Advances in Industrial Control

For other titles published in this series, go to
www.springer.com/series/1412

Other titles published in this series:

(*continued after Index*)

Béla Lantos · Lőrinc Márton

Nonlinear Control of Vehicles and Robots

 Springer

Prof. Béla Lantos
Dept. Control Engineering & Information
  Technology
Budapest University of Technology
  & Economics
Magyar Tudósok Körútja 2
1117 Budapest
Hungary
lantos@iit.bme.hu

Dr. Lőrinc Márton
Dept. Electrical Engineering
Sapientia Hungarian University
  of Transylvania
Soseaua Sighisoarei 1C
547367 Targu Mures
Romania
martonl@ms.sapientia.ro

ISSN 1430-9491
ISBN 978-1-4471-2618-8   ISBN 978-1-84996-122-6 (eBook)
DOI 10.1007/978-1-84996-122-6
Springer London Dordrecht Heidelberg New York

British Library Cataloguing in Publication Data
A catalogue record for this book is available from the British Library

Cover design: eStudio Calamar S.L.

Printed on acid-free paper

Springer is part of Springer Science+Business Media (www.springer.com)

Advances in Industrial Control

Professor (Emeritus) O.P. Malik
Department of Electrical and Computer Engineering
University of Calgary
2500, University Drive, NW
Calgary, Alberta
T2N 1N4
Canada

Professor K.-F. Man
Electronic Engineering Department
City University of Hong Kong
Tat Chee Avenue
Kowloon
Hong Kong

Professor G. Olsson
Department of Industrial Electrical Engineering and Automation
Lund Institute of Technology
Box 118
221 00 Lund
Sweden

Professor A. Ray
Department of Mechanical Engineering
Pennsylvania State University
0329 Reber Building
University Park
PA 16802
USA

Professor D.E. Seborg
Chemical Engineering
University of California Santa Barbara
3335 Engineering II
Santa Barbara
CA 93106
USA

Doctor K.K. Tan
Department of Electrical and Computer Engineering
National University of Singapore
4 Engineering Drive 3
Singapore 117576
Singapore

Professor I. Yamamoto
Department of Mechanical Systems and Environmental Engineering
Faculty of Environmental Engineering
The University of Kitakyushu
1-1, Hibikino, Wakamatsu-ku, Kitakyushu, Fukuoka, 808-0135
Japan

This book is dedicated to our Families.

Series Editors' Foreword

The series *Advances in Industrial Control* aims to report and encourage technology transfer in control engineering. The rapid development of control technology has an impact on all areas of the control discipline. New theory, new controllers, actuators, sensors, new industrial processes, computer methods, new applications, new philosophies..., new challenges. Much of this development work resides in industrial reports, feasibility study papers and the reports of advanced collaborative projects. The series offers an opportunity for researchers to present an extended exposition of such new work in all aspects of industrial control for wider and rapid dissemination.

Applications arising from moving vehicles and robotic arms that have in common a requirement to follow a path invariably involve co-ordinate systems and are described by nonlinear system models. Motivated by these common practical and theoretical issues, Béla Lantos and Lőrinc Márton have pursued an interesting agenda of generic unification, drawing together a modelling framework for describing these types of systems and then investigating the many concomitant control and applications issues. Their results are comprehensively presented in this new monograph *Nonlinear Control of Vehicles and Robots* for the *Advances in Industrial Control* series.

As might be expected from such a globally-oriented approach to nonlinear control systems, the authors pursue a variety of themes and readers from differing backgrounds will be interested in following different concepts as they read the monograph; however, to gain a perspective on the monograph's contents, two themes are considered in this Foreword, applications and control techniques.

The range of applications that the authors seek to model, analyse, and control within a unified framework is wide, and interesting, and includes:

- Robotic-arm systems—both multi-link and SCARA systems
- Automobiles—ground travel
- Fixed-wing aircraft—aerial travel
- Helicopters – indoor quad-rotorcraft
- Marine vessels—surface ships

- Underwater vessels—underwater autonomous vehicles and
- Control of formations of different vehicle types

The monograph also presents some specific nonlinear system issues for mechanical systems, and the reports of this work include:

- Studies of friction—static and dynamic models; and
- Studies of backlash—models and compensation

The theme of unified modelling for a wide range of vehicular and robotic systems is complemented by a second theme of nonlinear control systems design. The approach taken is similar to a toolbox approach, for the authors describe a range of nonlinear system control techniques and then demonstrate their use on the wide range of applications given above. Each nonlinear control method is selected for a particular application according to its appropriateness for that system. The suite of nonlinear control system methods includes:

- Nonlinear system stability methods
- Input-output linearization
- Flatness control
- Sliding -mode control; and
- Receding-horizon control

To support the development of the physical system modelling there is an appendix on the kinematics and dynamic-modelling fundamentals, and for the nonlinear control, there is an appendix on differential geometry that presents Lie algebra topics and discusses other subjects related to nonlinear systems.

The relevance of nonlinear control methods for industrial applications is growing. The first applications are destined to occur where there really is no alternative as in path following vehicular and robotic applications. Feasibility studies of the potential benefits for systems that are complex and require high control performance will also aid the penetration of these techniques into other industrial fields. The project of Béla Lantos and Lőrinc Márton in unifying models across a range of application areas and the use of a portfolio of nonlinear control methods is likely to be attractive to a wide range of readers. Typically, these will range from the industrial engineer seeking ways of to enhance existing process control performance to the control theorist and the control postgraduate interested in making nonlinear control accessible and usable.

The Series Editors are pleased to welcome this entry among a growing number of *Advances in Industrial Control* monographs in the nonlinear systems and nonlinear control field. Other recent entries in this field that might interest the reader include *Tandem Cold Metal Rolling Mill Control: Using Practical Advanced Methods* (ISBN 978-0-85729-066-3, 2010) by John Pittner and Marwan A Simaan; *Induction Motors* (ISBN 978-1-84996-283-4, 2010) by Riccardo Marino, Patrizio Tomei and Cristiano M. Verrelli; *Detection and Diagnosis of Stiction in Control Loops: State of the Art and Advanced Methods* (ISBN 978-1-84882-774-5, 2010) edited by Mohieddine Jelali and Biao Huang and *Control of Ships and Underwater Vehicles:*

Design for Underactuated and Nonlinear Marine Systems (ISBN 978-1-84882-729-5, 2009) by Khac Duc Do and Jie Pan.

Industrial Control Centre   M.J. Grimble
Glasgow   M.A. Johnson
Scotland, UK

Preface

Control techniques are indispensable in the design of robots and modern vehicles. Using feedback control the safety and efficiency of these mechanical systems can considerably be improved. In order to achieve good control performances, the mathematical model of the controlled mechanical system has to be taken into consideration during control algorithm design. The dynamic model of vehicles and robots are nonlinear.

First the book briefly outlines the most important nonlinear control algorithms that can be applied for the control of mechanical systems. The very first requirement of each control system is the closed loop stability. It is why the stability analysis methods for nonlinear systems are presented in detail. Basic nonlinear control methods (feedback linearization, backstepping, sliding control, receding horizon control) that can be applied for mechanical systems are also reviewed.

For efficient controller design it is inevitable the knowledge of the dynamic model of controlled mechanical system. A framework for the modeling of vehicles and robots are introduced. Staring from the dynamic model of rigid bodies, the mechanical model of robotic manipulators, ground, aerial and marine vehicles are presented. The nonlinear effects that appear in the model of different mechanical systems are discussed.

The control of robots and different type of vehicles are discussed in separate chapters. The model based tracking control of robotic manipulators is addressed in different approaches. Firstly it is assumed that the parameters of the mathematical model of the robotic system are known. For such systems the classical robot control methods are presented such as cascade control, nonlinear decoupling and hybrid position/force control. For the control of robots with unknown parameters selftuning adaptive control is proposed. If the robot prescribed path include sharp corners, backstepping control techniques are suggested.

The ground vehicles generally move in unknown environment with stationary or moving obstacles. Some control algorithms are proposed for these systems that take into consideration the static and dynamic obstacles based on input–output linearization and receding horizon control techniques.

Receding horizon control is also applied for the control of aircrafts. This control algorithm is extended with a robust disturbance observer. For the control of a quadrotor helicopter, backstepping control techniques are applied.

For nonlinear ship control the acceleration feedback can be combined with nonlinear PID control. Adaptive control techniques can be applied for ships with unknown parameters. The control of 6 degree of freedom ships is solved using backstepping control techniques.

For simultaneous control of a group of vehicles, formation control techniques can to be applied. In this work two approaches are suggested for vehicles that move on a surface: potential field method and passivity theory.

Non-smooth nonlinearities such as friction and backlash severely influence the control performances of mechanical systems. To solve the problem of friction compensation and identification in robotic systems, efficient friction modeling techniques are necessary. A piecewise linearly parameterized model is introduced to describe the frictional phenomenon in mechanical control system. The behavior of the control systems with Stribeck friction and backlash is analyzed in a hybrid system approach. Prediction and analysis methods for friction and backlash generated limit cycles are also presented. A friction identification method is introduced that can be applied for robotic manipulators driven by electrical motors and for hydraulic actuators as well. The piecewise linear friction model is also applied for robust adaptive friction and payload compensation in robotic manipulators.

The appendixes of the book are important for understanding other chapters. The kinematic and dynamic foundations of physical systems and the basis of differential geometry for control problems are presented. Readers who are familiar with these fundamentals may overstep the appendixes.

The reader of this book will become familiar with the modern control algorithms and advanced modeling techniques of the most common mechatronic systems: vehicles and robots. The examples that are included in the book will help the reader to apply the presented control and modeling techniques in their research and development work.

Budapest   Béla Lantos
Târgu Mureş (Marosvásárhely)   Lőrinc Márton

Acknowledgements

The research work related to this book was supported by the Hungarian National Research program under grant No. OTKA K71762 and by the Advanced Vehicles and Vehicle Control Knowledge Center under grant No. NKTH RET 04/2004, and by the Control System Research Group of the Hungarian Academy of Sciences. The second author's research was also supported by the János Bolyai Research Scholarship of the Hungarian Academy of Sciences.

The authors would like to thank for László Kis, Gergely Regula, Zoltán Prohászka, and Gábor Kovács, colleagues from the Department of Control Engineering and Information Technology, Budapest University of Technology and Economics, for their help.

The authors also recognize the contribution of Prof. Nariman Sepehri, University of Manitoba, Winnipeg, Canada, and Szabolcs Fodor, Sapientia Hungarian University of Transylvania to the section that deals with hydraulic actuators. Useful remarks and comments related to the control of the underactuated ball and beam system were given by Prof. John Hung, Auburn University, Alabama, USA.

Contents

Nomenclature

List of Abbreviations

(BM)	Backlash Mode
(CM)	Contact Mode
2D	Two Dimensional
3D	Three Dimensional
BLDC	Brushless Direct Current
CAN	Controller Area Network
CAS	Collision Avoidance System
CB	Center of Buoyancy
CCD	Charge Coupled Device
COG	Center Of Gravity
CPU	Central Processing Unit
DAC	Digital Analogue Converter
DC	Direct Current
DFP	Davidon Fletcher Power
DGA	Direct Geometric Approach
DH	Denavit Hartenberg
DOF	Degree Of Freedom
DSC	Digital Signal Controller
DSP	Digital Signal Processor
ECEF	Earth Centered Earth Fixed
ECI	Earth Centered Inertial
EKF	Extended Kalman Filter
emf	Electromotoric Force
FSTAB	Formation Stabilizing Controller
GAS	Globally Asymptotically Stable
GMS	Generalized Maxwell Slip
HLC	High Level Controller
IMU	Inertial Measurement Unit
ISS	Input to State Stability
LLC	Low Level Controller

LMI	Linear Matrix Inequality
LOS	Line of Sight
LQ	Linear Quadratic
LS	Least Square
LTI	Linear Time Invariant
LTV	Linear Time Varying
MIMO	Multiple Input Multiple Output
NED	North East Down
PC	Personal Computer
PD	Proportional Derivative
PI	Proportional Integrative
PID	Proportional Integral Derivative
PM	Pierson Moskowitz
PWM	Pulse Width Modulation
QP	Quadratic Programming
RHC	Receding Horizon Control
RPM	Rotation Per Minute
RPY	Roll Pitch Jaw
RTW	Real-Time Workshop
SI	International System
SISO	Single Input Single Output
SPI	Serial Peripheral Interface
TCP	Tool Center Point
UAV	Unmanned Aerial Vehicle
UGAS	Uniformly Globally Asymptotically Stable
UGS	Uniformly Globally Stable
UGV	Unmanned Ground Vehicle
UMV	Unmanned Marine Vehicle
wb	without bias
wfg	with finite gain

List of Notations

$\Rightarrow$	it follows
$\Leftrightarrow$	if and only if
$\in$	element of
$\notin$	not element of
$\geq$	partial ordering
$a := b$	a is defined by b
0	null element of linear space
$A = \{a :$ properties of $a\}$	definitions of set A
$\cap$	set intersection
$\cup$	set union
A^0	internal points of set A
$\bar{A}$	closure of set A
$\langle A \rangle, co(A)$	convex hull of set A
$\overline{\langle A \rangle}, \overline{co}(A)$	closed convex hull of set A

$A^\perp$	orthogonal complement of subspace A
$A \times B = \{(x, y) : x \in A,\ y \in B\}$	direct product of sets A and B
R^1, R	set of real numbers
C^1, C	set of complex numbers
R^n, C^n	real or complex Euclidean space
$F : A \to B$	mapping (function) F from A to B
$D(F)$	domain of F
$\mathrm{kernel}(F), \ker(F)$	kernel (null) space of F
$\mathrm{range}(F), R(F)$	range space of F
$\sin(\alpha), S_\alpha$	sine function
$\cos(\alpha), C_\alpha$	cosine function
$\tan(\alpha), T_\alpha$	tangent function
$\mathrm{atan}(x), \arctan(x)$	inverse tangent function
$\mathrm{sgn}(x), \mathrm{sign}(x)$	signum function
$\mathrm{sat}(x)$	saturation function
$F(\cdot, y)$	function $F(x, y)$ for fixed y
$F(y) \circ G(x)$	composition $F(G(x))$ of functions
$z, \tilde{z} = \bar{z}$	complex number and its conjugate
$\mathcal{A}, \mathcal{B}, \mathcal{C}$	linear mappings
$\mathcal{I}$	identity mapping
$a, b, c, \ldots$	vectors
$A, B, C, \ldots$	matrices
I	identity matrix
A^T, a^T	transpose of a real matrix or vector
$\langle a, b \rangle = a \cdot b = a^T b = b^T a$	scalar (inner) product of a and b
$a \times b$	vector product of a and b
$a \circ b$	dyadic product of a and b
$[a \times]$	matrix of vector product belonging to a
$[a \circ b]$	matrix $a\,b^T$ of dyadic product
$\mathrm{diag}(a, b, c, \ldots)$	diagonal matrix
$\mathrm{rank}(A)$	rank of matrix A
$\det(A)$	determinant of matrix A
$\mathrm{trace}(A)$	trace of matrix A
$\mathrm{Span}\{a, b, c, \ldots\}$	space spanned by $a, b, c, \ldots$
$A = U \Sigma V^T$	singular value decomposition of A
$A = QR$	QR decomposition of matrix A
A^+	Moore–Penrose pseudoinverse of A
$|x|$	absolute value of x
$\|x\|$	norm of x
$(X, \cdot, F)$	linear space over field F
$(E, \|\ \|)$	linear normed space, Banach space
$(H, \langle \cdot, \cdot \rangle)$	Hilbert space
$\langle f, g \rangle$	scalar (inner) product in Hilbert space
$C^n[0, T]$	space of continuous functions in R^n
$C^{(n)}[0, T]$	n-times differentiable functions

$C^{(\infty)}$	space of smooth functions
$L_p^n[0, T]$	in p-norm integrable functions in R^n
$L_\infty^n[0, T]$	essentially bounded functions in R^n
$L(E_1 \to E_2)$	space of linear mappings
$K(E_1 \to E_2)$	space of bounded linear mappings
A^*	adjoint operator in Hilbert space
$f'(x), f''(x)$	gradient and Hess matrix
$F(s) = L\{f(t)\}$	Laplace transform
$F(z) = Z\{f_n\}$	Z transform
l_2	space of infinite sequences
q, q^{-1}	shift operators
$G(q), H(q), H^{-1}(q)$	stable (bounded) operators over l_2
ξ	random variable
$E\xi$	expectation (mean) value
$x(t) = x(t, \omega)$	stochastic process
$R_{xy}(\tau)$	cross-covariance function
$R_{xx}(\tau)$	auto-covariance function
$\Phi_{xy}(\omega)$	cross-spectral density
$\Phi_{xx}(\omega)$	power spectral density
$x \in R^n, u \in R^m, y \in R^p$	state, input signal, output signal
$x(t) = \varphi(t, \tau, x, u(\cdot))$	state transition function
$y(t) = g(t, x(t), u(t))$	output mapping
$\dot{x}(t) = f(t, x(t), u(t))$	state equation of nonlinear system
$\dot{x}(t) = A(t)x(t) + B(t)u(t)$	state equation of linear system
$y(t) = C(t)x(t) + D(t)u(t)$	output mapping of linear system
$\Phi(t, \tau)$	fundamental matrix of linear system
$\dot{x} = Ax + Bu$	state equation of LTI system
$y = Cx + Du$	output of LTI system
e^{At}	exponential matrix
$G(s), H(s)$	transfer functions of LTI systems
p, z	pole, zero
$x_{i+1} = Ax_i + Bu_i$	discrete time linear time invariant system
$x_{i+1} = \Phi x_i + \Gamma u_i$	sampled continuous time linear system
$D(z)$	discrete time transfer function
$M_c = [B, AB, \ldots, A^{n-1}B]$	controllability matrix of linear system
$M_o = [C^T, A^T C^T, \ldots, (A^T)^{n-1} C^T]^T$	observability matrix of linear system
$u = -Kx$	linear state feedback
$\dot{\hat{x}} = F\hat{x} + Gy + Hu$	linear state observer
$\hat{x}_i = F\hat{x}_{i-1} + Gy_i + Hu_{i-1}$	actual linear state observer
$y(t) = \varphi^T(t)\vartheta$	linear parameter estimation problem
$\hat{\vartheta}$	estimate of ϑ
$\tilde{\vartheta}$	estimation error of ϑ
$P(t) = [\sum \lambda^{t-i} \varphi(i)\varphi^T(i)]^{-1}$	matrix playing role in parameter estimation
$\varepsilon(t) = y(t) - \varphi^T(t)\hat{\vartheta}(t-1)$	residual

$\hat{\vartheta}(t) = \hat{\vartheta}(t-1) + P(t)\varphi(t)\varepsilon(t)$	recursive parameter estimation
K	kinetic energy
P	potential energy
$L = K - P$	Lagrange function
G	Gibbs function
K_i	coordinate system, frame
T_{K_1,K_2}, T_{12}	homogeneous transformation
A_{K_1,K_2}, A_{12}	orientation
p_{K_1,K_2}, p_{12}	position
$Rot(z, \varphi)$	rotation around z by angle φ
φ, ϑ, ψ	Euler angles, RPY-angles
$q = (s, w)$	quaternion, $s \in R^1, w \in R^3$
$\tilde{q} = (s, -w)$	conjugate of quaternion $q = (s, w)$
$q_1 * q_2$	quaternion product
v, a	velocity, acceleration
ω, ε	angular velocity, angular acceleration
q_i	generalized coordinate, joint variable
C_{12}, S_{12}	$\cos(q_1 + q_2), \sin(q_1 + q_2)$
$^n d_{i-1}, {}^n t_{i-1}$	partial velocity, partial angular velocity
m	mass
$J(q)$	Jacobian of physical system
ρ_c	center of mass
$I_x, I_{xy}, \ldots$	inertia moments
$\mathbf{I}, I_c$	inertia matrix
$H(q), M(q)$	generalized inertia matrix
F, τ	force, torque
F_f, τ_f	friction generated force and torque
$\frac{\partial f}{\partial x}$	Jacobian of vector–vector function
$f : X \to R^n$	vector field over X, X is open, $f \in C^\infty$
$V(X)$	set of vector fields over X
$S(X)$	set of smooth functions $a : X \to R^1$
$(R^n)^*$	space of row vectors (covectors)
$h : X \to (R^n)^*$	form (covector field) over X, $h \in C^{(\infty)}$
$F(X)$	set of forms (covector fields) over X
$y = T(x)$	nonlinear coordinate transformation
$s_{f,t}(x_0)$	integral curve, $\dot{x}(t) = f(x(t)), x(0) = x_0$
$\nabla a, da$	gradient of $a \in S(X)$
$L_f a = \langle da, f \rangle$	Lie derivative of a; $a \in S(X)$, $f \in V(X)$
$L_f g = \frac{\partial g}{\partial x} f - \frac{\partial f}{\partial x} g = [f, g]$	Lie derivative (bracket) of g; $f, g \in V(X)$
$L_f h = f^T (\frac{\partial h^T}{\partial x})^T + h \frac{\partial f}{\partial x}$	Lie derivative of h; $h \in F(X)$, $f \in V(X)$
$ad_f^{i+1} g = [f, ad_f^i g]$	ad-operator, $ad_f^0 g = g, ad_f^1 g = [f, g]$
	etc.
$M \subset X$	submanifold
TM_x	tangent space of submanifold M

$\Delta : x \mapsto L^k(x)$ $\Delta(x) = \mathrm{Span}\{f_1(x), \ldots, f_k(x)\}$

$\dot{x} = f(x) + \sum_{i=1}^{m} g_i(x)u_i$ affine nonlinear system, $f, g_i \in V(x)$

$\mathrm{Span}\{ad_f^i g_j : 1 \le j \le m, 0 \le i \le n - 1\}$ reachability distribution

$r = (r_1, \ldots, r_m)$ relative degree vector

$u = S^{-1}(x)(-q(x) + v)$ linearizing static feedback, $\det S(x) \ne 0$

$\dot{z} = \beta(x, z, v), u = \alpha(x, z, v)$ endogenous state feedback

Chapter 1
Introduction

Overview   This chapter introduces some general definitions for autonomous vehicles and robots and it also gives a short historical review of the most important
steps of robot and vehicle control development. The motivation of the research work
summarized in this book is also presented here. The final part of the chapter shortly
outlines the structure of the book.

1.1 Basic Notions, Background

Vehicles and robots are designed for transportation and manipulation tasks, respectively. They can change the position or orientation of bodies in space. The transportation task can be formulated as the motions to some significant distance in comparison with the size of the moved bodies. Manipulation is to make any other change
of the position and orientation of the bodies. In this case, the dimensions of the manipulated object are comparable with the distance of the motion.

Vehicles are engineering systems used for transportation. They can be categorized in function of the environment where they perform the transportation, for example, on the ground, in the water, in the air. A car or ground vehicle is a wheeled
motor vehicle used for transporting passengers or cargo, which also carries its own
engine or drive motor. An aircraft is a vehicle which is able to fly by being supported by the air. The most important type of aircrafts are the airplanes, which are
fixed wing aircrafts capable of flight using forward motion that generates lift as the
wing moves through the air, and helicopters, in which the lift and thrust are supplied
by one or more engine driven rotors.

The manipulation of an object can be solved using robots. A robot is an automatically guided machine, able to do tasks on its own. According to the definition
of European Robotics Research Network, the robot is "an automatically controlled,
reprogrammable, multipurpose manipulator, programmable in three or more axes,
which may be either fixed in place or mobile for use in industrial automation applications". This definition is also used by the International Federation of Robotics.

B. Lantos, L. Márton, *Nonlinear Control of Vehicles and Robots*,
Advances in Industrial Control,
DOI 10.1007/978-1-84996-122-6_1, © Springer-Verlag London Limited 2011

According to the Robotics Institute of America, a robot is a "reprogrammable multifunctional manipulator designed to move materials, parts, tools, or specialized devices through variable programmed motions for the performance of a variety of tasks".

The bridge between the vehicles and robots is represented by the mobile robots which are robots that have the capability to move around in their environment and are not fixed to one physical location. Similarly to vehicles, they are categorized in the function of the environment in which they travel: land or home robots are most commonly wheeled or tracked, but also include legged robots. Aerial robots are usually referred to as unmanned aerial vehicles. Underwater robots are usually called autonomous underwater vehicles.

The control of vehicles and robots is a great challenge and it represents an important field of control engineering research in the past decades. Even the very first generations of robots that were used in industry applied automatic control. In the case of vehicles, even today, many control and navigation tasks are solved by the human operator. It is because the vehicles generally move in an unknown environment and the control system may not act correctly in any situation. Similar problems arise in the case of robots that have to perform manipulation tasks in an unknown environment.

In many recent transportation and manipulation tasks, high degree of autonomy is desirable. Autonomous engineering systems can perform desired tasks in partially known environments without continuous human supervision. One of the first successful approach to increase the autonomy of aerial and recently the water and ground vehicles was the introduction of autopilots. An autopilot is an electrical, mechanical, or hydraulic system used to guide a vehicle without assistance from a human being. Standard autopilots for airplanes can solve the navigation and control of airplanes during normal flights but more complex tasks such as take off and landing have be solved by the human pilots. Nowadays each passenger aircraft capable of transporting more than twenty passenger must be equipped with an autopilot. Autopilots are also intensively applied in the case of ships.

The possibility to increase the autonomy of vehicles and robotic systems are facilitated by the new results and tools offered by the related technologies. Nowadays almost all automatic control problems are solved using digital computers. Most of vehicle and robot control algorithms require high computational costs and the sampling periods in order of milliseconds or even less. Hence, such fast microprocessors should be applied that support floating point operations at hardware level. Many of these types of processors are available these days at relatively low cost. The industrial microcontrollers beside the capability of fast computation also offers on-chip interfaces for sensors and actuators that can be used in automation. For the cases when intensive processing is necessary, for example, in the case of image processing based control, parallel microprocessor units can be applied. For formation control of vehicles, reliable wireless communication protocols are needed.

There are a wide range of different type of sensors that facilitate the localization of vehicles in space such as the satellite based Global Positioning System based sensors, inertial sensors, laser or ultrasound based distance sensors, stereo vision systems. For precise localization of vehicles, the signals from different sensors can be combined using sensor fusion techniques. For example, inertial sensors are now usually combined with satellite based localization systems. The inertial system provides short term data, while the satellite system based sensor corrects accumulated errors of the inertial system.

The modern, computer controlled propulsion systems of vehicles together with finely controllable actuators are also necessary for automatic control of these systems.

In the case of robots, the high resolution encoders allow precise velocity and position measurement. These sensors, combined with electrical or hydraulic actuators with small time constants facilitate the implementation of complex robot control algorithms that allow tracking precision in the sub-millimeter domain in a wide range of speed domain.

1.2 A Short History

Design of automatic (self moving) machines were reported even in the middle ages, by making "robots" especially for entertainment, but the real brake through happened only in the 20th century when the technological development allowed the military and industrial applicability of these machines. Industrial robots are considered as a sufficient condition of competitive manufacturing in many areas of industry, which aims to combine high productivity, quality, and adaptability at minimal cost. In this view after the Second World War, a considerable effort was made. The first programmable robot was designed in 1954 by George Devol and it is called "Universal Automation".

The UNIMATE, produced at a pioneering robot company with same name from USA, becomes the first industrial robot in use in 1961. It was used in the General Motors factory in New Jersey plant to work with heated die-casting machines. The UNIMATE robots feature up to six fully programmable axes of motion and are designed for high-speed handling. It was conceived in 1956 at a meeting between inventors George Devol and Joseph Engelberger. It was considered a successful industrial equipment, which means that the robot worked reliably and ensured uniform quality.

In 1969, Victor Scheinman working in the Stanford Artificial Intelligence Lab creates the Stanford Arm, which was the first successful electrically-powered, computer-controlled robot arm. It was actually a test robot used to develop industrial assembly techniques for commercial robot arms.

In 1973, the company ASEA (now ABB) introduced the first industrial robot, the IRB-6, which allowed continuous path motion.

In 1975, based on the design of Victor Scheinman, UNIMATE developed the Programmable Universal Manipulation Arm (Puma). The goal of this development was to create such an arm that can perform similar operations as a human worker who stands at a production line. The result was a 6 Degree Of Freedom (DOF) robot with six rotational joints that was widely used for automobile assembly and a variety of other industrial tasks, for example, welding, painting.

In 1978, under the guidance of Hiroshi Makino, professor at the University of Yamanashi, developed the Selective Compliance Assembly Robot Arm (SCARA). This robot has four DOF and it is pliable X–Y-direction and the end effector can both rotate and translate around and along the Z-direction. This arm is generally applied for pick and place like problems and for manufacturing and assembly.

The robots are also applied for space exploration. The Canadarm 1 was used on the Space Shuttle to maneuver payload from the payload bay of the space shuttle to its deployment position and then release it. It was able to catch a free-flying payload and maneuver it into the space shuttle. The SPAR Aerospace Ltd., a Canadian company, designed, developed, tested and built it. It was first used in 1981. In 2001, MD Robotics of Canada built the Space Station Remote Manipulator System (SS-RMS) which is also called Canadarm 2. It was successfully launched and has begun operations to complete the assembly of the International Space Station.

An important class of modern robots are represented by the so called light weight robots, designed for mobility and interaction with a priori unknown environments and with humans. They always have low weight to payload ratio, hence they are energy efficient. In 2006, the company KUKA reached the 1:1 weight-payload ratio with a compact 7-DOF robot arm. Important research effort in this field is performed in the DLR Institute of Robotics and Mechatronics, Germany, where different generations of DLR Light-Weight Robots were created.

Vehicles were in use much earlier than industrial robots. But reliable semi- or fully autonomous vehicles appear only after the Second World War, almost parallel with appearance of industrial robots.

The German engineer Karl Benz generally is acknowledged as the inventor of the modern automobile. He finished his creation in 1885. The mass production of cars was facilitated by the introduction of production lines in the car manufacturing industry by Ransom Olds and Henry Ford in the first two decades of the twentieth century. The development of automotive technology was rapid. Early developments included electric ignition and the electric self-starter, independent suspension, and four-wheel brakes. The appearance of on-board computers facilitated the implementation of digital feedback loops for making cars more safe and efficient. Reliable on-board computers are also the key elements in the driverless car development.

The history of unmanned autonomous ground vehicles started in 1977 with the Tsukuba Mechanical Engineering Lab in Japan. On a dedicated, clearly marked course it achieved speeds of up to 30 km/h, by tracking white street markers. In the 1980s, a vision-guided Mercedes-Benz robot van, designed by Ernst Dickmanns and his team at the Universität der Bundeswehr München, Germany, achieved 100 km/h on streets without traffic. Also in the 1980s' the DARPA-funded Autonomous Land Vehicle in the United States achieved the first road-following demonstration that used laser, radar, computer vision and feedback control to make a driverless vehicle. In 1995, the Carnegie Mellon University Navlab project achieved 98.2% autonomous driving on a 5000 km distance. In 2002, the DARPA Grand Challenge competitions were announced. During this competition, international teams compete in fully autonomous vehicle races over rough, unpaved terrain and in a non-populated suburban setting. In 2007, DARPA urban challenge involved autonomous cars driving in an urban setting. Hence today, there is a great effort to obtain fully autonomous ground vehicles. However, it seems that the first commercial driverless models will require automated highway systems. These smart roads have markers or transmitters that can be sensed by the car to determine the precise position of the road center or the border of the road.

The unmanned aerial vehicles are already wide spread and they are generally used by military applications. These systems come in two varieties: some are controlled from a remote location, and others fly autonomously based on preprogrammed flight plans using more complex dynamic automation systems. The earliest Unmanned Aerial Vehicle was A.M. Low's "Aerial Target" developed during the First World War, an autonomous aircraft that already applied electrically driven gyros for its own stabilization. Unmanned airplanes were developed before, during and after the Second World War, but they were little more than remote-controlled airplanes until the sixties-seventies. After that the developments in digital control and sensor equipments allowed the self guidance and automatic control of this vehicles, but human supervision is still applied during their flights. These systems are generally applied for recognition, but they are also used for armed attack, transportation or scientific research.

The first remotely controlled water vehicle was designed by Albert Tesla in 1897. The craft were constructed of iron, powered by an electric battery and equipped with a radio-mechanical receiver that accepted commands from a wireless transmitter. After the Second World War, the real breakthrough occurs in the field of unmanned underwater vehicles. The first autonomous underwater vehicle was developed at the Applied Physics Laboratory at the University of Washington as early as 1957 by Stan Murphy, Bob Francois and Terry Ewar. Since then hundreds of different autonomous underwater vehicles have been designed. They are applied for scientific research, recognition, and underwater manipulation. Unmanned surface ships have been tested since Second World War. Nowadays, there are several types of unmanned boats in service with mainly military applications.

1.3 Control Systems for Vehicles and Robots, Research Motivation

A large amount of theoretical results in control are produced in different research and development institutions all over the world but only a little part of them are put in practice. There is a gap between theory and practice.

Our aim was twofold. First, we wanted to collect the fundamentals and the main directions of the control theory of robots and vehicles. Evidently, we had to use results from other authors, select them and unify the notations. In the selections of large collection of results, our principle was to prefer recently popular directions

Fig. 1.1 Vehicle or robot control system

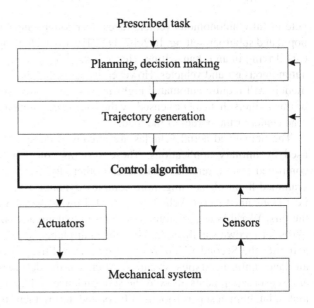

and those that have good chance for industrial applications and other real-time implementations.

The second aim was to complete these results with our own results in a unified approach of modeling of robots and vehicles, their control and the real-time implementation of the control algorithms.

The general structure of a control system for a vehicle or robot is shown in Fig. 1.1. As it is presented, the control system has clearly separable modules.

The planning and decision making module determines some basic parameters for the motion. For example, in the case of a transportation task, it generates the goal point coordinates and the intermediate points of the motion in such a way that obstacles are avoided and the solution is energy-efficient. Based on these parameters, the trajectory generation module calculates the prescribed motion that is generally given by the prescribed position, velocity, and acceleration functions in time. The planning and trajectory generation can be performed on-line or off-line. If the transportation and manipulation task is performed in a known environment that will not change over time, the off-line planning is enough. However, there are some cases when these modules rely on sensor information. For instance, if a mobile robot meets a moving obstacle, its trajectory has to be replanned on-line during the motion of the robot.

This book focuses on control algorithms (highlighted in Fig. 1.1) that calculate the control signals for the actuators of the robotic or vehicle system in function of the prescribed motion and senor information. The controlled system can be described with nonlinear models. Even a simple one DOF positioning system contains nonlinearity in its model, since the friction in the system can be described with nonlinear model. In more complex robotic or vehicle systems with many DOF, due to nonlinear forces and torques that inherently act on the system, the models that describe their motions are nonlinear. Some of them, as the centrifugal and Coriolis force, induce continuous nonlinearities in the system model. Other nonlinearities that are present in vehicles and robots, such as friction and backlash, are nonsmooth.

In order to obtain precise positioning and tracking in robot and vehicle control systems for a wide velocity range, these nonlinearities has to be taken into consideration during controller design.

In this view, firstly the nonlinear model of vehicle and robot systems have to be elaborated. This can be done starting from basic mechanical principles and laws. The complexity of the nonlinear models depends on their number of DOF. For example, in the case of ground vehicles, three DOF models can be applied but airplanes can be described with six DOF models. The influence of the environment on the motion cannot be omitted. For example, in the case of ships, the ocean currents and waves cannot be neglected.

The development of such control algorithms is necessary that can compensate the effect of these nonlinearities on the motion of the systems. If the nonlinearities are known, the control algorithms can be developed based on feedback linearization techniques. For vehicles and robots with unknown or partially known parameters, robust and adaptive control algorithms can be applied.

Different modeling and compensation approaches are necessary for nonsmooth nonlinearities (friction, backlash), that can severely influence the performances of mechanical control systems. The influence of these nonlinearities on control performances have to be studied. The development of such models is advisable that can easily be incorporated in the control algorithm. To obtain the model parameters online and off-line identification techniques have to be developed.

1.4 Outline of the Following Chapters

This book has twelve chapters, including this one, and two appendixes. The relations among the remaining chapters is presented in Fig. 1.2.

Chapter 2 gives an overview of the most important classes of nonlinear systems. Three examples illustrate the dynamic model of simple systems. A brief survey is offered about the most important stability theorems of nonlinear systems. Often used nonlinear compensation methods are presented starting from the theoretically important input-output linearization through the practical methods of flatness control, backstepping, sliding control, and receding horizon control. These methods are often used in later chapters both for single robot (vehicle) and formation control.

Chapter 3 derives detailed models of a large class of robots (industrial robots, cars, airplanes, and ships), the often used plants in single and formation control of autonomous systems.

Based on the modeling and basic control techniques introduced in Chaps. 2 and 3, the following five chapters present, how the control algorithms for nonlinear vehicles and robots can be designed.

Chapter 4 deals with the control of open chain rigid robots based on their nonlinear dynamic model. Firstly, a cascade control scheme is presented for decentralized robot control which is a linear method using only the average effective inertia

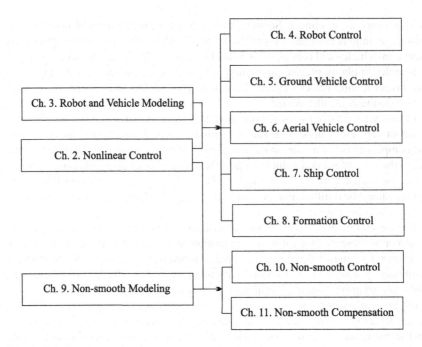

Fig. 1.2 Relations among the chapters

along the prescribed path computed from the nonlinear model. Decoupling methods both in joint and Cartesian space are presented together with the method of hybrid position and force control for compliance problems, and the method of parameter estimation using selftuning adaptive control. Backstepping control is used to solve precision control if the path contains sharp corners. Simulation results illustrate the control properties of the methods.

Chapter 5 presents methods for ground vehicle path design in case of static (lost load) and dynamic (incoming car) obstacles. For Collision Avoidance Systems two control methods are suggested, the first uses input affine approximation and input-output linearization while the second applies Receding Horizon Control. Simulation results demonstrate control efficiency.

Chapter 6 concentrates on two problems. The first deals with the longitudinal control of an airplane using receding horizon control. In order to improve stability, an internal disturbance observer based control is applied for stabilizing the system for varying Mach number and height parameters. The receding horizon control is a high level control which gives optimal reference signal for the internal control system. The second problem is the backstepping control of an indoor quadrotor helicopter supported with state estimation using inertial sensors and visual system which are also part of the embedded control realization.

Chapter 7 discusses the nonlinear control methods using acceleration feedback with nonlinear PID, adaptive feedback linearization and backstepping in 6-DOF. Some similarities with nonlinear decoupling of robots are shown. Constrained control allocations are taken into consideration. Simulation results illustrate the control properties of the methods.

Chapter 8 considers two problems in formation control. The first presents a stabilization method for ground vehicles (cars) based on potential field technique and guaranteed collision avoidance. The second deals with the stabilization of surface ships based on passivity theory. Simulation results illustrate the control properties of the methods.

Chapter 9 first introduces some definitions and theorems for the solution and stability of nonsmooth dynamical systems. Afterward, it is presented how the friction and backlash type nonlinearities influence the nonlinear model of mechanical systems. The chapter also contains a survey of the most important friction models introduced in the last years. It also presents a novel friction model which serves as a basis for identification and adaptive compensation of nonlinear friction.

The following two chapters are based on the control methods introduced in Chap. 2 and on modeling techniques presented in Chap. 9.

In Chap. 10, a hybrid model of mechanical systems with nonlinear friction and backlash is presented. Based on this model, the control of mechanical systems can be solved with guaranteed asymptotic stability in the simultaneous presence of these nonlinearities. Based on the friction model introduced in the previous chapter, a new easily applicable method is introduced to analyze the friction induced oscillatory behavior in mechanical control systems.

The first part of Chap. 11 presents a friction and a backlash identification method for mechanical systems that are based on measurements collected during controlled robot motion. The influences of other nonlinearities that appear in the system model were also taken into consideration during the identification procedure. Afterward, the problem of robust adaptive tracking control of robotic systems with unknown friction and payload is discussed. The effect of backlash on adaptive friction compensation is also analyzed.

Finally, Chap. 12 sums up the conclusions of this book and presents the future research directions.

Appendices A and B can help in understanding the other chapters for reader less familiar in control theory. In Appendix A, general modeling techniques are presented for mechanical systems and robots. Appendix B summarizes the basic notions of differential geometry that are necessary to understand the nonlinear control techniques.

Chapter 2
Basic Nonlinear Control Methods

Overview This chapter gives a survey of the basic nonlinear control methods. First, the main classes of nonlinear systems are summarized, then some examples are presented how to find the dynamic model of simple systems and an overview of the stability methods is given. It is followed by an introduction to the concepts of some often used nonlinear control methods like input/output linearization, flatness control, backstepping and receding horizon control.

2.1 Nonlinear System Classes

At the interface of a nonlinear system Σ, we can observe input and output signals. We can denote the input signal by $u(\cdot)$ and its value at time moment t by $u(t)$. Similarly, $y(\cdot)$ is the output signal and $y(t)$ is its value at time moment t. We can assume that the signals are elements of functions spaces which are usually linear spaces having an appropriate norm and some topological properties (convergence of Cauchy sequences of functions, existence of scalar product etc.). The chosen function space (Banach space, Hilbert space etc.) may depend on the control problem to be solved.

The state x of the nonlinear dynamic system is an abstract information which represents the entire past signals of the system observed at the interface, that is, $x(\tau) = x$ represents in abstract form all the signals $u(\sigma), y(\sigma), \forall \sigma \leq \tau$. The system can be characterized by the state transition function $x(t) = \phi(t, \tau, x, u(\cdot))$ and the output mapping $y(t) = g(t, x(t), u(t))$ where t is the actual time, τ and x are the initial time and initial state respectively, and for causal systems $u(\cdot)$ denotes the input signal between τ and t.

Specifically, for linear systems the transients can be written as $x(t) = \Phi(t, \tau)x + \Theta(t, \tau)u(\cdot)$ and $y(t) = C(t)x(t) + D(t)u(t)$. The state transition for linear systems can be thought as the superposition of the initial state transient and the result of the input excitation.

B. Lantos, L. Márton, *Nonlinear Control of Vehicles and Robots*,
Advances in Industrial Control,
DOI 10.1007/978-1-84996-122-6_2, © Springer-Verlag London Limited 2011

Fig. 2.1 Block scheme of the
nonlinear system

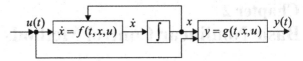

2.1.1 State Equation of Nonlinear Systems

If the state transition function satisfies some continuity conditions, then the state is
the solution of the nonlinear state differential equation with appropriately chosen
$f(t, x, u)$ and the system is called continuous time smooth nonlinear system:

$$\dot{x} = f(t, x, u),$$
$$y = g(t, x, u),$$
(2.1)

where "dot" denote time derivative and we used short forms in notation instead
of $\dot{x}(t) = f(t, x(t), u(t))$ and $y(t) = g(t, x(t), u(t))$. We shall consider dominantly
smooth systems in the book therefore "smooth" will be omitted and the system is
called continuous time nonlinear time variant system (NLTV). If f does not depend
explicitly on t, that is, $f(x, y)$, the nonlinear system is time invariant (NLTI). We
shell usually assume that $x \in R^n$, $u \in R^m$, $y \in R^p$. The block scheme of the nonlinear system is shown in Fig. 2.1.

Linear systems are special cases of nonlinear ones. Continuous time linear time
varying (LTV) systems have linear state differential equation:

$$\dot{x} = A(t)x + B(t)u,$$
$$y = C(t)x + D(t)u,$$
(2.2)

The solution of the state equation for LTV system can be expressed by the fundamental matrix $\Phi(t, \tau)$ which is the solution of the matrix differential equation

$$\dot{\Phi}(t, \tau) = A(t)\Phi(t, \tau), \qquad \Phi(\tau, \tau) = I.$$
(2.3)

The fundamental matrix has some important properties:

$$\Phi(t, \tau)\Phi(\tau, \theta) = \Phi(t, \theta), \qquad \Phi(t, \tau)\Phi(\tau, t) = I,$$
$$\frac{d\Phi(t, \tau)}{d\tau} = -\Phi(t, \tau)A(\tau).$$
(2.4)

Based on these properties, the solution of the state differential equation for linear
systems can be expressed by the fundamental matrix:

$$x(t) = \Phi(t, \tau)x + \int_{\tau}^{t} \Phi(t, \theta)B(\theta)u(\theta)\, d\theta.$$
(2.5)

The block scheme of the LTV system is shown in Fig. 2.2.

Fig. 2.2 Block scheme of the
LTV system

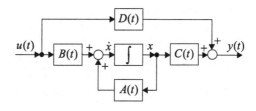

Fig. 2.3 Mathematical
scheme for solving nonlinear
state equation

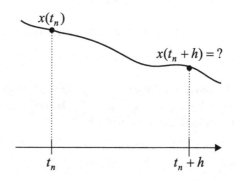

For continuous time linear time invariant (LTI) systems, A, B, C, D are constant matrices and the fundamental matrix is the exponential matrix $\Phi(t, \tau) = e^{A(t-\tau)}$, moreover the initial time can be chosen $\tau = 0$ because the system is time invariant. The LTI system has transfer function $G(s)$. The solution of the state ("differential" will be omitted) equation and the transfer function are

$$x(t) = e^{At}x + \int_0^t e^{A(t-\theta)}Bu(\theta)\,d\theta, \tag{2.6}$$

$$G(s) = C(sI - A)^{-1}B + D, \tag{2.7}$$

where $(sI - A)^{-1}$ is the Laplace-transform of e^{At}.

The state equation of nonlinear systems can usually be solved only numerically hence we show some numerical methods to find the approximate solution. Here, we assume that the control $u(t) := u_0(t)$ is known therefore $\dot{x}(t) = f(t, x(t), u_0(t)) =: f^*(t, x(t))$, and for simplicity we omit "star" in the notation. Assume $x(t_n)$ is known and we want to determine $x(t_n + h)$, see Fig. 2.3.

Euler method:

$$x(t_n + h) \approx x(t_n) + hx'(t_n) = x(t_n) + hf\big(t_n, x(t_n)\big). \tag{2.8}$$

Second order Taylor series:

$$x(t_n + h) \approx x(t_n) + hx'(t_n) + \frac{h^2}{2}x''(t_n)$$

$$= x(t_n) + hf\big(t_n, x(t_n)\big)$$

$$+ \frac{h^2}{2} \{ f_x'(t_n, x(t_n)) \cdot f(t_n, x(t_n)) + f_t'(t_n, x(t_n)) \}. \qquad (2.9)$$

Unfortunately, higher order Taylor series are hard to determine hence Runge and Kutta developed such methods which differ from the n-th order Taylor series only in $o(h^n)$. Now we show two of such methods.

Second order Runge–Kutta method:

$$x(t_n + h) \approx x(t_n) + h\{a_1 f(t_n, x(t_n)) + a_2 f(t_n + \alpha h, x(t_n)$$
$$+ \beta f(t_n, x(t_n)))\}, \qquad (2.10)$$

where the unknowns are a_1, a_2, α, β. In order to guarantee $o(h^2)$ (small ordo) precision $a_1 = 1 - a_2$, $\alpha = \frac{1}{2a_2}$, $\beta = \frac{h}{2a_2}$ should be satisfied. For the remaining parameter, two methods are wide-spread:

Heun's trapezoidal algorithm: $a_2 = \frac{1}{2}$.
Modified Euler–Cauchy algorithm: $a_2 = 1$.

Fourth order Runge–Kutta method:
The essence of the method is that the change of the function $x(t)$ will be estimated from the values of the derivatives determined at four points of the function (once at the left end of the interval, twice in the middle and once at right end), and take the average of them with weights assuring $o(h^4)$ approximation at the right end point of the interval:

$$k_1 = hf(t_n, x(t_n)),$$

$$k_2 = hf\left(t_n + \frac{h}{2}, x(t_n) + \frac{1}{2}k_1\right),$$

$$k_3 = hf\left(t_n + \frac{h}{2}, x(t_n) + \frac{1}{2}k_2\right), \qquad (2.11)$$

$$k_4 = hf(t_n + h, x(t_n) + k_3),$$

$$x(t_n + h) \approx x(t_n) + \frac{1}{6}(k_1 + 2k_2 + 2k_3 + k_4).$$

The computation is usually performed with variable step size. The goal is to find the solution as quick as possible while guaranteeing the necessary precision for which simple step change strategies are available.

Multistep methods: Other type of methods assume equidistant step size and try to find exact solution for polynomial form $x(t)$. The solution is assumed in the form

$$x_{n+1} = \sum_{i=0}^{k} a_i x_{n-i} + \sum_{i=-1}^{k} b_i hf(t_{n-i}, x_{n-i}). \qquad (2.12)$$

Performing the normalization of the running time $\ldots, t_{n-1}, t_n, t_{n+1}$ to normalized time $\ldots, -1, 0, 1$ the following conditions appear:

$$
1 = \sum_{i=0}^{k} a_i (-i)^j + \sum_{i=-1}^{k} b_i j (-i)^{j-1}, \quad j \in [1, m],
$$

(2.13)

$$
1 = \sum_{i=0}^{k} a_i .
$$

The accuracy can be influenced by m and k. Wide-spread methods are:

- *Adams–Bashforth method*: $k = m - 1$, $a_1 = \cdots = a_k = 0$, $b_{-1} = 0$.
- *Adams–Moulton method*: $k = m - 2$, $a_1 = \cdots = a_k = 0$, $b_{-1} \neq 0$.
- *Predictor–corrector method:* The first Adams–Bashforth (predictor) step determines the estimation x_{n+1} because now $b_{-1} = 0$ and thus x_{n+1} does not appear on the right side of the equation system. After this, the last x_{n+1} can cyclically be applied on the right side of the Adams–Moulton (corrector) steps gradually improving the approximation.

2.1.2 Holonomic and Nonholonomic Systems

Nonlinear systems have often to satisfy constraints that restrict the motion of the system by limiting the paths which the system can follow. For simplicity, we assume that the configuration space Q is an open subset of R^n with coordinates $q = (q_1, \ldots, q_n)^T$. More general configuration spaces can be handled by an appropriate choice of local coordinates.

A simple example is the case of two particles attached by a massless rod. The configuration of each particle is described by a point $r_i \in R^3$, but all trajectories of the particles have to satisfy

$$
\| r_1 - r_2 \|^2 = L^2,
$$

where L is the length of the rod. The constraints act through the application of *constraint forces* modifying the motion of the system such that the constraints are always satisfied. The constraint is an example of a holonomic constraint.

In the sequel, we shall define holonomic and nonholonomic constraints according to Murray and Li, and Sastry, see [103]. Then we investigate how the motion equation can be simplified by using coordinate transformation.

2.1.2.1 Holonomic Constraints

More generally, a constraint is called *holonomic* if it restricts the motion of the system to a smooth hypersurface in the unconstrained configuration space Q. Holo-

nomic constraints can be represented locally by algebraic constraints on the configuration space:

$$h_i(q) = 0, \quad i = 1, \ldots, k. \tag{2.14}$$

We shall assume that the constraints are linearly independent, that is, the matrix
$\frac{\partial h}{\partial q}$ has full row rank. Since holonomic constraints define a smooth hypersurface
in the configuration space, it is possible to eliminate the constraints by choosing a
set of coordinates for this surface. These new coordinates parametrize all allowable
motions of the system and are not subject to any further constraints.

The constraint forces are linear combinations of the gradients of the constraint
functions $h_i : Q \rightarrow R$. Let $h : Q \rightarrow R^k$ represent the vector valued constraint function then the constraint forces can be written as

$$F_c = \left[\frac{\partial h}{\partial q} \right]^T \lambda, \tag{2.15}$$

where $\lambda \in R^k$ is called Lagrange multiplier vector measuring the relative magnitudes of the constraint forces. No work is done if the system is moved along feasible
trajectories since

$$F_c^T \dot{q} = \lambda^T \frac{\partial h}{\partial q} \dot{q} = \lambda^T \frac{d}{dt} \big(h(q) \big) = 0.$$

A fundamentally *more general constraint* occurs in some robotics and mechanical applications (multifingered grasping, rolling disk etc.) where the constraint has
the form

$$A^T(q)\dot{q} = 0. \tag{2.16}$$

Such constraint is called *Pfaffian constraint*. Notice that $A^T(q)$ is $k \times n$ and $\dot{q}$ appears linearly in the constraint equations. This linearity gives chance to eliminate the
constraints using a nonlinear coordinate transformation based on the fundamental
results of differential geometry summarized in Appendix B. On the other hand, we
shall consider mechanical systems and will also refer to the results in Appendix A,
especially we shall use the Lagrange equation describing the dynamic model.

Since Pfaffian constraints restrict the allowable velocity of the system but not
necessarily the configurations, we cannot always represent it as an algebraic constraint on the configuration space. The Pfaffian constraint is called *integrable* if there
exists a vector valued function $h : Q \rightarrow R^k$ such that

$$A^T(q)\dot{q} = 0 \quad \Longleftrightarrow \quad \frac{\partial h}{\partial q} \dot{q} = 0. \tag{2.17}$$

Thus, an integrable Pfaffian constraint is equivalent to a holonomic constraint. No-
tice that we do not require that $A^T = \frac{\partial h}{\partial q}$, but only that they define the same subspace
of the allowable velocities at every $q \in Q$.

2.1.2.2 Nonholonomic Constraints

Pfaffian constraint that is not integrable is an example of *nonholonomic* constraint. For such constraints, the instantaneous velocities of the system are constrained to an $n - k$ dimensional subspace, but the set of reachable configurations is not restricted to some $n - k$ dimensional hypersurface in the configuration space. Since not all Pfaffian constraints are integrable, hence the motion equations have to be extended to this case.

It is still possible to speak about constraint forces. Such forces are generated by the Pfaffian constraints to insure that the system does not move in the directions given by the rows of the constraint matrix $A^T(q)$. The constraint forces at $q \in Q$ are defined by $F_c = A(q)\lambda$, where the Lagrange multiplier $\lambda \in R^k$ is the vector of relative magnitudes of the constraint forces. If the constraints are integrable, then this is identical to the holonomic case since $A(q)$ and $[\frac{\partial h}{\partial q}]^T$ have the same range space.

The constraint forces for Pfaffian constraints prevent motion of the system in directions which would violate the constraints. In order to include these forces in the dynamics, we have to add an additional assumption, namely forces generated by the constraints do no work on the system. This assumption is often referred as d'Alambert principle.

By using the Lagrangian $L(q, \dot{q})$ for the unconstrained system and assuming the constraints in the form $A^T(q)\dot{q} = 0$ the equation of motion can be written in the form

$$\frac{d}{dt}\frac{\partial L}{\partial \dot{q}} - \frac{\partial L}{\partial q} = A(q)\lambda + F, \tag{2.18}$$

where F represents the nonconservative and externally applied forces.

For systems having Lagrangian $L(q, \dot{q}) = \frac{1}{2}\dot{q}^T M(q)\dot{q} - P(q)$ (kinetic energy minus potential energy), the motion equations can be written as

$$M(q)\ddot{q} + C(q, \dot{q})\dot{q} + n(q, \dot{q}) = A(q)\lambda + F,$$

where $n(q, \dot{q})$ denotes nonconservative and potential forces and F is the external force. Differentiating the constraint equation yields

$$A^T(q)\ddot{q} + \dot{A}^T(q)\dot{q} = 0. \tag{2.19}$$

By solving the motion equation for $\ddot{q}$ and substituting it into the differentiated constraint equation, we obtain:

$$\ddot{q} = M^{-1}(A\lambda + F - C\dot{q} - n),$$
$$A^T M^{-1} A\lambda = A^T M^{-1}(C\dot{q} + n - F) - \dot{A}^T \dot{q}, \tag{2.20}$$
$$\lambda = (A^T M^{-1} A)^{-1}(A^T M^{-1}(C\dot{q} + n - F) - \dot{A}^T \dot{q}),$$

where we suppressed the dependence on q and $\dot{q}$.

2.1.2.3 Lagrange–d'Alambert Equations

It is useful to derive the equations without solving for the instantaneous constraint forces present in the system. It means that we project the motion of the system onto the feasible directions by ignoring the forces in the constrained directions. The resulting dynamics is in a form well suited for closed-loop control.

At a given configuration q the directions at which the system is allowed to move is given by the null space of $A^T(q)$. We adopt the classical notation of virtual displacement $\delta q \in R^n$ which satisfies $A^T(q)\delta q = 0$. If F is a generalized force applied to the system, then we call $\delta W = F \cdot \delta q$ the virtual work due to F acting along a virtual displacement δq, where "dot" denotes scalar product. D'Alambert's principle states that the constraint forces do no virtual work, hence

$$\big(A(q)\lambda\big) \cdot \delta q = 0 \quad \text{for } A^T(q)\dot q = 0. \tag{2.21}$$

Notice that δq is not the same as $\dot q$. The generalized velocity $\dot q$ satisfies both the velocity constraints and the motion equations, while the virtual displacement only satisfies the constraints.

In order to eliminate the constraint forces from the motion equations, we can project the motion onto the linear subspace generated by the null space of $A^T(q)$. Since $(A\lambda) \cdot \delta q = 0$, we obtain the so called *Lagrange–d'Alambert equations*:

$$\left(\frac{d}{dt}\frac{\partial L}{\partial \dot q} - \frac{\partial L}{\partial q} - F\right) \cdot \delta q = 0, \tag{2.22}$$

$$A^T(q)\delta q = 0. \tag{2.23}$$

Specifically, if $q = (q_1^T, q_2^T)^T$ and $A^T(q) = [A_1^T(q)\ A_2^T(q)]$, where $A_2^T(q)$ is square and invertible, then we can use the constraint $\dot q_2 = -(A_2^T)^{-1}A_1^T\dot q_1$ to eliminate $\dot q_2$ and $\ddot q_2$ from the motion equations. The method is illustrated in the following example.

2.1.2.4 Dynamics of Rolling Disk

Consider a disk rolling on the horizontal plane such that the disk plane remains orthogonal to the horizontal plane during motion. The configuration of the disk is given by the position (x, y) of the contact point, the heading angle θ, and the orientation of the disk with respect to the vertical, ϕ. We use the notation $q = (x, y, \theta, \phi)^T$. Let R denote the radius of the disk. We assume as input the driving torque on the wheel, τ_θ, and the steering torque about the vertical axis, τ_ϕ. Let m be the mass of the disk, I_1 its inertia about the horizontal (rolling) axis, and I_2 its inertia about the vertical axis. We assume that the disk rolls without slipping hence the following velocity constraints have to be satisfied:

$$\dot x - R\dot\phi C_\theta = 0, \quad \dot y - R\dot\phi S_\theta = 0 \quad \Longleftrightarrow \quad A^T(q)\dot q = \begin{bmatrix} 1 & 0 & 0 & -RC_\theta \\ 0 & 1 & 0 & -RS_\theta \end{bmatrix}\dot q = 0.$$

These constraints assume that the disk rolls in the direction in which it is heading. It is evident that the constraints are everywhere linearly independent.

The unconstrained Lagrangian is the kinetic energy associated with the disk:

$$L(q, \dot{q}) = \frac{1}{2}m(\dot{x}^2 + \dot{y}^2) + \frac{1}{2}I_1\dot{\theta}^2 + \frac{1}{2}I_2\dot{\phi}^2.$$

Denote $\delta q = (\delta x, \delta y, \delta\theta, \delta\phi)^T$ the virtual displacement of the system. The Lagrange–d'Alembert equations are given by

$$\left(\begin{bmatrix} m & 0 & 0 & 0 \\ 0 & m & 0 & 0 \\ 0 & 0 & I_1 & 0 \\ 0 & 0 & 0 & I_2 \end{bmatrix} \ddot{q} - \begin{bmatrix} 0 \\ 0 \\ \tau_\theta \\ \tau_\phi \end{bmatrix} \right) \cdot \delta q = 0, \quad \text{where} \quad \begin{bmatrix} 1 & 0 & 0 & -RC_\theta \\ 0 & 1 & 0 & -RS_\theta \end{bmatrix} \delta q = 0.$$

From the constraints, we can solve for δx and δy:

$$\delta x = RC_\theta \delta\phi, \qquad \delta y = RS_\theta \delta\phi.$$

The scalar product can be written as

$$\left(\begin{bmatrix} 0 & 0 \\ mRC_\theta & mRS_\theta \end{bmatrix} \begin{bmatrix} \ddot{x} \\ \ddot{y} \end{bmatrix} + \begin{bmatrix} I_1 & 0 \\ 0 & I_2 \end{bmatrix} \begin{bmatrix} \ddot{\theta} \\ \ddot{\phi} \end{bmatrix} - \begin{bmatrix} \tau_\theta \\ \tau_\phi \end{bmatrix} \right) \cdot \begin{bmatrix} \delta\theta \\ \delta\phi \end{bmatrix} = 0,$$

and since $\delta\theta$ and $\delta\phi$ are free, we obtain the motion equation

$$\begin{bmatrix} 0 & 0 \\ mRC_\theta & mRS_\theta \end{bmatrix} \begin{bmatrix} \ddot{x} \\ \ddot{y} \end{bmatrix} + \begin{bmatrix} I_1 & 0 \\ 0 & I_2 \end{bmatrix} \begin{bmatrix} \ddot{\theta} \\ \ddot{\phi} \end{bmatrix} = \begin{bmatrix} \tau_\theta \\ \tau_\phi \end{bmatrix}.$$

Differentiating the constraints, we obtain

$$\ddot{x} = RC_\theta \ddot{\phi} - RS_\theta \dot{\theta}\dot{\phi}, \qquad \ddot{y} = RS_\theta \ddot{\phi} + RC_\theta \dot{\theta}\dot{\phi},$$

and substituting the results into the motion equation we obtain the system dynamics as follows:

$$\begin{bmatrix} I_1 & 0 \\ 0 & I_2 + mR^2 \end{bmatrix} \begin{bmatrix} \ddot{\theta} \\ \ddot{\phi} \end{bmatrix} = \begin{bmatrix} \tau_\theta \\ \tau_\phi \end{bmatrix}.$$

Note that in this simple example the system of equations does not depend on x and y, but in general it is not valid.

The motion of the x- and y-position of the contact point can be obtained from the first order differential equations

$$\dot{x} = RC_\theta \dot{\phi}, \qquad \dot{y} = RS_\theta \dot{\phi},$$

which follow from the constraints. Thus, given the trajectory of θ and ϕ, we can determine the trajectory of the disk rolling on the horizontal plane. Dividing the motion equations into a set of second-order differential equations in a reduced set of variables plus a set of first-order differential equations representing the constraints is typical for many nonholonomic systems.

2.1.2.5 Holonomic Property Related to Frobenius Theorem

Consider the nonlinear system satisfying constraints in which beside the generalized coordinate $q \in R^n$ appears also its derivative $\dot{q} \in R^n$:

$$A^T(q)\dot{q} = 0, \quad \text{rank } A^T(q) = k < n. \tag{2.24}$$

Assume that $A^T(q)$ is $k \times n$ and denote its null (kernel) space $\ker A^T(q)$ at q_0.

Let us define the function $\Delta(q_0) := \ker A^T(q_0)$ which maps q_0 into a linear subspace $\ker A^T(q_0)$, hence it is a distribution. If this distribution is involutive, that is, it is closed for Lie bracket defined by $[f, g] = \frac{\partial g}{\partial q} f - \frac{\partial f}{\partial q} g$ for vector fields $f(q), g(q)$ then, based on Frobenius theorem, we may find local coordinates $\bar{q} = (\bar{q}_1, \ldots, \bar{q}_n)^T$ such that (2.24) can be written as

$$\dot{\bar{q}}_{n-k+1} = \cdots = \dot{\bar{q}}_n = 0 \quad \Leftrightarrow \quad \bar{q}_{n-k+1} = c_{n-k+1}, \ldots, \bar{q}_n = c_n \tag{2.25}$$

for appropriately chosen constants $c_{n-k+1}, \ldots, c_n$ determined by the initial conditions and thus we can eliminate the coordinates $\bar{q}_{n-k+1}, \ldots, \bar{q}_n$. Therefore, we call the constraints (2.24) *holonomic* if the distribution Δ is involutive, otherwise the constraints are *nonholonomic*. Notice that the application of Frobenius theorem needs to solve a set of partial differential equations which is a hard problem. Hence, we seek for other methods simplifying the motion equation in the presence of constraints.

2.1.2.6 Motion Equation in Hamiltonian Form

The equation of motion for the constrained mechanical system can be assumed in Lagrange form

$$\frac{d}{dt}\left(\frac{\partial L}{\partial \dot{q}}\right) - \left(\frac{\partial L}{\partial q}\right) = A(q)\lambda, \quad A^T(q)\dot{q} = 0, \tag{2.26}$$

where $A(q)\lambda$ describes the constrained forces, $\lambda \in R^k$ are the Lagrange multipliers. For simplicity of notation, the generalized external force was neglected. The Lagrangian $L(q, \dot{q})$ is defined by $L = K - P$ where K is the kinetic energy and P is the potential energy. The kinetic energy is the quadratic form $K = \frac{1}{2}\langle M(q)\dot{q}, \dot{q}\rangle$. Here, $M(q)$ is the generalized inertia matrix. We can assume that no simplifications were applied hence $M(q)$ is positive definite and hence

$$\det\left[\frac{\partial^2 L}{\partial \dot{q}_i \partial \dot{q}_j}\right] \neq 0. \tag{2.27}$$

In order to eliminate the Lagrange multipliers and find the state equation of the mechanical system, the Legendre transformation $p := \frac{\partial L}{\partial \dot{q}} = M(q)\dot{q}$ and the Hamiltonian $H = K + P = 2K - L$ will be applied. Because of $K = \frac{1}{2}\langle p, \dot{q}\rangle$, the Hamiltonian can be written as

$$H(q, p) = \sum_{i=1}^{n} p_i q_i - L(q, \dot{q}), \quad p_i = \frac{\partial L}{\partial \dot{q}_i}, \quad i = 1, \ldots, n. \quad (2.28)$$

It follows from the new form of the kinetic energy and the positive definiteness of $M(q)$ that p is an independent system, hence the motion equation of the constrained mechanical system can be written in the Hamiltonian form

$$\dot{q} = \frac{\partial H(q, p)}{\partial p}, \qquad \dot{p} = -\frac{\partial H(q, p)}{\partial q} + A(q)\lambda,$$
$$A^T(q)\frac{\partial H(q, p)}{\partial p} = 0. \quad (2.29)$$

The constraint forces $A(q)\lambda$ can be determined by differentiating the constraint equation by the time and expressing λ, that is,

$$\left[\frac{\partial A^T(q)}{\partial q}\frac{\partial H(q, p)}{\partial p}\right]^T \frac{\partial H(q, p)}{\partial p}$$
$$+ A^T(q)\frac{\partial^2 H(q, p)}{\partial p^2}\left[-\frac{\partial H(q, p)}{\partial q} + A(q)\lambda\right] = 0. \quad (2.30)$$

Notice that here a matrix A^T is differentiated by a vector q therefore it is a problem how to implement it. Fortunately not the derivative has to be determined but the derivative multiplied by a vector $\frac{\partial H}{\partial p}$ which is easier to obtain.

Let us consider the following mathematical problem. Let A be any matrix of type $m \times n$ and the vector x have dimension n. Then A can be assumed as $A = \sum_{i,j} a_{ij} E_{ij}$ where the only nonzero element of the matrix E_{ij} is in the intersection of the i-th row and the j-th column. Hence, differentiating by a_{ij} and applying the result for the vector x we obtain $E_{ij}x$ which is a column vector with the only nonzero element in the i-th row which is x_j. Repeating the process for all the elements of A we obtain $\frac{\partial A}{\partial x}x$ which is an $m \times (mn)$ type matrix each column containing the above result for one element of A.

Now we can apply this result for the matrix A^T and the vector $\frac{\partial H}{\partial p}$. Since rank A^T is k, hence it follows from (2.27) and (2.28)

$$\det\left[A^T(q)\frac{\partial^2 H(q, p)}{\partial p^2}A(q)\right] \neq 0, \quad (2.31)$$

thus λ can be expressed from (2.30) and put into (2.29) which results in the motion equation of the constrained (holonomic or nonholonomic) system.

2.1.2.7 Coordinate Transformation of Constrained Motion Equations

Here we try to eliminate the constraint equations without the solution of partial differential equations. The idea is to introduce a coordinate transformation $\tilde{p} = T(q)p$ and partition $\tilde{p}$ as $\tilde{p} = (\tilde{p}^{1T} \tilde{p}^{2T})^T$ where $\tilde{p}^2 = 0$ suits to the constraints. Notice that this is only possible if the constraints are holonomic, otherwise we obtain the motion equation in the variables q and the new $\tilde{p}$ and the constraints are not eliminated. Of course the goal is to eliminate the constraints and assure $\tilde{p}^2 = 0$.

By the assumptions $A^T(q)$ has type $k \times n$ and rank $A^T(q) = k$ thus there exists a matrix $S(q)$ of type $n \times (n-k)$ satisfying $A^T(q)S(q) = 0$ and having rank $S(q) = n - k$. Hence, we are able to define the coordinate transformation according to

$$A^T(q)S(q) = 0 \quad \Rightarrow \quad S(q) = \left[S^1(q) \ldots S^{n-k}(q) \right], \tag{2.32}$$

$$T(q) = \begin{bmatrix} S^T(q) \\ A^T(q) \end{bmatrix}, \qquad \tilde{p} = T(q)p \quad \Leftrightarrow \quad p = T^{-1}(q)\tilde{p}, \tag{2.33}$$

$$\tilde{H}(q, \tilde{p}) = H\left(q, T^{-1}(q)\tilde{p}\right). \tag{2.34}$$

In order to find the transformed motion equations, we have to determine the derivatives of $\tilde{H}$ and the relations to the derivatives of H. During this process, we can apply the rule about the derivative of any inverse matrix A^{-1} by a scalar variable a: $\frac{\partial A^{-1}}{\partial a} = -A^{-1}\frac{\partial A}{\partial a}A^{-1}$. Denote e_i the i-th standard unit vector and $\langle a, b \rangle = a^T b = b^T a = \langle b, a \rangle$ the scalar product of a and b, respectively. Then

$$\tilde{H}(q, \tilde{p}) = H\left(q, T^{-1}(q) \sum_j \tilde{p}_j e_j\right),$$

$$\frac{\partial \tilde{H}(q, \tilde{p})}{\partial \tilde{p}_i} = \left\langle \frac{\partial H(q, p)}{\partial p}, T^{-1}(q)e_i \right\rangle = \left\langle T^{-T}(q)\frac{\partial H(q, p)}{\partial p}, e_i \right\rangle,$$

$$\frac{\partial \tilde{H}(q, \tilde{p})}{\partial \tilde{p}} = T^{-T}(q)\frac{\partial H(q, p)}{\partial p} \quad \Leftrightarrow \quad \frac{\partial H(q, p)}{\partial p} = T^T(q)\frac{\partial \tilde{H}(q, \tilde{p})}{\partial \tilde{p}}, \tag{2.35}$$

$$\frac{\partial \tilde{H}(q, \tilde{p})}{\partial q_i} = \frac{\partial H(q, p)}{\partial q_i} + \left\langle \frac{\partial H(q, p)}{\partial p}, -T^{-1}(q)\frac{\partial T(q)}{\partial q_i}T^{-1}(q)\tilde{p} \right\rangle,$$

$$\frac{\partial \tilde{H}(q, \tilde{p})}{\partial q_i} = \frac{\partial H(q, p)}{\partial q_i} - \left\langle T^{-T}(q)\frac{\partial H(q, p)}{\partial p}, \frac{\partial T(q)}{\partial q_i}p \right\rangle,$$

$$\frac{\partial H(q, p)}{\partial q_i} = \frac{\partial \tilde{H}(q, \tilde{p})}{\partial q_i} + \left\langle \frac{\partial \tilde{H}(q, \tilde{p})}{\partial \tilde{p}}, \frac{\partial T(q)}{\partial q_i}p \right\rangle. \tag{2.36}$$

Let us consider now the time derivative of $\tilde{p}$. Since $\tilde{p} = T(q)p$, $\tilde{p}^1 = S^T(q)p$, $\tilde{p}^2 = A^T(q)p$, therefore

$$\dot{\tilde{p}} = \frac{dT(q)}{dt}p + T(q)\left(-\frac{\partial H(q, p)}{\partial q} + A(q)\lambda\right). \tag{2.37}$$

Since $S^T(q)A(q) = 0$, thus the constraint forces $A^T(q)A(q)\lambda$ appear only in $\dot{\tilde{p}}^2$. For simplicity of the next derivations, we suppress writing down the variables q, p. First, using (2.33) and (2.37), it follows

$$\dot{\tilde{p}}_i^1 = \frac{dS^{iT}}{dt}p - S^{iT}\frac{\partial H}{\partial q}. \tag{2.38}$$

Second, if the constraints are satisfied then, using (2.35) and (2.38), it yields

$$0 = A^T\dot{q} = A^T\frac{\partial H}{\partial p} = A^T[S\ A]\frac{\partial \tilde{H}}{\partial \tilde{p}} = A^T A\frac{\partial \tilde{H}}{\partial \tilde{p}^2} \quad \Leftrightarrow \quad \frac{\partial \tilde{H}}{\partial \tilde{p}^2} = 0, \tag{2.39}$$

$$\frac{dS^{iT}}{dt}p = \left\langle p, \frac{\partial S^i}{\partial q}\frac{\partial H}{\partial p}\right\rangle = p^T\frac{\partial S^i}{\partial q}[S^1(q)\ldots S^{n-k}(q)]\frac{\partial \tilde{H}}{\partial \tilde{p}^1}, \tag{2.40}$$

$$\left\langle \frac{\partial \tilde{H}}{\partial \tilde{p}}, \frac{\partial T}{\partial q_l}p\right\rangle = p^T\left[\frac{\partial S}{\partial q_l}\frac{\partial A}{\partial q_l}\right]\frac{\partial \tilde{H}}{\partial \tilde{p}} = p^T\frac{\partial S}{\partial q_l}\frac{\partial \tilde{H}}{\partial \tilde{p}^1}. \tag{2.41}$$

Third, using (2.36) the second term in (2.38) can be determined:

$$S^{iT}\frac{\partial H}{\partial q} = \sum_l S_l^i\frac{\partial H}{\partial q_l} = \sum_l S_l^i\left(\frac{\partial \tilde{H}}{\partial q_l} + p^T\frac{\partial S}{\partial q_l}\frac{\partial \tilde{H}}{\partial \tilde{p}^1}\right),$$

$$S^{iT}\frac{\partial H}{\partial q} = \sum_l S_l^i\frac{\partial \tilde{H}}{\partial q_l} + p^T\sum_l\left(\left[\frac{\partial S^1}{\partial q_l}\cdots\frac{\partial S^{n-k}}{\partial q_l}\right]S_l^i\right)\frac{\partial \tilde{H}}{\partial \tilde{p}^1}. \tag{2.42}$$

It is easy to show that, for example, $\sum_l \frac{\partial S_l^1}{\partial q_l}S_l^i = \frac{\partial S^1}{\partial q}S^i$ which together with the appropriate term from (2.40) and (2.38) results in the *Lie bracket* with negative sign

$$-[S^i, S^1] = -\left(\frac{\partial S^1}{\partial q}S^i - \frac{\partial S^i}{\partial q}S^1\right) \tag{2.43}$$

and in the state equations of $\tilde{p}_i^1, i = 1, \ldots, n - k$,

$$\dot{\tilde{p}}_i^1 = -S^{iT}\frac{\partial \tilde{H}}{\partial q} + \left(-p^T[S^i, S^1]\cdots - p^T[S^i, S^{n-k}]\right)\frac{\partial \tilde{H}}{\partial \tilde{p}^1}. \tag{2.44}$$

Using similar steps the state equations of $\tilde{p}_i^2, i = 1, \ldots, k$ can be obtained:

$$\dot{\tilde{p}}_i^2 = -A^{iT}\frac{\partial \tilde{H}}{\partial q} + \left(-p^T[A^i, S^1]\cdots - p^T[A^i, S^{n-k}]\right)\frac{\partial \tilde{H}}{\partial \tilde{p}^1}$$

$$+ A^{iT}(q)A(q)\lambda. \tag{2.45}$$

Introducing the notation

$$
\tilde{J}(q,\tilde{p}) = \begin{pmatrix} 0 & & S \\ & \begin{pmatrix} -p^T[S^1,S^1] & \cdots & -p^T[S^1,S^{n-k}] \\ -S^T & \vdots & \vdots & \vdots \\ -p^T[S^{n-k},S^1] & \cdots & -p^T[S^{n-k},S^{n-k}] \end{pmatrix} \\ -A^T & \begin{pmatrix} -p^T[A^1,S^1] & \cdots & -p^T[A^1,S^{n-k}] \\ \vdots & \vdots & \vdots \\ -p^T[A^k,S^1] & \cdots & -p^T[A^k,S^{n-k}] \end{pmatrix} \end{pmatrix} \tag{2.46}
$$

the dynamic model will be

$$
\begin{pmatrix} \dot{q} \\ \dot{\tilde{p}}^1 \\ \dot{\tilde{p}}^2 \end{pmatrix} = \tilde{J}(q,\tilde{p}) \begin{pmatrix} \frac{\partial \tilde{H}(q,\tilde{p})}{\partial q} \\ \frac{\partial \tilde{H}(q,\tilde{p})}{\partial \tilde{p}^1} \end{pmatrix} + \begin{pmatrix} 0 \\ 0 \\ A^T(q)A(q)\lambda \end{pmatrix}, \qquad \frac{\partial \tilde{H}(q,\tilde{p})}{\partial \tilde{p}^2} = 0. \tag{2.47}
$$

As can be seen, although $\frac{\partial \tilde{H}(q,\tilde{p})}{\partial \tilde{p}^2} = 0$, that is, $\tilde{H}$ does not depend on $\tilde{p}^2$, there is no guarantee that by using the above transformation the derivative of $\tilde{p}^2$ is zero and the constraint equation can be omitted. The reason is that, denoting by $\tilde{J}_r(q,\tilde{p}^1)$ the upper part of $\tilde{J}(q,\tilde{p})$,

$$
\tilde{J}_r(q,\tilde{p}^1) := \begin{pmatrix} 0 & & S \\ -S^T & \begin{pmatrix} -p^T[S^1,S^1] & \cdots & -p^T[S^1,S^{n-k}] \\ \vdots & \vdots & \vdots \\ -p^T[S^{n-k},S^1] & \cdots & -p^T[S^{n-k},S^{n-k}] \end{pmatrix} \end{pmatrix}, \tag{2.48}
$$

a bracket $\{\,,\,\}_r$ can be defined for scalar valued functions F_r, G_r mapping $(q,\tilde{p}^1)$ in R^1 by

$$
\{F_r, G_r\}_r(q,\tilde{p}^1) := \begin{pmatrix} \frac{\partial F_r^T}{\partial q} & \frac{\partial F_r^T}{\partial \tilde{p}^1} \end{pmatrix} \tilde{J}_r(q,\tilde{p}^1) \begin{pmatrix} \frac{\partial G_r}{\partial q} \\ \frac{\partial G_r}{\partial \tilde{p}^1} \end{pmatrix}. \tag{2.49}
$$

This bracket is skew-symmetric, satisfies the Leibniz rule, but usually does not satisfy the Jacobi identity. Is also the Jacobi identity satisfied then the bracket $\{\,,\,\}_r$ is a so called Poisson bracket, in which case the derivative of $\tilde{p}^2$ is zero and the constraint equations can be omitted, see [128].

2.1.3 Differentially Flat Systems

In Appendix B, we have shown that under special conditions, perhaps locally, the nonlinear system can be linearized using static or dynamic feedback and then it can be further compensated by using well known linear control methods. However, it remains a serious problem to obtain the symbolic solution of a system of partial differential equations (Frobenius equations) for which no general and efficient methods are available. All these offer reason for seeking special solutions in which range the flatness control is a possible approach.

The idea of differentially flat systems arose at the 90th years based on the works of David Hilbert and Elie Cartan at the beginning of the 20th century. We shall discuss differentially flat systems using the results of [39]. Fundamentals of differential geometry and novel results in flatness control can be found in [1] and [80], respectively.

First, the idea of dynamic system will be extended to infinite dimensional manifolds which can be obtained from conventional systems through prolongation of vector fields. For infinite dimensional system, the so called Lie–Bäcklund transformation will be defined which is equivalent to an endogenous dynamic feedback and state transformation performed on an appropriate finite dimensional system. Based on this result, the class of differentially flat systems will be defined in such a way that the systems belonging to it can be converted to the trivial series of integrators by the help of the Lie–Bäcklund transformation.

First of all, we start with a new interpretation of the vector field. In Appendix B, the idea of the manifold $X \in R^n$, its tangent space TX and the Lie derivative $L_f h$ of smooth scalar valued function h relative to a vector field f have already been defined. Considering $L_f h = \sum_{i=1}^{n} f_i \frac{\partial h}{\partial x_i}$ allow us to identify the vector field f by

$$f = (f_1, \ldots, f_n)^T \sim \sum_{i=1}^{n} f_i \frac{\partial}{\partial x_i} \qquad (2.50)$$

which can be assumed the componentwise or differential operator expression of the vector field f. This interpretation can easily be generalized for infinite dimension.

2.1.3.1 Prolongation of Vector Fields, the Cartan Fields

Let us consider the system given by the differential equation

$$\dot{x} = f(x, u) \qquad (2.51)$$

where x is a point of an n dimensional manifold M, $u \in R^m$ and f is $C^{(\infty)}$ vector field on M which can be described by the local coordinates $\frac{\partial}{\partial x_i}$ ($i = 1, 2, \ldots, n$) in the form $f = \sum_{i=1}^{n} f_i \frac{\partial}{\partial x_i}$. It is assumed $u(t) \in C^{(\infty)}$. Let us introduce the infinite dimensional vector

$$\xi(t) = [\xi_1 \ \xi_2 \ \xi_3 \ \cdots]^T = \left[x(t)^T \ u(t)^T \ \dot{u}(t)^T \ \cdots \right]^T$$

for which $\xi \in M \times R^m \times R^m \times \cdots = M \times R_\infty^m = \bar{M}$, that is, ξ is a point of an infinite dimensional manifold. Let $\bar{f}$ be a vector field over $M \times R_\infty^m = \bar{M}$ derived from f

by using

$$\dot{\xi} = \bar{f}(\xi) = \begin{bmatrix} f(\xi_1, \xi_2) \\ \xi_3 \\ \xi_4 \\ \vdots \end{bmatrix}. \tag{2.52}$$

Notice that, despite of the infinite dimensional vector field, every row depends only on a finite number of elements of ξ. The vector field $\bar{f}$ can be written in local coordinates

$$\bar{f} = \sum_{i=1}^{n} f_i \frac{\partial}{\partial x_i} + \sum_{j \geq 0} \sum_{i=1}^{m} u_i^{(j+1)} \frac{\partial}{\partial u_i^{(j)}} \tag{2.53}$$

which will be called a Cartan field. It follows from the construction of the vector field $\bar{f}$ that if a path $t \to (x(t), u(t))$ is a solution of (2.51) then the path $t \to (x(t), u(t), \dot{u}(t), \ddot{u}(t), \ldots)$ is a solution of (2.52).

In what follows, we shall understand under control system the ordered pair $(\bar{M}, \bar{f})$ where $\bar{M}$ is an infinite dimensional manifold and $\bar{f}$ is a Cartan field of the form (2.53) over $\bar{M}$.

Now consider the linear dynamic system consisting of the series of integrators:

$$x_{r_1}^{(\kappa_1)} = u_1$$
$$x_{r_2}^{(\kappa_2)} = u_2 \qquad r_1 = 1; r_i = r_{i-1} + \kappa_{i-1}; \sum_i \kappa_i = n.$$
$$\vdots$$
$$x_{r_m}^{(\kappa_m)} = u_m$$

In the above form, r_i denotes the dimension, κ_i is the order (number of integrators).

The infinite dimensional vector field in local coordinates belonging to this system can be written in the form

$$\bar{f}_{\text{triv}} = \sum_{j \geq 0} \sum_{i=1}^{m} x_i^{(j+1)} \frac{\partial}{\partial x_i^{(j)}}. \tag{2.54}$$

This vector field will be called the trivial vector field. Notice that there is a one-to-one relation between the integral curves $t \to (x(t), u(t))$ of the trivial system and the properly regular (differentiable) time functions $t \to (x_{r_1}(t), \ldots, x_{r_m}(t))$.

2.1.3.2 Lie–Bäcklund Transformation

Consider the control systems $(\bar{M}, \bar{f})$ and $(\bar{N}, \bar{g})$ derived from the differential equations $\dot{x} = f(x, u)$ and $\dot{y} = g(y, v)$ (dim u = dim v) where $\bar{M} = M \times R_\infty^m$

(dim $M = n$) and $\bar{N} = N \times R_\infty^m$ (dim $N = l$), that is, the traditional dimension number of the two systems are different, while the number of inputs is the same. Furthermore consider an invertible $C^{(\infty)}$ mapping which defines a coordinate transformation between $\bar{M}$ and $\bar{N}$: $\xi = \Phi(\zeta)$, $\xi \in \bar{M}$ and $\zeta \in \bar{N}$. Denote Ψ the inverse coordinate transformation which is also $C^{(\infty)}$ and let $\zeta = \Psi(\xi)$. Only such coordinate transformations will be considered for which each of their components depends only on a finite number of coordinates, similarly to the prolonged vector fields. Thus $\zeta_1 = \psi_1(\bar{\xi}_1)$, $\zeta_2 = \psi_2(\bar{\xi}_2), \ldots, \zeta_i = \psi_i(\bar{\xi}_i), \ldots$ where $\bar{\xi}_i = (\xi_{i,1}, \ldots, \xi_{i,r_i})$ and r_i are finite.

Now consider an integral curve $t \to \xi(t)$ of the vector field $\bar{f}$. Its image over the manifold $\bar{N}$ is $t \to \zeta(t) = \Psi(\xi(t))$. Denote $\Psi_* \bar{f}$ that vector field over the manifold $\bar{N}$ whose integral curve is $t \to \zeta(t) = \Psi(\xi(t))$. Differentiating the curve $\Psi(\xi(t))$ by the time yields

$$\frac{d}{dt}\Psi\big(\xi(t)\big) = \frac{\partial \Psi}{\partial \xi} \cdot \bar{f}(\xi) = L_{\bar{f}}\Psi(\xi),$$

but since $\Psi_* \bar{f} : \bar{N} \to T\bar{N}$ thus

$$\Psi_* \bar{f} = \left(L_{\bar{f}}\Psi\right) \circ \Phi = \left(L_{\bar{f}}\psi_1\right) \circ \Phi \frac{\partial}{\partial y} + \sum_{j \geq 0}\left(L_{\bar{f}}\psi_{j+1}\right) \circ \Phi \frac{\partial}{\partial v^{(j)}}. \tag{2.55}$$

Notice that in the above expression all terms are finite since the transformation and each element of the vector field depend on finite number of coordinates.

For transformation Ψ of general form, it is not satisfied that the resulting vector field (2.55) is a Cartan field, that is, (2.55) can be derived from a "traditional" form dynamic system over $\bar{N}$. To this it has to be required that the prolongation is a series of integrators, that is,

$$\frac{dv^{(j)}}{dt} = L_{\Psi_* \bar{f}}v^{(j)} = v^{(j+1)} = \psi_{j+2} \circ \Phi,$$

from which using (2.55) follows

$$L_{\Psi_* \bar{f}}v^{(j)} = \left(L_{\bar{f}}\psi_1\right) \circ \phi \frac{\partial v^{(j)}}{\partial y} + \sum_{k \geq 0}\left(L_{\bar{f}}\psi_{k+1}\right) \circ \Phi \frac{\partial v^{(j)}}{\partial v^{(k)}} = \left(L_{\bar{f}}\psi_{j+1}\right) \circ \Phi.$$

Comparing the right sides of the last two equations, we obtain that $\Psi_* \bar{f}$ is a Cartan field over $\bar{N}$, if and only if, $L_{\bar{f}}\psi_{j+1} = \psi_{j+2}$ is guaranteed for every positive j. Coordinate transformations mapping integral curves of Cartan fields into integral curves of Cartan fields will be called Lie–Bäcklund transformations.

The control systems $(\bar{M}, \bar{f})$ and $(\bar{N}, \bar{g})$ are called Lie–Bäcklund equivalent if there exists an invertible and $C^{(\infty)}$ coordinate transformation $\Psi : \bar{M} \to \bar{N}$, each element of which is function of only a finite number of coordinates and $\Psi_* \bar{f} = \bar{g}$.

The Lie–Bäcklund transformation saves the number of inputs and the equilibrium points. Consequently it follows that any two linear controllable systems of equal number of inputs are Lie–Bäcklund equivalent.

2.1.3.3 Endogenous State Feedback

The state feedback is dynamic if it has own state:

$$\dot{z} = \beta(x, z, v),$$
$$u = \alpha(x, z, v).$$
(2.56)

We shall deal only with such feedback where the number of inputs of the original and fed-back system does not change, that is, $\dim u = \dim v$. Using (2.56) for the original system (2.51), the differential equation of the closed loop will be

$$\dot{x} = f\big(x, \alpha(x, z, v)\big),$$
$$\dot{z} = \beta(x, z, v).$$
(2.57)

The dynamic feedback does not necessarily save the controllability properties. An example for this is the common use of state observer and state feedback for linear systems where the closed loop system obtained by using dynamic feedback because the states of the feedback are the state variables $\hat{x}$ of the observer. The system obtained so is not state controllable since the observer was designed to assure exponential decaying the estimation error according to the differential equation $(\dot{x} - \dot{\hat{x}}) = F(x - \hat{x})$ (where F is stable and quick), hence there is no input making the behavior of $\hat{x}$ independent of x. In order to eliminate such phenomenon, we restrict the investigations to so-called endogenous dynamic feedback defined as follows.

Let be given a system $(\bar{M}, \bar{f})$. The feedback given in the form (2.56) is called *endogenous dynamic feedback* if the prolongation of the closed loop system given by (2.57) is Lie–Bäcklund equivalent to the system $(\bar{M}, \bar{f})$.

Theorem 2.1 *Existence condition for endogenous state feedback.*
Consider the systems $(\bar{M}, \bar{f})$ and $(\bar{N}, \bar{g})$. If they are Lie–Bäcklund equivalent, then there exist an endogenous state feedback of type (2.56) and a traditional coordinate transformation (diffeomorphism) which transfer the system $\dot{x} = f(x, u)$ into the system prolonged with finite number of integrators:

$$\dot{y} = g(y, w),$$
$$w^{(r+1)} = v.$$

2.1.3.4 Flat Output Variable

The system $(\bar{M}, \bar{f})$ is differentially flat if it is Lie–Bäcklund equivalent to the trivial system given by the vector field

$$\bar{f}_{\text{triv}} = \sum_{j \geq 0} \sum_{i=1}^{m} y_i^{(j+1)} \frac{\partial}{\partial y_i^{(j)}}.$$
(2.58)

The variables of the vector y (their number is equal to the number of inputs of $(\bar{M}, \bar{f})$) will be called the flat outputs.

Corollary *Differentially flat systems can be linearized by dynamic state feedback and a coordinate transformation since (2.58) is arisen from a linear system prolonged with a series of integrators. According to this property, y is often called the linearizing output.*

Theorem 2.2 *Differentially flat system can be characterized by the flat output variable.*

Consider the differentially flat system $(\bar{M}, \bar{f})$ arisen from the prolongation of the system $\dot{x} = f(x, u)$. Denote Ψ the Lie–Bäcklund transformation which transfers $(\bar{M}, \bar{f})$ into a trivial system and denote Φ the inverse transformation. It follows from the definition of differential flatness that there is a finite q such that

$$x = \phi_1\left(y, \dot{y}, \ddot{y}, \ldots, y^{(q)}\right),$$
$$u = \phi_2\left(y, \dot{y}, \ddot{y}, \ldots, y^{(q+1)}\right),$$
(2.59)

and x and u identically satisfy the equation $\dot{x} = f(x, u)$ describing the system. Furthermore, there is a finite p yielding

$$y = \psi_1\left(x, u, \dot{u}, \ldots, u^{(p)}\right).$$
(2.60)

According to (2.59) and (2.60), the path of x and u can be determined from y and its derivatives and there is a one-to-one correspondence between the integral curves $t \to (x(t), u(t))$ satisfying $\dot{x} = f(x, u)$ and the properly regular time functions $t \to y(t)$. It means that in case of differentially flat systems there exists a set of variables inside of which the differential constraints defining the system disappear.

The system (2.51) is differentially flat around a point $\Leftrightarrow$ it is possible to find an output variable y, $\dim y = \dim u = m$, which smoothly depends around that point on x, u, and a finite number of derivatives of u, such that the resulting square system is left and right input–output invertible with *trivial zero dynamics* [39].

Corollary *In order to admit the differential flatness property, it is "enough" to find the output y and the functions ϕ_1, ϕ_2 and ψ_1 appearing in (2.59) and (2.60).*

2.1.3.5 Path Design for Differentially Flat Systems

Beside the linearizability (which not always can be exploited because many times not all the components of the state vector can be measured), differential flatness can play important role for *path design*. Namely, the path design can be performed in the space of the variables of y. Then the path of the state and the inputs can simply be determined using the time derivatives according to (2.59), without the integration of the differential equations. Hence, during path design it is sufficient to choose a properly regular path satisfying the goals of the control in the space of the variables of y using simple interpolation techniques (Lagrange polynomials etc.).

2.1.3.6 Application Possibilities for Different System Classes

Recently, there are strong necessary and sufficient conditions to decide whether the system is differentially flat or not (see, for example, [80, Theorem 6.7]), however the steps to be performed are too complicated for general nonlinear systems. Exceptions are SISO nonlinear systems for which necessary and sufficient conditions are available [56] and the linearizing feedback remains static (if it exists). There are also necessary and sufficient conditions for some kind of driftless systems [90]. For people dealing with robotics, it has been known for a long time that open chain, rigid and in all degrees of freedom actuated mechanical systems can be linearized using static nonlinear feedback (called nonlinear decoupling in robotics) which, as mentioned above, is equivalent to the differential flatness property. On the other hand, there exist also differentially flat mechanical systems that are underactuated or nonholonomic. A counter example is the inverted pendulum which is not differentially flat.

2.2 Dynamic Model of Simple Systems

Some simple examples will be shown for finding the dynamic model of simple mechanical systems using physical laws. We try to apply as simple knowledge as possible in the discussions. For example, the kinetic energy and the potential energy of a mass point are $K = \frac{1}{2}mv^2$ and $P = mgh$, respectively, where g is the gravity acceleration and h is the hight of the mass point over the horizontal plane. Similarly, the kinetic energy of a mass point rotating around an axis is $K = \frac{1}{2}\Theta\omega^2$ where ω is the angular velocity, $\Theta = ml^2$ is the inertia moment to the rotation axis and l is the length of the massless rod containing the mass point at its end.

For a bit more complicated case of a single rigid body, we can apply the results in Appendix A. For example, for a rigid body having velocity v_c, angular velocity ω and inertia matrix I_c belonging to the center of mass the kinetic energy is $K = \frac{1}{2}\langle v_c, v_c\rangle m + \frac{1}{2}\langle I_c\omega, \omega\rangle$. However, this result was also derived from the above relations because the rigid body is the limit of infinite many point masses. From the physical laws, we shall apply the Lagrange equations, see also Appendix A.

2.2.1 Dynamic Model of Inverted Pendulum

Consider a pendulum mounted on a moving car (see Fig. 2.4).

Fig. 2.4 The model of the
inverted pendulum

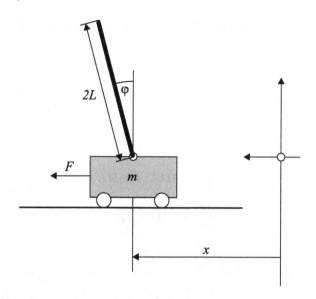

Let L be the length of the rod mounted on the car and M the mass of the rod then
the inertial moment of the rod belonging to the center of mass:

$$\Theta = \int_{-L}^{L} l^2 \, dm = \int_{-L}^{L} l^2 \frac{M}{2L} \, dl = \frac{M}{2L} \frac{l^3}{3} \bigg|_{-L}^{L} = \frac{M}{6L} 2L^3 = \frac{ML^2}{3}. \qquad (2.61)$$

The generalized coordinates are x and φ. The coordinates of the center of mass
of the rod are x_M and y_M, its velocity is v_M. For simplicity, let us introduce the
usual notations in robotics: $S_\varphi = \sin(\varphi)$, $C_\varphi = \cos(\varphi)$. Then

$$\begin{aligned}
x_M &= x + L S_\varphi \quad \Rightarrow \quad \dot{x}_M = \dot{x} + \dot{\varphi} L C_\varphi, \\
y_M &= L C_\varphi \quad \Rightarrow \quad \dot{y}_M = -\dot{\varphi} L S_\varphi, \\
v_M^2 &= \dot{x}_M^2 + \dot{y}_M^2 = \dot{x}^2 + 2L C_\varphi \dot{x} \dot{\varphi} + L^2 C_\varphi^2 \dot{\varphi}^2 + L^2 S_\varphi^2 \dot{\varphi}^2 \\
&= \dot{x}^2 + 2L C_\varphi \dot{x} \dot{\varphi} + L^2 \dot{\varphi}^2.
\end{aligned} \qquad (2.62)$$

The kinetic energy K and the potential energy P are respectively,

$$\begin{aligned}
K &= \frac{1}{2} m \dot{x}^2 + \frac{1}{2} M v_M^2 + \frac{1}{2} \Theta \dot{\varphi}^2 \\
&= \frac{1}{2} m \dot{x}^2 + \frac{1}{2} M \left(\dot{x}^2 + 2L C_\varphi \dot{x} \dot{\varphi} + L^2 \dot{\varphi}^2 \right) + \frac{1}{2} \Theta \dot{\varphi}^2, \\
P &= M g L C_\varphi.
\end{aligned} \qquad (2.63)$$

The derivatives, needed to Lagrange equations, are as follows:

$$\frac{\partial K}{\partial \dot{x}} = m\dot{x} + M\dot{x} + MLC_\varphi\dot{\varphi},$$

$$\frac{\partial K}{\partial x} = 0,$$

$$\frac{\partial P}{\partial x} = 0,$$

$$\frac{\partial K}{\partial \dot{\varphi}} = MLC_\varphi\dot{x} + ML^2\dot{\varphi} + \Theta\dot{\varphi},$$ (2.64)

$$\frac{\partial K}{\partial \varphi} = -MLS_\varphi\dot{x}\dot{\varphi},$$

$$\frac{\partial P}{\partial \varphi} = -MgLS_\varphi,$$

$$\frac{d}{dt}\frac{\partial K}{\partial \dot{x}} = (m+M)\ddot{x} + MLC_\varphi\ddot{\varphi} - MLS_\varphi\dot{\varphi}^2,$$

$$\frac{d}{dt}\frac{\partial K}{\partial \dot{\varphi}} = MLC_\varphi\ddot{x} - MLS_\varphi\dot{x}\dot{\varphi} + \left(\Theta + ML^2\right)\ddot{\varphi}.$$ (2.65)

If the car is pulled with force F, then the Lagrange equations yield:

$$\frac{d}{dt}\frac{\partial K}{\partial \dot{x}} - \frac{\partial K}{\partial x} + \frac{\partial P}{\partial x}$$

$$= F \quad \Rightarrow \quad (m+M)\ddot{x} + MLC_\varphi\ddot{\varphi} - MLS_\varphi\dot{\varphi}^2 = F,$$ (2.66)

$$\frac{d}{dt}\frac{\partial K}{\partial \dot{\varphi}} - \frac{\partial K}{\partial \varphi} + \frac{\partial P}{\partial \varphi}$$

$$= 0 \quad \Rightarrow \quad MLC_\varphi\ddot{x} + \left(\Theta + ML^2\right)\ddot{\varphi} - MgLS_\varphi = 0.$$ (2.67)

From here, it can be obtained firstly

$$\ddot{x} = \frac{F - MLC_\varphi\ddot{\varphi} + MLS_\varphi\dot{\varphi}^2}{m + M},$$ (2.68)

then after some algebraical manipulations

$$MLC_\varphi\frac{F - MLC_\varphi\ddot{\varphi} + MLS_\varphi\dot{\varphi}^2}{m+M} + ML^2\frac{4}{3}\ddot{\varphi} - MgLS_\varphi = 0,$$

$$L\left(\frac{4}{3} - \frac{MC_\varphi^2}{m+M}\right)\ddot{\varphi} = gS_\varphi - \frac{C_\varphi}{m+M}F - \frac{MLS_\varphi C_\varphi\dot{\varphi}^2}{m+M},$$

Fig. 2.5 Simplified model of
a car active suspension

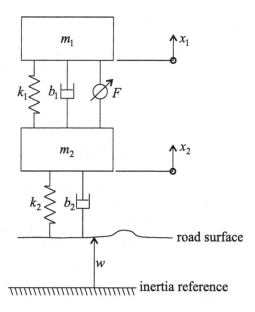

and finally

$$\ddot{\varphi} = \frac{g S_\varphi - \frac{C_\varphi}{m+M}(F + M L S_\varphi \dot{\varphi}^2)}{L(\frac{4}{3} - \frac{M C_\varphi^2}{m+M})}. \tag{2.69}$$

The state vector is $(x, \dot{x}, \varphi, \dot{\varphi})^T$ and the input signal is F. The two output signals
are x and φ. The system is underactuated because there are two output signals and
only one input signal.

2.2.2 Car Active Suspension Model

Let us consider a vehicle (car, bus etc.) divided into 4 parts in case of 4 wheels, each
part is a composite mechanical spring-damper system consisting of the quarter part
of the mass of the body (together with the passengers) and the mass of the wheel.
The vertical positions are described by upward directed x_1 and x_2, see Fig. 2.5. The
distance between the road surface and the wheel's contact point is the disturbance
w varying together with the road surface. The connections between the different
parts are flexible modeled by viscous friction and spring constant. The suspension
is active which means that the actuator produces the force F (control signal).

A good suspension system should have satisfactory road holding stability, while
providing good traveling comfort when riding over bumps and holes in the road.
When the car is experiencing any road disturbance, it should not have large oscillations and the oscillations should dissipate quickly.

Denote m_1 and m_2 the mass of the (quarter) body and the wheel respectively, while the flexible connections are described by the viscous damping factors (b_1, b_2) and the spring constants (k_1, k_2). The motion equations, using also the action-reaction (d'Alambert) principle, are as follows:

$$m_1\ddot{x}_1 = F - b_1(\dot{x}_1 - \dot{x}_2) - k_1(x_1 - x_2),$$
$$m_2\ddot{x}_2 = -F + b_1(\dot{x}_1 - \dot{x}_2) + k_1(x_1 - x_2) - b_2(\dot{x}_2 - \dot{w}) - k_2(x_2 - w).$$

As usual, the second derivatives will be expressed:

$$\ddot{x}_1 = -\frac{b_1}{m_1}(\dot{x}_1 - \dot{x}_2) - \frac{k_1}{m_1}(x_1 - x_2) + \frac{1}{m_1}F,$$

$$\ddot{x}_2 = \frac{b_1}{m_2}(\dot{x}_1 - \dot{x}_2) + \frac{k_1}{m_2}(x_1 - x_2) - \frac{b_2}{m_2}(\dot{x}_2 - \dot{w}) - \frac{k_2}{m_2}(x_2 - w) - \frac{1}{m_2}F,$$

$$y_1 := x_1 - x_2,$$

$$\ddot{x}_1 = -\frac{b_1}{m_1}\dot{y}_1 - \frac{k_1}{m_1}y_1 + \frac{1}{m_1}F,$$

$$\ddot{x}_2 = \frac{b_1}{m_2}\dot{y}_1 + \frac{k_1}{m_2}y_1 - \frac{b_2}{m_2}(\dot{x}_2 - \dot{w}) - \frac{k_2}{m_2}(x_2 - w) - \frac{1}{m_2}F,$$

$$\ddot{y}_1 = \ddot{x}_1 - \ddot{x}_2$$

$$= -\left(\frac{b_1}{m_1} + \frac{b_2}{m_2}\right)\dot{y}_1 - \left(\frac{k_1}{m_1} + \frac{k_2}{m_2}\right)y_1$$

$$+ \frac{b_2}{m_2}(\dot{x}_2 - \dot{w}) + \frac{k_2}{m_2}(x_2 - w) + \left(\frac{1}{m_1} + \frac{1}{m_2}\right)F.$$

In order to eliminate $\dot{w}$, the equation of $\ddot{y}_1$ will be integrated:

$$\dot{y}_1 = -\left(\frac{b_1}{m_1} + \frac{b_2}{m_2}\right)y_1 + \frac{b_2}{m_2}(x_2 - w)$$

$$+ \underbrace{\int\left\{-\left(\frac{k_1}{m_1} + \frac{k_2}{m_2}\right)y_1 + \frac{k_2}{m_2}(x_2 - w) + \left(\frac{1}{m_1} + \frac{1}{m_2}\right)F\right\}dt}_{y_2}.$$

Denote y_2 the integral and let us differentiate it:

$$\dot{y}_2 = -\left(\frac{k_1}{m_1} + \frac{k_1}{m_2}\right)y_1 + \frac{k_2}{m_2}(x_2 - w) + \left(\frac{1}{m_1} + \frac{1}{m_2}\right)F.$$

Perform the substitution $x_2 = x_1 - y_1$:

$$\dot{y}_2 = \frac{k_2}{m_2}x_1 - \left(\frac{k_1}{m_1} + \frac{k_1}{m_2} + \frac{k_2}{m_2}\right)y_1 + \left(\frac{1}{m_1} + \frac{1}{m_2}\right)F - \frac{k_2}{m_2}w,$$

$$\dot{y}_1 = \frac{b_2}{m_2}x_1 - \left(\frac{b_1}{m_1} + \frac{b_2}{m_2} + \frac{b_2}{m_2}\right)y_1 + y_2 - \frac{b_2}{m_2}w,$$

$$\ddot{x}_1 = -\frac{b_1}{m_1}\left\{-\left(\frac{b_1}{m_1} + \frac{b_2}{m_2}\right)y_1 + \frac{b_2}{m_2}(x_1 - y_1 - w) + y_2\right\} - \frac{k_1}{m_1}y_1 + \frac{1}{m_1}F$$

$$= \frac{b_1b_2}{m_1m_2}x_1 + \left\{\frac{b_1}{m_1}\left(\frac{b_1}{m_1} + \frac{b_1}{m_2} + \frac{b_2}{m_2}\right) - \frac{k_1}{m_1}\right\}y_1 - \frac{b_1}{m_1}y_2 + \frac{1}{m_1}F + \frac{b_1b_2}{m_1m_2}w.$$

State equation:

$$\dot{\bar{x}} = A\bar{x} + Bu, \qquad y = C\bar{x},$$

$$\bar{x} := (x_1, \dot{x}_1, y_1, \dot{y}_1)^T, \qquad u = (F, w)^T, \qquad y = x_1 - x_2 = y_1,$$

$$A = \begin{bmatrix} 0 & 1 & 0 & 0 \\ -\frac{b_1b_2}{m_1m_2} & 0 & \frac{b_1}{m_1}\left(\frac{b_1}{m_1} + \frac{b_1}{m_2} + \frac{b_2}{m_2}\right) - \frac{k_1}{m_1} & -\frac{b_1}{m_1} \\ \frac{b_2}{m_2} & 0 & -\left(\frac{b_1}{m_1} + \frac{b_1}{m_2} + \frac{b_2}{m_2}\right) & 1 \\ \frac{k_2}{m_2} & 0 & -\left(\frac{k_1}{m_1} + \frac{k_1}{m_2} + \frac{k_2}{m_2}\right) & 0 \end{bmatrix},$$

$$B = \begin{bmatrix} 0 & 0 \\ \frac{1}{m_1} & \frac{b_1b_2}{m_1m_2} \\ 0 & -\frac{b_2}{m_2} \\ \left(\frac{1}{m_1} + \frac{1}{m_2}\right) & -\frac{k_2}{m_2} \end{bmatrix},$$

$$C = \begin{bmatrix} 0 & 0 & 1 & 0 \end{bmatrix}.$$

Some typical values for a bus in SI units:

$$m_1 = 2500, \qquad b_1 = 350, \qquad k_1 = 80000,$$

$$m_2 = 320, \qquad b_2 = 15020, \qquad k_2 = 500000.$$

Typical requirements in case of bus:
In order to assure the traveling comfort, settling time shorter than 10 s and overshoot less that 5% is usually necessary. For example, when the bus runs onto a 10 cm high step the body will oscillate within a range of ± 5 mm (i.e., ± 0.005 m) and should return to smooth ride within 5 s.

2.2.3 The Model of the 2 DOF Robot Arm

Consider the two degree of freedom robot arm moving in the vertical plane, see the block scheme in Fig. 2.6.

The generalized coordinates are $q = (q_1, q_2)^T := (\varphi_1, \varphi_2)^T$ and the generalized forces are the driving torques $\tau := (\tau_1, \tau_2)^T$. The position vectors pointing from

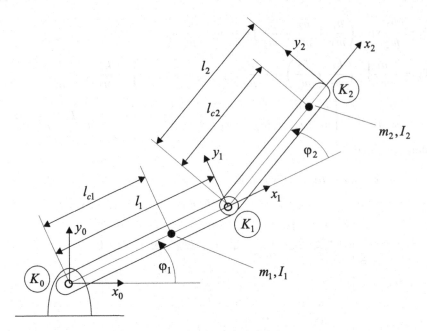

Fig. 2.6 The model of a 2 DOF robot moving in vertical plane

the origin of the fixed frame (coordinate system) K_0 into the origins of the moving frames K_1 and K_2 are respectively,

$$p_{01} = (l_1 C_1, l_1 S_1, 0)^T,$$
$$p_{02} = (l_1 C_1 + l_2 C_{12}, l_1 S_1 + l_2 S_{12}, 0)^T,$$
(2.70)

where the notations $C_1 := \cos(q_1), \ldots, S_{12} = \sin(q_1 + q_2)$ are used. The velocities in the basis of the fixed frame K_0 can be obtained by formal differentiation of the position vectors:

$$^0 v_1 = \begin{bmatrix} -l_1 S_1 \\ l_1 C_1 \\ 0 \end{bmatrix} \dot{q}_1,$$

$$^0 v_2 = \begin{bmatrix} -l_1 S_1 - l_2 S_{12} & -l_2 S_{12} \\ l_1 C_1 + l_2 C_{12} & l_2 C_{12} \\ 0 & 0 \end{bmatrix} \begin{pmatrix} \dot{q}_1 \\ \dot{q}_2 \end{pmatrix}.$$
(2.71)

The inertia matrix is constant in the frame of the moving body but is nonconstant in the fixed frame, hence we transform the velocities into the basis of the moving frame:

$$v_1 := {}^1v_1 = \begin{bmatrix} C_1 & S_1 & 0 \\ -S_1 & C_1 & 0 \\ 0 & 0 & 1 \end{bmatrix} {}^0v_1 = \begin{bmatrix} c0 \\ l_1 \\ 0 \end{bmatrix} \dot{q}_1,$$

$$v_2 := {}^2v_2 = \begin{bmatrix} C_{12} & S_{12} & 0 \\ -S_{12} & C_{12} & 0 \\ 0 & 0 & 1 \end{bmatrix} {}^0v_2 = \begin{bmatrix} l_1S_2 & 0 \\ l_1C_2+l_2 & l_2 \\ 0 & 0 \end{bmatrix} \begin{pmatrix} \dot{q}_1 \\ \dot{q}_2 \end{pmatrix}.$$

(2.72)

If each moving frame will be parallelly transferred to the origin of the center of mass, then the velocity (v_c) and the inertia matrix belonging to the center of mass have to be used in the motion equation. In the block scheme, I_1 and I_2 are already the right lower elements of the inertia matrices I_{c1} and I_{c2}, respectively. The velocities v_{c1}, v_{c2} can be immediately written down using simple geometrical considerations:

$$v_{c1} = \begin{bmatrix} 0 \\ l_{c1} \\ 0 \end{bmatrix} \dot{q}_1,$$

$$v_{c2} = \begin{bmatrix} l_1S_2 & 0 \\ l_1C_2+l_{c2} & l_{c2} \\ 0 & 0 \end{bmatrix} \begin{pmatrix} \dot{q}_1 \\ \dot{q}_2 \end{pmatrix}.$$

(2.73)

From the geometrical image, the angular velocities are simply

$$\omega_1 = (0, 0, \dot{q}_1)^T,$$
$$\omega_2 = (0, 0, \dot{q}_1 + \dot{q}_2)^T,$$

which can also be written in matrix-vector form:

$$\omega_1 = \begin{bmatrix} 0 \\ 0 \\ 1 \end{bmatrix} \dot{q}_1,$$

$$\omega_2 = \begin{bmatrix} 0 & 0 \\ 0 & 0 \\ 1 & 1 \end{bmatrix} \begin{pmatrix} \dot{q}_1 \\ \dot{q}_2 \end{pmatrix}.$$

(2.74)

Hence, the kinetic and potential energy of the robot arm are respectively,

$$K = \frac{1}{2}\langle v_{c1}, v_{c1}\rangle m_1 + \frac{1}{2}\langle I_1\omega_1, \omega_1\rangle$$
$$+ \frac{1}{2}\langle v_{c2}, v_{c2}\rangle m_2 + \frac{1}{2}\langle I_2\omega_2, \omega_2\rangle,$$

(2.75)

$$P = m_1gl_{c1}S_1 + m_2g(l_1S_1 + l_{c2}S_{12}).$$

After simple algebraic manipulations, the kinetic energy can be formulated in the quadratic form

$$K = \frac{1}{2}\langle H(q)\dot{q}, \dot{q}\rangle = \frac{1}{2}\sum_j \sum_k D_{jk}\dot{q}_j\dot{q}_k,$$

where

$$H(q) = \begin{bmatrix} D_{11}(q) & D_{12}(q) \\ D_{12}(q) & D_{22}(q) \end{bmatrix},$$

$$D_{11} = m_1 l_{c1}^2 + m_2\left(l_1^2 + l_{c2}^2 + 2l_1 l_{c2}C_2\right) + I_1 + I_2, \qquad (2.76)$$

$$D_{12} = m_2(l_1 C_2 + l_{c2})l_{c2} + I_2,$$

$$D_{22} = m_2 l_{c2}^2 + I_2.$$

The dynamic model of the robot arm can be found using the Lagrange equations. Performing the differentiations and introducing the notations

$$D_{112}(q) = -m_2 l_1 l_{c2} S_2,$$

$$D_1(q) = \frac{\partial P}{\partial q_1} = m_1 g l_{c1}C_1 + m_2 g (l_1 C_1 + l_{c2}C_{12}), \qquad (2.77)$$

$$D_2(q) = \frac{\partial P}{\partial q_2} = m_2 g l_{c2}C_{12},$$

we obtain the following dynamic model of the two degree of freedom robot arm:

$$D_{11}(q)\ddot{q}_1 + D_{12}(q)\ddot{q}_2 + D_{112}(q)\left(2\dot{q}_1\dot{q}_2 + \dot{q}_2^2\right) + D_1(q) = \tau_1,$$

$$D_{12}(q)\ddot{q}_1 + D_{22}(q)\ddot{q}_2 - D_{112}(q)\dot{q}_1^2 + D_2(q) = \tau_2. \qquad (2.78)$$

The state vector may be $x := (q_1, q_2, \dot{q}_1, \dot{q}_2)^T$, the control signal (the input of the model) is $u = (\tau_1, \tau_2)^T$. Since, from physical considerations, the matrix $H(q)$ is positive definite, thus invertible, therefore $\ddot{q}_1$ and $\ddot{q}_2$ can be expressed from (2.77) and (2.78). Hence, the state equation can be written in the input affine form

$$\dot{x} = f(x) + G(x)u. \qquad (2.79)$$

2.3 Stability of Nonlinear Systems

The stability of nonlinear systems will be discussed in state space, in signal spaces and in the class of dissipative systems. The discussion is based on the classical work of Lyapunov [86] and more recent results of LaSalle and Lefschetz [77], Vidyasagar [152], Khalil [62] and van der Schaft [127].

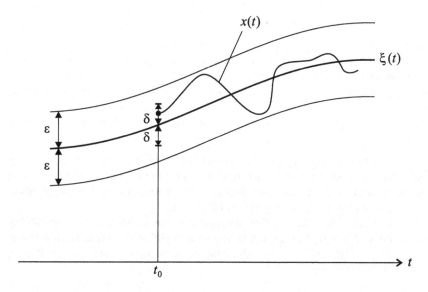

Fig. 2.7 Illustration of Lyapunov stability definition

2.3.1 Stability Definitions

Let the state equation of the *time varying* nonlinear system be

$$\dot{x} = f(t, x, u),$$
$$y = g(t, x, u). \tag{2.80}$$

Assuming $u(t) = 0$ or $u(t) = u_0(t)$ known input signal, the nonlinear system can be brought on the form $\dot{x} = f(t, x)$. In what follows we shall assume $f \in C_{t,x}^{(0,1)}([a, \infty) \times D_x)$, that is, f is continuous in t and differentiable in x, furthermore $a \in R^1$ and $D_x \subset R^n$ is open, for example $a = 0$ and $D_x = R^n$. Although different stability definitions exist for nonlinear systems, the most widespread is the Lyapunov stability definition.

Lyapunov stability
Let $\xi(t)$ be the solution of the differential equation $\dot{x} = f(t, x)$. We say that the solution $\xi(t)$ is stable in Lyapunov sense, if for every $t_0 \in [a, \infty)$ and for every $\varepsilon > 0$ there exists $\delta(\varepsilon, t_0) > 0$ such that if $\|x(t_0) - \xi(t_0)\| < \delta(\varepsilon, t_0)$ and $x(t)$ is the solution then $\|x(t) - \xi(t)\| < \varepsilon$ for every $t \in [t_0, \infty)$.

The definition can also be illustrated so that small perturbing the initial condition everywhere with respect to $\xi(t)$ and starting the system from the perturbed initial condition the new solution remains in the neighborhood of $\xi(t)$. A clear interpretation is shown in Fig. 2.7. The solution $\xi(t)$ may be an equilibrium point of the system, a limit cycle (periodical solution) or any other solution of the state equation (for example during chaos investigations).

The following terminology is usual:

(i) The solution $\xi(t)$ of the system is *unstable* if $\xi(t)$ is not stable.
(ii) The solution $\xi(t)$ of the system is *uniformly stable* if $\xi(t)$ is stable in Lyapunov sense and $\delta(\varepsilon, t_0) = \delta(\varepsilon)$ does not depend on t_0.
(iii) The solution $\xi(t)$ of the system is *asymptotically stable* if $\xi(t)$ is stable in Lyapunov sense and for every $t_0 \in [a, \infty)$ there exists $\delta_1(t_0) > 0$ such that if $\|x(t_0) - \xi(t_0)\| < \delta_1(t_0)$ and $x(t)$ is the solution then $\lim_{t \to \infty} \|x(t) - \xi(t)\| = 0$.

The definition of (for the practice particularly important) asymptotic stability means that small perturbing everywhere with respect to $\xi(t)$ and starting the system from the perturbed initial condition then the new solution remains in the neighborhood of $\xi(t)$ and actually approaches $\xi(t)$ if $t \to \infty$.

We shall show that the stability investigation can be reduced to the stability investigation of $\xi \equiv 0$ (equilibrium point) of a modified system satisfying a special condition. Namely, let us introduce the notation $\tilde{x}(t) := x(t) - \xi(t)$, that is, $x(t) = \xi(t) + \tilde{x}(t)$, then

$$\frac{d\tilde{x}(t)}{dt} = \frac{dx(t)}{dt} - \frac{d\xi(t)}{dt} = f\big(t, \xi(t) + \tilde{x}(t)\big) - f\big(t, \xi(t)\big) =: \tilde{f}\big(t, \tilde{x}(t)\big),$$

which means that the new form of the stability problem

$$\frac{d\tilde{x}(t)}{dt} = \tilde{f}\big(t, \tilde{x}(t)\big),$$

$$\tilde{\xi}(t) \equiv 0 \text{ (equilibrium point)} \quad \Leftrightarrow \quad \tilde{f}(t, 0) \equiv 0. \tag{2.81}$$

However, it should be stressed that the new system has to satisfy the condition $\tilde{f}(t, 0) \equiv 0$. Henceforth, we assume that the transformation has already been performed and omit the notation "$\sim$".

The stability is a local property in the definition, see $\delta(\varepsilon, t_0) > 0$. It does not say anything directly about the region of attraction for the equilibrium point $\xi \equiv 0$. Further results are needed to answer what is the region around the equilibrium point from which choosing the initial condition the trajectory (solution) remains in the neighborhood of the equilibrium point or even more it does approach the equilibrium point.

We shall speak about *global stability* if the domain of attraction is the whole space. According to these, we shall use the abbreviations GAS, UGAS etc. meaning that the system is globally asymptotically stable, uniformly globally asymptotically stable etc.

2.3.2 Lyapunov Stability Theorems

In the stability investigations, the idea of positive definite function plays an important role, in the applications it serves as Lyapunov function candidate.

(i) The scalar valued function $V(t, x)$ is *positive definite*, if there exists a scalar valued function $W(x) \in C$ such that $V(t, x) \geq W(x) > 0$ for every $\|x\| \neq 0$ and $V(t, 0) \equiv W(0) = 0$.
(ii) $V(t, x)$ is *negative definite* if $-V(t, x)$ is positive definite.

2.3.2.1 Lyapunov's Direct Method

Theorem 2.3 *Lyapunov's first theorem, direct method.*
Let $\xi \equiv 0$ an equilibrium point, that is,

$$\frac{dx}{dt} = f(t, x), \qquad f(t, 0) \equiv 0, \qquad f(t, x) \in C_{t,x}^{(0,1)}. \tag{2.82}$$

1. *If $\exists V(t, x) \in C_{t,x}^{(1,1)}([a, \infty) \times D_x)$ positive definite function (so called Lyapunov function) such that for every $x(t)$ solution along the trajectory yields*

$$\dot{V}(t, x) := \frac{dV(t, x(t))}{dt} \leq 0 \quad (\dot{V} \text{ negative semidefinite}), \tag{2.83}$$

then the equilibrium point $\xi \equiv 0$ is stable in Lyapunov sense.
2. *If additionally*

$$\dot{V}(t, x) := \frac{dV(t, x(t))}{dt} < 0 \quad (\dot{V} \text{ negative definite}), \tag{2.84}$$

then the equilibrium point $\xi \equiv 0$ is asymptotically stable in Lyapunov sense.

Notice that

$$\frac{dV}{dt} = \frac{\partial V}{\partial x}\frac{dx}{dt} + \frac{\partial V}{\partial t} = \frac{\partial V}{\partial x}f(t, x) + \frac{\partial V}{\partial t}. \tag{2.85}$$

The following simple proof can seldom be found in control books for this general time variant case. The proof is based on the fact that any continuous function over a compact (bounded and closed) set assumes its extremum value.

Proof (Stability part) Let $\varepsilon > 0$ and consider the spherical surface $S_\varepsilon = \{x : \|x\| = \varepsilon\}$ that is evidently compact. Then $W(x) \in C$ assumes its minimum over S_ε, that is, $\exists x^* \in S_\varepsilon$, such that $\inf_{x \in S_\varepsilon} W(x) = W(x^*) = \alpha > 0$. Let $t_0 \in (a, \infty)$, then, since $V(t, x)$ is continuous and $V(t, 0) = 0$, hence $\exists 0 < \delta(t_0, \varepsilon) < \varepsilon$ such that $\forall \{x : \|x\| < \delta\}$ yields $0 \leq V(t, x) < \alpha$. Let $\|x(t_0)\| < \delta$ and $x(t)$ (not identically zero) solution of the state equation. It can be shown that the trajectory $x(t)$ remains in the interior of the spherical surface $S_\varepsilon: \|x(t)\| < \varepsilon, \forall t \in [t_0, \infty)$. Otherwise, it would exist $t_1 \in [t_0, \infty)$, such that $\|x(t_0)\| < \delta < \varepsilon, \|x(t)\| < \varepsilon, \forall t < t_1$, but $\|x(t_1)\| = \varepsilon$. This leads to contradiction because $dv/dt := \dot{V}(t, x(t)) \leq 0$, thus $v(t)$ is monotone nonincreasing along the trajectory, hence $\alpha \leq W(x(t_1)) \leq$

$V(t_1, x(t_1)) \leq V(t_0, x(t_0)) < \alpha$. Therefore, $\|x(t)\| < \varepsilon$, $\forall t \in [t_0, \infty)$, and thus the equilibrium point $\xi \equiv 0$ is stable. $\hspace{2cm}$ $\square$

Proof (Asymptotic stability part) We already know that $\xi \equiv 0$ is stable. Let $\|x(t_0)\| < \delta_1(t_0)$ sufficiently small and $x(t)$ (not identically zero) solution. It has to be proven that $\lim_{t \to \infty} \|x(t)\| = 0$. Now we know that $dV/dt < 0$ (negative definite) along the trajectory.

First, we shall prove that, using the notation $\lim_{t \to \infty} V(t, x(t)) = \inf_t V(t, x(t))$ $=: \alpha$, it cannot be satisfied $\alpha > 0$. If, in contradiction, we would assume $\alpha > 0$, then it would follow from $\alpha > 0$ that $\|x(t)\| \geq \beta > 0$, $\forall t \in [t_0, \infty)$. Otherwise, that is, in case of $\beta = 0$, there would exist a sequence $\{t_1, t_2, \ldots, t_k, \ldots\}$ such that $\lim_{t_k \to \infty} V(t_k, x(t_k)) = 0$, which would be a contradiction because of $\lim_{t \to \infty} V(t, x(t)) = \alpha > 0$. Hence, $\alpha > 0 \Rightarrow \|x(t)\| \geq \beta > 0$, $\forall t \in [t_0, \infty)$.

After this, we shall show that $\alpha > 0$ leads also to contradiction. Namely, we know that the equilibrium point is stable, thus we can assume there exists $0 < \beta \leq \|x(t)\| < \varepsilon$, $\forall t \in [t_0, \infty)$. Since $\dot{V}(t, x) < 0$ (negative definite), there exists $W_1(x) \in C$ positive definite and continuous function such that $\dot{v}(t) := \dot{V}(t, x(t)) \leq -W_1(x(t)) < 0$. Let $\gamma := \inf_{\beta \leq \|x\| < \varepsilon} W_1(x)$, then $\gamma > 0$ and by integration

$$v(t) = v(t_0) + \int_{t_0}^{t} \dot{V}\big(\tau, x(\tau)\big) \, d\tau \leq v(t_0) - \int_{t_0}^{t} W_1(\tau) \, d\tau \leq v(t_0) - \gamma(t - t_0).$$

Hence, $\exists t$, $v(t) < 0$ which is a contradiction since V is positive definite, and thus $V(t, x(t)) \geq 0$ along the trajectory. Hence, in contradiction to the original assumption, $\alpha = 0$ and therefore $\lim_{t \to \infty} V(t, x(t)) = 0$.

It has yet to be shown that also the trajectory $x(t)$ approaches zero. But this follows from the fact that V is positive definite, $W \in C$ is continuous and positive definite, $V(t, x(t)) \geq W(x(t)) \geq 0$, thus $0 = \lim_{t \to \infty} V(t, x(t)) \geq \lim_{t \to \infty} W(x(t)) \geq 0 \Rightarrow \lim_{t \to \infty} W(x(t)) = 0$, and since $W \in C$ is positive definite thus $\lim_{t \to \infty} x(t) = 0$. Hence, the equilibrium point $\xi \equiv 0$ is *asymptotically stable* in Lyapunov sense. $\hspace{2cm}$ $\square$

In order to use Lyapunov's direct method, an appropriate Lyapunov function candidate has to be found from which we can deduce the stability of the equilibrium point. Positive functions like norm, energy functions etc. can be suggested as Lyapunov function candidates. In some simple cases other possibilities exist, for example, the question arises whether is it possible to deduce the stability of the nonlinear system from the stability of the linearized system.

2.3.2.2 Lyapunov's Indirect Method

Theorem 2.4 *Lyapunov's second theorem, indirect method.*

Let $\xi \equiv 0$ be an equilibrium point, i.e. $\frac{dx}{dt} = f(t, x)$, $f(t, 0) \equiv 0$, $f(t, x) \in C_{t,x}^{(0,1)}$, and let us introduce the notations $A(t) := f'_x(t, x)|_{x=0}$ and $f_1(t, x) := f(t, x) - A(t)x$.

If it is satisfied

(i) $\lim_{\|x\|\to 0} \sup_{t\geq 0} \frac{\|f_1(t,x)\|}{\|x\|} = 0$.
(ii) $A(\cdot)$ *is bounded.*
(iii) $\dot{x} = A(t)x$ *linear system is uniformly asymptotically stable,*

then the equilibrium point $\xi \equiv 0$ *of the original nonlinear system is uniformly asymptotically stable.*

The sketch of the proof is as follows. Consider the linearized system $\dot{x} = A(t)x$ and its fundamental matrix $\Phi(t, \tau)$. Let

$$P(t) := \int_t^\infty \Phi^T(\tau, t)\Phi(\tau, t)\, d\tau,$$

then $\langle P(t)x, x\rangle = \int_t^\infty \|\Phi(\tau, t)x\|^2\, d\tau > 0$. Hence, $\langle P(t)x, x\rangle$ is a Lyapunov function candidate for which the following conditions are satisfied:

(1) $P(t)$ is positive definite and $\exists \alpha, \beta > 0$, such that $\forall t, x : \alpha\|x\|^2 \leq \langle P(t)x, x\rangle \leq \beta\|x\|^2$, furthermore it yields $\dot{P}(t) + P(t)A(t) + A^T(t)P(t) = -I$.
(2) $V(t, x) := \langle P(t)x, x\rangle$ is positive definite function with $W(x) := \alpha\|x\|^2$.
(3) $\dot{V}(t, x) = -\|x\|^2 + 2\langle P(t)f_1(t, x), x\rangle < 0$, since $\|2\langle P(t)f_1(t, x), x\rangle\|$ can be made less than $\|x\|^2$ in some neighborhood of $x = 0$ because of assumption (i) and properties (1).

Notice that Lyapunov's indirect method does not give any information about the attraction domain of the equilibrium, hence only the local stability of the equilibrium has been proven.

For LTI systems, the classical stability idea is strongly related to the poles of the transfer function which are equal to the eigenvalues of the state matrix. Stability was defined for such systems so that the transients are decaying if the initial conditions are changed from zero to nonzero value. The following theorem shows the connection between classical and Lyapunov stability.

Theorem 2.5 *Lyapunov equation characterizing stable LTI system.*

Consider the continuous time linear time invariant system $\dot{x} = Ax$ *for which evidently* $\xi \equiv 0$ *equilibrium point. Let P be a constant, symmetrical and positive definite matrix* $(P > 0)$ *and let* $V(x) := \langle Px, x\rangle$ *be the Lyapunov candidate function. Perform the substitution* $\dot{x} = Ax$ *in the derivative of V:* $\dot{V} = \langle P\dot{x}, x\rangle + \langle Px, \dot{x}\rangle = \langle (PA + A^T P)x, x\rangle$. *Then* $\dot{V} < 0$ *(negative definite), if* $\exists Q > 0$ *positive definite matrix such that the Lyapunov equation is satisfied:*

$$PA + A^T P = -Q, \quad P > 0, \ Q > 0. \tag{2.86}$$

The following statements are equivalent:

(i) *For all eigenvalues,* s_i *of the matrix A yields* $\text{Re } s_i < 0$.
(ii) *For every* $Q > 0$, *the Lyapunov equation has the positive definite solution* $P := \int_0^\infty e^{A^T t} Q e^{At}\, dt > 0$.

The proof is simple see, for example, [152].

2.3.2.3 LaSalle's Method for Time Invariant Systems

For the nonlinear time invariant (autonomous) system class $\dot{x} = f(x)$, $f(0) = 0$ LaSalle [77] gave a stronger result than Lyapunov's direct method which is based on the idea of positive limit set and invariant set.

Positive limit set.
The positive limit set of the trajectory $x(t)$ is $\Gamma^+ := \{y : \exists\{t_n\}$ sequence, $t_n \to \infty$ as $n \to \infty$, such that $\lim_{n\to\infty} x(t_n) = y\}$. The positive limit set of the trajectory $x(t)$ may be for example a limit point $\lim_{t\to\infty} x(t)$ (if it exists), or a limit cycle (isolated, closed and periodical curve) to which the trajectory converges.

Invariant set.
The set M is invariant if each trajectory (solution of the differential equation $\dot{x} = f(x)$) going through any point of M lies entirely in M. If the set E is not invariant then we can speak about its maximal invariant set M contained in E: $M = \max\{H \subset E : x(0) \in H, x(t) \text{ solution} \Rightarrow \forall t : x(t) \in H\}$. The positive limit set Γ^+ of a bounded trajectory $x(t)$ is always nonempty, compact (bounded and closed) and invariant.

Theorem 2.6 *LaSalle's maximal invariant set theorem for autonomous nonlinear system.*

Assume the system $\dot{x} = f(x)$ has equilibrium point $0 = f(0)$ and $f(x) \in C_x^{(1)}$. Assume there exists $V(x)$ Lyapunov function which satisfies the following conditions for some fixed $r > 0$ over the compact set $\{x : |V(x)| \leq r\} =: \Omega_r$:

(i) *$V(x)$ is positive definite, i.e. $V(x) \geq 0$ and $V(x) = 0 \Leftrightarrow x = 0$.*
(ii) *$\dot{V}(x) = \langle \text{grad } V, f \rangle \leq 0$ (negative semidefinite).*
(iii) *$V(x) \in C_x^{(1)}$ (continuously differentiable).*

Let M be the maximal invariant set contained in $E := \{x \in \Omega_r : \dot{V}(x) = 0\}$. Then for every initial condition $x(0) \in \Omega_r$ the resulting trajectory $x(t)$ asymptotically approaches the maximal invariant set M.

More detailed it means that for every $\varepsilon > 0$ there exists T, such that for every $t > T$ the trajectory $x(t)$ remains in the ε-neighborhood of M: There exist $y(t, \varepsilon) \in M + \{x : \|x\| < \varepsilon\}$ such that $\|x(t) - y(t, \varepsilon)\| < \varepsilon$.

Notice that Ω_r is an attraction set of the equilibrium point although not necessarily the largest one.

Corollary *If $M = \{0\}$ is the maximal invariant set contained in $E = \{x \in \Omega_r : \dot{V}(x) = 0\}$, then for every $x(0) \in \Omega_r$ the corresponding $x(t)$ trajectory asymptotically approaches the equilibrium point $0 = f(0)$.*

Corollary *Let $x = (y^T, z^T)^T$ and assume that beside the conditions of LaSalle's theorem it is additionally satisfied:*

(i) *$\dot{y} = \varphi(y, z)$ and $\dot{z} = \psi(y, z)$.*

(ii) $\dot{V}(x) = \dot{V}(y, z) = 0 \Rightarrow y = 0$.
(iii) $\varphi(0, z) = 0 \Leftrightarrow z = 0$.

Then the equilibrium point $y = 0$, $z = 0$ *is asymptotically stable. It is evident namely, that according to (ii) and the definition of E in each point of* $M \subset E$ *yields* $y = 0$. *Since M is an invariant set and y is constant zero over it, therefore* $\dot{y} = 0$ *for each trajectory running in M, and thus by (iii) yields also* $z = 0$. *Hence, M consists only of a single point and any trajectory* $x(t)$ *starting in* Ω_r *asymptotically approaches the equilibrium point.*

2.3.2.4 *K* and *KL* Comparison Functions for Stability Investigations

In order to sharpen the stability theorems in the direction of global asymptotic stability, some further definitions are necessary.

(i) The function $\alpha : [0, a) \to R^+$ is called K-function if $\alpha(\cdot)$ is strictly monotone increasing and $\alpha(0) = 0$. Especially, if $a = \infty$ and $\alpha(r) \to \infty$ if $r \to \infty$ then $\alpha(r)$ is a K_∞-function.
(ii) The function $\beta : [0, a) \times R^+ \to R^+$ is called *KL*-function if $\beta(r, s)$ satisfies that $\beta(\cdot, s)$ is a K-function in the first variable for every fixed s, while $\beta(r, \cdot)$ is monotone decreasing in the second variable for every fixed r and $\beta(r, s) \to 0$ if $s \to \infty$.

It is clear that, for example, $\alpha(r) = \text{atan}(r)$ is a K-function, but not a K_∞-function, however $\alpha(r) = r^c$ is a K_∞-function for $c > 0$, while $\beta(r, s) = r^c e^{-s}$ is a *KL*-function for $c > 0$.

The following theorems are valid for $\dot{x} = f(t, x)$, the proofs can be found in [62].

Theorem 2.7 *Relations between stability and comparison functions.*
The equilibrium point $\xi \equiv 0$ *is*

(i) *Uniformly stable* $\Leftrightarrow$ *there exist* $\alpha(r)$ K-function and $c > 0$ independent of t_0 such that

$$\|x(t)\| \leq \alpha(\|x(t_0)\|), \quad \forall t \geq t_0 \geq 0, \ \forall \|x(t_0)\| < c.$$

(ii) *Uniformly asymptotically stable* $\Leftrightarrow$ *there exist* $\beta(r, s)$ *KL-function and* $c > 0$ *independent of* t_0 *such that*

$$\|x(t)\| \leq \beta(\|x(t_0)\|, t - t_0), \quad \forall t \geq t_0 \geq 0, \ \forall \|x(t_0)\| < c.$$

(iii) *Globally uniformly asymptotically stable* $\Leftrightarrow$ *there exist* $\beta(r, s)$ *KL-function such that*

$$\|x(t)\| \leq \beta(\|x(t_0)\|, t - t_0), \quad \forall t \geq t_0 \geq 0, \ \forall x(t_0) \in R^n.$$

Theorem 2.8 *Relations between Lyapunov function and qualified stability.*
Let $D \subset R^n$ be an open set, $0 \in D$, $V : [0, \infty) \times D \to R$ continuously differentiable function satisfying

$$W_1(x) \leq V(t, x) \leq W_2(x),$$

where $W_1(x)$, $W_2(x)$ are positive definite and continuous,

$$\frac{\partial V}{\partial x} f(t, x) + \frac{\partial V}{\partial t} \leq 0.$$

Then the equilibrium point $\xi \equiv 0$ is

(i) *Uniformly stable.*
(ii) *Uniformly asymptotically stable if $\frac{\partial V}{\partial x} f(t, x) + \frac{\partial V}{\partial t} \leq -W_3(x)$, where $W_3(x)$ is positive definite and continuous.*
(iii) *If beside the previous condition it is also satisfied that r and c are chosen so that $B_r := \{x : \|x\| \leq r\} \subset D$ and $0 < c < \min_{\|x\|=r} W_1(x)$, then every trajectory starting in $\{x \in B_r : W_2(x) \leq c\}$ satisfies*

$$\|x(t)\| \leq \beta(\|x(t_0)\|, t - t_0), \ \forall t \geq t_0 \geq 0$$

for some class KL-function. Specifically, if $D = R^n$ and $W_1(x)$ is radially unbounded then the equilibrium is globally uniformly asymptotically stable.

Remark The proof of the strong result iii) needs the construction of the class K-functions $\alpha_1(\|x\|) \leq W_1(x)$, $W_2(x) \leq \alpha_2(\|x\|)$, $\alpha_3(\|x\|) \leq W_3(x)$, then $\alpha = \alpha_3 \circ \alpha_2^{-1}$, $y(t) =: \sigma(y_0, t - t_0)$ is the solution of the scalar nonlinear differential equation $\dot{y} = -\alpha(y)$, $y(t_0) = V(t_0, x(t_0))$ and finally $\beta(x(t_0), t - t_0) := \alpha_1^{-1}(\sigma(\alpha_2(\|x(t_0)\|), t - t_0))$.

Till now, the input signal has not appeared in the stability investigations, hence the discussion should be extended to the input-state stability of nonlinear systems. Consider the nonlinear system $\dot{x} = f(t, x, u)$. We shall assume that the unexcited system has equilibrium point $\xi \equiv 0$, that is, $f(t, 0, 0) \equiv 0$, and we shall investigate the influence of the input signal $u(t)$ on the stability.

2.3.2.5 Input-to-State Stability (ISS)

The nonlinear system $\dot{x} = f(t, x, u)$ is called input-to-state stable (ISS), if there exist $\beta(r, s)$ class *KL*-function and $\gamma(r)$ class *K*-function such that for every initial condition $x(t_0)$ and every bounded input $u(t)$ there exits the solution $x(t)$, $\forall t \geq t_0$ satisfying the following inequality:

$$\|x(t)\| \leq \beta(\|x(t_0)\|, t - t_0) + \gamma\left(\sup_{t_0 \leq \tau \leq t} \|u(\tau)\|\right).$$

Notice that the above condition assures the boundedness of the trajectory $x(t)$ for every bounded input $u(t)$. If t is increasing the solution $x(t)$ is uniformly bounded by a K-function, in the argument of which stands $\sup_{t \geq t_0} \|u(t)\|$, furthermore if $t \to \infty$ yields $u(t) \to 0$ then $x(t) \to 0$ is also valid. Since for $u(t) \equiv 0$ the inequality reduces to $\|x(t)\| \leq \beta(\|x(t_0)\|, t - t_0)$, hence the equilibrium point $\xi(t) \equiv 0$ of the unexcited system is globally uniformly asymptotically stable. The following theorem gives a condition of ISS, the proof can be found in [62].

Theorem 2.9 *Sufficient condition for ISS stability.*
Let $V : [0, \infty) \times R^n \to R$ be a continuously differentiable function satisfying the conditions

$$\alpha_1(\|x\|) \leq V(t, x) \leq \alpha_2(\|x\|),$$

$$\frac{\partial V}{\partial x} f(t, x, u) + \frac{\partial V}{\partial t} \leq -W_3(x), \quad \forall \|x\| \geq \rho(\|u\|) > 0$$

for every $(t, x, u) \in [0, \infty) \times R^n \times R^m$, where α_1, α_2 are class K_∞-functions, ρ is class K-function and $W_3(x)$ is positive definite over R^n. Then $\dot{x} = f(t, x, u)$ is input-to-state stable with the choice $\gamma := \alpha_1^{-1} \circ \alpha_2 \circ \rho$.

In the expressions, -1 denotes the inverse function and circle denotes the composition of functions.

2.3.3 Barbalat Lemmas

Functions like $\sin(t)$, e^{-at}, $\log(t)$ etc. and combinations thereof often appear in dynamic system responses. Unfortunately, it is not an easy thing to deduce system behavior from such signals. Against typical engineering feeling, if $\dot{f} \to 0$ it does not follow that $f(t)$ has a limit as $t \to \infty$. For example, $f(t) = \sin(\log(t))$ has no limit as $t \to \infty$ while $\dot{f}(t) = \frac{\cos(\log(t))}{t} \to 0$ as $t \to \infty$. Similarly, the fact that $\lim_{t \to \infty} f(t)$ exists, does not imply that $\lim_{t \to \infty} \dot{f}(t) = 0$. For example, $f(t) = e^{-t} \sin(e^{2t})$ tends to zero while its derivative $\dot{f}(t) = -e^{-t} \sin(e^{2t}) + 2e^t \cos(e^{2t})$ is unbounded. On the other hand, if $f(t)$ is lower bounded and decreasing ($\dot{f}(t) \leq 0$), then it converges to a limit.

To find reliable results, consider first the definition of the limit value of a function. We know from standard calculus that $\lim_{t \to \infty} f(t) = a$ if for every $\varepsilon > 0$ there exists $T(\varepsilon) > 0$ such that for all $t \geq T$ follows $|f(t) - a| < \varepsilon$. As a consequence, if $\lim_{t \to \infty} f(t) \neq 0$, then there exists a $\varepsilon_0 > 0$ such that for all $T > 0$ we can find a $T_1 \geq T$ satisfying $|f(T_1)| \geq \varepsilon_0$.

Uniformly continuous functions are especially important. We call $f(t)$ uniformly continuous if for every $\varepsilon > 0$ there exists $\delta(\varepsilon) > 0$ such that $|f(t_2) - f(t_1)| < \varepsilon$ if $|t_2 - t_1| < \delta$. According to the mean value theorem, if the function $f : R \to R$ is differentiable, then $f(t_2) - f(t_1) = \dot{f}(\tau)(t_2 - t_1)$ where $\tau \in (t_1, t_2)$. If $\dot{f} \in L_\infty$, that is, $\dot{f}$ is essentially bounded, then f is uniformly continuous and δ can be chosen $\varepsilon / \|\dot{f}\|_\infty$.

Lemma 2.1 (Barbalat lemmas)

1. Let $f : R \rightarrow R$ be uniformly continuous on $[0, \infty)$. Suppose that $\lim_{t \rightarrow \infty} \int_0^t f(\tau) \, d\tau$ exists and is finite. Then $\lim_{t \rightarrow \infty} f(t) = 0$.
2. Suppose that $f \in L_\infty$, $\dot{f} \in L_\infty$ and $\lim_{t \rightarrow \infty} \int_0^t f(\tau) \, d\tau$ exists and is finite. Then $\lim_{t \rightarrow \infty} f(t) = 0$.
3. If $f(t)$ is differentiable and has finite limit $\lim_{t \rightarrow \infty} f(t) < \infty$, and if $\dot{f}$ is uniformly continuous, then $\dot{f}(t) \rightarrow 0$ as $t \rightarrow \infty$.

Proof

1. If the first statement is not true, then there is a $\varepsilon_0 > 0$ such that for all $T > 0$ we can find $T_1 \geq T$ satisfying $|f(T_1)| \geq \varepsilon_0$. Since $f(t)$ is uniformly continuous, there is a positive constant $\delta > 0$ such that $|f(t + \tau) - f(t)| < \varepsilon_0/2$ for all $t > 0$ and $0 < \tau < \delta$. Hence,

$$\left| f(t) \right| = \left| f(t) - f(T_1) + f(T_1) \right| \geq \left| f(T_1) \right| - \left| f(t) - f(T_1) \right| > \varepsilon_0 - \varepsilon_0/2 = \varepsilon_0/2$$

for every $t \in [T_1, T_1 + \delta]$. Therefore,

$$\left| \int_{T_1}^{T_1+\delta} f(t) \, dt \right| = \int_{T_1}^{T_1+\delta} \left| f(t) \right| dt > \varepsilon_0 \delta/2$$

where the equality holds because $f(t)$ does not change the sign for $t \in [T_1, T_1 + \delta]$. Thus, $\int_0^t f(t) \, dt$ cannot converge to a finite limit as $t \rightarrow \infty$, which is a contradiction. Hence, the first statement is valid.
2. The second statement follows from the fact that if $\dot{f} \in L_\infty$ then f is uniformly continuous and the first statement can be applied.
3. The third statement follows from the fact that because of

$$f(\infty) - f(0) = \int_0^\infty \dot{f}(t) \, dt$$

the right side exists and is finite. Since $\dot{f}$ is uniformly continuous, we can apply the first statement with $\dot{f}$ instead of f.   □

The following lemma gives an immediate corollary of Barbalat's lemma, which looks like an invariant set theorem in Lyapunov stability analysis of time varying systems.

Lemma 2.2 (Lyapunov-like form of Barbalat's lemma) *If a scalar valued function $V(x, t)$ satisfies the following three conditions:*

(i) $V(x, t)$ *is lower bounded.*
(ii) $\dot{V}(x, t)$ *is negative semidefinite.*
(iii) $\dot{V}(x, t)$ *is uniformly continuous in t.*

Then $\dot{V}(x, t) \rightarrow 0$ as $t \rightarrow \infty$.

Observe that $V(x, t)$ can simply be a lower bounded function instead of being positive definite. Notice that $\dot{V}(x, t)$ is uniformly continuous if $\ddot{V}(x, t)$ exists and is bounded.

2.3.4 Stability of Interconnected Passive Systems

Stability problems can be considered also in function spaces for fixed (typically zero) initial conditions. For $p \in [1, \infty)$, the Lebesgue measurable function $f : [0, \infty) \to R^1$ belongs to $L_p[0, \infty)$ if $\int_0^\infty |f(t)|^p \, dt < \infty$ and then, with the norm $\|f\|_p = (\int_0^\infty |f(t)|^p \, dt)^{1/p}$, the space $L_p[0, \infty)$ is a Banach space (linear normed space in which the Cauchy sequences are convergent). If $p = \infty$, then with the norm $\|f\|_\infty = \operatorname{ess\,sup} |f(t)| < \infty$ the space $L_\infty[0, \infty)$ is also a Banach space.

The definition can easily be extended to vector valued functions $f = (f_1, \ldots, f_n)^T$, $f_i \in L_p[0, \infty)$, $i = 1, \ldots, n$ where, with the norm $\|f\|_p := (\sum_{i=1}^n \|f_i\|_p^2)^{1/2}$, the space $L_p^n[0, \infty)$ is also a Banach space.

For $p = 2$, a scalar product $\langle f, g \rangle = \int_0^\infty \langle f(t), g(t) \rangle \, dt$ can be explained and the norm can be defined by $\|f\| = (\langle f, f \rangle)^{1/2}$. Then $L_2^n[0, \infty)$ is a Hilbert space.

Further extensions are possible for functions which are not necessarily bounded but satisfy the above conditions for every finite $T > 0$ if the function is substituted by zero for $\forall t > T$. Such extended spaces are denoted by L_{pe}^n. For simplicity, we shall consider only $L_p^n[0, \infty)$ signal spaces.

2.3.4.1 L_p-Stable Nonlinear Systems

Let $G : L_p^n \to L_p^m$ a *nonlinear* system mapping $u \in L_p^n[0, \infty)$ input signals into $y \in L_p^m[0, \infty)$ output signals. Then the nonlinear system is

(i) L_p-stable with finite gain (wfg) if there exist $\gamma_p, \beta_p \geq 0$ constants such that

$$x \in L_p^n \quad \Rightarrow \quad \|y\|_p \leq \gamma_p \|x\|_p + \beta_p.$$

(ii) L_p-stable with finite gain without bias (wb) if there exist $\gamma_p \geq 0$ constant such that

$$x \in L_p^n \quad \Rightarrow \quad \|y\|_p \leq \gamma_p \|x\|_p.$$

The *gain* of the nonlinear system is defined by $\gamma_p(G) = \inf\{\gamma_p\}$ satisfying (i) for wfg and (ii) for wb, respectively.

Let us consider now interconnected stable nonlinear systems according to Fig. 2.8 and ask for the stability of the closed loop. The proofs of the following two theorems are simple, see also [152].

Fig. 2.8 Scheme of
nonlinear control system

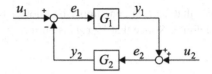

Theorem 2.10 *Small gain theorem.*

Consider the closed loop system according to Fig. 2.8 and let $p \in [1, \infty]$.
Assume that G_1, G_2 are causal and L_p-stable wb (or wfg) nonlinear systems,
$\gamma_{1p} := \gamma_{1p}(G_1)$ *and* $\gamma_{2p} := \gamma_{2p}(G_2)$. *Then the closed loop system is L_p-stable,*
if

$$\gamma_{1p} \cdot \gamma_{2p} < 1.$$

The inputs of nonlinear electrical circuits are voltage and current generators, the
outputs are voltages and currents and the instantaneous power is $\langle y(t), u(t) \rangle$. The
circuit is passive if the input energy (integral of the power) in the system does not
exceed the change of the stored energy. If the inductive and capacitive elements
are initially chargeless, then $\int_0^\infty \langle y(t), u(t) \rangle \, dt \geq 0$ for passive electrical circuits.
Passive systems generalize this principle.

Notice, scalar product needs that the dimension of input and output signals is the
same.

2.3.4.2 Passive Systems

Let $G : L_2^n \to L_2^n$ a nonlinear system. We say that

(i) G is passive, if $\exists \beta \in R^1$ such that $\langle u, Gu \rangle = \int_0^\infty \langle u(t), (Gu)(t) \rangle \, dt \geq \beta$,
$\forall u \in L_2$.

(ii) G is strictly input passive, if $\exists \alpha > 0, \beta \in R^1$ such that $\langle u, Gu \rangle \geq \alpha \|u\|_2^2 + \beta$,
$\forall u \in L_2$.

(iii) G is strictly output passive, if $\exists \alpha > 0, \beta \in R^1$ such that $\langle u, Gu \rangle \geq \alpha \|(Gu)\|_2^2 + \beta$, $\forall u \in L_2$.

Notice that since $u \equiv 0 \in L_2$ thus $0 \geq \beta$. It can easily be shown that strictly output
passive system is L_2-stable wfg.

Theorem 2.11 *Passivity theorem in Lp space.*

Consider the closed loop system in Fig. 2.8 with $G_1, G_2 \in L_2^n$.

(i) *If both G_1 and G_2 are strictly output passive and both u_1 and u_2 are active
then, for the output $\binom{y_1}{y_2}$, the closed loop is strictly output passive and L_2-stable
wfg.*

(ii) *If G_1 is strictly output passive, G_2 is passive and $u_2 = 0$ (i.e., only u_1 is active),
then for the output y_1, the closed loop is strictly output passive and L_2-stable
wfg.*

(iii) *If G_1 is passive, G_2 is strictly input passive and $u_2 = 0$ (i.e., only u_1 is active),
then for the output y_1, the closed loop is strictly output passive and L_2-stable
wfg.*

2.3.4.3 Dissipative Systems

Dissipativity can be seen as passivity in state space. The storage function serves as Lyapunov function. The number of inputs and outputs is the same, the input and output signals are from L_2.

Consider the nonlinear system $\Sigma : \dot{x} = f(x, u), y = h(x, u), x \in R^n, u, y \in R^m$. The real valued function $s(u, y)$ will be called *supply rate*. The system is called dissipative if there exist $S : R^n \to R^+$ so called *storage function* such that for every $x(t_0) \in R^n$ and for every $t_1 \geq t_0$ the following dissipative inequality yields:

$$S\big(x(t_1)\big) \leq S\big(x(t_0)\big) + \int_{t_0}^{t_1} s\big(u(t), y(t)\big) dt. \tag{2.87}$$

Specifically, if equality holds then the system is called lossless.

As an example, consider the supply rate $s(u, y) := \langle u, y \rangle$. Assuming $S \geq 0$, then

$$\int_0^T \langle u(t), y(t) \rangle dt \geq S\big(x(T)\big) - S\big(x(0)\big) \geq -S\big(x(0)\big) \tag{2.88}$$

for every $x(0)$, for every $T > 0$ and for every input $u(\cdot)$ which means that the input–output mapping $G_{x(0)}$ is passive with the choice $\beta = -S(x(0))$. Here, S can be considered to be the stored energy motivating the following definitions.

Passive dissipative systems
Assume the nonlinear system has the form $\Sigma : \dot{x} = f(x, u), y = h(x, u), x \in R^n$, $u, y \in R^m$, then the system

(i) Is *passive*, if it is dissipative with respect to the supply rate $s(u, y) = \langle u, y \rangle$.
(ii) Is *strictly input passive* if it is dissipative with respect to the supply rate $s(u, y) = \langle u, y \rangle - \delta \|u\|^2$ where $\delta > 0$.
(iii) Is *strictly output passive* if it is dissipative with respect to the supply rate $s(u, y) = \langle u, y \rangle - \varepsilon \|y\|^2$ where $\varepsilon > 0$.
(iv) Is *conservative* if it is lossless and $s(u, y) = \langle u, y \rangle$.
(v) Has L_2-gain $\leq \gamma$ if it is dissipative with respect to the supply rate $s(u, y) = \frac{1}{2}\gamma^2 \|u\|^2 - \frac{1}{2}\|y\|^2$ where $\gamma > 0$.

For example, in the last case

$$\frac{1}{2} \int_0^T \big(\gamma^2 \|u(t)\|^2 - \|y(t)\|^2\big) dt \geq S\big(x(T)\big) - S\big(x(0)\big) \geq -S\big(x(0)\big),$$

from which after regrouping follows

$$\int_0^T \|y(t)\|^2 dt \leq \gamma^2 \int_0^T \|u(t)\|^2 dt + 2S\big(x(0)\big),$$

which is equivalent to L_p-stability wfg $\leq \gamma$.

The dissipativity condition (2.87) can be substituted by its differentiated form which is sometimes easier to check. By dividing the inequality by $t_1 - t_0$ and letting $t_1 \to t_0$, we obtain the equivalent *differential dissipation inequality*

$$S'_x(x)f(x, u) \leq s(u, h(x, u)), \quad \forall x, u,$$

$$S'_x(x) = \left(\frac{\partial S}{\partial x_1}(x) \cdots \frac{\partial S}{\partial x_n}(x) \right). \tag{2.89}$$

Specifically, in case of input affine nonlinear system, that is, $\Sigma_a : \dot{x} = f(x) + g(x)u$, $y = h(x)$, and assuming the system is passive with respect to the supply rate $s(u, y) = \langle u, y \rangle = \langle u, h(x) \rangle$, then the differential form for Σ_a amounts to $S'_x(x)[f(x) + g(x)u] \leq u^T h(x)$, $\forall x, u$, or equivalently (choosing first $u = 0$) to the *Hill–Moylan conditions*

$$S'_x(x)f(x) \leq 0,$$

$$S'_x(x)g(x) = h^T(x). \tag{2.90}$$

Special cases:

(i) For LTI system, the conditions are equivalent to the Kalman–Yacubovitch–Popov conditions if the storage function is $S(x) = \frac{1}{2}x^T P x$, where $P = P^T \geq 0$, namely $A^T P + P A \leq 0$ and $B^T P = C$.

(ii) If the input affine system Σ_a is strictly output passive with respect to the supply rate $s(u, y) = \langle u, y \rangle - \varepsilon \|y\|^2$ where $\varepsilon > 0$, then (choosing first $u = 0$) yields

$$S'_x(x)f(x) \leq -\varepsilon h^T(x)h(x),$$

$$S'_x(x)g(x) = h^T(x). \tag{2.91}$$

(iii) If the input affine system Σ_a has finite L_2-gain with respect to the supply rate $s(u, y) = \frac{1}{2}\gamma \|u\|^2 - \frac{1}{2}\|y\|^2$, then

$$S'_x(x)[f(x) + g(x)u] - \frac{1}{2}\gamma^2 \|u\|^2 + \frac{1}{2}\|h(x)\|^2 \leq 0, \quad \forall x, u. \tag{2.92}$$

(iv) The maximum of the left side in the inequality (2.92) is reached at $u^* = \frac{1}{\gamma^2}g^T(x)S'^T_x(x)$ and substituting it back we obtain the *Hamilton–Jacobi–Bellman* inequality

$$S'_x(x)f(x) + \frac{1}{2\gamma^2}S'_x(x)g(x)g^T(x)S'^T_x(x) + \frac{1}{2}h^T(x)h(x) \leq 0, \quad \forall x. \tag{2.93}$$

Hence, Σ_a has L_2-gain $\leq \gamma$ if and only if (2.93) has continuously differentiable solution $S \geq 0$. We remark that the equality form is in strong connection with the Hamilton–Jacobi–Bellman equality of dynamic programming.

The following theorems were proven in [127].

Theorem 2.12 *Extremum property of S_a amongst storage functions.*
Consider the nonlinear system Σ. The system is dissipative with respect to the supply rate $s(u, y) \Leftrightarrow$ for every initial condition $x(0) = x$ yields

$$S_a(x) = \sup_{u(\cdot), T \geq 0} - \int_0^T s\big(u(t), y(t)\big)\, dt < \infty. \tag{2.94}$$

Furthermore, every other storage function satisfies $S_a(x) \leq S(x)$, $\forall x$. Thus, $S_a(x)$ can be considered the maximal energy which can be extracted from the system Σ if the initial condition is x.

If each state x of Σ can be reached from x_* within finite time, then Σ is dissipative $\Leftrightarrow S_a(x_*) < \infty$. Moreover, if G_{x_*} is passive wb ($\beta = 0$), or has L_2-gain $\leq \gamma$ wb, then $S_a(x_*) = 0$.

Zero-state detectability.
The nonlinear system Σ_a is called zero-state detectable if $u(t) \equiv 0$, $y(t) \equiv 0$, $\forall t \geq 0 \Rightarrow \lim_{t \to \infty} x(t) = 0$. $S_a(x_*) = 0$ is equivalent to the zero-state detectability condition.

Theorem 2.13 *Connection between passivity and asymptotic stability.*
Consider the zero-state detectable nonlinear system Σ_a. Let $S \geq 0$ with $S(0) = 0$ be solution to (2.91) in case of strict output passivity, or (2.92) in case of finite L_2-gain. Then $\xi \equiv 0$ is asymptotically stable equilibrium point. Moreover if $S(x)$ is proper, that is, $\{x : S(x) < c\}$ is compact, $\forall c > 0$, then the equilibrium point is globally asymptotically stable.

Consider the closed loop nonlinear system according to Fig. 2.8 but at the place of G_i stands the system $\Sigma_i : \dot{x}_i = f_i(x_i, e_i)$, $y_i = h_i(x_i, e_i)$, $i = 1, 2$. Assume that the strict dissipative conditions

$$S_i\big(x_i(t_1)\big) \leq S_i\big(x_i(t_0)\big) + \int_{t_0}^{t_1} \big(\langle e_i(t), y_i(t) \rangle - \varepsilon_i \|y_i(t)\|^2\big)\, dt, \quad i = 1, 2 \tag{2.95}$$

are satisfied.

Theorem 2.14 *Stability of interconnected dissipative systems.*

(i) *Suppose both Σ_1 and Σ_2 are passive (strictly output passive) and (u_1, u_2) external inputs are active, then the closed loop system with outputs (y_1, y_2) is output passive (strictly output passive).*

(ii) *Suppose S_i satisfies (2.95), is continuously differentiable and has (isolated) local minimum at x_i^*, $i = 1, 2$, furthermore no external inputs are active, that is, $u_1 = u_2 = 0$. Then (x_1^*, x_2^*) is stable equilibrium point of the closed loop system.*

(iii) *Suppose Σ_i is strictly output passive and zero-state detectable, S_i satisfying (2.95), is continuously differentiable and has (isolated) local minimum at*

$x_i^* = 0$, $i = 1, 2$, *furthermore* no external inputs *are active, that is,* $u_1 = u_2 = 0$. *Then* $(0, 0)$ *is asymptotically stable equilibrium point of the closed loop. If additionally* S_i *has global minimum at* $x_i^* = 0$, $i = 1, 2$, *then* $(0, 0)$ *is globally asymptotically stable equilibrium point of the closed loop system.*

Theorem 2.15 *Small gain theorem for interconnected dissipative systems.*

(i) *Suppose* Σ_i *is zero-state detectable, has finite gain* γ_i, $i = 1, 2$, *with* $\gamma_1 \cdot \gamma_2 < 1$. *Suppose* S_i *satisfies* (2.95), *is continuously differentiable and has (isolated) local minimum at* $x_i^* = 0$, $i = 1, 2$, *furthermore* no external inputs *are active, that is,* $u_1 = u_2 = 0$. *Then* $(0, 0)$ *is asymptotically stable equilibrium point of the closed loop. If additionally* S_i *has global minimum at* $x_i^* = 0$ *and is proper,* $i = 1, 2$, *then* $(0, 0)$ *is globally asymptotically stable equilibrium point of the closed loop system.*

(ii) *Suppose* Σ_{ai} *is input affine system, has finite gain* γ_i, $i = 1, 2$, *with* $\gamma_1 \cdot \gamma_2 < 1$. *Suppose* S_i *satisfies* (2.95), *is continuously differentiable and* $S_i(0) = 0$ *(therefore* $S_i(x)$ *is positive semidefinite), and let* $x_i^* = 0$ *be asymptotically stable equilibrium point of the unexcited system* $\dot{x}_i = f_i(x_i)$, $i = 1, 2$, *furthermore* no external inputs *are active, that is,* $u_1 = u_2 = 0$. *Then* $(0, 0)$ *is asymptotically stable equilibrium point of the closed loop system.*

2.4 Input–Output Linearization

It is a meaningful idea to control the nonlinear system in such a way that first the nonlinear system is linearized then the linearized system is compensated by using attractive methods of linear control theory. For linearizing the nonlinear system, state feedback and output feedback can be suggested, called input/state linearization and input/output linearization, respectively. Their theories are well developed, see for example [56, 57, 152], however their attractiveness is limited for low dimensional systems because of computational problems, mainly because of the need for solving partial differential equations, the so called Frobenius equations. The basic results were also surveyed in Appendix B and we assume the reader became acquainted with the main techniques from there.

Comparing the two approaches, input/state linearization has the advantage that the number of inputs and outputs may be different, but has the drawback that a large number of sets of Frobenius equations has to be solved.

On the other hand, input/output linearization has the disadvantage that the number of inputs and outputs must be equal, but the solution of Frobenius equation can be evaded, which is a benefit, except for SISO system if the input should be eliminated from the zero dynamics. Notice that eliminating the input from the zero dynamics for MIMO nonlinear systems is usually not possible because the arising distribution spanned by the fix part of the nonlinear coordinate transformation needed for linearization is usually not involutive and the Frobenius equation cannot be applied.

In both cases, the remaining part of the computations can be performed automatically by using symbolic computation tools, like MATLAB Extended Symbolic Toolbox, Mathematica, etc.

Now we summarize here the basic results regarding input/output linearization discussed in Appendix B. The system is assumed to be input affine, that is,

$$\dot{x} = f(x) + \sum_{i=1}^{m} u_i g_i(x),$$

$$y_i(x) = h_i(x), \quad i = 1, \ldots, m, \ h = (h_1, \ldots, h_m)^T,$$

$$f, g_i \in V(X), h_i \in S(X), \quad i = 1, \ldots, m,$$

where f, g_i are vector fields and h_i are real valued functions, all appropriately smooth.

As can be seen, the input does not appear in the output mappings hence the idea is to differentiate the outputs so many times that at the end the input appears in the results. This will define the components of the possible vector degree for the different outputs. For example, differentiating y_i by the time yields

$$\dot{y}_i = \nabla h_i \dot{x} = \left\langle dh_i, f + \sum_{i=1}^{m} g_i u_i \right\rangle = L_f h_i + \sum_{j=1}^{m} u_j L_{g_j} h_i.$$

If each $L_{g_j} h_i$ is zero, then the differentiation is continued and we get

$$\ddot{y}_i = \nabla(L_f h_i) \dot{x} = \left\langle d(L_f h_i), f + \sum_{i=1}^{m} g_i u_i \right\rangle = L_f^2 h_i + \sum_{j=1}^{m} u_j L_{g_j}(L_f h_i).$$

Let r_i be the smallest integer for which

$$y_i^{(r_i)} = L_f^{r_i} h_i + \sum_{j=1}^{m} L_{g_j}\left(L_f^{r_i-1} h_i\right) u_j.$$

The remaining problem is how to express the inputs from the formulas, which is related to the linear independence of the functions weighting the inputs after differentiation.

The MIMO input affine nonlinear system is said to have the *relative degree vector* $r = (r_1, \ldots, r_m)^T$ at the point x_0 if $\exists U(x_0)$ open set such that:

(i) $L_{g_j} L_f^k h_i(x) \equiv 0, \ \forall x \in U, j = 1, \ldots, m, k = 0, \ldots, r_i - 2.$

(ii) The matrix $S(x) = [s_{ij}(x)]_{m \times m}$ is nonsingular at x_0 where $s_{ij} = L_{g_j} L_f^{r_i-1} h_i$, i.e.

$$S(x) = \begin{bmatrix} L_{g_1} L_f^{r_1-1} h_1 & \cdots & L_{g_m} L_f^{r_1-1} h_1 \\ \vdots & \ddots & \vdots \\ L_{g_1} L_f^{r_m-1} h_m & \cdots & L_{g_m} L_f^{r_m-1} h_m \end{bmatrix}.$$

Fig. 2.9 Block scheme of the
MIMO input/output
linearized system

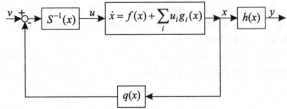

Since $S(x_0)$ is nonsingular hence $S(x)$ is also nonsingular in a sufficiently small neighborhood of x_0. The components $q_i(x) = L_f^{r_i} h_i(x), i = 1, \ldots, m$ and the column vector $q(x)$ will also be needed to internal compensation, see Fig. 2.9.

Theorem 2.16 *The MIMO nonlinear system can be input/output linearized in the neighborhood $U(x_0)$ if it has relative degree vector and $S(x_0)$ is invertible. There is a smooth nonlinear transformation $z = T(x)$, where $z = (z_0^T, z_u^T)^T$, z_o is locally observable while z_u is locally not observable. By using nonlinear static feedback $u(x) = S^{-1}(v - q(x))$, the observable part consists of decoupled integrators in Brunowsky canonical form. The not observable part remains nonlinear and has the form $\dot{z}_u = b_u(z_o, z_u) + A_u(z_o, z_u)u$. The observable part can be further compensated by external linear feedback. The full system stability is influenced by the stability of the unobservable part that is called the zero dynamics of the nonlinear system.*

In Appendix B, formulas can be found for computing the normal form of the input affine system after coordinate transformation $T(x)$. The system components of the observable and not observable parts are given in (B.51a) and (B.51b), respectively. The resulting observable system is given in (B.52a). The resulting zero dynamics is given in (B.52b) containing u and in (B.54) containing v, respectively.

The fixed parts of the components of the coordinate transformation $T(x)$ are given in (B.50), these have to be extended with further functions to make the transformation n-dimensional, that is, to a diffeomorphism. This step does not need the solution of any Frobenius equation, see Lemma B.1 in Appendix B, but the price is that the zero dynamics depends also on the input u.

An exception may be the SISO case where the distribution Δ spanned by the single vector g is evidently involutive. Hence, by using Frobenius theorem there are $\eta_1, \ldots, \eta_{n-1} \in S(X)$ functions whose differentials are independent at x_0 and satisfy $\langle d\eta_i, g(x) \rangle = 0$, $\forall x \in U$. Taking $n - r$ pieces from these functions, for example the first $n - r$, the coordinate transformation may be $T(x) = [dh^T, \ldots, (dL_f^{r-1}h)^T, (d\eta_1)^T, \ldots, (d\eta_{n-r})^T]$. By using this coordinate transformation, the input u will not appear in the zero dynamics, i.e. $\dot{z}_u = f_u(z_o, z_u)$.

Further details can also be found in Appendix B regarding the stability test of the zero dynamics and the extension of the minimum phase property to nonlinear systems based on the stability of the zero dynamics. Another possibility is the use of precompensator based on the dynamic extension algorithm if the vector relative degree does originally not exist. In this case, the compensator is a dynamic system, see (B.56) and (B.57).

2.5 Flatness Control

We illustrate here how can the flatness property be exploited in the path design of underactuated systems, for which not every path is realizable. Assuming underactuated system, it means that for a separately designed path $x(t)$ it may be possible that there is no appropriate control $u(t)$ which can realize the path.

On the other hand, for differentially flat systems we know that both the state and the control can be parametrized by the output variable and its derivatives. Hence, choosing a fine sequence of "corner points" for the path in the output variable and interpolating amongst them analytically, that is, determining also the derivatives in the necessary order, we can simultaneously find both the state trajectory and the corresponding nominal control.

Flatness properties of cable-driven crane.
We shall illustrate the flatness concept and its use in path design for a cable-driven crane. For simplicity, only the two dimensional (2D) planar version is considered. The scheme of the 2D crane is shown in Fig. 2.10. (Since the pulley's mass in point B has been neglected, hence L_s and its actuator motor is omitted in the discussion.) The planar crane is a 3-DOF system which is underactuated because the only inputs are the two motor torques moving the ropes L_1 and $L_2 + L_3$.

The model of the planar crane is differentially flat [69]. In simplified formulation, it means that if $\dot{x} = f(x, u)$ is the dynamic model of the system then there exist a new variable y (the flat output) and finite integers q and p such that $x = \phi_1(y, \dot{y}, \ddot{y}, \ldots, y^{(q)})$, $u = \phi_2(y, \dot{y}, \ddot{y}, \ldots, y^{(q+1)})$, $y = \psi_1(x, u, \dot{u}, \ldots, u^{(p)})$. For the crane, $y := (x, z)^T$ where x, z are the coordinates of the load (point C) in the plane of the crane. The coordinates of point B will be denoted by x_B, z_B. The flatness relations for the 2D crane are summarized as follows:

$$\theta = \arctan\left(-\ddot{x}/(\ddot{z} + g)\right),$$

$$d = \frac{\ddot{x}(z - (k+l)C_\alpha) - (\ddot{z} + g)(x - (k+l)S_\alpha)}{(\ddot{z} + g)S_a - \ddot{x}C_a},$$

$$\beta = \arcsin\left((1 - d/l)S_{\alpha+\theta}\right),$$

$$\gamma = \left(\pi + \beta - (\alpha + \theta)\right)/2,$$

$$L_1 = lS_\beta/S_\gamma,$$

$$L_2 = lS_{\gamma-\beta}/S_\gamma, \tag{2.96}$$

$$x_B = (k+l)S_\alpha - L_2 S_{\alpha-\beta},$$

$$z_B = (k+l)C_\alpha - L_2 C_{\alpha-\beta},$$

Fig. 2.10 Block scheme of
the cable-driven 2D crane

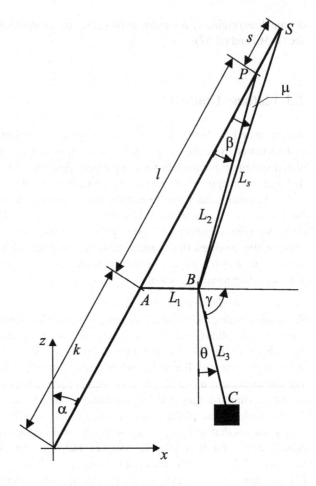

$$L_3 = \sqrt{(x - x_B)^2 + (z - z_B)^2},$$

$$T_3 = m_{\text{load}}\big((\ddot{z} + g)C_\theta - \ddot{x}S_\theta\big),$$

$$u_1 = T_1\rho_1 - \frac{J_1}{\rho_1}\ddot{L}_1,$$

$$u_2 = T_3\rho_2 - \frac{J_2}{\rho_2}(\ddot{L}_2 + \ddot{L}_3).$$

If $y, \dot{y}, \ldots, y^{(4)}$ are known, then all system variables can be computed from them except singularities. For prescribed initial y_0 and final y_1 and transient time T_F, five times differentiable $y(t)$ polynomials can be found connecting the points with straight line in the space from which all the system states $x(t)$ and inputs $u(t)$ can be reconstructed.

It follows from the equilibrium of the internal forces that the line AB in the direction of the horizontal rope L_1 is the bysectrix of the angle 2γ between PB and

BC. Hence, the magnitudes of the internal forces T_3 for L_2 and T_1 for L_1 are related by $T_1 = 2C_\gamma T_3$. By using the flat output all the variables describing the 2D crane can be determined by (2.96). In order to avoid the first and second order symbolic differentiation of the nonlinear expressions for L_1, L_2, L_3, numerical differentiation can be applied (see `spline`, `unmkpp`, `mkpp` in MATLAB). Finally, the flatness based input torques u_1, u_2 for the motors can be determined, where J_i and ρ_i denote the inertia and the radius of the winch, respectively.

Assuming zero derivatives until order five in the initial and final positions, the 11th order polynomial connecting (x_0, z_0) and (x_1, z_1) has the form

$$x(\tau) = x_0 + (x_1 - x_0)\tau^6 \times \left[a_6 + a_7\tau + a_8\tau^2 + a_9\tau^3 + a_{10}\tau^4 + a_{11}\tau^5\right], \quad (2.97)$$

where $\tau = t/T_F$ is the normalized time and the coefficients for both $x(\tau)$ and $z(\tau)$ are

$$a_6 = 0.462, \qquad a_7 = -1.980, \qquad a_8 = 3.465,$$
$$a_9 = -3.080, \qquad a_{10} = 1.386, \qquad a_{11} = -0.252. \qquad (2.98)$$

Neglecting the feedforward compensated friction and the centripetal and Coriolis effects (because of the slow angular velocities of typical cranes) the basic equations of the underactuated system are

$$\frac{J_1}{\rho_1}\ddot{L}_1 = T_1\rho_1 - u_1,$$
$$\frac{J_2}{\rho_2}(\ddot{L}_2 + \ddot{L}_3) = T_3\rho_2 - u_2. \qquad (2.99)$$

The system is submitted to constraints with respect to L_1 and L_2 hence Lagrange multipliers could have been used. Fortunately, instead of this relatively complicated method, the equilibrium between internal forces can be exploited, see Kiss [69], who developed a form of the dynamic model for the choice $x = (\gamma, \alpha - \beta, L_3, \dot{\gamma}, (\dot{\alpha} - \dot{\beta}), \dot{L}_3)^T$ where $\alpha = $ const. However, the measured outputs are $L_1, L_2 + L_3, \dot{L}_1, \dot{L}_2 + \dot{L}_3$ which depend on these state variables in a nonlinear way and would make necessary the use of nonlinear state estimation.

Hence, a new model has been developed using $x = (L_1, L_2, L_3, \dot{L}_1, \dot{L}_2, \dot{L}_3)^T$, see [74]. The idea is the use of the cosine theorem by which

$$C_\gamma = \frac{l^2 - L_1^2 - L_2^2}{2L_1L_2}, \qquad C_\beta = \frac{l^2 + L_2^2 - L_1^2}{2lL_2}. \qquad (2.100)$$

For L_1 and L_2, the expressions in (2.96) can be used. By elementary geometry and using

$$\varphi = (\pi/2) - (\gamma + (\alpha - \beta)), \qquad \theta = \pi - (2\gamma + (\alpha - \beta)),$$

it yields

$$x = kS_\alpha + L_1C_\varphi + L_3S_\theta = kS_\alpha + L_1S_{\gamma+(\alpha-\beta)} + L_3S_{2\gamma+(\alpha-\beta)},$$
$$z = kC_\alpha + L_1S_\varphi - L_3C_\theta = kC_\alpha + L_1C_{\gamma+(\alpha-\beta)} + L_3C_{2\gamma+(\alpha-\beta)}. \qquad (2.101)$$

The main problem is the computation of T_3 from the state variables which is needed in (2.99). This can be performed using $m\ddot{x} = -T_3 S_\theta$, $m(\ddot{z} + g) = T_3 C_\theta$ for the load, from which it follows

$$C_\theta \ddot{x} + S_\theta(\ddot{z} + g) = 0, \qquad m\left(-S_\theta \ddot{x} + C_\theta(\ddot{z} + g)\right) = T_3. \qquad (2.102)$$

Notice that the presented flatness based method computes only the nominal path and the nominal (open loop) control but does not solve the feedback control problem. However, the results can be used for feedforward compensation or linearization along the path.

2.6 Backstepping

Backstepping control is a popular method for controlling nonlinear systems given in strictly feedback form

$$\dot{x}_1 = G_1(\bar{x}_1)x_2 + f_1(\bar{x}_1),$$
$$\dot{x}_2 = G_2(\bar{x}_2)x_3 + f_2(\bar{x}_2),$$
$$\vdots \qquad\qquad\qquad\qquad (2.103)$$
$$\dot{x}_{n-1} = G_{n-1}(\bar{x}_{n-1})x_n + f_{n-1}(\bar{x}_{n-1}),$$
$$\dot{x}_n = G_n(\bar{x}_n)u + f_n(\bar{x}_n),$$
$$y = h(x_1),$$

where $x_i \in R^m$, $i = 1, \ldots, n$ are vector parts of the state vector, $y \in R^m$ is the output, $\bar{x}_i = (x_1^T, \ldots, x_i^T)^T$ is a short notation, h is an invertible function (i.e. its Jacobian h^{x_1} is an invertible matrix), G_i are invertible matrices, $i = 1, \ldots, n$ and all functions are appropriately smooth.

We assume the task is to realize a prescribed path $y_d(\theta) \in R^m$ for the output variable y. The path is parametrized in $\theta \in R^1$. Let $v_s(\theta, t) \in R^1$ be the prescribed speed of the path parameter to which a parameter speed error $\omega_s(\theta, \dot{\theta}, t) = v_s(\theta, t) - \dot{\theta}$ may exist. The result of the path design subtask is the path parameter θ, the path $y_d(\theta)$ and the desired path parameter speed $v_s(\theta, t)$. An extra problem is to make the parameter speed error ω_s going to zero using high level control.

A special case is $\theta = t$, $v_s = 1$ in which case $\omega_s \equiv 0$ is already satisfied and the desired path is $y_d(t)$.

We shall often need the total derivative by the time of parameters of the form $\alpha_{i-1}(\bar{x}_{i-1}, \theta, t)$, that is,

$$\frac{d\alpha_{i-1}}{dt} = \frac{\partial \alpha_{i-1}}{\partial x_1}\dot{x}_1 + \cdots + \frac{\partial \alpha_{i-1}}{\partial x_{i-1}}\dot{x}_{i-1} + \frac{\partial \alpha_{i-1}}{\partial \theta}\dot{\theta}$$

$$=: \sigma_{i-1} + \alpha_{i-1}^\theta \dot{\theta} = \sigma_{i-1} + \alpha_{i-1}^\theta(v_s - \omega_s). \qquad (2.104)$$

Observe that the derivatives $\dot{x}_i$ can be substituted by the right side of the state equations. The first term σ_{i-1} in the total time derivative $\dot{\alpha}_{i-1}$ is that part which does not contain $\dot{\theta}$. Notice that any variable in the upper index denote the partial derivative by the variable, see also h^{x_1}.

Now we present the backstepping compensation method similarly to Skjetne, Fossen and Kokotovich [130], but we neglect the disturbance effects for simplicity.

At every level, we design the virtual control α_{i-1}, for stabilization purposes we choose Hurwitz matrix A_i, that is, matrix whose eigenvalues are on the open left complex plane, choose positive definite matrix $Q_i > 0$ and solve the Lyapunov equation

$$P_i A_i + A_i^T P_i = -Q_i, \quad Q_i > 0, P_i > 0, A_i \text{ is Hurwitz,} \qquad (2.105)$$

to find the symmetric and positive definite matrix $P_i > 0$ which will be used in the construction of the Lyapunov function V_i. The Lyapunov function for the composite system will be $V = V_n$.

During the process, the original variables x_i will be changed to z_i and the equilibrium point will be $z = (z_1^T, \ldots, z_n^T)^T = 0$ whose stability has to be guaranteed. The *backstepping algorithm* consists of the following steps.

Step $i = 1$
In the first step, let $z_1 = y - y_d$ and $z_2 = x_2 - \alpha_1$. Then

$$\dot{z}_1 = h^{x_1} \dot{x}_1 - \dot{y}_d = h^{x_1}(G_1 x_2 + f_1) - \dot{y}_d$$
$$= h^{x_1} G_1(z_2 + \alpha_1) + h^{x_1} f_1 - y_d^\theta(v_s - \omega_s).$$

Define the first virtual control by

$$\alpha_1 := G_1^{-1}(h^{x_1})^{-1}[A_1 z_1 - h^{x_1} f_1 + y_d^\theta v_s]. \qquad (2.106)$$

Substituting α_1 into $\dot{z}_1$ then $\dot{z}_1 = h^{x_1} G_1 z_2 + h^{x_1} G_1 \alpha_1 - y_d^\theta(v_s - \omega_s)$ and canceling the appropriate terms with positive and negative sign yields

$$\dot{z}_1 = h^{x_1} G_1 z_2 + A_1 z_1 + y_d^\theta \omega_s. \qquad (2.107)$$

Define the first partial Lyapunov function by $V_1 = z_1^T P_1 z_1$ and take into consideration that $\dot{V}_1 = 2 z_1^T P_1 \dot{z}_1$ and

$$2 z_1^T P_1 A_1 z_1 = 2\langle z_1, P_1 A_1 z_1 \rangle = \langle z_1, (P_1 A_1 + A_1^T P_1) z_1 \rangle = -z_1^T Q_1 z_1$$

then

$$\dot{V}_1 = 2 z_1^T P_1[h^{x_1} G_1 z_2 + A_1 z_1 + y_d^\theta \omega_s]$$
$$= -z_1^T Q_1 z_1 + 2 z_1^T P_1 h^{x_1} G_1 z_2 + \underbrace{2 z_1^T P_1 y_d^\theta}_{\tau_1} \omega_s, \qquad (2.108)$$

$$\tau_1 = 2z_1^T P_1 y_d^\theta.$$

Step $i = 2$

Since the desired path appears only in the definition of z_1, therefore some differences will be in $\dot{z}_2$. In the second step, let $z_3 = x_3 - \alpha_2$. Then $\dot{z}_2 = \dot{x}_2 - \dot{\alpha}_1$ and thus

$$\dot{z}_2 = G_2(z_3 + \alpha_2) + f_2 - \sigma_1 - \alpha_1^\theta(v_s - \omega_s)$$
$$= G_2 z_3 + G_2 \alpha_2 + f_2 - \sigma_1 - \alpha_1^\theta(v_s - \omega_s).$$

Define the second virtual control by

$$\alpha_2 := G_2^{-1}\left[A_2 z_2 - P_2^{-1} G_1^T \left(h^{x_1} \right)^T P_1 z_1 - f_2 + \sigma_1 + \alpha_1^\theta v_s \right]. \qquad (2.109)$$

Substituting α_2 into $\dot{z}_2$ and canceling the appropriate terms with positive and negative sign yields

$$\dot{z}_2 = G_2 z_3 + A_2 z_2 - P_2^{-1} G_1^T \left(h^{x_1} \right)^T P_1 z_1 + \alpha_1^\theta \omega_s. \qquad (2.110)$$

Define the second partial Lyapunov function by $V_2 = V_1 + z_2^T P_2 z_2$ and take into consideration that $\dot{V}_2 = \dot{V}_1 + 2z_2^T P_2 \dot{z}_2$ and $2z_2^T P_2 A_2 z_2 = -z_2^T Q_2 z_2$ then

$$\dot{V}_2 = -z_1^T Q_1 z_1 + 2z_1^T P_1 h^{x_1} G_1 z_2 + \tau_1$$
$$+ 2z_2^T P_2 \left[G_2 z_3 + A_2 z_2 - P_2^{-1} G_1^T \left(h^{x_1} \right)^T P_1 z_1 + \alpha_1^\theta \omega_s \right]$$
$$= -z_1^T Q_1 z_1 - z_2^T Q_2 z_2 + 2z_2^T P_2 G_2 z_3 + \underbrace{\left(\tau_1 + 2z_2^T P_2 \alpha_1^\theta \right)}_{\tau_2} \omega_s, \quad (2.111)$$

$$\tau_2 = \tau_1 + 2z_2^T P_2 \alpha_1^\theta.$$

Steps $i = 3, \dots, n-1$

Notice that $\alpha_1, \dot{z}_1, \dot{V}_1, \alpha_2, \dot{z}_2$ contain h^{x_1} but $\dot{V}_2$ does not. Hence, in the next steps some modifications are necessary:

$$z_i = x_i - \alpha_{i-1},$$
$$\alpha_i = G_i^{-1}\left[A_i z_i - P_i^{-1} G_{i-1}^T P_{i-1} z_{i-1} - f_i + \sigma_{i-1} + \alpha_{i-1}^\theta v_s \right],$$
$$\dot{z}_i = G_i z_{i+1} + A_i z_i - P_i^{-1} G_{i-1}^T P_{i-1} z_{i-1} + \alpha_{i-1}^\theta \omega_s,$$

$$V_i = \sum_{j=1}^{i} z_j^T P_j z_j, \qquad (2.112)$$

$$\dot{V}_i = -\sum_{j=1}^{i} z_j^T Q_j z_j + 2z_i^T P_i G_i z_{i+1} + \tau_i \omega_s,$$

$$\tau_i = \tau_{i-1} + 2z_i^T P_i \alpha_{i-1}^\theta.$$

Step $i = n$

In the last step, let $z_n = x_n - \alpha_{n-1}$, $\dot{z}_n = G_n u + f_n - \sigma_{n-1} - \alpha_{n-1}^\theta$, from which the control law can be determined:

$$u = G_n^{-1}\left[A_n z_n - P_n^{-1} G_{n-1}^T P_{n-1} z_{n-1} - f_n + \sigma_{n-1} + \alpha_{n-1}^\theta v_s\right],$$

$$\dot{z}_n = A_n z_n - P_n^{-1} G_{n-1}^T P_{n-1} z_{n-1} + \alpha_{n-1}^\theta \omega_s,$$

$$V_n = \sum_{j=1}^{n} z_j^T P_j z_j,$$

$$\dot{V}_n = -\sum_{j=1}^{n} z_j^T Q_j z_j + \tau_n \omega_s,$$

$$\tau_n = \tau_{n-1} + 2 z_n^T P_n \alpha_{n-1}^\theta.$$

(2.113)

The closed loop system can be brought on the form

$$\dot{z} = A(\bar{x}_n, \theta, t) z + b(\bar{x}_{n-1}, \theta, t) \omega_s,$$

(2.114)

where

$A(\bar{x}_n, \theta, t)$

$$= \begin{bmatrix} A_1 & h^{x_1} G_1 & 0 & 0 & \cdots & 0 \\ -P_2^{-1} G_1^T (h^{x_1})^T P_1 & A_2 & G_2 & 0 & \cdots & 0 \\ 0 & -P_3^{-1} G_2^T P_2 & A_3 & G_3 & \cdots & 0 \\ \vdots & \vdots & \ddots & \ddots & \ddots & \vdots \\ 0 & 0 & \cdots & 0 & -P_n^{-1} G_{n-1}^T P_{n-1} & A_n \end{bmatrix},$$

$$b(\bar{x}_{n-1}, \theta, t) = \begin{pmatrix} y_d^\theta \\ \alpha_1^\theta \\ \alpha_2^\theta \\ \vdots \\ \alpha_{n-1}^\theta \end{pmatrix}.$$

(2.115)

Remark

- The system stability depends on $\tau_n \omega_s$ where ω_s is the speed error depending on the path parameter. However, if a high level controller makes ω_s convergent to zero then the derivative of the Lyapunov function $\dot{V} = \dot{V}_n$ becomes negative definite.
- If the path is designed as function of the time, that is, $\theta = t$ and the path is $y_d(t)$, then $\omega_s \equiv 0$, and the closed loop is globally asymptotically stable.

- The notation $\dot{\alpha}_{i-1} = \sigma_{i-1} + \alpha_{i-1}^{\theta} \dot{\theta}$ hides the complexity of the computation. In reality, the formulas are early exploding with increasing n. The use of symbolic computation tools is suggested to find the derivatives for $n > 2$.

2.7 Sliding Control

Sliding control is an important method in the class of variable structure control approaches [8, 150]. Its importance is justified by the fact that sliding controller does not need the precise model of the process to be controlled, it is enough to know a simplified model and some upper bounds of the modeling errors. On the other hand, important other methods exist which ever are not sliding controller but integrate in their algorithms the sliding control principle too. Such a method is for example self-tuning adaptive control. We try to introduce the sliding control concept as simple as possible.

Assume the state variable is $x = (y, y', \ldots, y^{(n-1)})^T$ and the SISO system has input affine model and error signal

$$y^{(n)} = f(x) + g(x)u, \tag{2.116}$$

$$\tilde{y}(t) = y_d(t) - y(t), \tag{2.117}$$

respectively. Suppose $x_d(0) = x(0)$ otherwise the error will be composed by integration. The switching surface $s(x,t) = 0$, called sliding surface, is defined by the rule

$$s(x,t) := \left(\frac{d}{dt} + \lambda \right)^{n-1} \tilde{y}(t), \tag{2.118}$$

where $\lambda > 0$. It is evident that if the trajectory remains on the sliding surface (sliding mode) then the error goes exponentially to zero. Notice that $\frac{1}{2}|s|^2$ can be considered as Lyapunov function hence it is useful to require

$$\frac{1}{2} \frac{d}{dt} |s|^2 = ss' \leq -\eta |s|, \tag{2.119}$$

where $\eta > 0$.

If the above conditions are satisfied but in the beginning the trajectory is not on the sliding surface, that is, $s(0) \neq 0$, then (since, for example, in case of $s(0) > 0$ yields $s' \leq \eta$) the trajectory surely reaches the sliding surface $s = 0$ within the time

$$t_s \leq \frac{|s(0|}{\eta}. \tag{2.120}$$

The time for approaching the sliding surface can be influenced by the parameter η while the time of decaying by the parameter λ, which simplify the design. In consequence, the main task of every control system, that is, eliminating the error, is conceptually fulfilled.

2.7.1 Sliding Control of Second Order Systems

The remaining question is how to choose the control law in order to satisfy the above conditions. We illustrate it through examples for second order systems of increasing complexity.

2.7.1.1 Proportional Control

Let the system to be controlled $\ddot{y} = f(x) + u$, the error signal is $\tilde{y}(t)$, $n = 2$, and by definition

$$s = \dot{\tilde{y}} + \lambda \tilde{y}, \tag{2.121}$$

$$s' = \ddot{y}_d - \ddot{y} + \lambda \dot{\tilde{y}} = \ddot{y}_d - f - u + \lambda \dot{\tilde{y}}. \tag{2.122}$$

Assume that instead of $f(x)$ only its estimation $\hat{f}(x)$ is available. Let the form of the control

$$u = \hat{u} + k(x)\,sign(s). \tag{2.123}$$

It is clear that on the sliding surface $s = 0$, $sign(0) = 0$ by definition, hence $u = \hat{u}$. Reaching the sliding surface the best concept is to remain on it since then the error goes exponentially to zero. But if $s = 0$ (constant), then $s' = 0$, too. The only information is $\hat{f}(x)$ therefore the strategy may be

$$s = 0, \quad \hat{f}(x) \quad \Rightarrow \quad s' = 0, \quad u = \hat{u}, \tag{2.124}$$

$$\hat{u} = \ddot{y}_d - \hat{f}(x) + \lambda \dot{\tilde{y}}. \tag{2.125}$$

However, if the sliding surface has not been reached yet, that is, $s \neq 0$, then in order to reach the sliding surface within finite time, $ss' \leq -\eta|s|$ has to be satisfied, hence

$$s' = \hat{f} - f - k(x)\,sign(s), \tag{2.126}$$

$$ss' = \left(f - \hat{f}\right)s - k(x)|s| \leq \left|f - \hat{f}\right||s| - k(x)|s| \leq -\eta|s|, \tag{2.127}$$

$$k(x) \geq \left|f(x) - \hat{f}(x)\right| + \eta. \tag{2.128}$$

Let the upper bound of the deviation between model and estimation be given by the known function $F(x)$,

$$\left|f(x) - \hat{f}(x)\right| \leq F(x) \tag{2.129}$$

then $k(x)$ can be chosen

$$k(x) = F(x) + \eta. \tag{2.130}$$

2.7.1.2 Integral Control

Let the system be again $\ddot{y} = f(x) + u$, but now the error is given by the integral $\int_0^t \tilde{y}(\tau)\,d\tau$, for which formally $n = 3$ can be assumed. Then by definition

$$s = \left(\frac{d}{dt} + \lambda\right)^2 \int_0^t \tilde{y}(\tau)\,d\tau = \dot{\tilde{y}} + 2\lambda\tilde{y} + \lambda^2 \int_0^t \tilde{y}(\tau)\,d\tau, \qquad (2.131)$$

$$s' = \ddot{y}_d - \ddot{y} + 2\lambda\dot{\tilde{y}} + \lambda^2\tilde{y} = \ddot{y}_d - f - u + 2\lambda\dot{\tilde{y}} + \lambda^2\tilde{y}. \qquad (2.132)$$

Hence, similarly to the previous case, the strategy may be $s = 0$, $\hat{f}(x) \Rightarrow s' = 0$, $u = \hat{u}$ yielding

$$u = \hat{u} + k(x)\,sign(s), \qquad (2.133)$$

$$\hat{u} = \ddot{y}_d - \hat{f}(x) + 2\lambda\dot{\tilde{y}} + \lambda^2\tilde{y}, \qquad (2.134)$$

$$k(x) = F(x) + \eta. \qquad (2.135)$$

It is an advantage of integral control that $s(0) = 0$ can also be assured, that is, the control can already start on the sliding surface eliminating t_s, if we choose

$$s(x,t) := \dot{\tilde{y}} + 2\lambda\tilde{y} + \lambda^2 \int_0^t \tilde{y}(\tau)\,d\tau - \dot{\tilde{y}}(0) - 2\lambda\tilde{y}(0). \qquad (2.136)$$

Remark The difference between proportional and integral control is in the computation of s and $\hat{u}$, however $k(x)$ is the same. Henceforth, we shall deal only with proportional control, and in case of integral control we modify the final s and $\hat{u}$ according to the above results.

2.7.1.3 Proportional Control in Case of Variable Gain

In the above examples, we had unit gain. Now, we consider again the second order system $\ddot{y} = f(x) + g(x)u$ but with state dependent gain having strictly positive sign. Suppose that the estimation $\hat{g}(x)$ of the state dependent gain and a parameter $\beta > 0$ (may also be state dependent) is known, furthermore the relation between real and estimated gain satisfies the special condition

$$\beta^{-1} \le \frac{\hat{g}}{g} \le \beta. \qquad (2.137)$$

By definition

$$s = \left(\frac{d}{dt} + \lambda\right)\tilde{y} = \dot{\tilde{y}} + \lambda\tilde{y}, \qquad (2.138)$$

$$s' = \ddot{y}_d - \ddot{y} + \lambda\dot{\tilde{y}} = \ddot{y}_d - f - gu + \lambda\dot{\tilde{y}}. \qquad (2.139)$$

Let the form of the controller

$$u = \hat{b}^{-1}\big[\hat{u} + k(x)\,sign(s)\big], \qquad (2.140)$$

and derive $\hat{u}$ from the strategy $s = 0,\ \hat{f}(x),\ \hat{b}(x) \Rightarrow s' = 0,\ u = \hat{u}$, then

$$\hat{u} = \ddot{y} - \hat{f}(x) + \lambda \tilde{y}. \qquad (2.141)$$

Considering the general case $s \neq 0$ and supposing the sliding surface has not been reached yet, then

$$s' = \hat{f} - f + \big(1 - g\hat{g}^{-1}\big)\hat{u} - g\hat{g}^{-1}k(x)\,sign(s), \qquad (2.142)$$

$$ss' \leq |f - \hat{f}||s| + |1 - g\hat{g}^{-1}||\hat{u}||s| - g\hat{g}^{-1}k(x)|s| \leq -\eta|s|, \qquad (2.143)$$

$$k(x) \geq \hat{g}g^{-1}|f - \hat{f}| + |\hat{g}g^{-1} - 1||\hat{u}| + \hat{g}g^{-1}\eta. \qquad (2.144)$$

According to the bounds of $\hat{g}g^{-1}$ the nonlinear gain $k(x)$ can be chosen

$$k(x) = \beta(F + \eta) + (\beta - 1)|\hat{u}|. \qquad (2.145)$$

Now let us consider the more practical case if minimal and maximal bounds of the gain are known:

$$g_{min} \leq g \leq g_{max}. \qquad (2.146)$$

This case can be transformed to the previous one by using the geometric average of the minimal and maximal gains:

$$\hat{g} := \sqrt{g_{min} \cdot g_{max}}, \qquad (2.147)$$

$$\beta := \sqrt{\frac{g_{max}}{g_{min}}}, \qquad (2.148)$$

$$\beta^{-1} \leq \frac{\hat{g}}{g} \leq \beta. \qquad (2.149)$$

Notice that b_{min}, b_{max} may be state dependent.

2.7.2 Control Chattering

Unfortunately, the realization of sliding control has an unpleasant problem. Since $k(x)$ is usually nonzero on the sliding surface and

$$sign(x) = \begin{cases} +1, & \text{if } s > 0, \\ 0, & \text{if } s = 0, \\ -1, & \text{if } s < 0 \end{cases}$$

Fig. 2.11 Boundary layer for
second order system

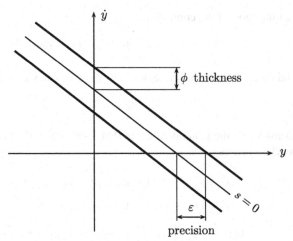

hence the control is discontinuous, the jump is $\pm k(x)$. When the trajectory reaches
the sliding surface $s(x,t) = 0$, the control should perform $\pm k(x)$ jump within zero
time, which is impossible. The trajectory steps over to the other side of the sliding
surface, then returns back to the sliding surface, but cannot stop within zero time
there, hence steps over to the first side, and repeats the process.

Meanwhile between the two sides of the sliding surface the change of the control
is $\pm k(x)$ hence the control is chattering with high frequency (control chattering).
This high frequency excitation may be harmfully, especially if the system has nonmodeled structural frequency ω_{struct} of resonance character near to the frequency of
chattering, because the magnitude may be amplified which can damage the system.

2.7.2.1 Compensation of Control Chattering Using Boundary Layer

For eliminating high frequency control chattering, first the *sign* function will be
changed to the *sat* function:

$$sat(s/\phi) = \begin{cases} +1, & \text{if } s > \phi, \\ s/\phi, & \text{if } -\phi \leq s \leq \phi, \\ -1, & \text{if } s < -\phi. \end{cases}$$

The price is that a boundary layer neighboring should be formed around the sliding
surface:

$$B(x,t) = \{x : |s(x,t)| \leq \phi\}.$$

Outside the boundary layer we save the original control while inside it u will be
interpolated by $sat(s/\phi)$. For example, in case of known gain, proportional control
and $y_d = const$ the boundary layer around the sliding surface $s = -\dot{y} + \lambda(y_d - y)$
can be characterized by its thickness ϕ and precision ε, see Fig. 2.11.

From theoretical point of view, it is an important consequence that both $\tilde{y}$ and $\tilde{y}^{(i)}$ are bounded within the boundary layer. Denoting the differential operator (Laplace transform variable) by $p = \frac{d}{dt}$ then for $x(0) = 0$ initial condition

$$\tilde{y} = \frac{1}{(p+\lambda)^{n-1}}s,$$

$$\tilde{y}^{(i)} = \frac{p^i}{(p+\lambda)^{n-1}}s = \frac{1}{(p+\lambda)^{n-1-i}}\left(1 - \frac{\lambda}{(p+\lambda)}\right)^i s.$$

By repeated application of the convolution theorem, it follows within the boundary layer

$$\left|\int_0^t e^{-\lambda\tau}s(t-\tau)\,d\tau\right| \le \phi/\lambda \quad \Rightarrow \quad |\tilde{y}| \le \frac{\phi}{\lambda^{n-1}} =: \varepsilon,$$

$$\left|1 - \int_0^t \lambda e^{-\lambda\tau}\,d\tau\right| \le 2,$$

$$|\tilde{y}^{(i)}| \le \frac{1}{\lambda^{n-1-i}}2^i\phi = (2\lambda)^i\frac{\phi}{\lambda^{n-1}} = (2\lambda)^i\varepsilon.$$

If the initial conditions differ from zero, then the error remains still bounded because of the stability of the filter: $x = x_d + \mathcal{O}(\varepsilon)$.

Now consider the design of the controller if a boundary layer is present. The thickness ϕ of the boundary layer will be chosen time varying in order to decrease the tracking error $\tilde{y}$. Hence, the condition for $\frac{1}{2}s^2$ will be modified to

$$\frac{1}{2}\frac{d}{dt}|s|^2 = ss' \le (\dot{\phi} - \eta)|s|, \tag{2.150}$$

which means that we prescribe stronger condition for $\dot{\phi} < 0$ than for $\dot{\phi} \ge 0$.

2.7.2.2 Boundary Layer and Time Varying Filter for Second Order System

Consider first *proportional control* with constant gain $g = 1$. Evidently $\dot{\phi}$ and η have opposite effects, hence, assuming $k(x)$ belongs to η without boundary layer, then in case of boundary layer it needs

$$\tilde{k}(x) = k(x) - \dot{\phi}, \tag{2.151}$$

$$u = \hat{u} + \tilde{k}(x)\,sat(s/\phi). \tag{2.152}$$

Therefore, within the boundary layer yields

$$s' = \hat{f} - f - \tilde{k}(x)s/\phi, \tag{2.153}$$

from which it follows

$$s' = -\tilde{k}(x_d)s/\phi + \Delta f(x_d) + \mathcal{O}(\varepsilon). \tag{2.154}$$

Notice that s' behaves as a first order linear system along the trajectory $x_d(t)$ under the effect of bounded disturbance. Let choose the time constant equal to λ then we obtain the *balance condition*

$$\frac{\tilde{k}(x_d)}{\phi} := \lambda, \tag{2.155}$$

from which it follows for the varying boundary layer

$$\dot{\phi} + \lambda\phi = k(x_d). \tag{2.156}$$

Consider now the case of *state dependent gain*. Then using (2.144) and $\eta := \eta - \dot{\phi}$, it follows

$$\tilde{k}(x) \geq k(x) - \frac{\hat{g}}{g}\dot{\phi}, \tag{2.157}$$

$$\dot{\phi} > 0: \quad \tilde{k}(x) \geq k(x) - \left(\frac{\hat{g}}{g}\right)_{\min}\dot{\phi} = k(x) - \beta^{-1}\dot{\phi},$$

$$\tilde{k}(x_d) = k(x_d) - \beta^{-1}\dot{\phi},$$

$$\dot{\phi} < 0: \quad \tilde{k} \geq k(x) + \left(\frac{\hat{g}}{g}\right)_{\max}(-\dot{\phi}) = k(x) - \beta\dot{\phi}, \tag{2.158}$$

$$\tilde{k}(x_d) = k(x_d) - \beta\dot{\phi}.$$

Prescribe for variable gain the balance condition

$$\frac{\tilde{k}(x_d)}{\phi}\beta := \lambda, \tag{2.159}$$

then we obtain for the thickness of the boundary layer the law

$$\dot{\phi} > 0: \quad \frac{\phi\lambda}{\beta} = k(x_d) - \beta^{-1}\dot{\phi} \quad \Rightarrow \quad \dot{\phi} + \lambda\phi = \beta k(x_d),$$

$$\dot{\phi} < 0: \quad \frac{\phi\lambda}{\beta} = k(x_d) - \beta\dot{\phi} \quad \Rightarrow \quad \dot{\phi} + \frac{\lambda}{\beta^2}\phi = \frac{1}{\beta}k(x_d). \tag{2.160}$$

2.7.3 Sliding Control of Robot

Sliding control elaborated for SISO system can be applied for robots considering each degree of freedom as a separate SISO system and modeling the coupling effects as disturbances [8].

Robot:$H\ddot{q} + h = \tau.$

Controller:$\tau = \hat{H}u + \hat{h},$

$u = ?$ sliding control.

Closed loop:$\ddot{q} = H^{-1}(\hat{h} - h) + H^{-1}\hat{H}u.$

All the functions are state dependent, $x = (q^T, \dot{q}^T)^T$ is the state, q is the joint variable. Suppose the estimation $\hat{H}, \hat{h}$ is good enough, especially the elements in the main diagonal of $H^{-1}\hat{H} = I + H^{-1}(\hat{H}^{-1} - H)$ are strictly positive (i.e., the gains are positive). Denote $(H^{-1})_i$ the i-th row of H^{-1}, let $\Delta h = \hat{h} - h$ and $x^+ = (|x_1|, \ldots, |x_n|)^T$ for any vector x. Develop for each subsystem a sliding controller:

$$\ddot{q}_i = \left(H^{-1}\right)_i \Delta h + \left(H^{-1}\hat{H}\right)_{ii} u_i + \sum_{i \neq j}\left(H^{-1}\hat{H}\right)_{ij} u_j,$$

$$\left.\begin{aligned} g_{i,\min} \leq \left(H^{-1}\hat{H}\right)_{ii} \leq g_{i,\max} \\ \beta_i = \sqrt{g_{i,\max}/g_{i,\min}} \\ G_i = \hat{g}_i^{-1} = 1/\sqrt{g_{i,\max} \cdot g_{i,\min}} \end{aligned}\right\} \text{ functions of } q.$$

Then the control law for the integrating sliding controller yields

$$u_i = G_i(q)\left\{\hat{u}_i + \tilde{k}_i(q, \dot{q}) \, sat(s_i/\phi_i)\right\},$$

$$\hat{u}_i = \ddot{q}_{di} + 2\lambda\dot{\tilde{q}}_i + \lambda^2 \tilde{q}_i,$$

$$s_i = \dot{\tilde{q}}_i + 2\lambda\tilde{q}_i + \lambda^2 \int_0^t \tilde{q}_i(\theta) \, d\theta - \dot{\tilde{q}}_i(0) - 2\lambda\tilde{q}_i(0),$$

$$k_i(q, \dot{q}) \geq \beta_i\left\{(1 - \beta_i^{-1})|\hat{u}_i| + \left(H^{-1}\right)_i^+ \Delta h^+ + \sum_{j \neq i} G_j|\ddot{q}_{dj}|\left|\left(H^{-1}\hat{H}\right)_{ij}\right| + \eta_i\right\}.$$

Notice that during the computation of k_i the approximation $u \approx G_j\ddot{q}_{dj}$ was applied in order to avoid the solution of an equation system. For computing $\tilde{k}_i$ and ϕ_i, the formulas elaborated above can be used.

2.8 Receding Horizon Control

Model Predictive Control (MPC) is a popular method especially in the process industry where relatively slow process models allow online optimization. Linear predictive control is well elaborated both in frequency (operator) domain [25] and state space [87]. Depending on the type of constraints, optimal prediction leads to

Quadratic (QP) or Nonlinear Programming (NP) which are well supported by existing softwares (e.g., MATLAB Optimization Toolbox).

2.8.1 Nonlinear Receding Horizon Control

For nonlinear systems, recent approaches are usually based on new optimum seeking methods suited for the predictive control problem or traditional analytical optimum conditions and gradient based optimization techniques [4, 65]. Basis for the later ones is a general form of the Lagrange multiplicator rule which is also valid in Banach spaces [73].

Here a version of the Receding Horizon Control (RHC) approach of model predictive control will be presented. Using RHC, the optimization is performed in the actual horizon in open loop, then the first element of the optimal control sequence will be performed in closed loop, the horizon is shifted to the right and the process is repeated for the new horizon. For simplification of the notation, the start of the moving horizon will be normalized to zero in every step.

It follows from the RHC principle that stability problems can arise if the objective function does not contain any information about the behavior after the horizon. To improve the stability properties, the techniques of Frozen Riccati Equation (FRE) and Control Lyapunov Function (CLF) can be suggested [161]. From engineering point of view, the increase of the horizon length N can improve the stability chance however at the same time the real-time optimization problem will be more complex.

Critical may be the presence of constraints in form of limited control set or even more in form of state variable constraints. Although efficient numerical optimization techniques exist for constrained optimization like example, active set method or interior point method, however in real-time environment simpler gradient based techniques are more realistic. They apply penalty function and control projection instead of the constraints and use conjugate gradient or Davidon–Fletcher–Power (DFP) method for unconstrained optimization. They are especially important for quick systems like robots and vehicles where the necessary sampling time is typically in the order of 10–20 ms for the (high level) control system. Hence, we will concentrate here only on quadratic objective function however the system may be nonlinear.

Typical finite horizon nonlinear predictive control problems in discrete time lead to optimization in finite dimensional space where the variables are $x = \{x_i\}_{i=0}^N$ and $u = \{u_i\}_{i=0}^{N-1}$, $Q_i \geq 0$ and $R_i > 0$ punish large states and control signals respectively, the optimality criterion to be minimized is

$$F_0(x, u) = \frac{1}{2} \sum_{i=0}^{N-1} \left[\langle Q_i x_i, x_i \rangle + \langle R_i u_i, u_i \rangle \right] + \frac{1}{2} \langle Q_N x_N, x_N \rangle,$$

$$F_0(x, u) =: \sum_{i=0}^{N-1} L_i(x_i, u_i) + \Phi(x_N),$$

(2.161)

the constraints are the discrete time state equation

$$\varphi(x_i, u_i) - x_{i+1} = 0, \tag{2.162}$$

the control set $u_i \in M$ and the initial condition $a - x_0 = 0$. If (x^*, u^*) is the optimal solution then

$$J'_x(x^*, u^*)x + J'_u(x^*, u^*)u$$

is the derivative of

$$J(x, u) = F_0(x, u) + \langle \lambda_0, a - x_0 \rangle + \langle \lambda_1, \varphi(x_0, u_0) - x_1 \rangle + \cdots$$
$$+ \langle \lambda_N, \varphi(x_{N-1}, u_{N-1}) - x_N \rangle. \tag{2.163}$$

By introducing the Hamiltonian functions

$$H_i = \langle \lambda_{i+1}, \varphi(x_i, u_i) \rangle + L_i(x_i, u_i),$$

and using the results of [73] the necessary condition of the optimality are obtained as

$$\lambda_N = Q_N x_N, \tag{2.164}$$

$$\lambda_i = \partial H_i / \partial x_i = Q_i x_i + (\partial \varphi / \partial x_i)^T \lambda_{i+1}, \tag{2.165}$$

$$\partial H_i / \partial u_i = R_i u_i + (\partial \varphi / \partial u_i)^T \lambda_{i+1}, \tag{2.166}$$

$$dJ = \sum_{i=0}^{N-1} \langle \partial H_i / \partial u_i, u_i^* - u_i \rangle \le 0, \quad \forall u_i \in M. \tag{2.167}$$

For the control design within the actual horizon, first the initial condition x_0 and the approximation of u are needed. The latter may be the solution in the previous horizon shifted to the left and supplying the missing term u_N (for example, repeating the last term etc.). Notice that for the very first horizon no initial control sequence is present yet thus special method is necessary to supply it.

The optimization repeats the following steps:

- Solution of the state equations in $x = \{x_i\}_{i=0}^{N}$.
- Computation of the Lagrange multipliers λ_i.
- Computation of the derivatives $\partial H_i / \partial u_i$.
- Numerical optimization based on gradient type methods (gradient, conjugate gradient, DFP etc.) to find $u = \{u_i\}_{i=0}^{N-1}$.

Nonpredictive design method should be used to find the initial approximation for the first horizon. If the original system is a continuous time one, then first it can be approximated by a discrete time one, for example,

$$\dot{x} = f_c(x, u) \quad \Rightarrow \quad x_{i+1} = x_i + T f_c(x_i, u_i) =: \varphi(x_i, u_i), \tag{2.168}$$

where T is the sampling time. If the full state can not be measured then x_0 can be approximated by using extended Kalman filter or other state estimator/observer [30].

If $y = Cx$ is the system output and $\tilde{y} = y_d - y$ is the error, then the cost function can be modified as

$$\tilde{y} = y_d - Cx, \tag{2.169}$$

$$2L_i(x_i, u_i) = \langle \tilde{Q}_i \tilde{y}_i, \tilde{y}_i \rangle + \langle S_i x_i, x_i \rangle + \langle R_i u_i, u_i \rangle, \tag{2.170}$$

$$2\Phi(x_N) = \langle \tilde{Q}_N \tilde{y}_N, \tilde{y}_N \rangle, \tag{2.171}$$

and the derivatives can be computed by

$$\partial L_i / \partial x_i = -C^T \tilde{Q}_i \tilde{y}_i + S_i x_i, \tag{2.172}$$

$$\partial \Phi / \partial x_N = -C^T \tilde{Q}_N \tilde{y}_N. \tag{2.173}$$

Input constraints are enforced by projecting u_i into the constraints set. State constraints can be taken into consideration as an additional penalty added to $L(x_i, u_i)$ in the cost function. The weighting in the term $\Phi(x_N)$ has great influence on the stability and dynamic behavior of the system under predictive control. The appropriate weighting term can be chosen in experiments.

The control design strategy can be summarized in the following steps:

- Development of the nonlinear dynamic model of the system.
- Optimal (suboptimal, flatness-based etc.) open loop control signal design used for initial approximation of the control sequence in the horizon.
- Elaboration of the disturbance model reduced on the system input.
- Development of the model based nonlinear predictive controller and its use in closed loop control. The first element of the control sequence in the actual horizon is completed by the feedforward compensations of the disturbance.

Remark As can be seen from (2.165) and (2.166), $\partial \varphi / \partial x_i$ and $\partial \varphi / \partial u_i$ are needed during numerical optimization therefore their computation time has large influence to the real-time properties. In mechanical systems often appear terms of the form $f(x) = A^{-1}(x)b(x)$ at the right side of the state equation. See, for example, robots where $\ddot{q} = -H^{-1}(q)h(q, \dot{q}) + H^{-1}(q)\tau$. The following formula can speed up the computation:

$$\frac{\partial f}{\partial x_i} = -A^{-1} \frac{\partial A}{\partial x_i} A^{-1} b + A^{-1} \frac{\partial b}{\partial x_i}. \tag{2.174}$$

2.8.2 Nonlinear RHC Control of 2D Crane

We can assume that the path and the nominal control for the differentially flat 2D crane have already been designed, see Sect. 2.1.3. Now we shall concentrate only on the steps which can accelerate the RHC computation based on the above remark. Since $\varphi(x_i, u_i) = x_i + T f_c(x_i, u_i)$, hence it is enough to find $f_c(x, u)$ and then solve its differentiation.

First, the Jacobian (dx) and Hessian $(D2x)$ have to be found for which carefully performed symbolic differentiation can be suggested. By using them the derivatives, $\dot{x}, \ddot{x}, \dot{z}, \ddot{z}$ can be determined. For example, in the case of $x(t)$ yields

$$\dot{x} = dx \cdot \left(\dot{L}_1 \dot{L}_2 \dot{L}_3\right)^T,$$
$$\ddot{x} = dx \cdot \left(\ddot{L}_1 \ddot{L}_2 \ddot{L}_3\right)^T + \left(\dot{L}_1 \dot{L}_2 \dot{L}_3\right) \cdot D2x \cdot \left(\dot{L}_1 \dot{L}_2 \dot{L}_3\right)^T. \tag{2.175}$$

The resulting continuous time dynamic model in MATLAB convention for *indexing* is the following:

$$A(1,:) = C_\theta \, dx + S_\theta \, dz,$$
$$b(1) = \left(\dot{L}_1 \dot{L}_2 \dot{L}_3\right)(-C_\theta D2x - S_\theta D2z)\left(\dot{L}_1 \dot{L}_2 \dot{L}_3\right)^T - S_\theta g,$$
$$A(2,:) = 2C_\gamma m(S_\theta \, dx - C_\theta \, dz)\rho_1,$$
$$A(2,1) = A(2,1) + J_1/\rho_1,$$
$$b(2) = \left(\dot{L}_1 \dot{L}_2 \dot{L}_3\right)\left(2C_\gamma m(-S_\theta D2x + C_\theta D2z)\rho_1\right)\left(\dot{L}_1 \dot{L}_2 \dot{L}_3\right)^T$$
$$\qquad + 2C_\gamma mC_\theta g\rho_1,$$
$$A(3,:) = m(S_\theta \, dx - C_\theta \, dz)\rho_2,$$
$$A(3,2:3) = A(3,2:3) + [J_2/\rho_2 J_2/\rho_2],$$
$$b(3) = \left(\dot{L}_1 \dot{L}_2 \dot{L}_3\right)\left(m(-S_\theta D2x + C_\theta D2z)\rho_2\right)\left(\dot{L}_1 \dot{L}_2 \dot{L}_3\right)^T + mC_\theta g\rho_2,$$
$$\left(\ddot{L}_1 \ddot{L}_2 \ddot{L}_3\right)^T = A^{-1}b - A^{-1}(0 u_1 u_2)^T.$$

For RHC A and b have to be differentiated once more, see the above remark. However, this can also be performed by using symbolic computation tools. Notice that symbolic computations can be performed offline, only the final formulas have to be implemented in real-time.

An important question remains the appropriate choice of the sampling time T so that the approximation of the continuous time system by the discrete time one using Euler formula is accurate enough. Do not forget that the sampling time has influence on the horizon length because the effective horizon length in time units is NT which has to be fitted to the speed of the system. For large N, the use of basis functions (splines etc.) can be suggested in order to decrease the number of variables to be optimized.

Numerical results with RHC control of a small size model of a real cable-driven crane of US Navy can be found in [74].

2.8.3 RHC Based on Linearization at Each Horizon

Till now we discussed RHC based on the nonlinear dynamic model. Of course there exist the possibility to linearize the system at the beginning of each horizon if the linear approximation is accurate enough in the entire horizon. Applying RHC for the LTV or LTI model, valid in the horizon, simple control implementations can be found. Especially, if the objective function is quadratic and the constraints are linear equalities and/or linear inequalities then in each horizon a Quadratic Programming (QP) problem has to be solved for which efficient active set or interior point methods are available. Especially, if there are no constraints then analytical solution is possible which enormously simplifies the real-time conditions.

2.9 Closing Remarks

The main goal of the chapter was to give a *survey* of the basic nonlinear control methods. Central question was the introduction of the main control ideas and the presentation of the results supporting the solutions of nonlinear control problems. In the presentation, a compromise was sought between theoretical precision and engineering view.

Nonlinear control systems have to be tested before application for which *simulation methods* can be applied. During simulation, the real system is substituted by its model. For modeling of continuous time dynamic systems, nonlinear and linear models can be used. Important problem is the computation of the transients for which in case of nonlinear systems numerical methods should be used. For this purpose, simple (Euler) and more accurate (Runge–Kutta and predictor–corrector) methods were shown. For linear systems, the computation can be simplified by using the fundamental matrix (LTV systems) or the exponential matrix or transfer function (LTI systems). Nonlinear systems can be divided into different classes.

Nonlinear systems often have to satisfy *kinematic constraints* in the space of generalized coordinates and their derivatives. The constraints can be divided into holonomic and nonholonomic ones. The kinematic constraints are usually given as a set of equations linear in the derivatives of the generalized coordinates but nonlinear in the generalized coordinates (Pfaffian constraints). If there exist a nonlinear coordinate transformation (diffeomorphism) after which the constraints can be eliminated and the number of variables can be reduced, then the system is *holonomic*, otherwise it is *nonholonomic*. In the dynamic model of nonholonomic system appear constraint forces, too, described by parts of the kinematic constraints and Lagrange multipliers. A method was shown how to determine the Lagrange multipliers from the dynamic model. Based on energy concept, the Hamiltonian form was introduced and an experiment was performed to simplify the Hamiltonian using a coordinate transformation which is related to the kernel space of the kinematic constraints. Conditions for omitting the kinematic constraints were given based on Poisson bracket.

An other nonlinear system class is the class of *differentially flat* systems which are Lie–Bäcklund equivalent to a trivial system. Differentially flat systems can be linearized by dynamic state feedback and a coordinate transformation. For such system, there is a flat output variable, having the same dimension as the system input, such that the state and the input can be expressed by the *flat output variable* and its derivatives of finite order. The resulting square system is left and right input–output invertible with trivial zero dynamics. Path design for differentially flat systems is a simple problem because it can be reduced to smooth interpolation between points in the space of the flat output variable. The result is not only the path but also the nominal control along the path. The nominal (open loop) control can be used for feedforward compensation or may be the basis for linearization along the path. The differential flatness property is especially useful for underactuated systems where not every path can be realized by control.

Modeling nonlinear systems is an often arising problem. Three simple examples were considered to illustrate how to find the dynamic model of physical systems. The first example was the dynamic modeling of an *inverted pendulum* mounted on a moving car. To find the model the kinetic and potential energy was found and the Lagrange equations were applied. The inverted pendulum is an underactuated system. The second example was the dynamic model of a *quarter car using active suspension*. The flexible connections between body and wheel were described by viscous damping factors and spring constants. The resulting model is an LTI system with active force input. The third model was a *2 DOF robot* having revolute joints and moving in vertical plane. First the position, velocity and angular velocity of the links were determined then the dynamic model was derived from the Lagrange equations. The coupling and centripetal effects were determined from the kinetic energy by using differentiation. The dynamic model is nonlinear.

Stability is a central problem of every control system, especially for nonlinear ones. First the stability, uniform stability and asymptotic stability of any solution of the time varying nonlinear state equation were defined. Then the solution was transformed to the equilibrium point at zero satisfying some conditions. The idea of positive definite Lyapunov function was introduced and based on it *Lyapunov's first theorem* (direct method) was formulated for stability and asymptotic stability. For time varying nonlinear systems, *Lyapunov's second theorem* (indirect method) was formulated allowing to conclude from the asymptotic stability of the linearized system to the asymptotic stability of the nonlinear one. For LTI system, the stability is equivalent to the solvability of the *Lyapunov equation*. For time invariant nonlinear system, *LaSalle's theorem* was formulated which gives also information about the attraction domain of the stability and brings into relation the asymptotic stability with the maximal invariant set. The classes of *K and KL functions* were introduced and two theorems were given for showing the relation to the uniformly asymptotically and the globally uniformly asymptotically stable systems. The *input-to-state stability* (ISS) was defined and a sufficient condition was formulated for ISS.

Three forms of the *Barbalat's lemma* were formulated which give sufficient conditions for zero limit value of scalar-valued time functions as the time goes to infinity. As an immediate corollary of Barbalat's lemma, a Lyapunov-like condition

was given which looks like an invariant set theorem in Lyapunov stability analysis
of time varying systems.

The system behavior in *Lp function spaces* was characterized with the nonlinear
gain. A small gain theorem was formulated for closed loop systems in such function
spaces. The discussions implicitly assumed zero initial conditions. *Passive systems,*
amongst them passive, strictly input passive and strictly output passive systems were
introduced and a passivity theorem in Lp spaces was formulated which allows to
conclude from the stability of the open loop subsystems to the stability of the closed
loop if only one external input is active.

Dissipative systems allow to integrate the Lyapunov theory and the stability in
L2 spaces in some sense. Special is that the number of inputs and outputs must
be equal because the scalar product plays a central role. Dissipativity is defined by
using the idea of *storage function* and *supply rate* which have to satisfy the dissipativity inequality. *Special classes* of dissipative systems are the passive, the strictly
input passive, the strictly output passive systems and the system having finite L2
gain. The dissipative property can be expressed also in differential form which can
be checked easier. *Three theorems* were formulated giving connection between passivity and asymptotic stability, stability conditions of interconnected systems without external input, and a small gain theorem for interconnected dissipative systems
without external inputs. These theorems will play important role in the stability investigations of multiagent systems moving in formation.

The control of nonlinear dynamic systems is a complex problem. In order to
simplify the problem, it is a useful idea to linearize the nonlinear system using appropriate nonlinear coordinate transformation and then compensate it further using
linear control methods. Sufficient condition for the *input–output linearization* of the
MIMO nonlinear input affine system is the existence of vector relative degree and
the nonsingularity of a special matrix composed from repeated Lie derivatives. If
the conditions are satisfied, then a nonlinear coordinate transformation has to be determined which can be formed by completing a set of functions appearing during
the process. Then an internal static nonlinear state feedback can be applied which
brings the system onto Brunovsky canonical form equivalent to decoupled integrators. However, beside of this linear system appears also a nonlinear one called zero
dynamics which is not observable (has no influence on the output). The *zero dynamics* has to be stable in order to assure closed loop stability. For SISO system, the
zero dynamics is independent from the system input, but in this case the Frobenius
equation has to be solved in order to construct the nonlinear coordinate transformation. For MIMO systems, this process cannot be applied and the zero dynamics
depends also on the system input. The number of inputs and outputs must be the
same in the construction, however the output may differ from the physical system
output. The weak point of the method is the large number of symbolic computations, the solution of the Frobenius (partial differential) equation in SISO case or
the computation of the nonlinear coordinate transformation in the MIMO case, and
the computation of the inverse coordinate transformation which is also needed to
find the zero dynamics.

Some other nonlinear control methods were also discussed in detail, especially
because of the complexity of input–output linearization. In *flatness control,* only the

path planning problem for differentially flat underactuated systems was considered. It was demonstrated for the planar (2D) cable-driven crane. The flat output variable was determined and it was shown that the state and the control can be computed from the flat output and its derivatives. Then the path planning was reduced to polynomial interpolation.

Backstepping control is a popular method for controlling nonlinear systems given in strictly feedback form. The dimension of the portions of the state vector, the input vector and the physical output must be the same and some invertability conditions have to be satisfied. It was assumed that the path was designed in the space and parametrized in a scalar path parameter which has to satisfy a parameter speed constraint, too. The design is decomposed in steps, in every step a nonlinear virtual control is developed, the state variable portion is changed to a new variable and a partial Lyapunov function is designed. At the end, the last Lyapunov function assures the *closed loop stability*, but high level control has to eliminate the parameter speed error. Especially, if the path is given as a function of time then the closed loop is already stable without high level control. Backstepping plays also important role in formation stabilization.

Sliding control is a kind of variable structure control. The sliding variable is the output of a stable filter with equal time constants driven by the output error. The sliding surface belongs to the zero value of the sliding variable. If the system remains on the sliding surface, then the error goes exponentially to zero. If the sliding variable is initially not on the sliding surface, then the sliding surface can be reached within limited time. A method was presented for the design of sliding controller satisfying the above conditions for SISO systems including proportional and integral control with constant or state dependent gain. About the functions in the system model only upper bounds were assumed. Since the method can cause high frequency *control chattering* in the neighborhood of the sliding surface therefore a *boundary layer* was introduced whose thickness determines the precision. Hence, the thickness was chosen state dependent in order to decrease the error. It was shown that with some modifications the sliding control developed for SISO systems can also be applied for robots considering coupling effects amongst the joints as disturbances.

Receding Horizon Control (RHC) is a kind of optimal control considered in discrete time in order to make the problem finite dimensional. Using RHC, the optimization is performed in the actual horizon in open loop, then the first element of the optimal control sequence is performed in closed loop, the horizon is shifted to the right and the process is repeated for the new horizon. It follows from the RHC principle that stability problems can arise if the objective function does not contain any information about the behavior after the horizon. It is clear that the increase of the horizon length can improve the stability chance however at the same time the real-time optimization problem becomes more complex. Critical is the presence of *constraints* for the control and/or state variables. In order to avoid time consuming constrained optimization in real-time, the method of penalty function, control projection and unconstrained optimization using gradient based techniques (conjugate gradient, DFP etc.) were suggested. Under these assumptions, the *RHC algorithm for nonlinear discrete time system* was elaborated. The critical steps of the computations were illustrated in the example of the RHC control of the planar (2D) crane which is differentially flat but underactuated.

Another RHC strategy may be the *linearization at the beginning of each horizon*. The linear approximation should be accurate enough in the entire horizon. Considering the actual LTV or LTI model and assuming the objective function is quadratic and the constraints are linear equalities and/or linear inequalities, then in each horizon a *Quadratic Programming* (QP) problem has to be solved for which efficient numerical methods are available. This approach can also be applied in safety critical systems where high level RHC computes optimal reference signals for the low level control system.

Chapter 3
Dynamic Models of Ground, Aerial and Marine Robots

Overview This chapter gives a survey of the dynamic modeling of the robots and vehicles in unified approach. First, the kinematic and dynamic model of the rigid body is discussed which is the basis for the further investigations. For robots, the dynamic model is developed using Appell's equation and Lagrange's equation. For ground cars, a nonlinear dynamic model, two nonlinear input affine approximations and a linearized model is derived. For airplanes, first the usual navigation frames are shown. Then the kinematic and dynamic equations are presented considering aerodynamic and gyroscopic effects. Finally, the main flying modes, the trimming and linearization principle and the concept of the parametrization of aerodynamic and trust forces for identification purposes are outlined. For surface and underwater ships, first the rigid body equations are developed, then the hydrodynamic and restoring forces and moments are shown. A short introduction is given for wind, wave and current models. Finally, the results are summarized in the dynamic model of ship both in body frame and NED frame. The chapter intensively builds on the results surveyed in Appendix A.

3.1 Dynamic Model of Rigid Body

Consider a rigid body moving relative to an inertial coordinate system K_I. To the rigid body, an own coordinate system K_B is fixed in the center of gravity (COG) which is moving together with the rigid body. The velocity, acceleration, angular velocity and angular acceleration are defined relative to the inertial frame. The differentiation rule in moving coordinate system gives the relation between the formal (componentwise) derivatives in the moving frame and the derivatives in the inertial frame. Especially $a = \dot{v} + \omega \times v$ where a, v, ω are the acceleration of COG, the velocity of COG and the angular velocity of the rigid body, respectively. The total external force F_{total} and the total external torque M_{total} are acting in COG.

B. Lantos, L. Márton, *Nonlinear Control of Vehicles and Robots*,
Advances in Industrial Control,
DOI 10.1007/978-1-84996-122-6_3, © Springer-Verlag London Limited 2011

3.1.1 Dynamic Model Based on Newton–Euler Equations

All vectors are expressed in the basis of the moving frame K_B which will be shown by the simplified notation $a, v, \omega, F_{\text{total}}, M_{\text{total}} \in K_B$. Their components in K_B are $a = (a_x, a_y, a_z)^T$, $v = (U, V, W)^T$, $\omega = (P, Q, R)^T$, $F_{\text{total}} = (F_X, F_Y, F_Z)^T$, $M_{\text{total}} = (L, M, N)^T$. The rigid body has mass m and inertia matrix I_c belonging to COG.

The dynamic model of the rigid body is given by Newton's force equation $ma = m(\dot{v} + \omega \times v) = F_{\text{total}}$, and Euler's moment equation $I_c \dot{\omega} + \omega \times (I_c \omega) = M_{\text{total}}$, from which follows

$$\dot{v} = -\omega \times v + \frac{1}{m} F_{\text{total}}, \tag{3.1}$$

$$\dot{\omega} = -I_c^{-1} \omega \times (I_c \omega) + I_c^{-1} M_{\text{total}}. \tag{3.2}$$

Let us compute first the inverse of the inertia matrix:

$$I_c = \begin{bmatrix} I_x & -I_{xy} & -I_{xz} \\ -I_{xy} & I_y & -I_{yz} \\ -I_{xz} & -I_{yz} & I_z \end{bmatrix} \Rightarrow I_c^{-1} = \begin{bmatrix} I_1 & I_2 & I_3 \\ I_2 & I_4 & I_5 \\ I_3 & I_5 & I_6 \end{bmatrix}. \tag{3.3}$$

$$\det(I_c) = I_x I_y I_z - 2 I_{xy} I_{xz} I_{yz} - I_x I_{yz}^2 - I_y I_{xz}^2 - I_z I_{xy}^2,$$

$$I_1 = (I_y I_z - I_{yz}^2)/\det(I_c),$$

$$I_2 = (I_{xy} I_z + I_{xz} I_{yz})/\det(I_c),$$

$$I_3 = (I_{xy} I_{yz} + I_{xz} I_y)/\det(I_c), \tag{3.4}$$

$$I_4 = (I_x I_z - I_{xz}^2)/\det(I_c),$$

$$I_5 = (I_x I_{yz} + I_{xy} I_{xz})/\det(I_c),$$

$$I_6 = (I_x I_y - I_{xy}^2)/\det(I_c).$$

Then we can continue with the computation of $-\omega \times (I_c \omega)$:

$$-\omega \times (I_c \omega)$$

$$= \begin{bmatrix} 0 & R & -Q \\ -R & 0 & P \\ Q & -P & 0 \end{bmatrix} \left(\begin{bmatrix} I_x \\ -I_{xy} \\ -I_{xz} \end{bmatrix} P + \begin{bmatrix} -I_{xy} \\ I_y \\ -I_{yz} \end{bmatrix} Q + \begin{bmatrix} -I_{xz} \\ -I_{yz} \\ I_z \end{bmatrix} R \right)$$

$$= \begin{pmatrix} (-RI_{xy} + QI_{xz})P + (RI_y + QI_{yz})Q + (-RI_{yz} - QI_z)R \\ (-RI_x - PI_{xz})P + (RI_{xy} - PI_{yz})Q + (RI_{xz} + PI_z)R \\ (QI_x + PI_{xy})P + (-QI_{xy} - PI_y)Q + (-QI_{xz} + PI_{yz})R \end{pmatrix}$$

$$= \begin{pmatrix} (0)PP + I_{xz}PQ - I_{xy}PR + I_{yz}QQ + (I_y - I_z)QR - I_{yz}RR \\ -I_{xz}PP - I_{yz}PQ + (I_z - I_x)PR + (0)QQ + I_{xy}QR + I_{xz}RR \\ I_{xy}PP + (I_x - I_y)PQ + I_{yz}PR - I_{xy}QQ - I_{xz}QR + (0)RR \end{pmatrix}.$$

Dynamic model of the rigid body

$$\dot{U} = RV - QW + F_X/m,$$

$$\dot{V} = -RU + PW + F_Y/m,$$

$$\dot{W} = QU - PV + F_Z/m,$$

$$\dot{P} = P_{pp}PP + P_{pq}PQ + P_{pr}PR + P_{qq}QQ + P_{qr}QR$$
$$+ P_{rr}RR + I_1\bar{L} + I_2M + I_3N,$$

$$\dot{Q} = Q_{pp}PP + Q_{pq}PQ + Q_{pr}PR + Q_{qq}QQ + Q_{qr}QR$$
$$+ Q_{rr}RR + I_2\bar{L} + I_4M + I_5N,$$

$$\dot{R} = R_{pp}PP + R_{pq}PQ + R_{pr}PR + R_{qq}QQ + R_{qr}QR$$
$$+ R_{rr}RR + I_3\bar{L} + I_5M + I_6N,$$

(3.5)

(3.6)

where the parameters P_{**}, Q_{**}, R_{**} are weighting factors for the products of the components of ω defined by

$$P_{pp} = -I_{xz}I_2 + I_{xy}I_3,$$

$$P_{pq} = I_{xz}I_1 - I_{yz}I_2 + (I_x - I_y)I_3,$$

$$P_{pr} = -I_{xy}I_1 + (I_z - I_x)I_2 + I_{yz}I_3,$$

$$P_{qq} = I_{yz}I_1 - I_{xy}I_3,$$

$$P_{qr} = (I_y - I_z)I_1 + I_{xy}I_2 - I_{xz}I_3,$$

$$P_{rr} = -I_{yz}I_1 + I_{xz}I_2,$$

(3.7)

$$Q_{pp} = -I_{xz}I_4 + I_{xy}I_5,$$

$$Q_{pq} = I_{xz}I_2 - I_{yz}I_4 + (I_x - I_y)I_5,$$

$$Q_{pr} = -I_{xy}I_2 + (I_z - I_x)I_4 + I_{yz}I_5,$$

$$Q_{qq} = I_{yz}I_2 - I_{xy}I_5,$$

$$Q_{qr} = (I_y - I_z)I_2 + I_{xy}I_4 - I_{xz}I_5,$$

$$Q_{rr} = -I_{yz}I_2 + I_{xz}I_4,$$

(3.8)

$$R_{pp} = -I_{xz}I_5 + I_{xy}I_6,$$

$$R_{pq} = I_{xz}I_3 - I_{yz}I_5 + (I_x - I_y)I_6,$$

$$R_{pr} = -I_{xy}I_3 + (I_z - I_x)I_5 + I_{yz}I_6,$$

$$R_{qq} = I_{yz}I_3 - I_{xy}I_6,$$

$$R_{qr} = (I_y - I_z)I_3 + I_{xy}I_5 - I_{xz}I_6,$$

$$R_{rr} = -I_{yz}I_3 + I_{xz}I_5.$$

(3.9)

The above equations do not consider the position and orientation of the rigid body, although many problems need also them, for example if a designed path has to be realized by the control. It means that $r_c \in K_I$ and the orientation of K_B relative to K_I are also important. Hence, we have to deal also with the kinematic part of the motion equation. For the description of the orientation, the Euler (roll, pitch, yaw) angles and the quaternions are especially important, see Appendix A.

3.1.2 Kinematic Model Using Euler (RPY) Angles

In case of Euler (RPY) angles, we can parametrize the relative orientation by the rotation angles Ψ, Θ, Φ around the z-, y-, x-axes, respectively:

$$
\begin{aligned}
A_{K_I, K_B} &= Rot(z, \Psi)\, Rot(y, \Theta)\, Rot(x, \Phi) \\[4pt]
&= \begin{bmatrix} C_\Psi & -S_\Psi & 0 \\ S_\Psi & C_\Psi & 0 \\ 0 & 0 & 1 \end{bmatrix}
\begin{bmatrix} C_\Theta & 0 & S_\Theta \\ 0 & 1 & 0 \\ -S_\Theta & 0 & C_\Theta \end{bmatrix}
\begin{bmatrix} 1 & 0 & 0 \\ 0 & C_\Phi & -S_\Phi \\ 0 & S_\Phi & C_\Phi \end{bmatrix} \\[4pt]
&= \begin{bmatrix} C_\Psi & -S_\Psi & 0 \\ S_\Psi & C_\Psi & 0 \\ 0 & 0 & 1 \end{bmatrix}
\begin{bmatrix} C_\Theta & S_\Theta S_\Phi & S_\Theta C_\Phi \\ 0 & C_\Phi & -S_\Phi \\ -S_\Theta & C_\Theta S_\Phi & C_\Theta C_\Phi \end{bmatrix} \Rightarrow
\end{aligned}
$$

$$
A_{K_I, K_B}(\Phi, \Theta, \Psi) = \begin{bmatrix}
C_\Psi C_\Theta & C_\Psi S_\Theta S_\Phi - S_\Psi C_\Phi & C_\Psi S_\Theta C_\Phi + S_\Psi S_\Phi \\
S_\Psi C_\Theta & S_\Psi S_\Theta S_\Phi + C_\Psi C_\Phi & S_\Psi S_\Theta C_\Phi - C_\Psi S_\Phi \\
-S_\Theta & C_\Theta S_\Phi & C_\Theta C_\Phi
\end{bmatrix}.
$$

$$(3.10)$$

Since Ψ, Θ, Φ can be considered as the joint variables of an RRR fictitious robot arm, hence the relation (A.115) between the angular velocity $\omega = \omega_x i_B + \omega_y j_B + \omega_z k_B$ and the joint velocities $\dot\Psi$, $\dot\Theta$, $\dot\Phi$ can be used, but in the formula the joints axes are $t_1 = k$ for $\dot\Psi$, $t_2 = j$ for $\dot\Theta$ and $t_3 = i$ for $\dot\Phi$, respectively. For the computation of the partial angular velocities, the partial products are already present in the above formulas, hence we can select the appropriate row and write it in the appropriate column of the "Jacobian":

$$
\begin{pmatrix} \omega_x \\ \omega_y \\ \omega_z \end{pmatrix} = \begin{bmatrix}
-S_\Theta & 0 & 1 \\
C_\Theta S_\Phi & C_\Phi & 0 \\
C_\Theta C_\Phi & -S_\Phi & 0
\end{bmatrix} \begin{pmatrix} \dot\Psi \\ \dot\Theta \\ \dot\Phi \end{pmatrix}.
$$

The determinant of the "Jacobian" matrix is $\det = -C_\Theta$ and its inverse is

$$
[\ ]^{-1} = -\frac{1}{C_\Theta} \begin{pmatrix}
0 & -S_\Phi & -C_\Phi \\
0 & -C_\Theta C_\Phi & C_\Theta S_\Phi \\
-C_\Theta & -S_\Theta S_\Phi & -S_\Theta C_\Phi
\end{pmatrix} = \begin{bmatrix}
0 & S_\Phi/C_\Theta & C_\Phi/C_\Theta \\
0 & C_\Phi & -S_\Phi \\
1 & T_\Theta S_\Phi & T_\Theta C_\Phi
\end{bmatrix}.
$$

Assume that the position of the COG is $r_c = X_c i_I + Y_c j_I + Z_c k_I$ in the basis of K_I. Then using the chosen parametrization $\omega = P i_B + Q j_B + R k_B$, $v = U i_B + V j_B + W k_B$ and the order $\dot{\Phi}, \dot{\Theta}, \dot{\Psi}$ we obtain the following kinematic equations:

$$\begin{pmatrix} \dot{X}_c \\ \dot{Y}_c \\ \dot{Z}_c \end{pmatrix} = A_{K_I, K_B}(\Phi, \Theta, \Psi) \begin{pmatrix} U \\ V \\ W \end{pmatrix}, \tag{3.11}$$

$$\begin{pmatrix} \dot{\Phi} \\ \dot{\Theta} \\ \dot{\Psi} \end{pmatrix} = \begin{bmatrix} 1 & T_\Theta S_\Phi & T_\Theta C_\Phi \\ 0 & C_\Phi & -S_\Phi \\ 0 & S_\Phi/C_\Theta & C_\Phi/C_\Theta \end{bmatrix} \begin{pmatrix} P \\ Q \\ R \end{pmatrix} =: F(\Phi, \Theta)\omega. \tag{3.12}$$

3.1.3 Kinematic Model Using Quaternion

The critical vertical pose of the rigid body at $\Theta = \pm\pi/2$ is singular configuration in Euler (RPY) angles hence some applications prefer the use of *quaternion* for orientation description.

It follows from the Rodriguez formula (A.13) and the properties (A.25) that $Rot(t, \varphi)$, $\|t\| = 1$, can be identified by an appropriately chosen unit quaternion q:

$$Rot(t, \varphi) = C_\varphi I + (1 - C_\varphi)[t \circ t] + S_\varphi[t \times] \leftrightarrow q = (C_{\varphi/2}, t S_{\varphi/2}),$$

$$q * (0, r) * \tilde{q} = (0, Rot(t, \varphi)r).$$

On the other hand, $t^2 = \langle t, t \rangle = 1$ hence $\langle t', t \rangle + \langle t, t' \rangle = 0 \Rightarrow t \perp t'$ thus $t, t \times t', t'$ is an orthogonal basis and $\omega = \lambda_1 t + \lambda_2 t \times t' + \lambda_3 t'$. Notice that prime denote time derivative here.

Lemma 3.1 $\omega = \varphi' t + (1 - C_\varphi) t' \times t + S_\varphi t'.$

Lemma 3.2 *If* $q = (s, w) = (C_{\varphi/2}, t S_{\varphi/2})$, $\|q\| = 1$ *then*

$$\omega = 2(-ws' + (-[w \times] + sI)w'). \tag{3.13}$$

The proofs of the above two lemmas can be found in [73].

Remarks

1. Since $\|q\|^2 = s^2 + w^2 = 1 \Rightarrow 2(w^T w' + ss') = 0$, hence

$$\begin{pmatrix} \omega \\ 0 \end{pmatrix} = 2 \begin{bmatrix} -w & -[w \times] + sI \\ s & w^T \end{bmatrix} \begin{pmatrix} s' \\ w' \end{pmatrix}. \tag{3.14}$$

2. The matrix in the above equation is orthogonal:

$$\begin{bmatrix} -w^T & s \\ [w \times] + sI & w \end{bmatrix} \begin{bmatrix} -w & -[w \times] + sI \\ s & w^T \end{bmatrix} = I. \tag{3.15}$$

3. The time derivative of the quaternion can be expressed by ω and q in bilinear form:

$$\begin{pmatrix} s' \\ w' \end{pmatrix} = \frac{1}{2} \begin{bmatrix} -w^T & s \\ [w\times] + sI & w \end{bmatrix} \begin{pmatrix} \omega \\ 0 \end{pmatrix} = \frac{1}{2} \begin{pmatrix} -\omega^T w \\ \omega s - \omega \times w \end{pmatrix},$$

$$\begin{pmatrix} s' \\ w' \end{pmatrix} = -\frac{1}{2} \begin{bmatrix} 0 & \omega^T \\ -\omega & [\omega\times] \end{bmatrix} \begin{pmatrix} s \\ w \end{pmatrix}. \tag{3.16}$$

The following lemma summarizes the results for the practice.

Lemma 3.3 *The orientation matrix $A_{K_I, K_B} =: B^T$ can be expressed by a quaternion in four dimensional vector form $q = (q_0, q_1, q_2, q_3)^T = (s, w^T)^T$. The time derivative $\dot{q}$ of the quaternion is bilinear in the angular velocity $\omega = (P, Q, R)^T$ of the rigid body and the quaternion q and yields*

$$\dot{q} = -\frac{1}{2} \begin{bmatrix} 0 & P & Q & R \\ -P & 0 & -R & Q \\ -Q & R & 0 & -P \\ -R & -Q & P & 0 \end{bmatrix} q =: -\frac{1}{2} \Omega_q q. \tag{3.17}$$

Remark Ω_q has the following properties:

$$\Omega_q^2 = -\|\omega\|^2 I,$$

$$\Omega_q^{2n} = (-1)^n \|\omega\|^{2n} I, \tag{3.18}$$

$$\Omega_q^{2n+1} = (-1)^n \|\omega\|^{2n} \Omega_q.$$

3.2 Dynamic Model of Industrial Robot

Many industrial robots are so called open chain rigid robots without branching. Basic relations for their kinematics and dynamics (kinetics) were summarized in Appendix A, the investigations here are based on them.

The robot consists of links connected by revolute (rotational R) or sliding (translational T) joints. The relation between links $i - 1$ and i can be characterized by the (original) Denavit–Hartenberg parameters $\vartheta_i, d_i, a_i, \alpha_i$ and the homogeneous transformation $T_{i-1,i}$ computed from them:

$$T_{i-1,i} = \begin{bmatrix} C_{\vartheta_i} & -S_{\vartheta_i} C_{\alpha_i} & S_{\vartheta_i} S_{\alpha_i} & a_i C_{\vartheta_i} \\ S_{\vartheta_i} & C_{\vartheta_i} C_{\alpha_i} & -C_{\vartheta_i} S_{\alpha_i} & a_i S_{\vartheta_i} \\ 0 & S_{\alpha_i} & C_{\alpha_i} & d_i \\ 0 & 0 & 0 & 1 \end{bmatrix}. \tag{3.19}$$

Joint variables are $q_i = \vartheta_i$ for rotational joint and $q_i = d_i$ for translational joint, respectively. For an m-DOF robot, the joint vector is $q = (q_1, \ldots, q_m)^T$. We often use abbreviations in robotics like $C_{\vartheta_i} = \cos(\vartheta_i)$ or $S_{12} = \sin(q_1 + q_2)$. The relation amongst the base frame K_B, the robot reference frame K_0, the frame of the last link K_m and the end effector frame K_E can be expressed by homogeneous transformations:

$$T_{0,m}(q) = T_{0,1}(q_1)T_{1,2}(q_2)\cdots T_{m-1,m}(q_m), \tag{3.20}$$

$$T_{B,E} = T_{B,0}T_{0,m}(q)T_{m,E}. \tag{3.21}$$

The velocity $^m v_m$ and the angular velocity $^m \omega_m$ can be expressed by the partial velocities $^m d_{i-1}$ and the partial angular velocities $^m t_{i-1}$ which define the configuration dependent Jacobian matrix $J_m(q)$ of the robot:

$$\begin{pmatrix} ^m v_m \\ ^m \omega_m \end{pmatrix} = {}^m J_m(q)\dot{q} = \begin{bmatrix} ^m d_0 & ^m d_1 & \cdots & ^m d_{m-1} \\ ^m t_0 & ^m t_1 & \cdots & ^m t_{m-1} \end{bmatrix}\dot{q}. \tag{3.22}$$

From the indexes, the right lower one (together with the name) identifies the quantity and the left upper one the basis in which the quantity was written down. The partial velocity and partial angular velocity can be computed by (A.115)–(A.118).

The velocity and angular velocity of the end effector can be determined from $^m J_m$, namely

$$T_{m,E} = \begin{bmatrix} A_{m,E} & p_{m,E} \\ 0^T & 1 \end{bmatrix} \quad \Rightarrow \quad {}^E J_E = \begin{bmatrix} A_{m,E}^T & ([p_{m,E}\times]A_{m,E})^T \\ 0 & A_{m,E}^T \end{bmatrix} {}^m J_m. \tag{3.23}$$

3.2.1 Recursive Computation of the Kinematic Quantities

We shall bring the dynamic model of the robot on the form $H\ddot{q} + h = \tau$, therefore the kinematic quantities will similarly be parametrized:

$$\begin{aligned} ^i \omega_i &= \Gamma_i \dot{q}, & ^i v_i &= \Omega_i \dot{q}, \\ ^i \varepsilon_i &= \Gamma_i \ddot{q} + \phi_i, & ^i a_i &= \Omega_i \ddot{q} + \theta_i, \end{aligned} \tag{3.24}$$

where $^i \omega_i$, $^i v_i$, $^i \varepsilon_i$ and $^i a_i$ are respectively, the angular velocity, velocity, angular acceleration and acceleration of the origin of the frame K_i fixed to the link i, all expressed in the basis of K_i. Using (A.50)–(A.51) with the choice $\rho := p_{i-1,i}$ and applying the differentiation rule in moving frame, it follows

$$\rho' = (1 - \kappa_i)t_{i-1}\dot{q}_i + \kappa_i t_{i-1}\dot{q}_i \times p_{i-1,i} =: d_{i-1}\dot{q}_i,$$

$$\rho'' = d_{i-1}\ddot{q}_i + \kappa_i t_{i-1}\dot{q}_i \times d_{i-1,i}.$$

However, the results are obtained in the basis of K_{i-1} which should be transformed to the basis of K_i by $A_{i-1,i}^{-1} = A_{i-1,i}^T$, hence

$$^i \omega_i = A_{i-1,i}^T {}^{i-1}\omega_{i-1} + {}^i t_{i-1}\dot{q}_i,$$

$$^i v_i = A_{i-1,i}^T \left(^{i-1} v_{i-1} + ^{i-1}\omega_{i-1} \times p_{i-1,i}\right) + ^i d_{i-1}\dot{q}_i,$$

$$^i \varepsilon_i = A_{i-1,i}^T \, ^{i-1}\varepsilon_{i-1} + ^i t_{i-1}\ddot{q}_i + ^i \omega_i \times ^i t_{i-1}\dot{q}_i,$$

$$^i a_i = A_{i-1,i}^T \left\{^{i-1} a_{i-1} + ^{i-1}\varepsilon_{i-1} \times p_{i-1,i} + ^{i-1}\omega_{i-1} \times \left(^{i-1}\omega_{i-1} \times p_{i-1,i}\right)\right\}$$

$$+ 2^i \omega_i \times ^i d_{i-1}\dot{q}_i - ^i t_{i-1}\dot{q}_i \times ^i d_{i-1}\dot{q}_i + ^i d_{i-1}\ddot{q}_i.$$

Here $^i d_{i-1}$, $^i t_{i-1}$ can be determined by using (A.111)–(A.114) for $m = i$. From the above formulas, the recursive relations for the computation of $\Gamma_i, \phi_i, \Omega_i, \theta_i$ can be read out. In order to simplify the formulas, the zero columns can be left out, that is, in the i-th step Γ_i, Ω_i have i columns and $\dim \dot{q} = i$:

$$\Gamma_i = \left[A_{i-1,i}^T \Gamma_{i-1} \mid ^i t_{i-1}\right],$$

$$^i \omega_i = \Gamma_i \dot{q},$$

$$\phi_i = A_{i-1,i}^T \phi_{i-1} + ^i \omega_i \times ^i t_{i-1}\dot{q}_i,$$

$$\Omega_i = \left[A_{i-1,i}^T \left\{\Omega_{i-1} - [p_{i-1,i}\times]\Gamma_{i-1}\right\} \mid ^i d_{i-1}\right], \tag{3.25}$$

$$\theta_i = A_{i-1,i}^T \left\{\theta_{i-1} + \phi_{i-1} \times p_{i-1,i} + ^{i-1}\omega_{i-1} \times \left(^{i-1}\omega_{i-1} \times p_{i-1,i}\right)\right\}$$

$$+ 2^i \omega_i \times ^i d_{i-1}\dot{q}_i - ^i t_{i-1}\dot{q}_i \times ^i d_{i-1}\dot{q}_i.$$

If ρ_{ci} denotes the center of gravity of link i in the basis of frame K_i, then

$$\Omega_{ci} = \Omega_i - [\rho_{ci}\times]\Gamma_i,$$

$$\theta_{ci} = \theta_i + \phi_i \times \rho_{ci} + ^i \omega_i \times \left(^i \omega_i \times \rho_{ci}\right), \tag{3.26}$$

$$^i a_{ci} = \Omega_{ci}\ddot{q} + \theta_{ci}.$$

The parametrization of the kinematic quantities is closely connected with the Jacobian matrix of the robot. The relation is sensitive on the choice of the basis vectors of the frames, it is different in the fixed frame K_0 and in the moving frame K_m. If prime ($'$) denotes the formal (componentwise) derivative by the time, then

$$\begin{pmatrix} ^\circ a_m \\ ^\circ \varepsilon_m \end{pmatrix} = ^\circ J_m \ddot{q} + \frac{d^\circ J_m}{dt}\dot{q} = \begin{bmatrix} ^\circ \Omega_m \\ ^\circ \Gamma_m \end{bmatrix}\ddot{q} + \begin{bmatrix} ^\circ \Omega_m' \\ ^\circ \Gamma_m' \end{bmatrix}\dot{q}, \tag{3.27}$$

$$\begin{pmatrix} ^m a_m \\ ^m \varepsilon_m \end{pmatrix} = ^m J_m \ddot{q} + \begin{pmatrix} ^m \Omega_m'\dot{q} + (^m \Gamma_m\dot{q}) \times (^m \Omega_m\dot{q}) \\ ^m \Gamma_m'\dot{q} \end{pmatrix}$$

$$= \begin{bmatrix} ^m \Omega_m \\ ^m \Gamma_m \end{bmatrix}\ddot{q} + \begin{bmatrix} ^m \Omega_m' + [(^m \Gamma_m\dot{q})\times]^m \Omega_m \\ ^m \Gamma_m' \end{bmatrix}\dot{q}. \tag{3.28}$$

3.2.2 Robot Dynamic Model Based on Appell's Equation

Using (A.83) and (A.85), we are able to derive the robot dynamic model based on the numerical results of the recursive forward kinematics. The method uses the "acceleration" energy or more precise Gibbs function.

Let us consider first a single link. Suppose the origin of the frame is in the COG then the Gibbs function is

$$\mathcal{G} = \frac{1}{2}\langle a_c, a_c \rangle m + \frac{1}{2}\langle I_c \varepsilon - 2(I_c \omega) \times \omega, \varepsilon \rangle,$$

where a_c is the acceleration, ε and ω are the angular acceleration and angular velocity, respectively, m is the mass of the link and I_c is the inertia matrix belonging to the COG in the frame fixed to the link.

We shall assume that the axes of the frame are parallel to the Denavit–Hartenberg frame hence the kinematic quantities can be computed by (3.25) and (3.26), furthermore the inertia parameters of the link and the non-rotating motor (or the end effector) mounted on the link are already merged according to the Huygens–Steiner formula (A.60)–(A.61). Applying Appell's equation for robots yields

$$\frac{\partial \mathcal{G}}{\partial \ddot{q}_i} + \frac{\partial P}{\partial q_i} = \tau_i, \quad i = 1, 2, \ldots, m, \tag{3.29}$$

where τ_i is the driving torque for link i. Since the Gibbs function of the robot is the sum of the Gibbs function of the links hence it is enough to clear the effect of a single link having index s and add them later. In order to simplify the formulas, we add the necessary number of zero columns to Ω_{cs} and Γ_s so that they will be $3 \times m$ matrices in case of m-DOF robot. Then

$$\frac{\partial \mathcal{G}_s}{\partial \ddot{q}_i} = \frac{1}{2}\left\langle \frac{\partial a_{cs}}{\partial \ddot{q}_i}, a_{cs} \right\rangle m_s + \frac{1}{2}\left\langle a_{cs}, \frac{\partial a_{cs}}{\partial \ddot{q}_i} \right\rangle m_s$$

$$+ \frac{1}{2}\left\langle I_{cs}\frac{\partial \varepsilon_s}{\partial \ddot{q}_i}, \varepsilon_s \right\rangle + \frac{1}{2}\left\langle I_{cs}\varepsilon_s - 2(I_{cs}\omega_s) \times \omega_s, \frac{\partial \varepsilon_s}{\partial \ddot{q}_i} \right\rangle$$

$$= \left\langle \frac{\partial a_{cs}}{\partial \ddot{q}_i}, a_{cs} \right\rangle m_s + \left\langle \frac{\partial \varepsilon_s}{\partial \ddot{q}_i}, I_{cs}\varepsilon_s - (I_{cs}\omega_s) \times \omega_s \right\rangle, \tag{3.30}$$

$$\frac{\partial a_{cs}}{\partial \ddot{q}_i} = \frac{\partial}{\partial \ddot{q}_i}(\Omega_{cs}\ddot{q} + \theta_{cs}) = \Omega_{cs,i}, \tag{3.31}$$

$$\frac{\partial \varepsilon_s}{\partial \ddot{q}_i} = \frac{\partial}{\partial \ddot{q}_i}(\Gamma_s\ddot{q} + \phi_s) = \Gamma_{s,i}, \tag{3.32}$$

where $\Omega_{cs,i}$ and $\Gamma_{s,i}$ are the i-th column of Ω_{cs} and Γ_s, respectively. Writing the scalar product in matrix product form, it follows

$$\frac{\partial \mathcal{G}_s}{\partial \ddot{q}_i} = m_s \Omega_{cs,i}^T(\Omega_{cs}\ddot{q} + \theta_{cs}) + \Gamma_{s,i}^T\left[I_{cs}(\Gamma_s\ddot{q} + \phi_s) - (I_{cs}\omega_s) \times \omega_s\right]$$

$$= \left(\Omega_{cs,i}^T \Omega_{cs} m_s + \Gamma_{s,i}^T I_{cs} \Gamma_s \right) \ddot{q} + \Omega_{cs,i}^T \theta_{cs} m_s$$
$$+ \Gamma_{s,i}^T \left[I_{cs} \phi_s - (I_{cs} \omega_s) \times \omega_s \right].$$

Adding the effect of the different links of the m-DOF robot, we obtain the robot dynamic model in the form

$$H(q)\ddot{q} + h_{cc}(q,\dot{q}) + h_g(q) = \tau, \tag{3.33}$$

$$H = \sum_{s=1}^{m} \left\{ \Omega_{cs}^T \Omega_{cs} m_s + \Gamma_s^T I_{cs} \Gamma_s \right\}, \tag{3.34}$$

$$h_{cc} = \sum_{s=1}^{m} \left\{ \Omega_{cs}^T \theta_{cs} m_s + \Gamma_s^T \left[I_{cs} \phi_s - (I_{cs}\omega_s) \times \omega_s \right] \right\}, \tag{3.35}$$

where h_{cc} is the centripetal and Coriolis effect and h_g is the gravity effect.

However, we have to find a computation rule also for the gravity effect. If m is a mass particle whose position vector is r in the fixed frame K_0 and its height over the horizontal plane is h, then its potential energy is mgh which can also be written as

$$-m\langle g_0, r \rangle = \left(-g_0^T \; 0 \right) \begin{pmatrix} r \\ 1 \end{pmatrix} m, \tag{3.36}$$

where g_0 is the gravity acceleration in the basis of K_0. If ρ_{cs} is the COG of link s in the Denavit–Hartenberg frame K_s, then the potential energy P of the robot and the gravity effect in Appell's equation are

$$P = \sum_{s=1}^{m} \left(-g_0^T \; 0 \right) T_{0,s} \begin{pmatrix} \rho_{cs} \\ 1 \end{pmatrix} m_s,$$

$$\frac{\partial P}{\partial q_i} = \sum_{s=1}^{m} \left(-g_0^T \; 0 \right) \frac{\partial T_{0,s}}{\partial q_i} \begin{pmatrix} \rho_{cs} \\ 1 \end{pmatrix} m_s \tag{3.37}$$

$$= \sum_{s=i}^{m} \left(-g_0^T \; 0 \right) T_{0,i-1} \frac{\partial T_{i-1,s}}{\partial q_i} \begin{pmatrix} \rho_{cs} \\ 1 \end{pmatrix} m_s.$$

Since by (A.119a)–(A.120b), $\partial T_{i-1,s} / \partial q_i = \Delta_{i-1} T_{i-1,s}$ hence

$$\frac{\partial P}{\partial q_i} = \left(-g_0^T \; 0 \right) T_{0,i-1} \Delta_{i-1} \sum_{s=i}^{m} m_s T_{i-1,s} \begin{pmatrix} \rho_{cs} \\ 1 \end{pmatrix} = G_i^T \begin{pmatrix} R_i \\ M_i \end{pmatrix}, \tag{3.38}$$

$$G_i^T = \left(-g_0^T \; 0 \right) T_{0,i-1} \Delta_{i-1}, \tag{3.39}$$

$$M_i = \sum_{s=i}^{m} m_s, \tag{3.40}$$

$$R_i = \sum_{s=i}^{m} m_s (A_{i-1,s} \rho_{cs} + p_{i-1,s}). \tag{3.41}$$

In the original Denavit–Hartenberg form, the joint axis is $t_{i-1} = k_{i-1}$ therefore:

(i) For rotational joint (R):

$$\Delta_{i-1} = \begin{bmatrix} 0 & -1 & 0 & 0 \\ 1 & 0 & 0 & 0 \\ 0 & 0 & 0 & 0 \\ 0 & 0 & 0 & 0 \end{bmatrix},$$

$$T_{0,i-1} \Delta_{i-1} = \begin{bmatrix} m_{0,i-1} & -l_{0,i-1} & 0 & 0 \\ 0 & 0 & 0 & 0 \end{bmatrix}, \tag{3.42}$$

$$G_i^T = (-g_0 \cdot m_{0,i-1}, g_0 \cdot l_{0,i-1}, 0, 0),$$

(ii) For translational joint (T):

$$\Delta_{i-1} = \begin{bmatrix} 0 & 0 & 0 & 0 \\ 0 & 0 & 0 & 0 \\ 0 & 0 & 0 & 1 \\ 0 & 0 & 0 & 0 \end{bmatrix},$$

$$T_{0,i-1} \Delta_{i-1} = \begin{bmatrix} 0 & 0 & 0 & n_{0,i-1} \\ 0 & 0 & 0 & 0 \end{bmatrix}, \tag{3.43}$$

$$G_i^T = (0, 0, 0, -g_0 \cdot n_{0,i-1}).$$

On the other hand

$$R_i = \sum_{s=i+1}^{m} m_s (A_{i-1,i} A_{i,s} \rho_{cs} + p_{i-1,i} + A_{i-1,i} p_{i,s}) + m_i (A_{i-1,i} \rho_{ci} + p_{i-1,i})$$

$$= A_{i-1,i} \left\{ \sum_{s=i+1}^{m} m_s (A_{i,s} \rho_{cs} + p_{i,s}) + m_i \rho_{ci} \right\} + \left(\sum_{s=i}^{m} m_s \right) p_{i-1,i},$$

hence the following *backward recursive algorithm* can be given for the computation of R_i and M_i:

$$R_{m+1} = 0, \qquad M_{m+1} = 0, \qquad M_i = M_{i+1} + m_i, \tag{3.44}$$

$$R_i = A_{i-1,i} (R_{i+1} + m_i \rho_{ci}) + M_i p_{i-1,i}. \tag{3.45}$$

Finally, the gravity effect in the robot dynamic model yields:

$$h_{g,i} = G_i^T \begin{pmatrix} R_i \\ M_i \end{pmatrix}. \tag{3.46}$$

3.2.3 Robot Dynamic Model Based on Lagrange's Equation

We have shown in Appendix A that the kinetic energy of the robot link can be computed by (A.79) if the origin of the frame is in COG. Assume the axes of the frame in COG are parallel to the axes of the Denavit–Hartenberg frame, then the kinetic energy of the robot is

$$
\begin{aligned}
K &= \sum_{s=1}^{m} \left\{ \frac{1}{2} \langle v_{cs}, v_{cs} \rangle m_s + \frac{1}{2} \langle I_{cs} \omega_s, \omega_s \rangle \right\} \\
&= \frac{1}{2} \sum_{s=1}^{m} \left\{ \langle \Omega_{cs} \dot{q}, \Omega_{cs} \dot{q} \rangle m_s + \langle I_{cs} \Gamma_s \dot{q}, \Gamma_s \dot{q} \rangle \right\} \\
&= \frac{1}{2} \sum_{s=1}^{m} \left\{ \langle \Omega_{cs}^T \Omega_{cs} m_s \dot{q}, \dot{q} \rangle + \langle \Gamma_s^T I_{cs} \Gamma_s \dot{q}, \dot{q} \rangle \right\} \quad \Rightarrow
\end{aligned}
\tag{3.47}
$$

$$
K = \frac{1}{2} \left\langle \left\{ \sum_{s=1}^{m} \{ \Omega_{cs}^T \Omega_{cs} m_s + \Gamma_s^T I_{cs} \Gamma_s \} \dot{q}, \dot{q} \right\} \right\rangle =: \frac{1}{2} \langle H(q) \dot{q}, \dot{q} \rangle.
$$

Therefore, the kinetic energy of the robot is the quadratic form of the symmetric and positive definite matrix $H(q)$ in the joint velocities $\dot{q}$. Introduce the well spread notation

$$
H(q) = \left[D_{jk}(q) \right]_{m \times m},
\tag{3.48}
$$

and use the Lagrange's equation (A.77) and the relation $D_{ij} = D_{ji}$ then

$$
\frac{\partial K}{\partial \dot{q}_i} = \frac{1}{2} \sum_{j=1}^{m} D_{ji} \dot{q}_j + \frac{1}{2} \sum_{k=1}^{m} D_{ik} \dot{q}_k = \sum_{j=1}^{m} D_{ij} \dot{q}_j,
\tag{3.49}
$$

$$
\frac{d}{dt} \frac{\partial K}{\partial \dot{q}_i} = \sum_{j=1}^{m} D_{ij} \ddot{q}_j + \sum_{j=1}^{m} \sum_{k=1}^{m} \frac{\partial D_{ij}}{\partial q_k} \dot{q}_j \dot{q}_k.
\tag{3.50}
$$

On the other hand

$$
axy + bxy = \frac{a+b}{2} xy + \frac{a+b}{2} xy,
$$

hence the second term can be made symmetric:

$$
\frac{d}{dt} \frac{\partial K}{\partial \dot{q}_i} = \sum_{j=1}^{m} D_{ij} \ddot{q}_j + \sum_{j=1}^{m} \sum_{k=1}^{m} \frac{1}{2} \left(\frac{\partial D_{ij}}{\partial q_k} + \frac{\partial D_{ik}}{\partial q_j} \right) \dot{q}_j \dot{q}_k.
\tag{3.51}
$$

Similarly, it can be obtained

$$\frac{\partial K}{\partial q_i} = \frac{1}{2} \sum_{j=1}^{m} \sum_{k=1}^{m} \frac{\partial D_{jk}}{\partial q_i} \dot{q}_j \dot{q}_k, \tag{3.52}$$

$$D_i := \frac{\partial P}{\partial q_i} = G_i^T \begin{pmatrix} R_i \\ M_i \end{pmatrix}. \tag{3.53}$$

Hence, the robot dynamic model in Lagrange form

$$\sum_{j=1}^{m} D_{ij} \ddot{q}_j + \sum_{j=1}^{m} \sum_{k=1}^{m} D_{ijk} \dot{q}_j \dot{q}_k + D_i = \tau_i, \quad i = 1, \dots, m, \tag{3.54}$$

$$D_{ijk} = \frac{1}{2} \left(\frac{\partial D_{ij}}{\partial q_k} + \frac{\partial D_{ik}}{\partial q_j} - \frac{\partial D_{jk}}{\partial q_i} \right), \tag{3.55}$$

where the following physical interpretation can be given:

D_{ii}	effective inertia,
D_{ij}	coupling inertia ($i \neq j$),
D_{ijj}	centripetal effect,
D_{ijk}	Coriolis effect ($j \neq k$),
D_i	gravity effect.

We have already merged the inertia parameter of the nonrotating motor (or the end effector) and the carrying link, but we have not considered the kinetic energy of the rotating rotor which is $K_{ri} = \frac{1}{2} \Theta_{ri} \dot{\varphi}_i^2 = \frac{1}{2} \Theta_{ri} v_i^2 \dot{q}_i^2$, where Θ_{ri} is the inertia moment and $\dot{\varphi}_i$ is the angular velocity of the rotor and $v_i = \dot{\varphi}_i / \dot{q}_i$ is the gear reduction. Since

$$\frac{d}{dt} \frac{\partial K_{ri}}{\partial \dot{q}_i} = \Theta_{ri} v_i^2 \ddot{q}_i, \tag{3.56}$$

hence an approximation of the rotor effect may be

$$D_{ii} := D_{ii} + \Theta_{ri} v_i^2. \tag{3.57}$$

Notice that D_{ij}, D_{ijk}, D_i are functions of q and they are varying during the motion of the robot. On the other hand, if D_{ij} (and thus the kinetic energy K) is known in symbolic form then D_{ijk} can be determined using partial differentiations. Similarly, if the potential energy P is known in symbolic form, then the gravity effect D_i can also be determined using partial differentiations.

Differentiation can be performed using symbolic computational tools like MATLAB Extended Symbolic Toolbox or Mathematica. Hence, beside the recursive numerical methods, we have also tools to determine offline the robot dynamic model in symbolic form and implement the simplified final set of formulas for real-time computations in the robot control algorithms. During this, it is useful to know that some functions are zero or can be computed from other functions in case of open chain rigid robots, see [133], namely

$$\frac{\partial D_{ik}}{\partial q_j} = 0, \quad \text{if } i, k \geq j \quad \Rightarrow \quad D_{ijk} = -D_{kji}, \quad \text{if } i, k \geq j, \quad (3.58)$$

$$D_{iji} = 0 \quad \text{if } i \geq j. \quad (3.59)$$

The results are illustrated for the $m = 4$ DOF case. In D^i, we have collected the D_{ijk} terms in matrix form for running j, k.

$$D^1 = \begin{bmatrix} 0 & D_{112} & D_{113} & D_{114} \\ * & D_{122} & D_{123} & D_{124} \\ * & * & D_{133} & D_{134} \\ * & * & * & D_{144} \end{bmatrix},$$

$$D^2 = \begin{bmatrix} -D_{112} & 0 & D_{213} & D_{214} \\ * & 0 & D_{223} & D_{224} \\ * & * & D_{233} & D_{234} \\ * & * & * & D_{244} \end{bmatrix},$$

$$\qquad (3.60)$$

$$D^3 = \begin{bmatrix} -D_{133} & -D_{213} & 0 & D_{314} \\ * & -D_{223} & 0 & D_{324} \\ * & * & 0 & D_{334} \\ * & * & * & D_{344} \end{bmatrix},$$

$$D^4 = \begin{bmatrix} -D_{114} & -D_{214} & -D_{314} & 0 \\ * & -D_{224} & -D_{324} & 0 \\ * & * & -D_{334} & 0 \\ * & * & * & 0 \end{bmatrix}.$$

3.2.4 Dynamic Model of SCARA Robot

SCARA robot is an RRTR type robot, see Fig. 3.1, its Denavit–Hartenberg parameters are collected in Table 3.1. All the joint axes are parallel and remain vertical during motion.

Assume the links have mass m_i, the center of gravity is $\rho_{ci} = (\rho_{xi}, \rho_{yi}, \rho_{zi})^T$, and the right lower element of the inertia matrix is $I_{ci,z} = I_{zi}$, $i = 1, \ldots, 4$. The other elements of the inertia matrices will have no influence on the dynamic model.

According to the Denavit–Hartenberg parameters, the homogeneous transformations between neighboring links are as follows:

$$T_{0,1} = \begin{bmatrix} C_1 & -S_1 & 0 & a_1 C_1 \\ S_1 & C_1 & 0 & a_1 S_1 \\ 0 & 0 & 1 & d_1 \\ 0 & 0 & 0 & 1 \end{bmatrix}, \qquad T_{1,2} = \begin{bmatrix} C_2 & -S_2 & 0 & a_2 C_2 \\ S_2 & C_2 & 0 & a_2 S_2 \\ 0 & 0 & 1 & d_2 \\ 0 & 0 & 0 & 1 \end{bmatrix},$$

Fig. 3.1 Simplified sketch of
the RRTR-type SCARA robot
with the coordinate systems

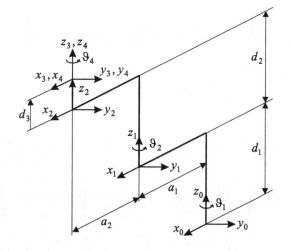

Table 3.1
Denavit–Hartenberg
parameters of SCARA robot

i	q_i	ϑ_i	d_i	a_i	α_i
1	ϑ_1	ϑ_1	d_1	a_1	$0°$
2	ϑ_2	ϑ_2	d_2	a_2	$0°$
3	d_3	$0°$	d_3	0	$0°$
4	ϑ_4	ϑ_4	0	0	$0°$

$$T_{2,3} = \begin{bmatrix} 1 & 0 & 0 & 0 \\ 0 & 1 & 0 & 0 \\ 0 & 0 & 1 & d_3 \\ 0 & 0 & 0 & 1 \end{bmatrix}, \qquad T_{3,4} = \begin{bmatrix} C_4 & -S_4 & 0 & 0 \\ S_4 & C_4 & 0 & 0 \\ 0 & 0 & 1 & 0 \\ 0 & 0 & 0 & 1 \end{bmatrix}.$$

The kinematic quantities can be obtained by using forward recursion:

$$\Gamma_1 = \begin{pmatrix} 0 \\ 0 \\ 1 \end{pmatrix}, \qquad \phi_1 = 0, \qquad \Omega_{c1} = \begin{pmatrix} -\rho_{y1} \\ a_1 + \rho_{x1} \\ 0 \end{pmatrix}, \qquad \theta_{c1} = \begin{pmatrix} -a_1 - \rho_{x1} \\ -\rho_{y1} \\ 0 \end{pmatrix} \dot{q}_1^2,$$

$$\Gamma_2 = \begin{pmatrix} 0 & 0 \\ 0 & 0 \\ 1 & 1 \end{pmatrix}, \qquad \phi_2 = 0, \qquad \Omega_{c2} = \begin{pmatrix} a_1 S_2 - \rho_{y2} & -\rho_{y2} \\ a_2 + a_1 C_2 + \rho_{x2} & a_2 + \rho_{x2} \\ 0 & 0 \end{pmatrix},$$

$$\theta_{c2} = \begin{pmatrix} -a_2 - a_1 C_2 - \rho_{x2} \\ a_1 S_2 - \rho_{y2} \\ 0 \end{pmatrix} \dot{q}_1^2 + 2 \begin{pmatrix} -a_2 - \rho_{x2} \\ -\rho_{y2} \\ 0 \end{pmatrix} \dot{q}_1 \dot{q}_2 + \begin{pmatrix} -a_2 - \rho_{x2} \\ -\rho_{y2} \\ 0 \end{pmatrix} \dot{q}_2^2,$$

$$\Gamma_3 = \begin{bmatrix} 0 & 0 & 0 \\ 0 & 0 & 0 \\ 1 & 1 & 0 \end{bmatrix}, \qquad \phi_3 = 0, \qquad \Omega_{c3} = \begin{bmatrix} a_1 S_2 - \rho_{y3} & -\rho_{y3} & 0 \\ a_2 + a_1 C_2 + \rho_{x3} & a_2 + \rho_{x3} & 0 \\ 0 & 0 & 1 \end{bmatrix},$$

$$\theta_{c3} = \begin{pmatrix} -a_2 - a_1 C_2 - \rho_{x3} \\ a_1 S_2 - \rho_{y3} \\ 0 \end{pmatrix} \dot{q}_1^2 + 2 \begin{pmatrix} -a_2 - \rho_{x3} \\ -\rho_{y3} \\ 0 \end{pmatrix} \dot{q}_1 \dot{q}_2 + \begin{pmatrix} -a_2 - \rho_{x3} \\ -\rho_{y3} \\ 0 \end{pmatrix} \dot{q}_2^2,$$

$$\Gamma_4 = \begin{bmatrix} 0 & 0 & 0 & 0 \\ 0 & 0 & 0 & 0 \\ 1 & 1 & 0 & 1 \end{bmatrix}, \qquad \phi_4 = 0,$$

$$\Omega_{c4} = \begin{bmatrix} a_2 S_4 + a_1 S_{24} - \rho_{y4} & a_2 S_4 - \rho_{y4} & 0 & -\rho_{y4} \\ a_2 C_4 + a_1 C_{24} - \rho_{x4} & a_2 C_4 + \rho_{x4} & 0 & \rho_{x4} \\ 0 & 0 & 1 & 0 \end{bmatrix},$$

$$\theta_{c4} = \begin{pmatrix} -a_2 C_4 - a_1 C_{24} - \rho_{x4} \\ a_2 S_4 + a_1 S_{24} - \rho_{y4} \\ 0 \end{pmatrix} \dot{q}_1^2 + \begin{pmatrix} -a_2 C_4 - \rho_{x4} \\ a_2 S_4 - \rho_{y4} \\ 0 \end{pmatrix} \dot{q}_2^2 + \begin{pmatrix} -\rho_{x4} \\ -\rho_{y4} \\ 0 \end{pmatrix} \dot{q}_4^2$$

$$+ 2 \begin{pmatrix} -a_2 C_4 - \rho_{x4} \\ a_2 S_4 - \rho_{y4} \\ 0 \end{pmatrix} \dot{q}_1 \dot{q}_2 + 2 \begin{pmatrix} -\rho_{x4} \\ -\rho_{y4} \\ 0 \end{pmatrix} \dot{q}_1 \dot{q}_4 + 2 \begin{pmatrix} -\rho_{x4} \\ -\rho_{y4} \\ 0 \end{pmatrix} \dot{q}_2 \dot{q}_4.$$

The dynamic model will be determined by using Appell's equation. Taking into consideration the form of Γ_s, the terms originated from $\Gamma_s^T I_{cs} \Gamma_s$ can be simplified in $H_{ij} = D_{ij}$:

$$[0 \; 0 \; \Gamma_{si,z}] \begin{bmatrix} * & * & * \\ * & * & * \\ * & * & I_{cs,z} \end{bmatrix} \begin{bmatrix} 0 \\ 0 \\ \Gamma_{sj,z} \end{bmatrix} = \Gamma_{si,z} \Gamma_{sj,z} I_{cs,z}.$$

Since the columns of Γ_s and ω_s are parallel and $\phi_i = 0$, hence the second term in h_{cc} is zero:

$$\Gamma_s^T \left[I_{cs} \phi_s - (I_{cs} \omega_s) \times \omega_s \right] = 0.$$

From h_{cc}, the centripetal and Coriolis effects D_{ijk} can be sorted. The potential energy of the robot is evidently $P = (m_3 + m_4) g q_3$ from which the gravity effect $D_i = \frac{\partial P}{\partial q_i}$ can easily be determined. Finally, the motion equation can be written in the form:

$$\sum_{j=1}^{4} D_{ij}(q) \ddot{q}_j + \sum_{j=1}^{4} \sum_{k=1}^{4} D_{ijk}(q) \dot{q}_j \dot{q}_k + D_i(q) = \tau_i, \quad i = 1, \dots, 4. \qquad (3.61)$$

It is useful to introduce the parameters P_i, $i = 1, \dots, 11$, and the functions F_i, $i = 1, \dots, 7$. The parameters can be computed offline.

$$P_1 = m_1\{\rho_{y1}^2 + (a_1 + \rho_{x1})^2\} + I_{z1}$$
$$+ m_2\{a_1^2 + \rho_{y2}^2 + (a_2 + \rho_{x2})^2\} + I_{z2}$$
$$+ m_3\{a_1^2 + (a_2 + \rho_{x3})^2 + \rho_{y3}^2\} + I_{z3}$$
$$+ m_4\{a_1^2 + a_2^2 + \rho_{x4}^2 + \rho_{y4}^2\} + I_{z4},$$

$$P_2 = m_4 a_2 \rho_{y4},$$

$$P_3 = m_4 a_2 \rho_{x4},$$

$$P_4 = m_2 a_1 \rho_{y2} + m_3 a_1 \rho_{y3},$$

$$P_5 = m_2 a_1 (a_2 + \rho_{x2}) + m_3 a_1 (a_2 + \rho_{x3}) + m_4 a_1 a_2, \qquad (3.62)$$

$$P_6 = m_4 a_1 \rho_{y4},$$

$$P_7 = m_4 a_1 \rho_{x4},$$

$$P_8 = m_2\{(a_2 + \rho_{x2})^2 + \rho_{y2}^2\} + I_{z2} + m_3\{(a_2 + \rho_{x3})^2 + \rho_{y3}^2\}$$
$$+ I_{z3} + m_4\{a_2^2 + \rho_{x4}^2 + \rho_{y4}^2\} + I_{z4},$$

$$P_9 = m_4\{\rho_{x4}^2 + \rho_{y4}^2\} + I_{z4},$$

$$P_{10} = m_3 + m_4,$$

$$P_{11} = (m_3 + m_4)g_z,$$

$$F_1 = 1, \quad F_2 = S_4, \quad F_3 = C_4, \quad F_4 = S_2, \quad F_5 = C_2, \quad F_6 = S_{24}, \quad F_7 = C_{24}.$$
$$(3.63)$$

By using the above parameters and functions, and considering (3.60), the non-trivial functions of the motion equations can be obtained in the following compact form for real-time implementation:

$$D_{11} = F_1 P_1 - 2F_2 P_2 + 2F_3 P_3 - 2F_4 P_4 + 2F_5 P_5 - 2F_6 P_6 + 2F_7 P_7,$$

$$D_{12} = F_1 P_8 - 2F_2 P_2 + 2F_3 P_3 - F_4 P_4 + F_5 P_5 - F_6 P_6 + F_7 P_7,$$

$$D_{13} = 0,$$

$$D_{14} = F_1 P_9 - F_2 P_2 + F_3 P_3 - F_6 P_6 + F_7 P_7,$$

$$D_{22} = F_1 P_8 - 2F_2 P_2 + 2F_3 P_3, \qquad (3.64)$$

$$D_{23} = 0,$$

$$D_{24} = F_1 P_9 - F_2 P_2 + F_3 P_3,$$

$$D_{33} = F_1 P_{10},$$

$$D_{34} = 0,$$

$$D_{44} = F_1 P_9;$$

$$D_{112} = -F_4 P_5 - F_5 P_4 - F_6 P_7 - F_7 P_6,$$
$$D_{113} = 0,$$
$$D_{114} = -F_2 P_3 - F_3 P_2 - F_6 P_7 - F_7 P_6,$$
$$D_{122} = -F_4 P_5 - F_5 P_4 - F_6 P_7 - F_7 P_6,$$
$$D_{123} = 0, \tag{3.65}$$
$$D_{124} = -F_2 P_3 - F_3 P_2 - F_6 P_7 - F_7 P_6,$$
$$D_{133} = 0,$$
$$D_{134} = 0,$$
$$D_{144} = -F_2 P_3 - F_3 P_2 - F_6 P_7 - F_7 P_6,$$

$$D_{213} = 0,$$
$$D_{214} = -F_2 P_3 - F_3 P_2,$$
$$D_{223} = 0,$$
$$D_{224} = -F_2 P_3 - F_3 P_2, \tag{3.66}$$
$$D_{233} = 0,$$
$$D_{234} = 0,$$
$$D_{244} = -F_2 P_3 - F_3 P_2;$$

$$D_{314} = 0, \qquad D_{324} = 0, \qquad D_{334} = 0, \qquad D_{344} = 0; \tag{3.67}$$

$$D_1 = 0, \qquad D_2 = 0, \qquad D_3 = -F_1 P_{11}, \qquad D_4 = 0. \tag{3.68}$$

The influence of the rotor inertia Θ_{ri} and the gear reduction $v_i = \dot{\phi}_i / \dot{q}_i$ can be taken into consideration by the corrections $D_{ii} := D_{ii} + \Theta_{ri} v_i^2, i = 1, \ldots, 4$.

3.3 Dynamic Model of Car

Consider the sketch of a car moving in horizontal plane, see Fig. 3.2. The rear wheels are driven and the front wheels can be steered. They are referred in the dynamic model by R and F, respectively. Important frames are K_{CoG} (fixed to the center of gravity of the car), K_W (fixed to the wheel) and K_{In} (inertial coordinate system). The origin of K_{CoG} has coordinates X and Y in K_{In}.

The simplified car model supposes that only the front wheels are steered and the left and right sides of the car are symmetrical, and therefore the two halves can be merged to a single unit in the model.

Fig. 3.2 Simplified sketch of car moving in horizontal plane

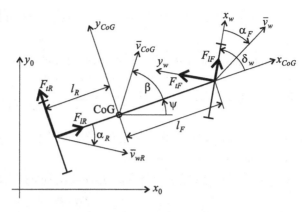

3.3.1 Nonlinear Model of Car

Let x_{CoG} be the longitudinal axle of the car, the axle y_{CoG} is pointed to the left from it and z_{CoG} is directed upwards. Denote l_F and l_R, respectively, the distance of the wheels from COG. Let the origin of the front wheel frame be at the x_{CoG}-axis, the x_W-axis is in the plane of the wheel and the turning angle (the angle between x_W and x_{CoG}) is δ_W. Denote $\dot{\psi}$ the angular velocity and v_{CoG} the absolute value of the velocity of COG, respectively. Let β be the angle between the velocity vector and x_{CoG}-axis. Denote v_{WF} the velocity vector of the front wheel and let α_F be the angle between the velocity vector of the front wheel and x_W-axis (then $\delta_W - \alpha_F$ is the angle between v_{WF} and the x_{CoG}-axis). The rear wheel velocity is v_{WR} and the angle to the x_{CoG}-axis is α_R.

The angles $\beta, \alpha_F, \alpha_R$ are usually called the vehicle body side slip angle, the tire side slip angle front and the tire side slip angle rear, respectively.

If the front wheel is steered to the left relative to the longitudinal axes x_{CoG} of the car, then v_{CoG} and v_{WF} vectors are at the left side of x_{CoG}, while the velocity vector v_{WR} is at the right side of x_{CoG}. The rotation angle of the wheel and the slide slip angle are considered positive in this case.

The forces acting at the origin of the coordinate system of the front wheel are the longitudinal force F_{lF} and the transversal force F_{tF}. It is assumed that the transversal component is $F_{tF} = c_F \alpha_F$ where c_F is constant. Similarly, the forces acting at the origin of the frame of the rear wheel are the longitudinal force F_{lR} and the transversal force F_{tR}, and the latter is $F_{tR} = c_R \alpha_R$ where c_R is constant.

The velocity of the front wheel is equal to the velocity of COG increased by the effect of the angular velocity $\dot{\psi}$ of the car, that is, the velocity of the point in circular motion on a radius l_F. The absolute value of the velocity vector will be denoted by v_G. Consider the components of the velocity of the front wheel in y_{CoG} and x_{CoG} directions, then using the above positive directions of the angles yields

$$v_{WF} \sin(\delta_W - \alpha_F) = l_F \dot{\psi} + v_G \sin(\beta),$$

$$v_{WF} \cos(\delta_W - \alpha_F) = v_G \cos(\beta), \tag{3.69}$$

$$\tan(\delta_W - \alpha_F) = \frac{l_F \dot{\psi} + v_G \sin(\beta)}{v_G \cos(\beta)}.$$

Similarly, the components of the rear wheel velocity in y_{CoG} and x_{CoG} directions, under the above positive directions, are respectively

$$v_{WR} \sin(\alpha_R) = l_R \dot{\psi} - v_G \sin(\beta),$$

$$v_{WR} \cos(\alpha_R) = v_G \cos(\beta),$$

from which follows

$$\tan(\alpha_R) = \frac{l_R \dot{\psi} - v_G \sin(\beta)}{v_G \cos(\beta)}. \tag{3.70}$$

Assuming small $\delta_W - \alpha_F$, α_R and β and using the approximations $\tan(\delta_W - \alpha_F) \approx \delta_W - \alpha_F$, $\tan(\alpha_R) \approx \alpha_R$, $\sin(\beta) \approx \beta$ and $\cos(\beta) \approx 1$, it follows

$$\alpha_F = \delta_W - \beta - \frac{l_F \dot{\psi}}{v_G}, \tag{3.71}$$

$$\alpha_R = -\beta + \frac{l_R \dot{\psi}}{v_G}. \tag{3.72}$$

Applying the usual notations $\cos(\beta) = C_\beta$, $\sin(\beta) = S_\beta$, and the differentiation rule in moving frames, then the kinematic model and based on it the dynamic model of the car can be derived.

$$\bar{v}_{COG} = \begin{pmatrix} C_\beta & S_\beta & 0 \end{pmatrix}^T v_G, \tag{3.73}$$

$$\bar{a}_{COG} = \dot{\bar{v}}_{COG} + \omega \times \bar{v}_{COG}$$

$$= \begin{pmatrix} C_\beta \\ S_\beta \\ 0 \end{pmatrix} \dot{v}_G + \begin{pmatrix} -S_\beta \\ C_\beta \\ 0 \end{pmatrix} \dot{\beta} v_G + \begin{bmatrix} 0 & -\dot{\psi} & 0 \\ \dot{\psi} & 0 & 0 \\ 0 & 0 & 0 \end{bmatrix} \begin{pmatrix} C_\beta \\ S_\beta \\ 0 \end{pmatrix} v_G$$

$$= \begin{bmatrix} C_\beta & -v_G S_\beta \\ S_\beta & v_G C_\beta \\ 0 & 0 \end{bmatrix} \begin{pmatrix} \dot{v}_G \\ \dot{\beta} \end{pmatrix} + \begin{pmatrix} -S_\beta \\ C_\beta \\ 0 \end{pmatrix} \dot{\psi} v_G. \tag{3.74}$$

Taking into account the direction of the forces, dividing by the mass m_v of the car, and considering only the nontrivial components of $\bar{a}_{COG}$, the acceleration is obtained:

$$\bar{a}_{COG} = \frac{1}{m_v} \begin{pmatrix} F_x \\ F_y \end{pmatrix} = \frac{1}{m_v} \begin{pmatrix} F_{lF} C_{\delta_w} - F_{tF} S_{\delta_w} + F_{lR} \\ F_{lF} S_{\delta_w} + F_{tF} C_{\delta_w} + F_{tR} \end{pmatrix}. \tag{3.75}$$

Some further steps are needed to find the motion equations.

$$\begin{bmatrix} C_\beta & -v_G S_\beta \\ S_\beta & v_G C_\beta \end{bmatrix}^{-1} = \frac{1}{v_G} \begin{pmatrix} v_G C_\beta & v_G S_\beta \\ -S_\beta & C_\beta \end{pmatrix}, \tag{3.76}$$

$$\begin{pmatrix} \dot{v}_G \\ \dot{\beta} \end{pmatrix} = \frac{1}{v_G} \begin{bmatrix} v_G C_\beta & v_G S_\beta \\ -S_\beta & C_\beta \end{bmatrix} \frac{1}{m_v} \begin{pmatrix} F_x \\ F_y \end{pmatrix} + \underbrace{\frac{1}{v_G} \begin{bmatrix} v_G C_\beta & v_G S_\beta \\ -S_\beta & C_\beta \end{bmatrix} \begin{pmatrix} -S_\beta \\ C_\beta \end{pmatrix} v_G \dot{\psi}}_{-\binom{0}{1}\dot{\psi}},$$

$$\frac{1}{v_G} \begin{bmatrix} v_G C_\beta & v_G S_\beta \\ -S_\beta & C_\beta \end{bmatrix} \frac{1}{m_v} \begin{pmatrix} F_{lF} C_{\delta_w} - F_{tF} S_{\delta_w} + F_{lR} \\ F_{lF} S_{\delta_w} + F_{tF} C_{\delta_w} + F_{tR} \end{pmatrix}$$

$$= \frac{1}{m_v} \begin{pmatrix} F_{lF} C_{\delta_w-\beta} - F_{tF} S_{\delta_w-\beta} + F_{lR} C_\beta + F_{tR} S_\beta \\ \frac{1}{v_G}[F_{lF} S_{\delta_w-\beta} + F_{tF} C_{\delta_w-\beta} - F_{lR} S_\beta + F_{tR} C_\beta] \end{pmatrix}, \tag{3.77}$$

$$F_{tF} = c_F \alpha_F = c_F \left(\delta_w - \beta - \frac{l_F \dot{\psi}}{v_G} \right),$$

$$F_{tR} = c_R \alpha_R = c_R \left(-\beta + \frac{l_R \dot{\psi}}{v_G} \right).$$

According to Fig. 3.2, the driving torque is

$$\tau_z = l_F \{ F_{lF} \sin(\delta_W) + F_{tF} \sin(90° + \delta_w) \} - l_R F_{tR}, \tag{3.78}$$

hence $\ddot{\psi} = \tau_z / I_z$, where I_z is the inertia moment.

Differential equations of the translational motion:

$$\dot{v}_G = \frac{1}{m_v} \Bigg\{ F_{lF} \cos(\delta_w - \beta) - c_F \left(\delta_w - \beta - \frac{l_F \dot{\psi}}{v_G} \right) \sin(\delta_w - \beta)$$

$$+ c_R \left(-\beta + \frac{l_R \dot{\psi}}{v_G} \right) \sin(\beta) + F_{lR} \cos(\beta) \Bigg\}, \tag{3.79}$$

$$\dot{\beta} = -\dot{\psi} + \frac{1}{m_v v_G} \Bigg\{ F_{lF} \sin(\delta_W - \beta) - F_{lR} \sin(\beta)$$

$$+ c_F \left(\delta_W - \beta - \frac{l_F \dot{\psi}}{v_G} \right) \cos(\delta_W - \beta)$$

$$+ c_R \left(-\beta + \frac{l_R \dot{\psi}}{v_G} \right) \cos(\beta) \Bigg\}. \tag{3.80}$$

Differential equation of the rotational motion:

$$\ddot{\psi} = \frac{1}{I_z} \Bigg\{ l_F F_{lF} \sin(\delta_W) + l_F c_F \left(\delta_w - \beta - \frac{l_F \dot{\psi}}{v_G} \right) \cos(\delta_w)$$

$$- l_R c_R \left(-\beta + \frac{l_R \dot{\psi}}{v_G} \right) \cos(\beta) \Bigg\}. \tag{3.81}$$

Kinetic equations:

$$\dot{X} = v_G \cos(\psi + \beta), \tag{3.82}$$

$$\dot{Y} = v_G \sin(\psi + \beta). \tag{3.83}$$

State equation:

$\dot{x} = f(x, u)$ nonlinear, typically $F_{lF} = 0$,

$x = (\beta, \psi, \dot{\psi}, v_G, X, Y)^T,$ $u = (\delta_w, F_{lR})^T,$ $y = (X, Y)^T.$ $\tag{3.84}$

3.3.2 Input Affine Approximation of the Dynamic Model

It follows from (3.77) that for small angles $\dot{v}_G = F_x/(m_v) \approx F_{lR}/(m_v)$ and $\dot{\beta} + \dot{\psi} = F_y/(m_v v_G) \approx (F_{tF} + F_{tR})/(m_v v_G)$. Based on these approximations two input affine models can be derived.

In the first input affine model, it is useful to introduce the rear $(S_h = F_{tR})$ and front $(S_v = F_{tF})$ side forces where

$$S_h = c_R(-\beta + l_R \dot{\psi}/v_G), \tag{3.85}$$

$$S_v = c_F(\delta_w - \beta - l_F \dot{\psi}/v_G). \tag{3.86}$$

It is clear that S_v can be considered to be an internal input because δ_w can be determined from it by

$$\delta_w = \frac{S_v}{c_F} + \beta + \frac{l_F \dot{\psi}}{v_G}. \tag{3.87}$$

Assuming small angles the *first input affine model* arises in the form

$$\dot{x} = \begin{pmatrix} -x_3 + S_h/(m_v x_4) \\ x_3 \\ -S_h l_R/I_z \\ 0 \\ x_4 C_{12} \\ x_4 S_{12} \end{pmatrix} + \begin{bmatrix} 1/(m_v x_4) & -x_1/(m_v x_4) \\ 0 & 0 \\ l_F/I_z & 0 \\ 0 & 1/m_v \\ 0 & 0 \\ 0 & 0 \end{bmatrix} u, \tag{3.88}$$

$\dot{x} = A(x) + B(x)u,$ $y = (x_5, x_6)^T = C(x),$

$x = (\beta, \psi, \dot{\psi}, v_G, X, Y)^T,$ $u = (S_v, F_{lR})^T,$ $y = (X, Y)^T.$

In the second input affine model, the notations $\alpha := F_{lR}/m_v$ and $r := \dot{\psi}$ will be introduced. On the other hand

$$F_{tF} + F_{tR} = -\frac{c_F + c_R}{v_G}\beta + \frac{c_R l_R - c_F l_F}{v_G}\dot{\psi} + \frac{c_F}{v_G}\delta_w,$$

hence, using the notations

$$a_{11} = -\frac{c_F + c_R}{m_v}, \qquad a_{12} = \frac{c_R l_R - c_F l_F}{m_v}, \qquad b_1 = \frac{c_F}{m_v},$$

$$a_{21} = \frac{c_R l_R - c_F l_F}{I_z}, \qquad a_{22} = -\frac{c_R l_R^2 - c_F l_F^2}{I_z}, \qquad b_2 = \frac{c_F l_F}{I_z} \tag{3.89}$$

and the new state variables $\bar{x} = (X, Y, \beta + \psi, v_G, \beta, r)^T$, the *second input affine model* is obtained

$$\dot{\bar{x}} = \begin{pmatrix} \bar{x}_4 C_3 \\ \bar{x}_4 S_3 \\ (a_{11}/\bar{x}_4)\bar{x}_5 + (a_{12}/\bar{x}_4^2)\bar{x}_6 \\ 0 \\ (a_{11}/\bar{x}_4)\bar{x}_5 + ((a_{12}/\bar{x}_4^2) - 1)\bar{x}_6 \\ a_{21}\bar{x}_5 + (a_{22}/\bar{x}_4)\bar{x}_6 \end{pmatrix} + \begin{bmatrix} 0 & 0 \\ 0 & 0 \\ b_1/\bar{x}_4 & 0 \\ 0 & 1 \\ b_1/\bar{x}_4 & 0 \\ b_2 & 0 \end{bmatrix} \bar{u}, \tag{3.90}$$

$$\dot{\bar{x}} = \bar{A}(\bar{x}) + \bar{B}(\bar{x})\bar{u}, \qquad \bar{y} = (\bar{x}_1, \bar{x}_2)^T = \bar{C}(x),$$

$$\bar{x} = (X, Y, \beta + \psi, v_G, \beta, r)^T, \qquad \bar{u} = (\delta_w, \alpha)^T, \qquad \bar{y} = (X, Y)^T.$$

The input affine car models will be used later for predictive control and formation control.

3.3.3 Linearized Model for Constant Velocity

If the velocity is constant, then the only input is the turning angle δ_w. Assuming small angles the dynamic model arises in linear form:

$$\dot{x} = Ax + Bu, \qquad y = Cx, \qquad x = (\beta, \dot{\psi})^T, \qquad u = \delta_W, \qquad y = \dot{\psi}, \tag{3.91}$$

$$\begin{pmatrix} \dot{\beta} \\ \ddot{\psi} \end{pmatrix} = \begin{bmatrix} -\frac{c_F + c_R}{m_v v_G} & \frac{c_R l_R - c_F l_F}{m_v v_G^2} - 1 \\ \frac{c_R l_R - c_F l_F}{I_{zz}} & -\frac{c_R l_R^2 + c_F l_F^2}{I_{zz} v_G} \end{bmatrix} \begin{pmatrix} \beta \\ \dot{\psi} \end{pmatrix} + \begin{bmatrix} \frac{c_F}{m_v v_G} \\ \frac{c_F l_F}{I_{zz}} \end{bmatrix} \delta_W. \tag{3.92}$$

The *characteristic equation* $\det(sI - A) = 0$ of the linearized model:

$$s^2 + \frac{c_F(I_{zz} + m_v l_F^2) + c_R(I_{zz} + m_v l_R^2)}{m_v v_G I_{zz}} s$$

$$+ \frac{(c_F + c_R)(l_F + l_R)^2 + m_v v_G^2(c_R l_R - c_F l_F)}{m_v v_G^2 I_{zz}} = 0.$$

The linearized system is stable if $c_R l_R > c_F l_F$ which is usually satisfied for production cars. However the cornering stiffnesses c_R, c_F depend on the road conditions.

The static gain is obtained for $s = 0$ by

$$A_{\text{stat}} = v_G \frac{c_F c_R (l_F + l_R)}{(c_F + c_R)(l_F + l_R)^2 + m_v v_G^2 (c_R l_R - c_F l_F)}. \tag{3.93}$$

In the practice, the characteristic velocity v_{char} is introduced where

$$v_{\text{char}}^2 := \frac{c_F c_R (l_f + l_R)^2}{m_v (c_R l_R - c_F l_F)} \tag{3.94}$$

which may be also negative (if $c_R l_R - c_F l_F < 0$) and is in strong relation with the static gain:

$$A_{\text{stat}} = \frac{1}{l_F + l_R} \cdot \frac{v_G}{1 + v_G^2 / v_{\text{char}}^2}. \tag{3.95}$$

For typical cars, v_{char} is between 68 and 112 km/h. Small characteristic velocity refers to under-steering.

The accuracy of the linearized model is usually satisfactory for lateral acceleration less than $0.4g$, see Kiencke and Nielsen [64]. More complex models of cars including four wheels driven models can be found in [64].

3.4 Dynamic Model of Airplane

Dynamic modeling of airplanes is a well elaborated field in aeronautics and control, see, for example, the excellent books of Cook [31], and Stevens and Lewis [138].

Before starting the discussion of the dynamic model of airplane, the different navigation coordinate systems (frames) playing important role in the modeling and control of vehicles will be reviewed.

3.4.1 Coordinate Systems for Navigation

The often used coordinate systems are ECI, ECEF, NED and ABC = BODY, see Fig. 3.3.

3.4.1.1 Earth Centered Inertial Frame

The ECI frame K_{ECI} is the favored inertial frame in near-earth environment. The origin of ECI is at the center of gravity of the earth in a fixed time moment, the x-axis is in the direction of the vernal equinox, the z-axis is in the rotation axis (north

Fig. 3.3 Coordinate systems
used in navigation

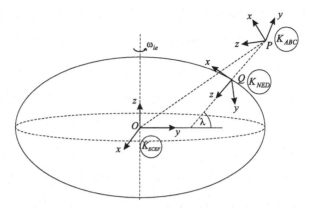

polar axis) of the earth and the y-axis makes the triplet to a right-handed orthogonal coordinate system. ECI is not moving. The fundamental laws of classical mechanics like Newton–Euler Equations are valid in inertial system. ECI is especially important for interplanetary path design.

A reference frame ECI that is a less perfect inertial frame, but much more convenient in terms of visualizing the trajectory of the vehicle, specifying its coordinates and earth's gravity, is a coordinate system with its origin at the center of the earth, translating with the earth, but with a fixed orientation relative to the sun (or to the average position of the fixed stars). Henceforth, this ECI approximation of the inertial frame will be assumed. It means that ECI and ECEF have common origin, ECI is not rotating while ECEF is rotating around the z-axis of ECI.

3.4.1.2 Earth Centered Earth Fixed Frame

The ECEF frame K_{ECEF} has its origin at the center of gravity of the earth in the actual time moment, the z-axis is the rotation axis of the earth, the x-axis is the intersection line of the equatorial plane of the earth and the main meridian through the observatory of Greenwich, UK, and the y-axis makes the triplet to a right-handed orthogonal coordinate system. ECEF is moving together with the earth. Its sidereal rotation rate relative to ECI is

$$\omega_E = \frac{(1 + 365.25) \text{ cycles}}{365.25 \cdot 24 \text{ h}} \cdot \frac{2\pi \text{ rad/cycle}}{3600 \text{ s/h}} = 7.2921151467 \times 10^{-5} \text{ rad/s}. \quad (3.96)$$

ECEF is specifically important for long distance path design between continents and for the path design and control of satellites and carrier rockets.

The earth can be approximated by a rotational ellipsoid according to WGS-84 standard which is the basis for Global Positioning System (GPS). The halve axes of the ellipse are $a = 6388137.0$ m and $b = 6356752.3142$ m, $(a > b)$. The rotational ellipsoid is arising from the rotation of the ellipse around the z-axis. The flatness f

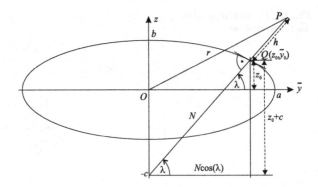

Fig. 3.4 Computational principle of spatial coordinates in ECEF frame

and eccentricity e are defined respectively, by

$$f = \frac{a-b}{a} = 0.0034, \qquad e = \sqrt{f(2-f)} = 0.0818.$$

In ECEF frame, each point P can be characterized by a vector $r = (x, y, z)^T$ from the origin to the point. However, in ECEF the point is identified by the geodetic coordinates $(\lambda, \varphi, h)^T$, which are the latitude, longitude and hight, respectively, and $(x, y, z)^T$ can be determined from them using the following method.

First, the intersection of the plane through P and the z-axis with the rotational ellipsoid will be determined. The intersection is an exemplar of the rotated ellipse, its axes have the direction of z and $\bar{y}$. The plane cuts out a section $\bar{y}_0$ from the equatorial plane whose projections onto the x-, y-axes of the ellipsoid can be determined based on the angle φ between the x-axis and $\bar{y}_0$, namely the projections are $x = \bar{y}_0 C_\varphi$ and $y = \bar{y}_0 S_\varphi$. The distance of P from the $\bar{y}$-axis of the ellipse is immediately the z-coordinate of the point P. The hight h of the point P is the shortest distance from the ellipse which defines the point $Q = (z_0, \bar{y}_0)$ on the ellipse. In point Q, the tangent of the ellipse can be determined. The line through Q orthogonal to the tangent intersects the z-axis in the point $R = (0, -c)$. The QR section is the normal to the tangent, its length is N. The length of the normal between Q and the intersection point with the $\bar{y}$-axis is $\bar{N}$. The angle between the normal and the $\bar{y}$-axis is λ. The relations are shown in Fig. 3.4.

Conversion from geodetic to rectangular coordinates:

$$(\lambda, \varphi, h)^T \rightarrow N(\lambda) = a/\sqrt{1 - e^2 S_\lambda^2},$$

$$x = (N + h)C_\lambda C_\varphi,$$

$$y = (N + h)C_\lambda S_\varphi, \qquad\qquad (3.97)$$

$$z = \left(N(1 - e^2) + h\right)S_\lambda.$$

Conversion from rectangular to geodetic coordinates:

$$(x, y, z)^T \rightarrow (\lambda, \varphi, h)^T$$

Initialization: $\quad h := 0, \qquad N := a, \qquad p := \sqrt{x^2 + y^2}, \qquad T_\varphi := y/x \xrightarrow{\text{atan2}} \varphi.$

Cycle: $\qquad S_\lambda := \dfrac{z}{N(1 - e^2) + h}, \qquad T_\lambda := \dfrac{z + e^2 N S_\lambda}{p} \xrightarrow{\text{atan}} \lambda,$

$\qquad\qquad N(\lambda) := a / \sqrt{1 - e^2 S_\lambda^2}, \qquad h = \dfrac{p}{C_\lambda} - N.$

$$(3.98)$$

By experiences, the convergence is quicker if λ is computed by atan instead of asin, convergence to cm accuracy requires 25 iterations. Alternatively, λ and h can be computed in closed form consisting of 15 formulas including 5 square roots and 1 atan [38].

3.4.1.3 North–East–Down Coordinate System

The NED frame can be found if the nearest point Q from the point P is determined on the rotational ellipsoid which will be the origin of the K_{NED} frame. Then the tangential plane at the point Q will be determined in which plane the orthonormal coordinate axes x_{NED}, y_{NED} and z_{NED} points to local north, east and down, respectively. Notice that z_{NED} points downwards, inside the rotational ellipsoid.

The orientation $A_{\text{ECEF,NED}}$ of the K_{NED} frame relative to K_{ECEF} can be found by using two rotations. The first rotation around z_{ECEF} by φ rotates the x-axis of the ECEF frame through the main meridian to the plane through P and z_{ECEF}. The rotated x-axis will be $\bar{y}$, see Fig. 3.4. The rotated y-axis is parallel to local east in Q. Hence, a second rotation around the rotated y-axis by the angle $-(\frac{\pi}{2} + \lambda)$ sets the axes of the K_{NED} frame:

$$A_{\text{ECEF,NED}}(\lambda, \varphi) = \begin{bmatrix} C_\varphi & -S_\varphi & 0 \\ S_\varphi & C_\varphi & 0 \\ 0 & 0 & 1 \end{bmatrix} \begin{bmatrix} C_{-(\frac{\pi}{2}+\lambda)} & 0 & S_{-(\frac{\pi}{2}+\lambda)} \\ 0 & 1 & 0 \\ -S_{-(\frac{\pi}{2}+\lambda)} & 0 & C_{-(\frac{\pi}{2}+\lambda)} \end{bmatrix}$$

$$= \begin{bmatrix} C_\varphi & -S_\varphi & 0 \\ S_\varphi & C_\varphi & 0 \\ 0 & 0 & 1 \end{bmatrix} \begin{bmatrix} -S_\lambda & 0 & -C_\lambda \\ 0 & 1 & 0 \\ C_\lambda & 0 & -S_\lambda \end{bmatrix}, \qquad (3.99)$$

$$A_{\text{ECEF,NED}}(\lambda, \varphi) = \begin{bmatrix} -S_\lambda C_\varphi & -S_\varphi & -C_\lambda C_\varphi \\ -S_\lambda S_\varphi & C_\varphi & -C_\lambda S_\varphi \\ C_\lambda & 0 & -S_\lambda \end{bmatrix},$$

$$B_G(\lambda, \varphi) := A_{\text{ECEF,NED}}^T = \begin{bmatrix} -S_\lambda C_\varphi & -S_\lambda S_\varphi & C_\lambda \\ -S_\varphi & C_\varphi & 0 \\ -C_\lambda C_\varphi & -C_\lambda S_\varphi & -S_\lambda \end{bmatrix} \qquad (3.100)$$

where λ is the geodetic latitude and φ is the longitude. The transformation rules for the same vector in different basis systems are $v_{\text{ECEF}} = A_{\text{ECEF,NED}} v_{\text{NED}}$ and $v_{\text{NED}} = B_G v_{\text{ECEF}}$.

3.4.1.4 Aircraft-Body Coordinate System

The origin of the ABC (or shortly B = BODY) coordinate system is in the center of gravity of the body (point P). Its x-, y-, z-axes are assumed to be aligned, respectively, forward, starboard, and down in the aircraft. The relative orientation between K_{NED} and K_B can be described by the Euler (RPY) angles ϕ, θ, ψ:

$$B_B(\phi, \theta, \psi) = \begin{bmatrix} C_\theta C_\psi & C_\theta S_\psi & -S_\theta \\ S_\phi S_\theta C_\psi - C_\phi S_\psi & S_\phi S_\theta S_\psi + C_\phi C_\psi & S_\phi C_\theta \\ C_\phi S_\theta C_\psi + S_\phi S_\psi & C_\phi S_\theta S_\psi - S_\phi C_\psi & C_\phi C_\theta \end{bmatrix}. \qquad (3.101)$$

The resulting orientation between K_{ECI} and K_B is

$$B = B_B(\phi, \theta, \psi) B_G(\lambda, \varphi). \qquad (3.102)$$

Notice that B can also be parametrized by the Euler (RPY) angles Φ, Θ, Ψ immediately between K_{ECI} and K_B, that is,

$$B(\Phi, \Theta, \Psi) = \begin{bmatrix} C_\Theta C_\Psi & C_\Theta S_\Psi & -S_\Theta \\ S_\Phi S_\Theta C_\Psi - C_\Phi S_\Psi & S_\Phi S_\Theta S_\Psi + C_\Phi C_\Psi & S_\Phi C_\Theta \\ C_\Phi S_\Theta C_\Psi + S_\Phi S_\Psi & C_\Phi S_\Theta S_\Psi - S_\Phi C_\Psi & C_\Phi C_\Theta \end{bmatrix}. \qquad (3.103)$$

These Euler angles appeared in the dynamic model of the rigid body. The Euler (RPY) angles between K_{NED} and K_B differ from them and can be determined from

$$B_B(\phi, \theta, \psi) := B(\Phi, \Theta, \Psi)\big[B_G(\lambda, \varphi)\big]^T \qquad (3.104)$$

using the inverse RPY method in Appendix A. The angular rates P, Q, R, which appear in the dynamic model of rigid body, are the elements of the absolute angular velocity vector ω_B.

3.4.2 Airplane Kinematics

If it does not give rise to misunderstanding, the simplified notations $K_{\text{ECI}} =: K_E$ and $K_{ABC} =: K_B$ will be used. The position of the airplane is identified by the vector p from the origin of K_E pointing into the COG of the airplane. The absolute velocity of the airplane is $v_{abs} = \dot{p}$, while the absolute acceleration is $a_{abs} = \ddot{p} = \dot{v}_{abs}$ in the inertia frame. The atmosphere in which the airplane is moving rotates by constant angular velocity ω_E and the airplane maneuvers in this medium. With robotics analogy, it can be imagined that the atmosphere is a link between K_E and

K_B which is moved by a rotational joint. The joint axis is the rotational axis of the earth whose direction is the z_E-axis of K_E. In order to change the orientation of the airplane relative to the atmosphere, the angular velocity has to be modified by the angular velocity ω_R so that the absolute angular velocity with respect to the inertia frame will be $\omega_B = \omega_E + \omega_R$. The relative velocity of the airplane sensed by people sitting in the airplane is v_B. By the differentiation rule in moving coordinate system, it follows in basis independent form

$$\omega_B = \omega_E + \omega_R,$$

$$v_{abs} = \dot{p} = v_B + \omega_E \times p,$$

$$\dot{v}_{abs} = \dot{v}_B + \omega_B \times v_B + \omega_E \times \dot{p}$$

$$= \dot{v}_B + \omega_B \times v_B + \omega_E \times (v_B + \omega_E \times p)$$

$$= \dot{v}_B + (\omega_B + \omega_E) \times v_B + \omega_E \times (\omega_E \times p)$$

$$= \dot{v}_B + (\omega_R + 2\omega_E) \times v_B + \omega_E \times (\omega_E \times p).$$

Since the inertia matrix of the airplane is constant only in the frame fixed to the airplane, hence it is useful to write down the moment equation in K_B. Since p is defined in K_E, therefore the force equation can be formulated in K_E. According to the chosen convention in notation $r_E = A_{K_E,K_B} r_B = B^T r_B \Leftrightarrow r_B = A_{K_E,K_B}^T r_E = B r_E$, the kinematic equations in basis dependent form yield:

$$\omega_B = B\omega_E + \omega_R,$$

$$\dot{\omega}_B = \varepsilon_B,$$

$$B\dot{p} = v_B + B(\omega_E \times p) \quad \Leftrightarrow \quad \dot{p} = B^T v_B + [\omega_E \times] p,$$

$$Ba_{abs} = \dot{v}_B + (\omega_B + B\omega_E) \times v_B + B\{\omega_E \times (\omega_E \times p)\} \quad \Leftrightarrow$$

$$\dot{v}_B = Ba_{abs} - B[\omega_E \times]^2 p - \{[\omega_B \times] + B[\omega_E \times]B^T\}v_B.$$

Remarks

(i) In K_E are computed: $p, \dot{p}, a_{abs}$.
(ii) In K_B are computed: $\omega_B, \varepsilon_B, v_B$.
(iii) From Newton's force equation is computed a_{abs}.
(iv) From Euler's moment equation is computed $\varepsilon_B = \dot{\omega}_B$ (angular acceleration).
(v) The computation of the absolute angular velocity can be based on Euler (RPY) angles (three variables) or quaternion (four variables).

3.4.3 Airplane Dynamics

Round-Earth-Equations
Denote F_B and T_B the external forces (except gravity effect) and torques in the body frame, respectively, and $g(p)$ the gravity acceleration in the inertial frame,

then for long distance flight the state equations with quaternion representation of the orientation are

$$
\begin{pmatrix} \dot{p} \\ \dot{v}_B \\ \dot{\omega}_B \\ \dot{q} \end{pmatrix} = \begin{bmatrix} [\omega_E \times] & B^T & 0 & 0 \\ -B[\omega_E \times]^2 & -([\omega_B \times] + B[\omega_E \times]B^T) & 0 & 0 \\ 0 & 0 & -I_c^{-1}[\omega_B \times]I_c & 0 \\ 0 & 0 & 0 & -\frac{1}{2}\Omega_q \end{bmatrix}
$$
$$
\times \begin{pmatrix} p \\ v_B \\ \omega_B \\ q \end{pmatrix} + \begin{pmatrix} 0 \\ \frac{F_B}{m} + Bg(p) \\ I_c^{-1}T_B \\ 0 \end{pmatrix}, \tag{3.105}
$$

where the state vector is $x = (p^T, v_B^T, \omega_B^T, q^T)^T \in R^{13}$. The state equation is non-linear because the state matrix contains $B, [\omega_B \times], \Omega_q$ which are state dependent matrices. The Round-Earth-Equations can be used for simulation and control of long distance flight and high velocity motion.

Alternatively, the orientation can be represented by Euler (RPY) angles, in which case, using the notation $\bar{\Phi} = (\Phi, \Theta, \Psi)^T$, appears $\dot{\bar{\Phi}} = F(\Phi, \Theta)\omega_B$ instead of the quaternion, see (3.12).

Flat-Earth-Equations
For short distance maneuvering, K_{NED} can be considered an inertial frame neglecting its acceleration to the acceleration of the airplane. The position vector p_{NED} points from the origin of K_{NED} into the COG of the airplane. The angular velocity ω_E of the earth is neglected and the gravity effect may be approximated by the mean value $\|g\| = 9.8054$ m/s^2 belonging to 45° latitude at sea-level. The Flat-Earth-Equations are as follows:

$$
\begin{pmatrix} \dot{v}_B \\ \dot{\omega}_B \\ \dot{\bar{\Phi}} \\ \dot{p}_{\text{NED}} \end{pmatrix} = \begin{pmatrix} -[\omega_B \times]v_B \\ -I_c^{-1}[\omega_B \times]I_c\omega_B \\ F(\varphi, \vartheta)\omega_B \\ B_B^T v_B \end{pmatrix} + \begin{pmatrix} \frac{F_B}{m} + B_B g_0 \\ I_c^{-1}T_B \\ 0 \\ 0 \end{pmatrix}, \tag{3.106}
$$

where the state vector is $x = (v_B^T, \omega_B^T, \underbrace{\varphi, \vartheta, \psi}_{\bar{\Phi}^T}, p_{\text{NED}}^T)^T$.

If the wind can be considered constant in a relatively larger region and has components W_N, W_E, W_D, then the velocity of the airplane relative to the air current is $v_R = v_B - B_B(W_N, W_E, W_D)^T$. This equation can be added to the Flat-Earth-Equations. The wind components can be considered as inputs and for the computation of the aerodynamic forces and moments v_R can be used instead of v_B.

Notice that the external forces and external torques acting to the airplane were not elaborated yet for use in the above equations. They are the gravity force, the aerodynamic forces/torques and the trust force. The trust is assumed to be a force in starboard direction through the COG.

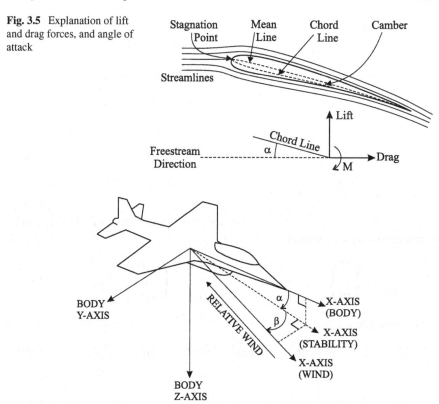

Fig. 3.5 Explanation of lift and drag forces, and angle of attack

Fig. 3.6 Interpretation of angle of attack and sideslip angle with respect to relative wind

3.4.4 Wind-Axes Coordinate System

Partly from aerodynamic, partly from control engineering reasons, some further coordinate systems have to be introduced beside K_E and K_B.

Figure 3.5 shows the cross section of a wing surface in the free air current having some deviation from free-stream direction where the deviation is in strong relation with the motion direction of the aircraft (see the direction of chord line). Two forces are acting to the wing: *lift* (L) and *drag* (D). The lift force is orthogonal to the free-stream direction while the drag force is parallel to it. The angle of attack α is the angle between chord line and free-stream direction.

The aerodynamic forces are caused by the relative motion of the airplane to the air. In case of free-stream, they are invariant to rotation around the velocity of the air current, hence the aerodynamic forces depend only on the angle of attack α and the sideslip angle β. Their interpretation to relative wind is shown in Fig. 3.6. The first rotation defines the stability axis, and the angle of attack is the angle between the body-fixed x-axis and the stability x-axis.

The x_W- and z_W-axes of the frame K_W have, respectively, $-D$- and $-L$-direction, furthermore y_W is the direction of sideforce. The orientation matrix (in the usual convention of robotics) is

$$A_{K_B,K_W} = Rot(y, -\alpha) Rot(z, \beta) \quad \Rightarrow$$

$$A_{K_B,K_W} = \begin{bmatrix} C_\alpha & 0 & -S_\alpha \\ 0 & 1 & 0 \\ S_\alpha & 0 & C_\alpha \end{bmatrix} \begin{bmatrix} C_\beta & -S_\beta & 0 \\ S_\beta & C_\beta & 0 \\ 0 & 0 & 1 \end{bmatrix} = \begin{bmatrix} C_\alpha C_\beta & -C_\alpha S_\beta & -S_\alpha \\ S_\beta & C_\beta & 0 \\ S_\alpha C_\beta & -S_\alpha S_\beta & C_\alpha \end{bmatrix},$$

$$S := A_{K_B,K_W}^T = \begin{bmatrix} C_\alpha C_\beta & S_\beta & S_\alpha C_\beta \\ -C_\alpha S_\beta & C_\beta & -S_\alpha S_\beta \\ -S_\alpha & 0 & C_\alpha \end{bmatrix},$$

$$r_B = A_{K_B,K_W} r_W, \qquad r_W = S r_B.$$

$$(3.107)$$

The true airspeed v_T has only x_W component hence the speed v_B of the airplane to be manipulated by the pilot is

$$v_B = \begin{pmatrix} U \\ V \\ W \end{pmatrix} = S^T v_W = A_{K_B,K_W} \begin{pmatrix} v_T \\ 0 \\ 0 \end{pmatrix} = \begin{pmatrix} C_\alpha C_\beta v_T \\ S_\beta v_T \\ S_\alpha C_\beta v_T \end{pmatrix}, \qquad (3.108)$$

from which follows

$$v_T = \sqrt{U^2 + V^2 + W^2}, \qquad \tan\alpha = W/U, \qquad \sin\beta = V/v_T. \qquad (3.109)$$

3.4.5 Gravity Effect

By the law of gravitational pull, the earth exert a force F_g to the airplane

$$F_g = -G \frac{Mm}{\|p\|^2} \frac{p}{\|p\|} = -GMm \frac{p}{\|p\|^3}, \qquad (3.110)$$

where G is a constant, M is the mass of the earth, m is the mass of the airplane and p is the vector pointing into the center of gravity of the airplane, furthermore $GM = 3.98599927 \times 10^{14}$ m^3/s^2. The vector of gravity acceleration is

$$g = -\frac{GM}{\|p\|^3} p. \qquad (3.111)$$

Its value varies with p, for example $\|g\|$ is at the equatorial plane on sea-level 9.81425 m/s^2, but at the poles 9.83193 m/s^2.

In NED frame, the gravity vector is $g = g_{scalar}(0, 0, 1)^T$ hence the gravity force in BODY frame is $A_{K_{NED},K_B}^T g = B_B g$, that is,

$$F_{Bg} = \begin{pmatrix} -S_\theta \\ S_\phi C_\theta \\ C_\phi C_\theta \end{pmatrix} g_{scalar} =: \begin{pmatrix} g_{B1} \\ g_{B2} \\ g_{B3} \end{pmatrix}. \qquad (3.112)$$

On the other hand, the gravity effect in wind-axes coordinate system is $F_{Wg} = S F_{Bg}$, that is,

$$
\begin{aligned}
F_{Wg} &= \begin{bmatrix} C_\alpha C_\beta & S_\beta & S_\alpha C_\beta \\ -C_\alpha S_\beta & C_\beta & -S_\alpha S_\beta \\ -S_\alpha & 0 & C_\alpha \end{bmatrix} \begin{pmatrix} -S_\theta \\ S_\phi C_\theta \\ C_\phi C_\theta \end{pmatrix} g_{scalar} \\
&= \begin{pmatrix} -C_\alpha C_\beta S_\theta + S_\beta S_\phi C_\theta + S_\alpha C_\beta C_\phi C_\theta \\ C_\alpha S_\beta S_\theta + C_\beta S_\phi C_\theta - S_\alpha S_\beta C_\phi C_\theta \\ S_\alpha S_\theta + C_\alpha C_\phi C_\theta \end{pmatrix} g_{scalar} \\
&=: \begin{pmatrix} g_{W1} \\ g_{W2} \\ g_{W3} \end{pmatrix}.
\end{aligned}
\tag{3.113}
$$

3.4.6 Aerodynamic Forces and Torques

The external force F_B (without gravity effect) and the external torque T_B can be decomposed into F_{BA}, T_{BA} aerodynamic (A) and F_{BT}, T_{BT} trust (T) components in the coordinate system of the airplane, and similarly in the wind-frame:

$$
F_B = \begin{pmatrix} F_x \\ F_y \\ F_z \end{pmatrix} = F_{BA} + F_{BT}, \qquad T_B = \begin{pmatrix} \bar{L} \\ M \\ N \end{pmatrix} = T_{BA} + T_{BT}, \tag{3.114}
$$

$$
F_W = S F_B = \begin{pmatrix} -D \\ Y \\ -L \end{pmatrix} + S F_{BT} = F_{WA} + F_{WT}, \tag{3.115}
$$

$$
T_W = S T_B = T_{WA} + T_{WB}.
$$

The (nongravity) forces and torques depend on the wing reference area S_{wa}, the free-stream dynamic pressure $\bar{q} = \frac{1}{2} \rho v_T^2$, different dimensionless coefficients $C_D, C_L, \ldots, C_n$ and, in case of the torques, on the wing span b and the wing mean geometric chord:

$$
\begin{aligned}
D_{stab} &= \bar{q} S_{wa} C_D \quad \text{drag,} \\
L_{stab} &= \bar{q} S_{wa} C_L \quad \text{lift,} \\
Y &= \bar{q} S_{wa} C_Y \quad \text{sideforce,} \\
\bar{L} &= \bar{q} S_{wa} b C_l \quad \text{rolling moment,} \\
M &= \bar{q} S_{wa} \bar{c} C_m \quad \text{pitching moment,} \\
N &= \bar{q} S_{wa} b C_n \quad \text{yawing moment.}
\end{aligned}
\tag{3.116}
$$

Fig. 3.7 The frame fixed to
the airplane with the
kinematic and force/torque
variables

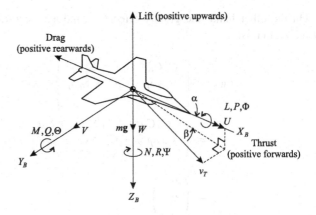

Fig. 3.8 Control surfaces of
conventional airplane

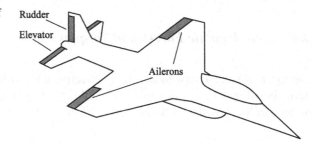

The $C_D, C_L, \ldots, C_n$ dimensionless coefficients depend in first line on the angles
α and β, the control surfaces and the Mach-number. The Mach-number at a point
by air convention is the local air speed divided by the speed of sound in case of
freestream of air. Notice that C_D and C_L are defined in the stability frame, not in
the wind-axes frame.

In Fig. 3.7, the interpretations of forces/torques are summarized in the three coordinate directions of K_B which are used during the derivations of the kinematic
and dynamic models. The orientation is described by the angles Φ (roll), Θ (pitch)
and Ψ (yaw), while the absolute angular velocity has components P, Q, R in the
basis of K_B. Figure 3.8 shows the control surfaces of a conventional airplane, called
elevator, aileron and rudder. For secondary flight control, additional flaps are used.
The control inputs belonging to elevator, aileron, rudder and flap may be denoted by
$\delta_e, \delta_a, \delta_r$ and δ_f, respectively.

The following relations will be important for finding the state equations of v_T, α
and β:

$$\omega_W = S\omega_B = \begin{bmatrix} C_\alpha C_\beta & S_\beta & S_\alpha C_\beta \\ -C_\alpha S_\beta & C_\beta & -S_\alpha S_\beta \\ -S_\alpha & 0 & C_\alpha \end{bmatrix} \begin{pmatrix} P \\ Q \\ R \end{pmatrix} \quad \Rightarrow$$

$$\begin{pmatrix} P_w \\ Q_w \\ R_w \end{pmatrix} = \begin{pmatrix} C_\alpha C_\beta P + S_\beta Q + S_\alpha C_\beta R \\ -C_\alpha S_\beta P + C_\beta Q - S_\alpha S_\beta R \\ -S_\alpha P + C_\alpha R \end{pmatrix},$$

$$S(\dot{v}_B + \omega \times v_B) = S\frac{d}{dt}\left(S^{-1}v_W\right) + (S\omega_B) \times (Sv_B)$$

$$= S\left(-S^{-1}\dot{S}S^{-1}\right)v_w + \dot{v}_W + \omega_W \times v_W$$

$$= -\dot{S}S^{-1}v_W + \dot{v}_W + \omega_w \times v_W$$

$$= SF_{BA} + SF_{BT} + SF_{Bg},$$

$$-\dot{S}S^{-1} = \begin{bmatrix} 0 & 0 & -C_\beta \\ 0 & 0 & S_\beta \\ C_\beta & -S_\beta & 0 \end{bmatrix}\dot{\alpha} + \begin{bmatrix} 0 & -1 & 0 \\ 1 & 0 & 0 \\ 0 & 0 & 0 \end{bmatrix}\dot{\beta},$$

$$\dot{v}_w - \dot{S}S^{-1}v_w + \omega_W \times v_W = \begin{pmatrix} \dot{v}_T \\ 0 \\ 0 \end{pmatrix} + \begin{pmatrix} 0 \\ \dot{\beta} \\ \dot{\alpha}C_\beta \end{pmatrix}v_T + \begin{pmatrix} 0 \\ R_w \\ -Q_w \end{pmatrix}v_T,$$

$$\begin{pmatrix} m\dot{v}_T \\ m\dot{\beta}v_T \\ m\dot{\alpha}v_T C_\beta \end{pmatrix} = \begin{pmatrix} -D \\ Y \\ -L \end{pmatrix} + \begin{pmatrix} C_\alpha C_\beta \\ -C_\alpha S_\beta \\ -S_\alpha \end{pmatrix}F_T + \begin{pmatrix} g_{W1} \\ g_{W2} \\ g_{W3} \end{pmatrix}$$

$$+ \begin{pmatrix} 0 \\ -R_w \\ Q_w \end{pmatrix}mv_T.$$

Assume the resulting external forces have the form in K_B

$$F_x = \bar{q}S_{wa}C_X + F_T - S_\Theta mg_{\text{scalar}},$$

$$F_y = \bar{q}S_{wa}C_Y + S_\Phi C_\Theta mg_{\text{scalar}},$$

$$F_z = \bar{q}S_{wa}C_Z + C_\Phi C_\Theta mg_{\text{scalar}},$$

and compare $S(C_X, C_Y, C_Z)^T$ and $(-C_{Dw}, C_{Yw}, -C_{Lw})^T$ then it follows

$$-C_{Dw} = C_\alpha C_\beta C_X + S_\beta C_Y + S_\alpha C_\beta C_Z,$$

$$C_{Yw} = -C_\alpha S_\beta C_X + C_\beta C_Y - S_\alpha S_\beta C_Z,$$

$$-C_{Lw} = -S_\alpha C_X + C_\alpha C_Z.$$

On the other hand, C_D and C_L are defined in the stability frame hence

$$\begin{pmatrix} -C_D \\ C_Y \\ -C_L \end{pmatrix} = Rot(y, -\alpha)^T \begin{pmatrix} C_X \\ C_Y \\ C_Z \end{pmatrix} = \begin{pmatrix} C_\alpha C_X + S_\alpha C_Z \\ C_Y \\ -S_\alpha C_X + C_\alpha C_Z \end{pmatrix},$$

from which the following aerodynamic coefficients can be obtained:

$$
\begin{aligned}
C_D &:= -C_X C_\alpha - C_Z S_\alpha, \\
C_L &:= -C_Z C_\alpha + C_X S_\alpha = C_{Lw}, \\
C_{Dw} &:= C_D C_\beta - C_Y S_\beta, \\
C_{Yw} &:= C_Y C_\beta + C_D S_\beta.
\end{aligned}
\tag{3.117}
$$

3.4.7 Gyroscopic Effect of Rotary Engine

If the airplane has a spinning rotor, then its angular moment has also to be taken into consideration. Assume the angular moment of the spinning rotor is $h = (h_x, h_y, h_z)^T$ and constant. Typically, $h = (I_p \Omega_p, 0, 0)^T$ where I_p is the inertia moment and Ω_p is the angular velocity of the spinning motor. The influence of the spinning rotor to the motion equation is $-I_c^{-1}(\omega \times h)$ where

$$
\begin{bmatrix} 0 & -R & Q \\ R & 0 & -P \\ -Q & P & 0 \end{bmatrix} \begin{pmatrix} h_x \\ h_y \\ h_z \end{pmatrix} = \begin{pmatrix} -Rh_y + Qh_z \\ Rh_x - Ph_z \\ -Qh_x + Ph_y \end{pmatrix},
$$

$$
\begin{aligned}
-I_c^{-1}(\omega \times h) &= - \begin{bmatrix} I_1 & I_2 & I_3 \\ I_2 & I_4 & I_5 \\ I_3 & I_5 & I_6 \end{bmatrix} \begin{pmatrix} -Rh_y + Qh_z \\ Rh_x - Ph_z \\ -Qh_x + Ph_y \end{pmatrix} \\
&= \begin{pmatrix} I_1(Rh_y - Qh_z) + I_2(Ph_z - Rh_x) + I_3(Qh_x - Ph_y) \\ I_2(Rh_y - Qh_z) + I_4(Ph_z - Rh_x) + I_5(Qh_x - Ph_y) \\ I_3(Rh_y - Qh_z) + I_5(Ph_z - Rh_x) + I_6(Qh_x - Ph_y) \end{pmatrix} \\
&=: \begin{pmatrix} P_h' \\ Q_h' \\ R_h' \end{pmatrix}.
\end{aligned}
\tag{3.118}
$$

If the airplane has $[x, y]$-plane symmetry, then $I_{xy}, I_{yz} = 0$, consequently $I_2, I_5 = 0$, and if $h = (I_p \Omega_p, 0, 0)^T$, that is, $h_x = I_p \Omega_p$, then $P_h' = I_3 Q h_x$, $Q_h' = -I_4 R h_x$, $R_h' = I_6 Q h_x$.

3.4.8 State Equations of Airplane

Using the motion equations of rigid body and taking into consideration the gravity, trust and aerodynamic effects the state equations of the airplane can be obtained in compact form:

Force equations in body frame:

$$\dot{U} = RV - QW + \frac{\bar{q}S_{wa}}{m}C_X + \frac{F_T}{m} + g_{B_1},$$

$$\dot{V} = PW - RU + \frac{\bar{q}S_{wa}}{m}C_Y + g_{B_2}, \tag{3.119}$$

$$\dot{W} = QU - PV + \frac{\bar{q}S_{wa}}{m}C_Z + g_{B_3}.$$

Force equations in wind-axes frame:

$$\dot{v}_T = -\frac{\bar{q}S_{wa}}{m}C_{Dw} + \frac{F_T}{m}C_\alpha C_\beta + g_{W_1},$$

$$\dot{\alpha} = -\frac{\bar{q}S_{wa}}{mv_T C_\beta}C_L + Q - T_\beta(C_\alpha P + S_\alpha R) - \frac{F_T}{mv_T C_\beta}S_\alpha + \frac{1}{v_T C_\beta}g_{W_3}, \tag{3.120}$$

$$\dot{\beta} = \frac{\bar{q}S_{wa}}{mv_T}C_{Yw} + S_\alpha P - C_\alpha R - \frac{F_T}{mv_T}C_\alpha S_\beta + \frac{1}{v_T}g_{W_2}.$$

Torque equations:

$$\dot{P} = P_{pp}PP + P_{pq}PQ + P_{pr}PR + P_{qq}QQ + P_{qr}QR + P_{rr}RR + P'_h$$
$$+ \bar{q}S_{wa}(bI_1C_l + \bar{c}I_2C_m + bI_3C_n),$$

$$\dot{Q} = Q_{pp}PP + Q_{pq}PQ + Q_{pr}PR + Q_{qq}QQ + Q_{qr}QR + Q_{rr}RR + Q'_h$$
$$+ \bar{q}S_{wa}(bI_2C_l + \bar{c}I_4C_m + bI_5C_n),$$

$$\dot{R} = R_{pp}PP + R_{pq}PQ + R_{pr}PR + R_{qq}QQ + R_{qr}QR + R_{rr}RR + R'_h$$
$$+ \bar{q}S_{wa}(bI_3C_l + \bar{c}I_5C_m + bI_6C_n).$$

$$\tag{3.121}$$

Kinematic equations:

$$\dot{\Phi} = P + T_\Theta(S_\Phi Q + C_\Phi R),$$

$$\dot{\Theta} = C_\Phi Q - S_\Phi R, \tag{3.122}$$

$$\dot{\Psi} = \frac{S_\Phi Q + C_\Phi R}{C_\Theta}.$$

Navigation equations:

$$p = (x_e, y_e, -H)^T \quad \Rightarrow \quad \dot{p} = B^T(U, V, W)^T = B^T(v_T C_\alpha C_\beta, v_T S_\beta, v_T S_\alpha C_\beta)^T$$

$$\dot{x}_N = C_\Psi C_\Theta U + (C_\Psi S_\Theta S_\Phi - S_\Psi C_\Phi)V + (C_\Psi S_\Theta C_\Phi + S_\Psi S_\Phi)W,$$

$$\dot{y}_E = S_\Psi C_\Theta U + (S_\Psi S_\Theta S_\Phi + C_\Psi C_\Phi)V + (S_\Psi S_\Theta C_\Phi - C_\Psi S_\Phi)W,$$

$$\dot{H} = S_\Theta U - C_\Theta S_\Phi V - C_\Theta C_\Phi W.$$

$$\tag{3.123}$$

Acceleration sensor equations:

$$a_x = \dot{U} - RV + QW - g_{B_1} \rightarrow a_x = \frac{\bar{q} S_{wa}}{m} C_X + \frac{F_T}{m},$$

$$a_y = \dot{V} - PW + RU - g_{B_2} \rightarrow a_y = \frac{\bar{q} S_{wa}}{m} C_Y, \qquad (3.124)$$

$$a_z = \dot{W} - QU + PV - g_{B_3} \rightarrow a_z = \frac{\bar{q} S_{wa}}{m} C_Z.$$

Remark In the above state equations, it is assumed that orientation, angular velocity etc. are defined relative to the ECI frame (see the use of capital letters). However, in many applications the NED frame, typically fixed at the start, can be considered as an approximate inertia frame, in which case the appropriate state variables will be denoted by lower case letters. For simplicity, the absolute value of velocity is denoted by $V := v_T$.

3.4.9 Linearization of the Nonlinear Airplane Model

Assume the NED frame can be considered an approximate inertia frame. It is useful to divide the Flat-Earth Flight into different sections in which the nonlinear model may be linearized around steady-state conditions and the perturbations can be controlled using linear control techniques. Steady-State Flight means $\dot{p}, \dot{q}, \dot{r} = 0$ and $\dot{u}, \dot{v}, \dot{w} = 0$ or $\dot{V}, \dot{\alpha}, \dot{\beta} = 0$, that is, accelerations are zero, furthermore the control vector $\delta = const$ where δ contains all control inputs. For Steady-State Flight, four typical cases can be distinguished.

1. *Steady Wings-Level Flight*: $\phi, \dot{\phi}, \dot{\theta}, \dot{\psi} = 0$ and θ_0, ψ_0 have nominal values.
2. *Steady Turning Flight*: $\dot{\phi}, \dot{\theta} = 0$, $\dot{\psi} = r$ is the turn rate and ϕ_0, θ_0 have nominal values.
3. *Steady Pull Up*: $\phi, \dot{\phi}, \dot{\psi} = 0$, $\dot{\theta} = q$ is the pull-up rate and ψ_0 has nominal value.
4. *Steady Roll*: $\dot{\theta}, \dot{\psi} = 0$, $\dot{\phi} = p$ is the roll rate and θ_0, ψ_0 have nominal values.

If a nonlinear system model is available in Simulink, then to find operation points x_0, u_0, y_0 the `trim` function of Simulink can be used. Some initial guess values have to be passed to trim in order to support Sequential Quadratic Programming (SQP) to find optimum equilibrium point. Initial guesses can contain flight level, speed, trust power etc. depending on the problem. The goal is to perform equilibrium (zero state derivatives) or coming near to it if no exact solution exists.

The function `trim('sys',x0,u0,y0,ix,iu,iy)` finds the trim point closest to `x0,u0,y0` that satisfies a specified set of state, input, and/or output conditions. The integer vectors `ix,iu,iy` select the values in `x0,u0,y0` that must be satisfied. If `trim` cannot find an equilibrium point that satisfies the specified set of conditions exactly, it returns the nearest point that satisfies the conditions, namely `abs([x(ix)-x0(ix);u(iu)-u0(iu);y(iy)-y0(iy)])`.

Use $[x,u,y,dx]$ =trim('sys',x0,u0,y0,ix,iu,iy,dx0,idx) to find specific nonequilibrium points, that is, points at which the system's state derivatives have some specified nonzero value. Here, dx0 specifies the state derivative values at the search's starting point and idx selects the values in dx0 that the search must satisfy exactly.

After having found the trim conditions, the linmod Simulink function can be used to perform the linearization of the system at the trim state x0 and input vector u0. The call slin=linmod('sys',x0,u0,...) returns a structure containing the state-space matrices, state names, operating point, and other information about the linearized model. Linearized models are valid for small perturbations around the operational points.

Important linearized models are the longitudinal and lateral models. Consider an airplane having piston engine truster characterized by the manifold pressure p_z and the engine speed in rpm.

For longitudinal motion, the active state variables are $x_{\text{long}} = (V, \alpha, r, \theta, H)^T$, the control inputs are $u_{\text{long}} = (\delta_e, p_z)^T$ and the output may be $y_{\text{long}} = (V, H)^T$.

For lateral motion $x_{\text{lat}} = (\beta, p, r, \psi, \phi, x_e, y_e)^T$ and $u_{\text{lat}} = (\delta_a, \delta_r)^T$ are the state variables and control inputs, respectively, while the outputs may be $y_{\text{lat}}^h = (\beta, \psi)^T$ for heading or $y_{\text{lat}}^r = (\beta, \phi)^T$ for roll.

For trim as initial values the velocity, altitude, heading (yaw), engine speed and manifold pressure can be specified.

3.4.10 Parametrization of Aerodynamic and Trust Forces and Moments

Without information about the dependence of the aerodynamic and trust coefficients on the states and control inputs the dynamic model is not defined because of the missed information about $C_X, C_Y, C_Z, C_l, C_m, C_n$ or C_D, C_L. Different approaches are known for their parametrization, however polynomial forms are the simplest ones. Of course, the coefficients of the parametrization have to be determined based on wind-tunnel testing and test-flight measurements. For identification of the parameters, well elaborated techniques [70] and the MATLAB function collection SIDPAC can be suggested, see for example:
http://www.aiaa.org/publications/supportmaterials

For illustration, consider the DHC-2 Beaver aircraft in the FDC 1.4 Simulink Toolbox for Flight Dynamics and Control Analysis, see for example:
http://www.xs4all.nl/~rauw/fdcreport/FDC14_preview_007.pdf
[120, 146].

The aerodynamic forces/torques are defined as $F_{\text{aero}} = d \cdot p_1(x, \dot{x}, u_{\text{aero}})$, where $d = q_{\text{dyn}} S_{wa} \text{diag}([1, 1, 1, b, \bar{c}, b])$, $q_{\text{dyn}} = (1/2)\rho V^2$ and $u_{\text{aero}} = (\delta_e, \delta_a, \delta_r, \delta_f)^T$. The force/torque coefficients of the dynamic model are polynomials:

$$C_{X_a} = C_{X_0} + C_{X_\alpha}\alpha + C_{X_{\alpha^2}}\alpha^2 + C_{X_{\alpha^3}}\alpha^3 + C_{X_q}\frac{q\bar{c}}{V} + C_{X_{\delta_r}}\delta_r + C_{X_{\delta_f}}\delta_f$$
$$+ C_{X_{\alpha\delta_f}}\alpha\delta_f,$$

$$C_{Y_a} = C_{Y_0} + C_{Y_\beta}\beta + C_{Y_p}\frac{pb}{2V} + C_{Y_r}\frac{rb}{2V} + C_{Y_{\delta_a}}\delta a + C_{Y_{\delta_r}}\delta r + C_{Y_{\delta_r a}}\delta r\alpha$$
$$\qquad + C_{Y_{\dot\beta}}\frac{\dot\beta b}{2V},$$

$$C_{Z_a} = C_{Z_0} + C_{Z_\alpha}\alpha + C_{Z_{\alpha^3}}\alpha^3 + C_{Z_q}\frac{q\bar{c}}{V} + C_{Z_{\delta_e}}\delta e + C_{Z_{\delta_e\beta^2}}\delta e\beta^2 + C_{Z_{\delta_f}}\delta f$$
$$\qquad + C_{Z_{\alpha\delta_f}}\alpha\delta f,$$

$$C_{l_a} = C_{l_0} + C_{l_\beta}\beta + C_{l_p}\frac{pb}{2V} + C_{l_r}\frac{rb}{2V} + C_{l_{\delta_a}}\delta a + C_{l_{\delta_a\alpha}}\delta a\alpha,$$

$$C_{m_a} = C_{m_0} + C_{m_\alpha}\alpha + C_{m_{\alpha^2}}\alpha^2 + C_{m_q}\frac{q\bar{c}}{V} + C_{m_{\delta_e}}\delta e + C_{m_{\beta^2}}\beta^2 + C_{m_r}\frac{rb}{2V}$$
$$\qquad + C_{m_{\delta_f}}\delta f,$$

$$C_{n_a} = C_{n_0} + C_{n_\beta}\beta + C_{n_p}\frac{pb}{2V} + C_{n_r}\frac{rb}{2V} + C_{n_{\delta_a}}\delta a + C_{n_{\delta_r}}\delta r + C_{n_q}\frac{q\bar{c}}{V} + C_{n_{\beta3}}\beta^3.$$

Notice that the state variables $V = v_T, \alpha, \beta, p, q, r$ together with the control surfaces appear in the polynomial approximations. Since C_Y depends also on $\dot\beta$; therefore, the state equation becomes explicit.

The engine power P varies also with the altitude due to changes in air density. The propeller can be assumed to be an ideal pulling disc. The nondimensional pressure increase in the propeller slipstream dp_t is related to the engine power P by the formulas

$$dp_t = \frac{\Delta p_t}{\frac{1}{2}\rho V^2} = c_1 + c_2\left(\frac{P}{\frac{1}{2}V^3}\right),$$

$$P = c_3\left\{-c_4 + \left[c_5(p_z + c_6)(n + c_7) + (n + c_7) + (c_8 - c_9 n)\left(1 - \frac{\rho}{\rho_0}\right)\right]\right\},$$

where $c_1, \ldots, c_9$ are known constants, p_z is the manifold pressure, n is the engine speed (RPM), ρ is the air-density and ρ_0 is the known air-density at sea level. The propulsive effects are assumed in the form $F_{\text{prop}} = d \cdot p_2(x, dp_t)$ containing the polynomials for the propulsive force and torque coefficients:

$$C_{X_p} = C_{X_{dp_t}}dp_t + C_{X_{\alpha dp_t^2}}\alpha(dp_t)^2,$$

$$C_{Y_p} = 0,$$

$$C_{Z_p} = C_{Z_{dp_t}}dp_t,$$

$$C_{l_p} = C_{l_{\alpha^2 dp_t}}\alpha^2 dp_t,$$

$$C_{m_p} = C_{m_{dp_t}}dp_t,$$

$$C_{n_p} = C_{n_{dp_t^3}}(dp_t)^3.$$

The identification of the model parameters can be performed using test-flight data and robust identification techniques (Bayes estimator, Least-Squares, nonlinear state estimator etc.), see [70].

The usual form of linear parameter estimation problem is $z(t) = \varphi^T(t)\vartheta + v(t)$ where $z(t)$ is the output measurement, ϑ is the parameter vector to be identified, $\varphi(t)$ contains computable weighting functions for the components of the parameter vector, and $v(t)$ is additive noise. The dynamic model identification of the airplane can be divided into subproblems, in which $C_{X_a}, \ldots, C_{n_a}$ and $C_{X_p}, \ldots, C_{n_p}$ play the role of the different $z(t)$, and the parameters to be identified are subsets of the parameters $C_{X_0}, C_{X_\alpha}, \ldots, C_{n_{dp_t^3}}$. Hence, for identification purposes, the test-flight data can be considered in the form

$$d^{-1}\begin{pmatrix} ma_{\text{sensor}} \\ I_c\dot{\omega} + \omega \times (I_c\omega) \end{pmatrix} = \begin{pmatrix} C_{X_a} \\ \vdots \\ C_{n_a} \end{pmatrix} + \begin{pmatrix} C_{X_p} \\ \vdots \\ C_{n_p} \end{pmatrix},$$

from which $C_{X_a}, \ldots, C_{n_a}$ and $C_{X_p}, \ldots, C_{n_p}$ should be determined. Robust numerical differentiation of ω is necessary in order to find $\dot{\omega}$ making the left side computable. Usual technique is to design test-flight sections of constant engine power (propeller slipstream, revolution etc.) so that the effect of engine power can be taken into consideration in the constant terms $C_{X_0}, \ldots, C_{n_0}$, reducing the problem to the identification of $C_{X_0}, \ldots, C_{n_{\beta^3}}$.

Similar approaches can be suggested for parametrization and identification of unknown small-scale airplane models.

3.5 Dynamic Model of Surface and Underwater Ships

For ships en route between continents, a star-fixed reference frame is used as inertia frame. For ships operating in local area, an earth-fixed tangent plane on the surface can be used for navigation. Due to the low speed of the ship relative to the earth angular velocity, the NED frame moving with the ship is assumed as inertia frame. The following discussions are limited to the latter case.

The BODY frame K_B is fixed to the ship. The x_B, y_B, z_B motion directions are called surge, sway and heave directions, respectively. Special is in the practice that the origin of the BODY frame K_B is usually different from the COG of the ship.

3.5.1 Rigid Body Equation of Ship

The dynamic model of ship satisfies the rigid body equations with added hydrodynamical forces and moments and environmental effects. Denote r_g the position vector from the origin of K_B to the COG and I_B the inertia matrix belonging to the origin of K_B. Let a_g, a, v and ω respectively, the acceleration of the COG, the acceleration of the origin of K_B, the velocity of the origin of K_B and the angular velocity of K_B, all relative to an inertia frame. Then, using the differentiation rule, the Huygens–Steiner formula and the Newton–Euler equations in Appendix A, it follows

$$a = \dot{v} + \omega \times v,$$

$$ma_g = m\big(\dot{v} + \omega \times v + \dot{\omega} \times r_g + \omega \times (\omega \times r_g)\big)$$

$$= m\big(\dot{v} - r_g \times \dot{\omega} - v \times \omega + (r_g \times \omega) \times \omega\big) = F_B,$$

$$I_B \dot{\omega} + \omega \times (I_B \omega) + m r_g \times (\dot{v} + \omega \times v) = M_B,$$

where F_B is the total force and M_B is the total moment acting in the origin of K_B. Adding $v \times v = 0$, the left sides of the last two equations can be rewritten as

$$\begin{bmatrix} mI_3 & -m[r_g \times] \\ m[r_g \times] & I_B \end{bmatrix} \begin{pmatrix} \dot{v} \\ \dot{\omega} \end{pmatrix}$$

$$+ \begin{bmatrix} 0 & -[(mv - mr_g \times \omega) \times] \\ -[(mv - mr_g \times \omega) \times] & [(mv \times r_g - I_B \omega) \times] \end{bmatrix} \begin{pmatrix} v \\ \omega \end{pmatrix}.$$

Hence, introducing the notations

$$M_{\text{RB}} = \begin{bmatrix} mI_3 & -m[r_g \times] \\ m[r_g \times] & I_B \end{bmatrix} = \begin{bmatrix} M_{11} & M_{12} \\ M_{21} & M_{22} \end{bmatrix}, \tag{3.125}$$

$$C_{\text{RB}}(v, \omega) = \begin{bmatrix} 0_{3 \times 3} & -[(mv - mr_g \times \omega) \times] \\ -[(mv - mr_g \times \omega) \times] & [(mv \times r_g - I_B \omega) \times] \end{bmatrix}$$

$$= \begin{bmatrix} 0_{3 \times 3} & -[(M_{11}v + M_{12}\omega) \times] \\ -[(M_{11}v + M_{12}\omega) \times] & -[(M_{21}v + M_{22}\omega) \times] \end{bmatrix}, \tag{3.126}$$

the rigid body (RB) part of the dynamic model can be written in the compact form

$$M_{\text{RB}} \begin{pmatrix} \dot{v} \\ \dot{\omega} \end{pmatrix} + C_{\text{RB}}(v, \omega) \begin{pmatrix} v \\ \omega \end{pmatrix} = \begin{pmatrix} F_B \\ M_B \end{pmatrix} =: \tau_{\text{RB}}, \tag{3.127}$$

$$M_{\text{RB}} = M_{\text{RB}}^T > 0 \quad \text{(constant)},$$

$$C_{\text{RB}}(v, \omega) = -C_{\text{RB}}^T(v, \omega), \quad \forall v, \omega. \tag{3.128}$$

Here M_{RB} is the generalized constant inertia matrix and C_{RB}, together with the linear and angular velocities, describe the centrifugal and Coriolis effects.

The following notation is adopted from Fossen [41]:

$$\eta := (x, y, z, \phi, \theta, \psi)^T,$$

$$\nu := (u, v, w, p, q, r)^T, \tag{3.129}$$

$$\tau := (X, Y, Z, \bar{L}, M, N)^T,$$

where $(x, y, z)^T$ is the position from the NED frame to the origin of K_B while the linear and angular velocities, forces and torques are expressed in the basis of K_B.

It is common to assume that the forces and moments τ_{RB} are separated into additive components of hydrodynamic forces due to radiation-induced and damping forces and moments ($\tau_H = \tau_R + \tau_D$), control forces and moments (τ), and environmental forces and moments (τ_{env}) due to wind, waves and currents.

3.5.2 Hydrodynamic Forces and Moments

The radiation-induced forces and moments include the added mass due to the surrounding world. It is useful to introduce the added mass inertia matrix M_A and the matrix of hydrodynamic centrifugal and Coriolis terms $C_A(v)$. In contrast to submerged volumes that have constant added mass, for surface vessels the added mass effect depends on the frequency of motion caused by water surface effects. At zero speed, the added mass matrix is expressed using the hydrodynamic derivatives,

$$M_A = \begin{bmatrix} A_{11} & A_{12} \\ A_{21} & A_{22} \end{bmatrix} = - \begin{bmatrix} X_{\dot u} & X_{\dot v} & X_{\dot w} & X_{\dot p} & X_{\dot q} & X_{\dot r} \\ Y_{\dot u} & Y_{\dot v} & Y_{\dot w} & Y_{\dot p} & Y_{\dot q} & Y_{\dot r} \\ Z_{\dot u} & Z_{\dot v} & Z_{\dot w} & Z_{\dot p} & Z_{\dot q} & Z_{\dot r} \\ L_{\dot u} & L_{\dot v} & L_{\dot w} & L_{\dot p} & L_{\dot q} & L_{\dot r} \\ M_{\dot u} & M_{\dot v} & M_{\dot w} & M_{\dot p} & M_{\dot q} & M_{\dot r} \\ N_{\dot u} & N_{\dot v} & N_{\dot w} & N_{\dot p} & N_{\dot q} & N_{\dot r} \end{bmatrix}, \tag{3.130}$$

where, for example, $X = -X_{\dot u} \dot u - \cdots - X_{\dot r} \dot r$ is the added mass force due to accelerations in different directions and $X_{\dot u} = -\frac{\partial X}{\partial \dot u}, \ldots, X_{\dot r} = -\frac{\partial X}{\partial \dot r}$. Henceforth, frequency independent model and $M_A = M_A^T > 0$ will be assumed.

The centripetal and Coriolis matrix $C_A(v)$ is defined by replacing M_{RB} with M_A in C_{RB}:

$$C_A(v) = \begin{bmatrix} 0_{3 \times 3} & -[(A_{11}v_1 + A_{12}v_2) \times] \\ -[(A_{11}v_1 + A_{12}v_2) \times] & -[(A_{21}v_1 + A_{22}v_2) \times] \end{bmatrix}. \tag{3.131}$$

The total damping matrix $D(v)$, mainly caused by potential, viscous, skin friction effects and wave drift damping, is approximated by the sum of a constant linear and a varying nonlinear term $D(v) = D + D_n(v)$ where the linear term is

$$D = \begin{bmatrix} D_{11} & D_{12} \\ D_{21} & D_{22} \end{bmatrix} = - \begin{bmatrix} X_u & X_v & X_w & X_p & X_q & X_r \\ Y_u & Y_v & Y_w & Y_p & Y_q & Y_r \\ Z_u & Z_v & Z_w & Z_p & Z_q & Z_r \\ L_u & L_v & L_w & L_p & L_q & L_r \\ M_u & M_v & M_w & M_p & M_q & M_r \\ N_u & N_v & N_w & N_p & N_q & N_r \end{bmatrix}. \tag{3.132}$$

For ships operating around zero speed the linear term dominates, while quadratic damping dominates at higher speed. The damping forces are always dissipative, that is, $v^T (D(v) + D(v)^T) v > 0, \forall v \neq 0$.

The hydrodynamic forces and moments can be assumed in the form

$$\tau_H = -M_A\dot{v} - C_A(v)v - D(v)v - g(\eta). \tag{3.133}$$

By introducing the notations $M = M_{\text{RB}} + M_A$ and $C(v) = C_{\text{RB}}(v) + C_A(v)$, the dynamic model can be written as

$$M\dot{v} + C(v)v + D(v)v + g(\eta) = \tau + \tau_{\text{env}}, \tag{3.134}$$

where $M = M^T > 0$ and $C(v) = -C(v)^T$. This latter property means that the quadratic form is identically zero, that is, $x^T C(v)x \equiv 0, \forall x$, which may be important for stability investigations.

3.5.3 Restoring Forces and Moments

Underwater vehicles and surface vessels are also affected by gravity and buoyancy forces. In hydrodynamic terminology, they are referred as restoring forces collected in the term $g(\eta)$. Fossen [41] treats underwater vehicles are different than surface vessels.

Underwater vehicles
Denote $r_b = (x_b, y_b, z_b)^T \in K_B$ and $r_g = (x_g, y_g, z_g)^T \in K_B$, respectively, the center of buoyancy (CB) and the center of gravity (CG). The gravity force is $W = mg \in K_{\text{NED}}$ and the buoyancy is $B = \rho g \nabla \in K_{\text{NED}}$ where ∇ is the displaced water volume. The forces can be converted to K_B by $f_g = (-S_\phi, C_\theta S_\phi, C_\theta C_\phi)^T W$ and $f_b = -(-S_\phi, C_\theta S_\phi, C_\theta C_\phi)^T B$. Then after easy computation follows

$$g(\eta) = -\begin{pmatrix} f_g + f_b \\ r_g \times f_g + r_b \times f_b \end{pmatrix}$$

$$= \begin{pmatrix} (W - B)S_\theta \\ -(W - B)C_\theta S_\phi \\ -(W - B)C_\theta C_\phi \\ -(y_g W - y_b B)C_\theta C_\phi + (z_g W - z_b B)C_\theta S_\phi \\ (z_g W - z_b B)S_\theta + (x_g W - x_b B)C_\theta C_\phi \\ -(y_g W - y_b B)S_\theta - (x_g W - x_b B)C_\theta S_\phi \end{pmatrix}. \tag{3.135}$$

A neutrally buoyant underwater vehicle will satisfy $W = B$. However, it is convenient to design underwater vehicles with $B > W$ such that the vehicle will surface automatically in emergency situation. An even simpler situation is if CB an CG are aligned vertically on the z_B-axis.

Surface vessels and semi-submersible restoring forces, surface ship
Restoring forces of surface vessels and semi-submersibles depend on the metacenter, the center of gravity (CG), the center of buoyancy (CB), and the geometry of the water plane.

The metacenter is the theoretical point at which an imaginary vertical line through CB intersects another vertical line (with respect to the horizontal plane of

Fig. 3.9 Concept of transverse metacentric stability

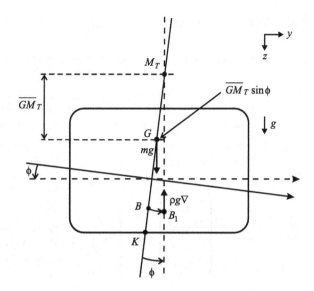

the ship) through a new CB created when the ship is displaced, or tilted, in the water, see Fossen [41].

For surface vessels at rest buoyancy and weight are in balance, that is, $mg = \rho g \nabla$. Let $z = 0$ denote the equilibrium position corresponding to nominal displayed water volume ∇. The hydrostatic force in heave will be $Z = -\rho g \delta \nabla(z)$ where $\delta \nabla(z) = \int_0^z A_{wp}(\zeta) \, d\zeta$ and $A_{wp}(\zeta)$ is the water plane area as a function of the heave position. For small perturbations, Z is linear in z, that is, $Z \approx -\rho g A_{wp}(0) z$, which is equivalent to a spring with stiffness and displacement. The restoring force is $\delta f_r = -\rho g (-S_\phi, C_\theta S_\phi, C_\theta C_\phi)^T \int_0^z A_{wp}(\zeta) \, d\zeta \in K_B$. The force pair is $W = B = \rho g \nabla$. Denoting the metacenters by

$$\overline{GM}_T = \text{transverse metacentric hight (m)},$$

$$\overline{GM}_L = \text{longitudinal metacentric hight (m)},$$

the force arm will be $r_r = (-\overline{GM}_L S_\theta, \overline{GM}_T S_\phi, 0)^T \in K_B$ and the restoring moment is $m_r = r_r \times f_r \in K_B$. The concept of transverse metacentric stability is illustrated in Fig. 3.9. Similar figure can be given to show lateral metacentric stability.

Neglecting the moment contribution to δf_r, the restoring forces and moments can be written in the form:

$$g(\eta) = -\begin{pmatrix} \delta f_r \\ m_r \end{pmatrix} = \begin{pmatrix} \rho g \int_0^z A_{wp}(\zeta) \, d\zeta (-S_\theta) \\ \rho g \int_0^z A_{wp}(\zeta) \, d\zeta \, C_\theta S_\phi \\ \rho g \int_0^z A_{wp}(\zeta) \, d\zeta \, C_\theta C_\phi \\ \rho g \nabla \overline{GM}_T S_\phi C_\theta C_\phi \\ \rho g \nabla \overline{GM}_L S_\theta C_\theta C_\phi \\ \rho g \nabla (-\overline{GM}_L C_\theta + \overline{GM}_T) S_\phi S_\theta \end{pmatrix}. \qquad (3.136)$$

3.5.4 Ballast Systems

In addition to restoring forces $g(\eta)$, the vessel can be pretrimmed by pumping water between ballast tanks of the vessel. Let z_d, ϕ_d, θ_d be the desired pretrimming values, then an equilibrium for $\dot{v}, v = 0$ can be satisfied by solving $g(\eta_d) = g_0 + w$ where g_0 is the effect of ballast tanks and w may be the actual disturbance. If $V_i(h_i) = \int_0^{h_i} A_i(h)\, dh$ is the actual volume of the ballast tank located in $r_i = (x_i, y_i, z_i)^T \in K_B$, $V_i \le V_{i,\max}$, $i = 1, \ldots, n$, then, because of $Z_{\text{ballast}} = \sum_{i=1}^n (W_i - B_i) = \rho g \sum_{i=1}^n (V_i - (V_{i,\max} - V_i)) = \rho g \sum_{i=1}^n (2V_i - V_{i,\max})$, it follows

$$
g_0 = \rho g \begin{pmatrix} 0 \\ 0 \\ Z_{\text{ballast}} \\ L_{\text{ballast}} \\ M_{\text{ballast}} \\ 0 \end{pmatrix} = \rho g \begin{pmatrix} 0 \\ 0 \\ \sum_{i=1}^n (2V_i - V_{i,\max}) \\ \sum_{i=1}^n y_i (2V_i - V_{i,\max}) \\ -\sum_{i=1}^n x_i (2V_i - V_{i,\max}) \\ 0 \end{pmatrix}. \tag{3.137}
$$

Assume $\int_0^z A_{wp}(\zeta)\, d\zeta \approx A_{wp}(0)z$ and introduce the notations $-Z_z = \rho g A_{wp}(0)$, $-Z_\theta = \rho g \int \int_{A_{wp}} x\, dA$, $-L_\phi = \rho g \nabla \overline{GM}_T$, and $-M_\theta = \rho g \nabla \overline{GM}_L$, then the ballast volumes $V_1, \ldots, V_n$ can be determined from

$$
\begin{bmatrix} -Z_z & 0 & -Z_\theta \\ 0 & -L_\phi & 0 \\ -Z_\theta & 0 & -M_\theta \end{bmatrix} \begin{pmatrix} z_d \\ \phi_d \\ \theta_d \end{pmatrix} = \rho g \begin{pmatrix} \sum_{i=1}^n (2V_i - V_{i,\max}) \\ \sum_{i=1}^n y_i (2V_i - V_{i,\max}) \\ -\sum_{i=1}^n x_i (2V_i - V_{i,\max}) \end{pmatrix} \tag{3.138}
$$

by using pseudo-inverse technique. In case of $[yz]$-symmetry $Z_\theta = 0$ yields.

Automatic ballast trimming can easily be realized by using PID-type feedback from z, ϕ, θ also in the presence of unknown disturbances [41].

3.5.5 Wind, Wave and Current Models

Wind and wave effects can be considered as additive disturbances. They are especially important for surface vessels and can be discussed in 3D, that is, $(x, y, \psi)^T$. The effect of ocean current can be taken into consideration through the current velocity vector v_c by introducing the relative velocity vector $v_r = v - v_c$.

3.5.5.1 Wind Models

Denote V_w the absolute value of the wind velocity and ψ_w the wind direction in K_{NED} relative to the North direction x_{NED}. Similarly, let ψ the direction of x_B-axis relative to the North direction, then $\gamma_r = \psi_w - \psi$ can be interpreted as the wind relative to the ship bow. The wind speed and direction can be measured with a wind sensor. The wind velocity vector can be assumed as $v_W = (u_w, v_w)^T$ where $V_w = \sqrt{u_r^2 + v_r^2}$, $\gamma_r = \operatorname{atan}(v_r/u_r) = \psi_w - \psi$, $u_r = V_r C_{\gamma_r} - u$ and $v_r = V_r S_{\gamma_r} - v$. The resulted force and torque will be denoted by $\tau_w = (X_w, Y_w, N_w)^T$.

To find the local velocity $V_w(h)$ above the sea level in hight h, a boundary-layer profile can be assumed, see, for example, $V_w(h) = V_w(10)(h/10)^{1/7}$ [24].

Wind forces and moments can be written in the forms

$$X_w = \frac{1}{2} C_X(\gamma_r) \rho_a V_r^2 A_T,$$

$$Y_w = \frac{1}{2} C_Y(\gamma_r) \rho_a V_r^2 A_L, \tag{3.139}$$

$$N_w = \frac{1}{2} C_N(\gamma_r) \rho_a V_r^2 A_L L,$$

and the coefficients C_X, C_Y, C_N can be further parametrized as

$$C_X = a_0 + a_1 \frac{2A_L}{L^2} + a_2 \frac{2A_T}{B^2} + a_3 \frac{L}{B} + a_4 \frac{S}{L} + a_5 \frac{C}{L} + a_6 M,$$

$$C_Y = -\left(b_0 + b_1 \frac{2A_L}{L^2} + b_2 \frac{2A_T}{B^2} + b_3 \frac{L}{B} + b_4 \frac{S}{L} + b_5 \frac{C}{L} + b_6 \frac{A_{SS}}{A_L} \right), \tag{3.140}$$

$$C_N = -\left(c_0 + c_1 \frac{2A_L}{L^2} + c_2 \frac{2A_T}{B^2} + c_3 \frac{L}{B} + c_4 \frac{S}{L} + c_5 \frac{C}{L} \right).$$

Here L is the overall length, B is the beam, A_L is the lateral projected area, A_T is the transversal projected area, A_{SS} is the lateral projected area of superstructure, S is the length of perimeter excluding masts and ventilators, C is the distance from bow of centroid of lateral projected area, and M is the number of masts or king posts seen in lateral projection [55].

3.5.5.2 Wind Generated Waves

Waves can be described by stochastic techniques using spectral formulation. For control development and simulation, the spectral formulation can be approximated by linear system derived from the peak value of the spectrum and other parameters.

Wave generation due to wind starts with small wavelets appearing on the water surface that increases the drag force which in turn allows short waves to grow. It is observed that a developing sea starts with high frequencies creating a spectrum with a peak at relatively high frequency. A long time storm creates a fully developed sea. After the wind has stopped, a low frequency decaying swell is formed with a low peak frequency. The wave spectrum can have two peak frequencies if a decaying swell of one storm interacts with the waves from another storm.

The often used Pierson–Moskowitz (PM) spectrum [115] is a two parameter wave spectrum

$$S(\omega) = A\omega^{-5} \exp\left(-B\omega^{-4}\right) \; [\text{m}^2\,\text{s}] \tag{3.141}$$

for fully developed sea, where $A = 8.1 \times 10^{-3} g^2$, $B = 0.74(g/V_{19.4})^4 = 3.11/H_s^2$ and $V_{19.4}$ is the wind speed at height 19.4 m over sea level. The significant wave height H_s is proportional to the square of the wind speed that is based on Gaussian assumption and $S(\omega)$ is narrow-banded. Based on the wave hight, Sea State codes and Beaufort numbers can be defined [116]. The peak (modal) frequency can be determined by making the derivative of $S(\omega)$ to zero, for which the model frequency is $\omega_0 = \sqrt[4]{4B/5}$, the model period is $T_0 = 2\pi/\omega_0$, and the maximum value is $S_{\max} = S(\omega_0) = \frac{5A}{4B\omega_0}\exp(-5/4)$.

Improved spectra are the modified PM-spectrum where $A = 4\pi^3 H_s^2/T_z^4$ and $B = 16\pi^3/T_z^4$, and T_z is the average zero-crossings period (or alternatively, $T_0 = T_z/0.710$, $T_1 = T_z/0.921$), the JONSWAP Spectrum by the ITTC (1984) standard satisfying

$$S(\omega) = 155 \frac{H_s^2}{T_1^4}\omega^{-5} \exp\left(\frac{-944}{T_1^4}\omega^{-4}\right)\gamma^Y \ [\text{m}^2\,\text{s}],$$

$$Y = \exp\left[-\left(\frac{0.191\omega T_1 - 1}{\sqrt{2}\sigma}\right)^2\right], \qquad \gamma = 3.3, \qquad (3.142)$$

$$\sigma = \begin{cases} 0.07 & \text{for } \omega \le 5.24/T_1, \\ 0.09 & \text{for } \omega > 5.24/T_1 \end{cases}$$

and the empirical Torsethaugen spectrum having two peaks [149].

For controller design and simulation, the spectral density functions may be approximated by simple linear systems like

$$G(s) = \frac{K_w s}{s^2 + 2\lambda\omega_0 s + \omega_0^2}, \qquad (3.143)$$

where λ is the damping coefficient, ω_0 is the dominating wave frequency. A convenient choice is $K_w = 2\lambda\omega_0\sigma$ where σ describes the wave intensity. Assuming $y(s) = G(s)w(s)$ where $w(s)$ is zero mean Gaussian white noise process with unit power across the spectrum, that is, $P_{ww}(\omega) = 1$, then the power density of $y(s)$ is $P_{yy}(\omega) = |G(j\omega)|^2$, especially for the above approximation

$$P_{yy}(\omega) = \frac{4(\lambda\omega_0\omega)^2\sigma^2}{(\omega_0^2 - \omega^2)^2 + 4(\lambda\omega_0\omega)^2}. \qquad (3.144)$$

Since $P_{yy}(\omega_0) = S(\omega_0)$ is the maximum value, hence $\sigma^2 = \max_{0<\omega<\infty} S(\omega)$. Specifically, for PM-spectrum $\sigma = \sqrt{\frac{A}{\omega_0^5}\exp(-\frac{B}{\omega_0^4})}$.

The above $G(s)$ can be cascaded or transfer functions with more parameters can be used in order to have better approximation of the spectrum. An alternative approach is to use nonlinear least-squares to compute λ so that $P_{yy}(\omega)$ fits $S(\omega)$ in LS-sense along the ω-axis. Transfer functions can easily be converted to state space form for simulation purposes.

For surface vessel moving with forward speed U, the peak frequency has to be modified for $G(s)$ according to

$$\omega_e(U, \omega_0, \beta) = \left| \omega_0 - \omega_0^2 \frac{U}{g} C_\beta \right|, \tag{3.145}$$

where β is the angle between the heading and the direction of the wave.

Control methods for surface vessels can be simulated and tested in 3D under the effect of environmental disturbances caused by wind forces and moments

$$X_{\text{wave}} = \frac{K_{w1}s}{s^2 + 2\lambda_1 \omega_{e1} s + \omega_{e1}^2} w_1 + d_1,$$

$$Y_{\text{wave}} = \frac{K_{w2}s}{s^2 + 2\lambda_2 \omega_{e2} s + \omega_{e2}^2} w_2 + d_2, \tag{3.146}$$

$$N_{\text{wave}} = \frac{K_{w3}s}{s^2 + 2\lambda_3 \omega_{e3} s + \omega_{e3}^2} w_3 + d_3,$$

where w_i are Gaussian white noise processes and d_i are drift processes modeled by Wiener processes according to $\dot{d}_i = w_{3+i}$, $i = 1, 2, 3$. Extra saturations can guarantee $|d_i| \leq d_{i,\max}$ for physical reality.

3.5.5.3 Ocean Current

Ocean currents are circulation systems of ocean waters produced by gravity, wind friction, water density variation and heat exchange. The oceans are divided into warm and cold water spheres and the Coriolis force try to turn the major currents to the East (northern hemisphere) or to the West (southern hemisphere). Additional effects are tidal components caused by planetary interactions like gravity. In coastal regions, tidal components can have high speeds.

Denote V_c the actual current speed whose direction is defined by the angle of attack α_c and the sideslip angle β_c. The velocity can be modeled by Gauss–Markov process $\dot{V}_c + \mu V_c = w$ where w is Gaussian white noise process. Specifically, for $\mu = 0$ it is a random walk process. Saturating elements can assure $V_{c,\min} \leq V_c(t) \leq V_{c,\max}$. The current direction can be fixed by α and β resulting in

$$\begin{pmatrix} u_c \\ v_c \\ w_c \end{pmatrix} = Rot(y, \alpha_c)^T Rot(z, -\beta_c)^T \begin{pmatrix} V_c \\ 0 \\ 0 \end{pmatrix}$$

$$= \begin{bmatrix} C_{\alpha_c} & 0 & -S_{\alpha_c} \\ 0 & 1 & 0 \\ S_{\alpha_c} & 0 & C_{\alpha_c} \end{bmatrix} \begin{bmatrix} C_{\beta_c} & -S_{\beta_c} & 0 \\ S_{\beta_c} & C_{\beta_c} & 0 \\ 0 & 0 & 1 \end{bmatrix} \begin{pmatrix} V_c \\ 0 \\ 0 \end{pmatrix}$$

$$= \begin{pmatrix} V_c C_{\alpha_c} C_{\beta_c} \\ V_c S_{\beta_c} \\ V_c S_{\alpha_c} C_{\beta_c} \end{pmatrix}. \tag{3.147}$$

For illustration, consider the dynamic model of a surface vessel in dynamic positioning, that is slowly maneuvering in sway and yaw satisfying $U \approx 0$, $C \approx 0$ and M, D are constant:

$$
\begin{bmatrix} m_{11} & m_{12} & 0 \\ m_{21} & m_{22} & 0 \\ 0 & 0 & 1 \end{bmatrix} \begin{bmatrix} \dot{v} \\ \dot{r} \\ \dot{\psi} \end{bmatrix} + \begin{bmatrix} d_{11} & d_{12} & 0 \\ d_{21} & d_{22} & 0 \\ 0 & -1 & 0 \end{bmatrix} \begin{bmatrix} v - v_c \\ r \\ \psi \end{bmatrix}
$$

$$
= \begin{bmatrix} b_1 \\ b_2 \\ 0 \end{bmatrix} \delta + \begin{bmatrix} Y_{\text{wind}} \\ N_{\text{wind}} \\ 0 \end{bmatrix} + \begin{bmatrix} Y_{\text{wave}} \\ N_{\text{wave}} \\ 0 \end{bmatrix},
$$

where δ is the rudder angle and v_c is the transverse current velocity. Substituting $v_c = V_c S_{\beta_c}$ and the Gauss–Markov model of V_c into the motion equations, the system with augmented current model is obtained:

$$
\begin{bmatrix} m_{11} & m_{12} & 0 & 0 \\ m_{21} & m_{22} & 0 & 0 \\ 0 & 0 & 1 & 0 \\ 0 & 0 & 0 & 1 \end{bmatrix} \begin{bmatrix} \dot{v} \\ \dot{r} \\ \dot{\psi} \\ \dot{V_c} \end{bmatrix} + \begin{bmatrix} d_{11} & d_{12} & 0 & -d_{11} S_{\beta_c} \\ d_{21} & d_{22} & 0 & -d_{21} S_{\beta_c} \\ 0 & -1 & 0 & 0 \\ 0 & 0 & 0 & -\mu \end{bmatrix} \begin{bmatrix} v \\ r \\ \psi \\ V_c \end{bmatrix}
$$

$$
= \begin{bmatrix} b_1 \\ b_2 \\ 0 \\ 0 \end{bmatrix} \delta + \begin{bmatrix} Y_{\text{wind}} \\ N_{\text{wind}} \\ 0 \\ 0 \end{bmatrix} + \begin{bmatrix} Y_{\text{wave}} \\ N_{\text{wave}} \\ 0 \\ 0 \end{bmatrix} + \begin{bmatrix} 0 \\ 0 \\ 0 \\ 1 \end{bmatrix} w. \tag{3.148}
$$

3.5.6 Kinematic Model

Since x, y, z are the coordinates of the ship positions in K_{NED} frame hence the velocity expressed in K_B can be transformed into velocity (the derivative of the position) in K_{NED} by using $A_{\text{NED},B} = Rot(z, \psi)Rot(y, \theta)Rot(x, \phi) = B^T =: R(\bar{\Phi})$ where $\bar{\Phi} = (\phi, \theta, \psi)^T$. Similarly, $(\dot{\phi}, \dot{\theta}, \dot{\psi})^T = F(\phi, \theta)\omega$ yields by (3.12). Therefore, introducing the Jacobian $J(\eta)$ of the ship, the kinematics can be obtained:

$$
J(\eta) = \begin{bmatrix} R(\bar{\Phi}) & 0 \\ 0 & F(\phi, \theta) \end{bmatrix}, \tag{3.149}
$$

$$
\dot{\eta} = J(\eta)v. \tag{3.150}
$$

The Jacobian matrix $J(\eta)$ is invertible if $\theta \neq \pm \pi/2$.

3.5.7 Dynamic Model in Body Frame

Collecting the previous results the 6D dynamic model in body frame can be written in the following form:

$$\dot{\eta} = J(\eta)v,$$

$$M\dot{v} + C(v)v + D(v)v + g(\eta) = \tau + g_0 + w. \tag{3.151}$$

Here, $\eta = (x, y, z, \phi, \theta, \psi)^T$ is the position and orientation, $v = (u, v, w, p, q, r)^T$ is the velocity and angular velocity, $M = M_{RB} + M_A$ is the mass matrix, $C(v) = C_{RB}(v) + C_A(v)$ is the matrix of centripetal and Coriolis effects, $D(v) = D + D_n(v)$ is the damping matrix consisting of linear and nonlinear parts, $g(\eta)$ is the vector of gravity and buoyancy effect, τ is the control force and torque, g_0 is the ballast force and torque, and finally w is the force and torque of environmental effects caused by wind, waves and current.

Important properties are $M = M^T > 0$ (symmetric and positive definite) in *ideal fluid* and $C(v) = -C(v)^T$ (skew symmetric). Hence, M is invertible and the quadratic form of C is identically zero in any variable. It should be noted that for *surface ships* moving with large speed in waves $M_A \neq M_A^T$.

The 12-dimensional state vector is $x = (\eta^T, v^T)^T$, the control input is τ, and the motion equations can be brought to the standard state space form $\dot{x} = f(x, u)$ because M is invertible. On the other hand $D > 0$, however $D \neq D^T$. The matrix $D(v)$ covers the potential damping, skin friction, wave drift damping and vortex shedding damping affects.

3.5.8 Dynamic Model in NED Frame

Avoiding singular configurations $\theta = \pm\pi/2$, the Jacobian matrix $J(\eta)$ is invertible and $v = J^{-1}(\eta)\dot{\eta}$. Hence $\dot{v} = J^{-1}(\eta)\ddot{\eta} - J^{-1}(\eta)\dot{J}(\eta)J^{-1}(\eta)\dot{\eta}$. By introducing the notations

$$M^*(\eta) = J^{-T}(\eta)MJ^{-1}(\eta),$$

$$C^*(\eta, \dot{\eta}) = J^{-T}(\eta)\left[C(v) - MJ^{-1}(\eta)\dot{J}(\eta)\right]J^{-1}(\eta)$$

$$= J^{-T}(\eta)\left[C(J^{-1}(\eta)\dot{\eta}) - MJ^{-1}(\eta)\dot{J}(\eta)\right]J^{-1}(\eta),$$

$$D^*(\eta, \dot{\eta}) = J^{-T}(\eta)D(v)J^{-1}(\eta) \tag{3.152}$$

$$= J^{-T}(\eta)D\left(J^{-1}(\eta)\dot{\eta}\right)J^{-1}(\eta),$$

$$g^*(\eta) = J^{-T}(\eta)g(\eta),$$

the motion equation can be written in the form

$$M^*(\eta)\ddot{\eta} + C^*(\eta, \dot{\eta})\dot{\eta} + D^*(\eta, \dot{\eta})\dot{\eta} + g^*(\eta) = J^{-T}(\eta)(\tau + g_0 + w). \tag{3.153}$$

Choosing $x^* = (\eta^T, \dot{\eta}^T)^T$ as state vector, and $u^* = J^{-T}(\eta)(\tau + g_0 + w)$ as input vector, then the kinematic and motion equations can be brought to the standard state space form $\dot{x}^* = f^*(x^*, u^*)$ by avoiding singular configurations $\theta = \pm\pi/2$.

3.6 Closing Remarks

The main goal of the chapter was to give an overview of the dynamic modeling of the different types of vehicles and robots in a unified approach.

First, the kinematic and dynamic models of the *rigid body* were derived which is the basis for further investigations. Two frames are considered, the inertial frame and the body frame. The differential equations for translational and rotational motion are formulated in body frame. General inertia matrix is assumed and for its inverse and the other coefficients in the differential equations closed formulas are given. The computation of the total force and total torque depends on the vehicle type, hence it was not discussed here. For the orientation of the body frame, relative to the inertial frame, two approaches are shown, one using Euler (RPY) angles and a second based on quaternion. The kinematic equations are differential equations describing the dependence of the orientation on the angular velocity. The quaternion description avoids the singularity in pitch angle in vertical pose.

For *industrial robots*, a forward recursive numerical algorithm has been formulated which determines the (linear and angular) velocities and accelerations of each robot link. For this purpose, such a parametrization is chosen which is compatible with the final form of the dynamic model. Based on this parametrization, it was possible to give a direct numerical method for the computation of the robot dynamic model by using Appell's equation. For the computation of the gravity effect, a backward recursive algorithm is given. Alternatively, the Lagrange's equation is used to find the dynamical model from the kinetic and potential energy based on partial differentiations which can be supported by symbolic computation tools. Formulas are presented which give the minimal number of centripetal and Coriolis terms needed in the dynamic model. As an illustration example, the dynamic model of the RRTR-type SCARA robot is presented in closed form for real-time implementation.

For *ground cars* moving in horizontal plane, the dynamic model has been derived by using Newton–Euler equations. The nonlinear dynamic model assumes that the transversal forces can be computed from the cornering stiffness and the tire side slip angle, while the control inputs are the steering angle and the rear wheel longitudinal force. The nonlinear dynamic model is not input affine which is a disadvantage for input/output linearization. However, for small (orientation, steering and wheel) angles the nonlinear dynamic model can be approximated by input affine nonlinear ones. Two input affine models are presented for later applications. One of them considers the front wheel transversal force as control input instead of the steering angle, however the steering angle can be computed from it for real-time implementation. In the second input affine model, the steering angle remains a control input. Since there are only two control inputs, while the degree of freedom is three, hence the car is underactuated.

For *airplanes*, first the usual navigation frames (ECI, ECEF, NED, BODY) are introduced. The position of a moving object can be parametrized by geodetic and rectangular coordinates, and formulas are presented for the conversion between them. For long distance flight, the motion equations are formulated in closed form by using quaternion description of orientation and the earlier results for rigid body. For short distance maneuvering the Flat-Earth Equations are presented, considering the NED frame as quasi-inertial frame, and using Euler (RPY) angles for orientation description. For the discussion of aerodynamic effects the wind-axes frame is introduced. Then the derivatives of the absolute value of the velocity, the angle of attack and the sideslip angle are determined which are important state variables. Conversion rules between wind-axes frame and body frame are given for the gravity effect and the aerodynamic forces and moments. For rotary engines the computation rule of gyroscopic effect is derived. The state equations of airplane are summarized in form of force equations (both in body and wind-axes frames), torque equations, kinematic equations, navigation equations and acceleration sensor equations. Finally, the main flying modes (Steady Wings-Level Flight etc.), the trimming and linearization principle and the concept of the parametrization of aerodynamic and trust forces for identification purposes are outlined. As an illustration example, a polynomial type parametrization of the aerodynamic and engine forces and moments is shown for the DHC-2 Beaver aircraft which is used in the FDC 1.4 Simulink toolbox.

For *surface and underwater ships*, first the rigid body equations are developed, then the hydrodynamic and restoring forces and moments are shown. Amongst them, the radiation induced forces and moments include the added mass due to the surrounding world. It can be modeled by the added mass inertia matrix and the hydrodynamic centrifugal and Coriolis term. Rigid body and added mass inertia matrices can be merged, and similarly rigid body and added mass centrifugal and Coriolis matrices. The first is positive definite, while the second is skew symmetric, which may be important in stability investigations of ship control algorithms. The total damping matrix is caused by potential, viscous and skin effects, and can be divided into constant linear and velocity dependent nonlinear terms. Restoring forces are caused by gravity and buoyancy forces. For their modeling formulas are given both for underwater ships and surface vessels. In dynamic positioning ballast tanks can play an important role. For ballast trimming, a simple strategy exists which can be realized by PID-type feedback. A short introduction is given for wind, wave and current models. These environmental effects are disturbances, their models are important for testing ship control algorithms using simulation. Finally, the results are summarized in the kinematic and the dynamic models of ships, both in body frame and NED frame.

Chapter 4
Nonlinear Control of Industrial Robots

Overview This chapter gives a brief survey of the most important control methods of industrial robots. From the control methods not using the dynamic model of the robot, first the decentralized three-loop cascade control is presented. This classical approach tries to apply linear control methods but its precision and allowable speed is limited. Then advanced control methods follow based on the robot dynamic model for high precision applications. The computed torque technique is practically nonlinear decoupling in joint space. In Cartesian space, first the transformation of sensory information and the realization of spatial generalized force by joint torques is discussed. It is followed by the nonlinear decoupling of the free motion in Cartesian space. The results of free motion are generalized in hybrid position and force control by using the operational space method. For the online identification of the robot parameters, the self-tuning adaptive control method is suggested. For the realization of nonsmooth path with sharp corners, a robust backstepping method is presented. The methods are illustrated in simulation results.

4.1 Decentralized Three-Loop Cascade Control

For relatively small load and speed, the results of classical linear control theory can be applied. The method considers the nonlinear effects as disturbance signal whose influence has to be decreased or eliminated. However, for larger load and quick motion, the precision is deteriorating and the control has to be succeeded by advanced control techniques.

4.1.1 Dynamic Model of DC Motor

For simplicity, consider a motor whose stator consists of a permanent magnet and the rotor is rotating in homogeneous magnetic field, see Fig. 4.1.

B. Lantos, L. Márton, *Nonlinear Control of Vehicles and Robots*,
Advances in Industrial Control,
DOI 10.1007/978-1-84996-122-6_4, © Springer-Verlag London Limited 2011

Fig. 4.1 Simplified sketch of
DC motor

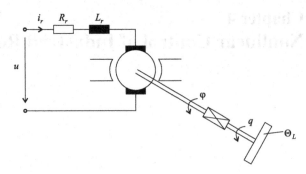

Denote R_r and L_r the resistance and inductivity of the rotor, respectively, and
let i_r be the rotor current and u the input voltage. By Lenz's induction law, there
is a back emf $c_1\dot\varphi$ which acts opposite to the input voltage of the motor where $\dot\varphi$ is
the angular velocity of the motor. On the other hand, by Biot–Savart's law, a force
proportional to the rotor current comes into being which causes a torque $\tau_m = c_2 i_r$
on the motor axis. Here, c_1, c_2 are motor constants. Let Θ_r be the inertia moment
of the rotor belonging to the motor axis. The robot can be assumed as a load inertia
Θ_L driven by the motor through a gear described by the gear reduction v which
decreases the angular velocity of the motor. At the link side after the gear reduction,
let $\dot q$ be the angular velocity, then $v = \dot\varphi/\dot q$. If the gear is lossless, then the powers
are equal on both sides of the gear reduction. Since the power N is the product
of torque and angular velocity, hence $N_m = N \Rightarrow \tau_m \dot\varphi = \tau \dot q \Rightarrow \tau = v\tau_m$. Notice
that while the gear reduction decreases the angular velocity at the load side, it also
increases the torque at the load side.

Suppose the loss is caused by viscous friction proportional to the angular velocity. Then the friction torque at the motor side is $f_m\dot\varphi$, while the friction torque
at the load side (reduced after the gear) is $f_{\mathrm{gear}}\dot q$. Assume the load is variable and
can be taken into consideration by a disturbance signal τ^*. Based on the above, the
dynamic behavior of the DC motor can be described by the following equations:

$$\Theta_L \ddot q + f_{\mathrm{gear}} \dot q + \tau^* + v\Theta_r \ddot\varphi + v f_m \dot\varphi = v c_2 i_r,$$

$$R_r i_r + L_r \frac{di_r}{dt} = u - c_1 \dot\varphi. \tag{4.1}$$

Dividing the first equation by v and transforming the variables to the motor side
($\dot q = \dot\varphi/v, \ddot q = \ddot\varphi/v$), then

$$\left(\Theta_r + \frac{\Theta_L}{v^2}\right)\ddot\varphi + \left(f_m + \frac{f_{\mathrm{gear}}}{v^2}\right)\dot\varphi = c_2 i_r - \frac{\tau^*}{v}. \tag{4.2}$$

By introducing the notations

$$\Theta := \Theta_r + \Theta_L/v^2 \quad \text{and} \quad f := f_m + f_{\mathrm{gear}}/v^2 \tag{4.3}$$

Fig. 4.2 Block scheme of
DC motor

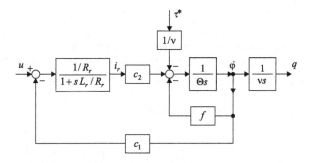

the dynamic model of DC motor can be brought to the following form (u is input, φ
is output, and τ^* is disturbance):

$$\Theta\ddot{\varphi} + f\dot{\varphi} = c_2 i_r - \frac{\tau^*}{\nu},$$

$$R_r i_r + L_r \frac{di_r}{dt} = u - c_1\dot{\varphi}. \tag{4.4}$$

Laplace-transforming and introducing the operator impedance $R_r + sL_r$ the
block scheme of the DC motor is obtained, see Fig. 4.2.

Let $\tilde{x} = (\varphi, \dot{\varphi}, i_r)^T$ be the state vector, $y = \varphi$ is the output signal, while the useful
input is the motor voltage u and the disturbance input is τ^*, i.e. $\tilde{u} = (u, \tau^*)^T$, then
it follows

$$\frac{d\tilde{x}}{dt} = \begin{bmatrix} 0 & 1 & 0 \\ 0 & -\frac{f}{\Theta} & \frac{c_2}{\Theta} \\ 0 & -\frac{c_1}{L_r} & -\frac{R_r}{L_r} \end{bmatrix} \tilde{x} + \begin{bmatrix} 0 & 0 \\ 0 & -\frac{1}{\nu\Theta} \\ \frac{1}{L_r} & 0 \end{bmatrix} \tilde{u} = A\tilde{x} + B\tilde{u},$$

$$y = \begin{bmatrix} 1 & 0 & 0 \end{bmatrix} \tilde{x} = C\tilde{x}. \tag{4.5}$$

The characteristic polynomial of the DC motor is

$$\det(sI - A) = \begin{vmatrix} s & -1 & 0 \\ 0 & s + (f/\Theta) & -c_2/\Theta \\ 0 & c_1/L_r & s + (R_r/L_r) \end{vmatrix}$$

$$= s\left[s^2 + \frac{fL_r + \Theta R_r}{\Theta L_r} s + \frac{fR_r + c_1 c_2}{\Theta L_r} \right],$$

the eigenvalues (poles) are

$$s_{1,2} = \frac{-(fL_r + \Theta R_r) \pm \sqrt{(fL_r + \Theta R_r)^2 - 4(fR_r + c_1 c_2)\Theta L_r}}{2\Theta L_r}$$

$$= \frac{-(T_m + T_e) \pm \sqrt{(T_m + T_e)^2 - 4(1 + \frac{c_1 c_2}{R_r f})T_m T_e}}{2T_m T_e},$$

$s_3 = 0$.

Here, $T_m = \Theta/f$ denotes the mechanical and $T_e = L_r/R_r$ the electrical time constants, respectively. Notice that c_1, c_2 are construction data of the motor and the eigenvalues (poles) may also be conjugate complex.

Let us introduce the notation $a := \frac{c_2/\Theta}{\det(sI - A)}$ then the transfer function between input voltage and angle output is

$$
G_{\varphi,u} = \begin{bmatrix} 1 & 0 & 0 \end{bmatrix} (sI - A)^{-1} \begin{bmatrix} 0 \\ 0 \\ 1/L_r \end{bmatrix}
$$

$$
= \begin{bmatrix} 1 & 0 & 0 \end{bmatrix} \begin{bmatrix} * & * & a \\ * & * & * \\ * & * & * \end{bmatrix} \begin{bmatrix} 0 \\ 0 \\ 1/L_r \end{bmatrix}
$$

$$
= \begin{bmatrix} * & * & a \end{bmatrix} \begin{bmatrix} 0 \\ 0 \\ 1/L_r \end{bmatrix} = \frac{a}{L_r}
$$

$$
\Rightarrow \quad G_{\varphi,u} = \frac{\frac{c_2}{\Theta L_r}}{\det(sI - A)}. \tag{4.6}
$$

It follows that the transfer function of the DC motor for angular velocity output $\dot{\varphi}$ and voltage input u yields

$$
G_{\dot{\varphi},u} = \frac{\frac{c_2}{\Theta L_r}}{(s - s_1)(s - s_2)} = \frac{A}{1 + 2\xi T s + T^2 s^2}, \tag{4.7}
$$

which leads to a second order $T\xi$ system, with to time constants if $\xi \geq 1$, or a damped system with conjugate complex poles if $\xi < 1$. In the practice both cases are possible, although only the first case will be considered in the sequel.

4.1.2 Design of Three-Loop Cascade Controller

The three-loop cascade control consists of internal current control (i), intermediate velocity control (revolution n) and external position control (rotation angle q), see Fig. 4.3. We will not discuss the power electronic realization, however it will be assumed that the current control is based on pulse width modulation (PWM) whose chopper frequency is at least 10 kHz. Hence, the effect of PWM can be neglected during the compensation if the transient time of the current control will not be made quicker than 0.5 ms.

During the design analog controllers will be assumed, but later the external position controller will be substituted by a sampled data controller (Direct Digital Controller, DDC). The classical results of linear control theory (crossover frequency,

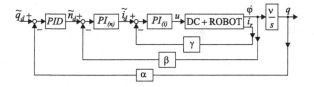

Fig. 4.3 Structure of three loops cascade controller

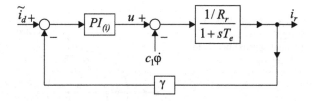

Fig. 4.4 Approximating method for current control design

phase margin etc.) will be applied. Since the analog controllers are using operational amplifiers, hence, the sensors provide voltage outputs denoted by "tilde": $\tilde{i} = \gamma i_r$, $\tilde{n} = \beta n$, $\tilde{q} = \alpha q$. The realization of the sampled data PID controller will be shown later.

4.1.2.1 Design of Internal Current Loop

The design of the PI controller of the current loop will be performed as if the back emf would be "disturbance". By this assumption, the goal of the PI controller is to speed up the first order system with one time constant which is an easy problem, see Fig. 4.4. This is of course only an approximation because back emf is one of the fundamentals of DC motor principles. The behavior of the closed loop with the designed controller, but without the approximation, will be investigated later.

Using the above assumption, the open loop transfer function is

$$G_{i0}(s) = \frac{A_{pi}}{T_{Ii}} \cdot \frac{1 + sT_{Ii}}{s} \cdot \frac{1/R_r}{1 + sT_e} \cdot \gamma, \qquad (4.8)$$

where $T_e = L_r/R_r$ is the electrical time constant. It is typical to choose $T_{Ii} := T_e$ so that the open loop becomes an integrator and the closed loop is a first order system with one time constant:

$$G_{i0}(s) = \frac{A_{pi}\gamma}{T_{Ii}R_r} \cdot \frac{1}{s} =: \frac{K_i}{s},$$

$$G_i(s) = \frac{1/\gamma}{1 + s/K_i} =: \frac{1/\gamma}{1 + sT_i} = \frac{1/\gamma}{1 + sa_iT_e}. \qquad (4.9)$$

In order to speed up the system, $T_i := a_iT_e < T_e$ should be satisfied, where $a_i < 1$, however the speeding up is limited by the PWM chopper frequency. The gain of the PI controller is $A_{pi} = \frac{R_r}{a_i\gamma}$.

Fig. 4.5 Block scheme of
current control with the
designed PI controller
without approximation

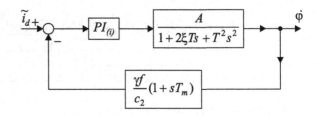

Fig. 4.6 Root locus of
PI-type current control
without approximation for
velocity output

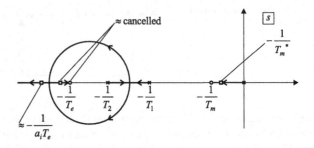

Consider the current control with the designed PI controller, but now without the above approximation. Since by Fig. 4.2

$$\dot{\varphi} = c_2 \frac{1/f}{1+sT_m} i_r \quad \Rightarrow \quad i_r = \frac{1+sT_m}{c_2/f}, \tag{4.10}$$

hence, choosing $\dot{\varphi}$ output, the internal loop can be brought to the form of Fig. 4.5, where the $T\xi$ term is the transfer function of the DC motor.

The root locus is shown in Fig. 4.6 for varying gain. Assume the approximating design results in large gain then the closed loop poles are at the places denoted by squares on the root locus. On the other side, $-1/T_e$ is a zero of the closed loop as well (since it is in the forward branch), while $-1/T_m$ is not a zero of the closed loop (since it is in the feedback branch). Hence, the zero at $-1/T_e$ and the pole (marked with square) in its neighborhood are almost canceling and can be neglected. The pole at the asymptote going to infinity along the negative real axis (marked also with square) will be $\approx -\frac{1}{a_i T_e}$, which is the above designed pole. The zero at $-1/T_m$ is not a closed loop zero, therefore the pole $-1/T_m^*$ is not canceled (marked also with square), and it will be the dominating pole of the closed loop for angular velocity output.

Summarizing the above, the resulting inner loop can be well approximated by the transfer function

$$G_{\dot{\varphi}\tilde{i}_d}(s) \approx \frac{A_i}{(1+sT_m^*)(1+sT_i)}, \quad A_i := \frac{c_2}{\gamma f}. \tag{4.11}$$

4.1.2.2 Design of Intermediate Velocity Loop

For the revolution control of the intermediate loop, a PI controller has to be designed for a second order system with two time constants which is also an easy problem. The open loop and closed loop transfer functions are respectively,

$$G_{n0}(s) \approx \frac{A_{pn}}{T_{In}} \cdot \frac{1 + sT_{In}}{s} \cdot \frac{A_i}{(1 + sT_m^*)(1 + sT_i)} \cdot \beta = \frac{K_n}{s(1 + sT_i)},$$

$$G_n(s) \approx \frac{A_n}{(1 + sT_1)(1 + sT_2)}, \quad A_n = \frac{1}{\beta}.$$

(4.12)

The characteristic equation of the closed loop is $1 + G_{n0}(s) = 0 \Rightarrow s^2 T_i + s + K_n = 0$. Setting aperiodic limit case it follows

$$s_{1,2} = \frac{-1 \pm \sqrt{1 - 4K_n T_i}}{2T_i} \quad \Rightarrow \quad K_n = \frac{1}{4T_i} \quad \Rightarrow \quad T_{1,2} = 2T_i.$$

4.1.2.3 Design of External Position Control Loop

First, an analog PID controller will be designed then it will be substituted by its approximation by a discrete time (sampled data) PID controller. However, the discrete time PID controller is followed by a D/A converter (DAC) whose gain should be taken into consideration in the open loop transfer function. Suppose the DAC converter has voltage domain $\pm U_{DA}$ and DAbits number of bits. Then the gain of the DAC converter is

$$A_{DA} = \frac{U_{DA}}{2^{DAbits-1}}.$$

(4.13)

The open loop transfer function of the position control is the product of the transfer functions of the PID controller, the gain of the DAC converter, the resulting transfer function of the two internal loops, the integral term $\frac{1}{vs}$ and the sensor transfer coefficient α:

$$G_0(s) = \frac{A_p}{T_I} \cdot \frac{(1 + s\tau_1)(1 + s\tau_2)}{s(1 + sT)} \cdot A_{DA} \cdot \frac{A_n}{(1 + sT_1)(1 + sT_2)} \cdot \frac{1}{vs} \cdot \alpha,$$

$$G_0(s) = K \frac{(1 + s\tau_1)(1 + s\tau_2)}{s^2(1 + sT_1)(1 + sT_2)(1 + sT)},$$

(4.14)

$$K := \frac{A_p A}{T_I}, \quad A := \frac{A_{DA} A_n \alpha}{v}.$$

Suppose $T_1 \geq T_2$ and (because of aperiodic limit case at the previous level) $T_1 < 10T_2$. The goal is to assure phase margin $\varphi \geq 75°$ which is suggested for stable real poles of the closed loop. Stable real poles eliminate oscillations in the closed loop transients. For large phase margin, one controller zero should be placed in the low frequency domain while the second zero can cancel the pole belonging to T_1. Hence,

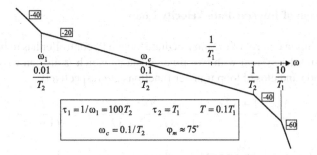

Fig. 4.7 Open loop Bode diagram of PID-type position control

the approximated amplitude function of the open loop Bode diagram in Fig. 4.7 can be chosen with the controller parameters τ_1, τ_2 and ω_c given there, which assures the desired large phase margin:

$$\varphi_t = 180° + \varphi(\omega_c) \approx 180° - 90° - 90° + 85° - 5° - 5° \approx 75°. \qquad (4.15)$$

The integration and differentiation times of the PID controller can be determined by

$$\tau_1 + \tau_2 = T_I + T \quad \Rightarrow \quad T_I = \tau_1 + \tau_2 - T,$$
$$\tau_1 \tau_2 = T_I(T_D + T) \quad \Rightarrow \quad T_D = \frac{\tau_1 \tau_2}{T_I} - T. \qquad (4.16)$$

The controller gain can be determined according to

$$\frac{A_1}{1} = \frac{\omega_c}{\omega_1} \quad \Rightarrow \quad A_1 = \omega_c/\omega_1,$$
$$\frac{K}{\omega_1^2} = A_1 \quad \Rightarrow \quad K = A_1 \omega_1^2 = \omega_c \omega_1 = \frac{A_p A}{T_I} \quad \Rightarrow \quad A_p = \frac{\omega_c \omega_1 T_I}{A}. \qquad (4.17)$$

The root locus of the entire control system is shown in Fig. 4.8. It can be seen that because of $\varphi_t \approx 75°$ the closed loop will contain also an almost canceling low frequency P/Z pair causing a slowly decaying transient such that the step response of the closed loop contains a small overshoot, although the closed loop poles are all real and there are no oscillations. However the robot path is rarely step-form, much more continuous, so that this effect may be also an advantage for robot control. The two integrators in the open loop assure asymptotically error-less tracking not only for constant but also for linear sections of the path.

The continuous time PID controller $G_{PID}(s)$ can be easily converted to discrete time PID controller $D_{PID}(z^{-1})$ saving the step response of the analog controller at

Fig. 4.8 Root locus of three loop cascade control with the PID-type position controller

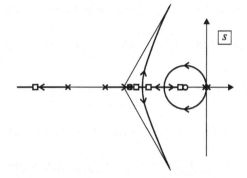

the integer multiples of the sampling time T_S (step response equivalence):

$$G_{\text{PID}}(s) = A_p\left(1 + \frac{1}{T_I s} + \frac{sT_D}{1+sT}\right)$$

$$\Rightarrow \quad D_{\text{PID}}(z^{-1}) = \frac{q_0 + q_1 z^{-1} + q_2 z^{-2}}{p_0 + p_1 z^{-1} + p_2 z^{-2}}, \tag{4.18}$$

$$q_0 = A_p\left(1 + \frac{T_D}{T}\right), \quad p_0 = 1,$$

$$q_1 = -A_p\left(1 + e^{-T_S/T} - \frac{T_S}{T_I} + \frac{2T_D}{T}\right), \quad p_1 = -\left(1 + e^{-T_S/T}\right), \tag{4.19}$$

$$q_2 = A_p\left[\left(1 - \frac{T_S}{T_I}\right)e^{-T_S/T} + \frac{T_D}{T}\right], \quad p_2 = e^{-T_S/T},$$

$$u_k := -p_1 u_{k-1} - p_2 u_{k-2} + q_0 e_k + q_1 e_{k-1} + q_2 e_{k-2}. \tag{4.20}$$

Here, e_k is the current input of the controller (the current error) and u_k is the current output of the controller (the current actuator input). The previous values should be stored in the controller memory. Never forget that before sending the controller output to the DAC converter, it should be tested for saturation and corrected if it would leave the input domain of DAC. For example, if the DAC has 12 bits then $u_K := 2048$ if $u_k \geq 2048$, and similarly $u_K := -2048$ if $u_k \leq -2048$.

4.1.3 Approximation of Load Inertia and Disturbance Torque

In robotic applications, the joint torques should cover the torques needed to the robot motion. If the robot path $q_d(t)$ is known, then it is useful to compute an average effective inertia along the path according to $\bar{D}_{ii} := \overline{D_{ii}(q_d(t))}$ which can play the role of $\Theta_{L,i}$. Notice that the effective inertia is positive because the diagonal elements of a positive definite matrix are always positive. Hence, to take the average is a clever conception. On the other hand, using Lagrange equations, we can write

$$\sum_j D_{ij}(q)\ddot{q}_j + \sum_j \sum_k D_{ijk}(q)\dot{q}_j \dot{q}_k + D_i(q) + v_i^2 \Theta_{r,i} \ddot{q}_i + v_i^2 f_{r,i} \dot{q}_i + f_{\text{gear},i} \dot{q}_i$$

$$= v_i c_{2i} i_{ri},$$

$$\bar{D}_{ii} \ddot{q}_i + v_i^2 \Theta_{r,i} \ddot{q}_i + v_i^2 f_{r,i} \dot{q}_i + f_{\text{gear},i} \dot{q}_i + \tau_i^* = v_i c_{2i} i_{ri},$$

from which follows that the disturbance torque yields

$$\tau_i^* := \sum_j D_{ij}(q)\ddot{q}_j + \sum_j \sum_k D_{ijk}(q)\dot{q}_j \dot{q}_k + D_i(q) - \bar{D}_{ii} \ddot{q}_i. \tag{4.21}$$

The results may be useful for testing three loop cascade control by using simulation.

4.2 Computed Torque Technique

The method of computed torque technique (CTT) is a nonlinear control method [8]. The controller consists of a centralized part intensively using the nonlinear dynamic model of the robot, and a decentralized part which is the desired joint acceleration plus PID correction for the error. If the model is exactly known, then the method realizes nonlinear decoupling in joint space. As will be shown, the centralized algorithm without PID correction is a kind of input/output linearization in joint space, that is, the system breaks up to decoupled double integrators without zero dynamics.

Robot: $H(q)\ddot{q} + h(q,\dot{q}) = \tau,$ (4.22)

Nonlinear controller: $\tau := H(q)u + h(q,\dot{q}),$ (4.23)

Closed loop: $H(q)\ddot{q} + h(q,\dot{q}) = H(q)u + h(q,\dot{q}), \quad \exists H^{-1},$ (4.24)

$\ddot{q} = u \quad \Leftrightarrow \quad \ddot{q}_i = u_i$ (decoupled double integrators),   (4.25)

$$u_i := \ddot{q}_{d,i} + k_{Pi}(q_{di} - q_i) + k_{Ii} \int_0^t (q_{di} - q_i)\, d\vartheta + k_{Di}(\dot{q}_{di} - \dot{q}_i). \tag{4.26}$$

Denoting by $e_i = q_{di} - q_i$ the error of the i-th joint control subsystem, then it follows from $\ddot{q}_i = u_i$ and (4.26)

$$\ddot{e}_i + k_{Di}\dot{e}_i + k_{Pi}e_i + k_{Ii} \int_0^t e_i\, d\vartheta = 0 \quad \Rightarrow \quad \dddot{e}_i + k_{Di}\ddot{e}_i + k_{Pi}\dot{e}_i + k_{Ii}e_i = 0. \tag{4.27}$$

Hence, the closed loop characteristic equation is

$$s^3 + k_{Di}s^2 + k_{Pi}s + k_{Ii} = 0, \tag{4.28}$$

and the closed loop is stable if the roots have negative real parts.

It is an easy way to find appropriate PID parameters. Assume the characteristic equation has three equal real roots:

$$(1 + sT)^3 = 1 + 3sT + 3s^2T^2 + s^3T^3 = 0 \quad \Leftrightarrow \quad s^3 + \frac{3}{T}s^2 + \frac{3}{T^2}s + \frac{1}{T^3} = 0.$$
(4.29)

Comparing the coefficients of the characteristic equations results in following choice of the controller parameters:

$$k_{Pi} := \frac{3}{T^2}, \qquad k_{Ii} := \frac{1}{T^3}, \qquad K_{Di} := \frac{3}{T}. \tag{4.30}$$

The time constant T of the closed loop can be chosen depending on the limit of the actuator signals (rotor current is proportional to driving torque). For example, in case of a 6-DOF robot for the more robust first three joints $T = 50$ ms can be chosen, and for the last three joints with smaller mass and inertia the value $T = 25$ ms can be suggested. Using simulation, it can be checked whether the maximal allowed rotor current is approached, in which case T should be increased before implementation. The decentralized PID controllers can be extended with integrator antiwindup.

In the realization, it should be taken into consideration that the robot and the model may be different. Often the load is unknown or there are nonmodeled nonlinearities. The available robot model $\hat{H}(q)\ddot{q} + \hat{h}(q,\dot{q}) = \tau$ may be different from the real robot, furthermore the control law can use only the available model:

Centralized control law: $\tau := \hat{H}(q)u + \hat{h}(q,\dot{q}),$ (4.31)

Decentralized control law: $u_i := \ddot{q}_{di} + \text{PID}.$ (4.32)

As a consequence, the closed loop system is only approximately decoupled:

$$H(q)\ddot{q} + h(q,\dot{q}) = \hat{H}(q)u + \hat{h}(q,\dot{q}), \tag{4.33}$$

$$\ddot{q} = H^{-1}(q)\big(\hat{h}(q,\dot{q}) - h(q,\dot{q})\big) + H^{-1}(q)\hat{H}(q)u. \tag{4.34}$$

Notice that the desired joint torque, computed by the centralized control law, can be realized by a low level current control loop after conversion of the torque to the dimension of the reference signal of the current control. For this purpose, the internal current control loop of the three loop cascade control can also be used, the other loops can be switched out or omitted.

4.3 Nonlinear Decoupling in Cartesian Space

Although we want to discuss here the nonlinear decoupling of the free motion, the control philosophy will be formulated in the space, hence transformation rule is needed how to implement it at the level of joint torques, since the actuators belong to the joints. From didactic point of view, we shall deal also here with transformation of the force/torque sensory information, which will be used later in hybrid position and force control.

If not only the motion of the robot but also the force and torque in the contact point between robot and environment have to be controlled then a force/torque sensor is needed. A typical six-component force/torque sensor is built in between the last link and the end effector (tool, gripper etc.), that is, at a place different from the tool center point (TCP), for which we usually formulate the control algorithm. On the other hand, in case of spatial control the control algorithm designs the force and torque to be exerted in the space, but their realization happens at the level of the joints, hence a transformation is needed between the desired generalized force at space level and the torque to be exerted at joint level.

4.3.1 Computation of Equivalent Forces and Torques

Suppose the force/torque sensor measures f_s force and m_s torque (moment) vectors in its own K_s sensor frame and we want to determine the equivalent force and torque in the tool center point (TCP) or in the contact point. The two cases are similar, since the contact point can be considered to be the TCP of a logical tool where the physical tool is augmented with the object. Hence, we consider only the TCP case.

Place a zero force in the form of $+f_s$ and $-f_s$ at the TCP. Denote p_{sE} the vector from the origin of K_s to the origin K_E of the end effector, that is, to the TCP. Then a force f_s appears at the TCP and the moment is increased by the moment of the couple of forces defined by $+f_s$ at the origin of K_s and $-f_s$ at the origin of K_E. The arm between them with appropriately chosen φ is $|p_{sE}|\cos(\varphi) = |p_{sE}|\sin(90°+\varphi)$, hence the moment vector of the couple of forces is $f_s \times p_{sE}$.

Basis independent form:

$$f_E = f_s,$$
$$m_E = m_s + f_s \times p_{sE}.$$

(4.35)

Basis dependent form:

$$K_s \xrightarrow{T_{s,E}} K_E,$$
$$f_s \qquad f_E = A_{sE}^T f_s,$$
$$m_s \qquad m_E = A_{sE}^T (m_s + f_s \times p_{sE}).$$

(4.36)

After simple manipulations, we obtain:

$$m_E = A_{sE}^T m_s - A_{sE}^T [p_{sE} \times] f_s = A_{sE}^T m_s + A_{sE}^T [p_{sE} \times]^T f_s$$
$$= A_{sE}^T m_s + \left([p_{sE} \times] A_{sE}\right)^T f_s,$$

from which the following *transformation rule* yields:

$$\begin{pmatrix} f_E \\ m_E \end{pmatrix} = \begin{bmatrix} A_{sE}^T & 0 \\ ([p_{sE}\times]A_{sE})^T & A_{sE}^T \end{bmatrix} \begin{pmatrix} f_s \\ m_s \end{pmatrix}. \tag{4.37}$$

The transformation rule can be generalized for the case of any two frames K_1, K_2 and the homogeneous transformation $T_{1,2}$ between them, if the point of attack of the force is changed:

$$\begin{pmatrix} f_2 \\ m_2 \end{pmatrix} = \begin{bmatrix} A_{12}^T & 0 \\ ([p_{12}\times]A_{12})^T & A_{12}^T \end{bmatrix} \begin{pmatrix} f_1 \\ m_1 \end{pmatrix}. \tag{4.38}$$

If the point of attack of the force is not changed, then there is no couple of forces and only A_{12}^T appears in the block-diagonal caused by the coordinate transformation (change of the basis).

4.3.2 Computation of Equivalent Joint Torques

Concentrate the force/torque specified in TCP in a generalized force F, the searched equivalent joint torque in τ, and the position and the orientation (according to Rodriguez formula) in the vector x, respectively:

$$F = \begin{pmatrix} f_E \\ m_E \end{pmatrix} \in R^6, \qquad \tau = \begin{pmatrix} \tau_1 \\ \tau_2 \\ \vdots \\ \tau_m \end{pmatrix} \in R^m, \qquad x = \begin{pmatrix} p \\ t\varphi \end{pmatrix} \in R^6. \tag{4.39}$$

Let J be the Jacobian matrix of the robot from K_0 to K_E, then, by using $\dot{x} = J\dot{q} \Rightarrow \delta x = J\delta q$ and assuming equivalent work at space level and at joint level, it follows

$$\langle F, \delta x \rangle = \langle \tau, \delta q \rangle = \langle F, J\delta q \rangle = \left(J^T F, \delta q \right). \tag{4.40}$$

Hence, the equivalent joint torque which belongs to the generalized force F in Cartesian space yields:

$$\tau = J^T F. \tag{4.41}$$

4.3.3 Robot Dynamic Model in Cartesian Space

The robot dynamic model in joint variables is $H\ddot{q} + h = \tau$. Let F be the specified force/torque at the TCP. Denote J the robot Jacobian to the TCP. Perform the following conversions and meanwhile introduce the vector α in order to simplify the notations:

$$\dot{x} = J\dot{q},$$

$$\ddot{x} = J\ddot{q} + \frac{dJ}{dt}\dot{q} = J\ddot{q} + \alpha,$$

$$\ddot{q} = J^{-1}\left(\ddot{x} - \frac{dJ}{dt}\dot{q}\right) = J^{-1}(\ddot{x} - \alpha),$$

$$HJ^{-1}(\ddot{x} - \alpha) + h = J^T F,$$

$$J^{-T}HJ^{-1}\ddot{x} + J^{-T}h - J^{-T}HJ^{-1}\alpha = F.$$

By introducing the notations

$$H^* := J^{-T}HJ^{-1}, \qquad h^* := J^{-T}h - H^*\alpha, \tag{4.42}$$

the robot dynamic model in Cartesian space is obtained in the following form:

$$H^*\ddot{x} + h^* = F. \tag{4.43}$$

The kinetic energy of the robot remains further on the quadratic form of H^* but in the variable $\dot{x}$, therefore H^* is positive definite and hence $\exists(H^*)^{-1}$ if the robot Jacobian is invertible.

4.3.4 Nonlinear Decoupling of the Free Motion

The control law will be developed in Cartesian space, while it will be realized through the joint torque (rotor current). The method is similar to the method of computed torque technique, but now the decoupling is performed in Cartesian space. It will be assumed that $\exists H^{*-1}$. This method will be the basis also for hybrid position/force control after appropriate corrections.

Robot in Cartesian space: $\qquad H^*\ddot{x} + h^* = F.$ $\qquad\qquad$ (4.44)

Nonlinear controller: $\qquad F := H^*u^* + h^*.$ $\qquad\qquad$ (4.45)

Closed loop: $\qquad H^*\ddot{x} + h^* = H^*u^* + h^*.$ $\qquad\qquad$ (4.46)

$$\ddot{x} = u^* \quad \Leftrightarrow \quad \ddot{x}_i = u_i^*, \quad i = 1,\ldots,6. \tag{4.47}$$

It can be seen that the compensated system is decomposed into six double integrators. The six directions are respectively the translations and the rotations around the axes of the base coordinate system, moreover these motions are decoupled in Cartesian space. Since the system has to be moved along a prescribed path in the space and should remain robust against perturbations of the load and the system parameters, hence the internal signals can be generated by the following decentralized controllers:

Decentralized controllers:

$$u_i^* = \ddot{x}_{di} + PID = \ddot{x}_{di} + k_{Pi}(x_{di} - x_i) + k_{Ii}\int(x_{di} - x_i)\,dt + k_{Di}(\dot{x}_{di} - \dot{x}_i). \quad (4.48)$$

Notice that during path design not only the path but also the first and second derivatives are planed. Hence, the signals $x_{di}, \dot{x}_{di}, \ddot{x}_{di}$ needed for the decentralized controllers are available, while the values of $x_i, \dot{x}_i$ are either measured by the sensors or computed from joint level sensory information based on the kinematic (geometry and velocity) model of the robot.

The choice of the decentralized controllers still remains to be done. We can set out from

$$\ddot{x}_i = u_i^* = \ddot{x}_{di} + k_{Pi}(x_{di} - x_i) + k_{Ii}\int(x_{di} - x_i)\,dt + k_{Di}(\dot{x}_{di} - \dot{x}_i), \quad (4.49)$$

from which it follows after reordering and differentiation

$$(x_{di} - x_i)''' + k_{Di}(x_{di} - x_i)'' + k_{Pi}(x_{di} - x_i)' + k_{Ii}(x_{di} - x_i) = 0. \quad (4.50)$$

Therefore, the system is a nonexcited linear system in the error signal $e_i = (x_{di} - x_i)$ whose characteristic equation is

$$s^3 + k_{Di}s^2 + k_{Pi}s + k_{Ii} = 0, \quad (4.51)$$

which has to be stabilized and sped up.

We can choose a reference model $(1 + sT)^3 = 0$ whose time constant T can be set to the necessary operation speed. For translations larger ($T = 50$ ms) while for rotations smaller ($T = 25$ ms) time constants can be suggested. The details are similar to computed torque technique.

Although the control law was developed in Cartesian space, it should be implemented at joint level taking also into consideration that only the nominal model of the robot is available.

Implementation:

$$\tau = J^T F = J^T(\hat{H}^* u^* + \hat{h}^*) = \hat{H}J^{-1}u^* + \hat{h} - HJ^{-1}\alpha$$

$$\Rightarrow \quad \tau = \hat{H}J^{-1}(u^* - \alpha) + \hat{h}. \quad (4.52)$$

Notice that $\alpha = \frac{dJ}{dt}\dot{q}$, which can also be computed as $(\Phi_E^T, \Theta_E^T)^T$ by using the results of the forward kinematics and simple corrections if $K_6 \neq K_E$.

4.4 Hybrid Position and Force Control

Let us consider the typical case as a prescribed motion in a subspace has to be realized while in the complementary subspace a prescribed force has to be exerted.

The method discussed here is originated from Khatib [63]. Other approaches can be found in [156]. As an example, consider the case as a character has to be written on a blackboard with a chalk, and at the same time a force has to be performed orthogonal to the surface of the blackboard, in order not to come off the blackboard and write in the air if there is a control error in the motion, however the force cannot be too large otherwise the chalk will break.

4.4.1 Generalized Task Specification Matrices

Suppose the end effector gripped an object and in the TCP a prescribed force f_d has to be exerted while a motion has to be realized in the space orthogonal to the force. Since the f_d direction is general in the K_o operation space (i.e., in the world frame), and similarly the complement space of motion, too, hence the method will be called operation space method. In the space orthogonal to f_d the motion, while in f_d direction the force will be controlled.

The task can be formalized in the following way. Choose a K_f frame whose origin is in the TCP but its z_f-axis is parallel to f_d, i.e. $(z_f \| f_d)$. Define a diagonal position specification matrix Σ_f such that

$$\Sigma_f = \begin{bmatrix} \sigma_x & 0 & 0 \\ 0 & \sigma_y & 0 \\ 0 & 0 & \sigma_z \end{bmatrix}, \tag{4.53}$$

where $\sigma_x = 1$, if free motion is possible in x_f direction, etc. ($\sigma_z = 0$ except free motion which will not be excluded). Notice that Σ_f is a mask matrix, where 1 stands for motion.

Similarly, a complement mask $\tilde{\Sigma}_f$ can be defined for force control:

$$\tilde{\Sigma}_f = I - \Sigma_f. \tag{4.54}$$

The relation between the coordinate systems (assuming only coordinate transformation, i.e., basis change) can be described as follows:

$$K_f \xrightarrow{A_f} K_o, \qquad r_f = A_f \rho_o, \qquad \rho_o = A_f^{-1} r_f. \tag{4.55}$$

The procedure can be repeated for the rotations around axes from a subspace and prescribed moment (torque) in the complementary subspace. Denote m_d the specified desired moment, K_m is the frame, Σ_m is the diagonal orientation specification matrix, $\tilde{\Sigma}_m = I - \Sigma_m$ is the diagonal moment specification matrix and A_m is the orientation matrix between K_m and K_o.

Generalized task specification matrices:

$$S = \begin{bmatrix} A_f^{-1} \Sigma_f A_f & 0 \\ 0 & A_m^{-1} \Sigma_m A_m \end{bmatrix}, \qquad \tilde{S} = \begin{bmatrix} A_f^{-1} \tilde{\Sigma}_f A_f & 0 \\ 0 & A_m^{-1} \tilde{\Sigma}_m A_m \end{bmatrix}. \tag{4.56}$$

The following steps are important subtasks in hybrid force/torque control:

1. Measuring the $x_d - x$ position/orientation and the $F_d - F$ force/torque errors, respectively, in the frame K_o.
2. Transformation of the errors into K_f and K_m, respectively.
3. Eliminating (zeroing) the errors in the noncontrolled subspaces.
4. Back-transformation of the remaining errors into K_o.

The four steps can easily be realized by $S(x_d - x)$ and $\tilde{S}(F_d - F)$.

4.4.2 Hybrid Position/Force Control Law

The control law will be divided into high-level nonlinear centralized and low-level linear decentralized parts according to the physical principles. The realization constraints of the motion and force controls will be taken into consideration in the appropriate subspaces by the generalized task specification matrices S and $\tilde{S}$, respectively.

The decentralized controllers do not consider the physical constraints, they are working in such a way as if the entire motion or force could be realized without any constraints.

The force control will be divided into active and damping part. The controller of the active part is PI controller, since the often appearing oscillation would be considerably magnified in the presence of D component. In order to damp the effect of oscillations, velocity proportional damping will be applied. Since the kinetic energy is the quadratic form of $\hat{H}^*$, hence it takes place in the weighting of the damping.

Nonlinear centralized controller:

$$F = F_{\text{motion}} + F_{\text{ccgf}} + F_{\text{active}},$$

$$F_{\text{motion}} = \hat{H}^* S u^*_{\text{motion}},$$

$$F_{\text{ccgf}} = \hat{h}^*, \tag{4.57}$$

$$F_{\text{active}} = \tilde{S} u^*_{\text{active}} + \hat{H}^* \tilde{S} u^*_{\text{damping}}.$$

Linear decentralized controllers:

$$u^*_{\text{motion}} = \ddot{x}_d + \text{PID} = \ddot{x}_d + k_{Pi}(x_d - x_i) + k_{Ii} \int (x_d - x_i)\, dt + k_{Di}(\dot{x}_d - \dot{x}_i),$$

$$u^*_{\text{active}} = F_d + \text{PI} = F_d + k_{PF}(F_d - F) + K_{IF} \int (F_d - F)\, dt, \tag{4.58}$$

$$u^*_{\text{damping}} = -K_{VF}\dot{x}.$$

Implementation:

$$\tau = J^T F = \hat{H} J^{-1} \big\{ S u^*_{\text{motion}} + \tilde{S} u^*_{\text{damping}} - \alpha \big\} + J^T \tilde{S} u^*_{\text{active}} + \hat{h}_{\text{ccgf}}. \tag{4.59}$$

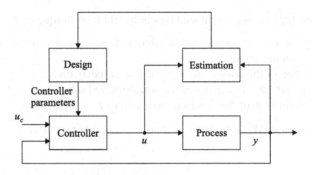

Fig. 4.9 Block scheme of self-tuning adaptive control

The torque τ can be realized also in this case by using decentralized current controllers at the lowest hierarchical level which obtain $\tilde{i}_d$ reference signal proportional to τ.

4.5 Self-Tuning Adaptive Control

The inertia parameters (masses, center of gravities, inertia moments) in the dynamic model of the robot are usually unknown hence we need their estimation in the realization of robot control algorithms based on the dynamic model. One possible solution is the use of self-tuning adaptive control, see Fig. 4.9. The self-tuning adaptive control method presented here is originated from Slotine and Li [131]. More about the analytical determination of the independent parameters of robots can be found in [133].

4.5.1 Independent Parameters of Robot Dynamic Model

Independent parameters can usually be chosen which appear linearly in the robot dynamic model. For example, in case of the 2-DOF RR-type robot arm (with the notations used in Sect. 2.2.3), the independent parameters are as follows:

$$
\begin{aligned}
\alpha_1 &= m_1 l_{c1}^2 + m_2 (l_1^2 + l_{c2}^2) + I_1 + I_2, \\
\alpha_2 &= m_2 l_1 l_{c2}, \\
\alpha_3 &= m_2 l_{c2}^2 + I_2, \\
\alpha_4 &= g(m_1 l_{c1} + m_2 l_1), \\
\alpha_5 &= g m_2 l_{c2}.
\end{aligned}
\tag{4.60}
$$

These parameters appear linearly in the dynamic model of the robot arm in Lagrange form:

$$D_{11} = \alpha_1 + 2C_2\alpha_2,$$

$$D_{12} = \alpha_3 + C_2\alpha_2 = D_{21},$$

$$D_{22} = \alpha_3,$$

$$D_{111} = 0,$$

$$D_{112} = -S_2\alpha_2 = D_{121} = D_{122}, \qquad (4.61)$$

$$D_{211} = S_2\alpha_2,$$

$$D_{212} = D_{221} = D_{222} = 0,$$

$$D_1 = C_1\alpha_4 + C_{12}\alpha_5,$$

$$D_2 = C_{12}\alpha_5.$$

Here we applied the usual notation $C_{12} = \cos(q_1 + q_2)$, where q_i is the i-th joint variable etc. Write down the dynamic model in the form

$$\sum_k D_{ik}\ddot{z}_k + \sum_k \left(\sum_j D_{ijk}\dot{q}_j \right)\dot{z}_k + D_i = \tau_i, \qquad (4.62)$$

where z_k will be chosen later. With this choice, the dynamic model can be brought to the following form:

$$\tau_1 = D_{11}\ddot{z}_1 + D_{12}\ddot{z}_2 + (D_{111}\dot{q}_1 + D_{121}\dot{q}_2)\dot{z}_1 + (D_{112}\dot{q}_1 + D_{122}\dot{q}_2)\dot{z}_2 + D_1$$

$$= (\alpha_1 + 2C_2\alpha_2)\ddot{z}_1 + (\alpha_3 + C_2\alpha_2)\ddot{z}_2 + (-S_2\alpha_2)(\dot{q}_1\dot{z}_2 + \dot{z}_1\dot{q}_2)$$

$$+ (-S_2\alpha_2)\dot{q}_2\dot{z}_2 + (C_1\alpha_4 + C_{12}\alpha_5), \qquad (4.63)$$

$$\tau_2 = D_{21}\ddot{z}_1 + D_{22}\ddot{z}_2 + (D_{211}\dot{q}_1 + D_{221}\dot{q}_2)\dot{z}_1 + (D_{212}\dot{q}_1 + D_{222}\dot{q}_2)\dot{z}_2 + D_2$$

$$= (\alpha_3 + C_2\alpha_2)\ddot{z}_1 + \alpha_3\ddot{z}_2 + (S_2\alpha_2)\dot{q}_1\dot{z}_1 + C_{12}\alpha_5. \qquad (4.64)$$

The so obtained model can already be written as the product of a matrix Y, consisting of elements Y_{ij} which are parameter independent signals, multiplied by the vector of independent parameters.

$$\begin{bmatrix} Y_{11} & Y_{12} & Y_{13} & Y_{14} & Y_{15} \\ Y_{21} & Y_{22} & Y_{23} & Y_{24} & Y_{25} \end{bmatrix} \begin{pmatrix} \alpha_1 \\ \alpha_2 \\ \alpha_3 \\ \alpha_4 \\ \alpha_5 \end{pmatrix} = \begin{pmatrix} \tau_1 \\ \tau_2 \end{pmatrix}. \qquad (4.65)$$

4.5.2 Control and Adaptation Laws

We shall use the robot dynamic model in three different forms during the derivation of the control and adaptation laws:

$$H(q)\ddot{z} + h(q, \dot{q}, \dot{z}) = \tau, \tag{4.66}$$

$$H(q)\ddot{z} + C(q, \dot{q})\dot{z} + D(q) = \tau,$$

$$\text{where } C = [C_{ik}], C_{ik} = \sum_{j} D_{ijk}(q)\dot{q}_j, \tag{4.67}$$

$$Y(q, \dot{q}, \dot{z}, \ddot{z})\alpha = \tau. \tag{4.68}$$

Choose the constant Λ matrix such that the s_i roots of $\det(sI + \Lambda) = 0$ satisfy $Re s_i < 0$. Denote $q_d(t), \dot{q}_d(t), \ddot{q}_d(t)$ the set point function and its derivatives belonging to the robot path. If the error $e = q_d - q$ satisfies the differential equation $\dot{e} + \Lambda e = 0$ then $e(t) \to 0$ exponentially and the speed of convergence can be influenced by the choice of Λ. Now we define the signal $q_r(t)$ to be the desired value corrected by the error integral weighted by Λ and call it the *reference signal*:

$$q_r := q_d + \Lambda \int (q_d - q) \, dt, \tag{4.69}$$

$$\dot{q}_r = \dot{q}_d + \Lambda(q_d - q), \tag{4.70}$$

$$\ddot{q}_r = \ddot{q}_d + \Lambda(\dot{q}_d - \dot{q}). \tag{4.71}$$

Observe that

$$s := \dot{q}_r - \dot{q} = \dot{q}_d - \dot{q} + \Lambda(q_d - q) = \dot{e} + \Lambda e. \tag{4.72}$$

It is satisfactory to assure by the control law that $s = 0$ comes to stay, because then $e(t) \to 0$ with exponential speed, which means that the difference between $q_d(t)$ and $q(t)$ disappears.

We shall prove that the following control law, which can be imagined to be the fusion of three control philosophies, the computed torque technique (nonlinear decoupling in joint space), the sliding control and the PID control, can assure $s(t) \equiv 0$ after a transient time. Remember that although the control law contains immediately only PD-component, it is merely apparent, because the I-component is present in q_r.

The control law has torque output which can be realized by current controllers of quick transient time at the lowest hierarchical control level (in case of DC motors the rotor current is proportional to the motor torque). The control law applies the actual value of the estimated robot parameter vector $\hat{\alpha}$.

Control law:

$$\tau := \hat{H}(q)\ddot{q}_r + \hat{C}(q, \dot{q})\dot{q}_r + \hat{D}(q) + K_p(q_r - q) + K_D s$$

$$= Y(q, \dot{q}, \dot{q}_r, \ddot{q}_r)\hat{\alpha} + K_p(q_r - q) + K_D s,$$

$$(K_p > 0, K_D > 0). \tag{4.73}$$

The closed loop with the above control law is

$$H(q)\ddot{q} + C(q, \dot{q})\dot{q} + D(q) = \hat{H}(q)\ddot{q}_r + \hat{C}(q, \dot{q})\dot{q}_r + \hat{D}(q) + K_p(q_r - q) + K_D s,$$

from which follows

$$H\ddot{q} = \hat{H}\ddot{q}_r + \hat{C}\dot{q}_r + \hat{D} + K_p(q_r - q) + K_D s - C\dot{q} - D + \underbrace{(C\dot{q}_r - C\dot{q}_r)}_{0}$$

$$= \hat{H}\ddot{q}_r + (\hat{C} - C)\dot{q}_r + (\hat{D} - D) + C(\underbrace{\dot{q}_r - \dot{q}}_{s})$$

$$+ K_p(q_r - q) + K_D s. \tag{4.74}$$

It will be shown that the closed loop is stable in Lyapunov sense. Let the Lyapunov function be the "kinetic and potential energy of the error" relative to q_r based on the analogy that the kinetic energy of the robot is $\frac{1}{2}\langle H\dot{q}, \dot{q}\rangle$, and the potential energy of a spring is $\frac{1}{2}kx^2$ where k is the spring constant:

$$V := \frac{1}{2}\langle H(q)s, s\rangle + \frac{1}{2}\langle K_p(q_r - q), q_r - q\rangle + \frac{1}{2}\langle \Gamma(\alpha - \hat{\alpha}), \alpha - \hat{\alpha}\rangle. \tag{4.75}$$

The Lyapunov function penalizes the parameter estimation error $\alpha - \hat{\alpha}$ by the weighting matrix $\Gamma > 0$. The derivative of the Lyapunov function is

$$\frac{dV}{dt} = \langle H\dot{s}, s\rangle + \frac{1}{2}\langle \dot{H}s, s\rangle + \langle K_p(\dot{q}_r - \dot{q}), q_r - q\rangle$$

$$+ \langle \Gamma(\dot{\alpha} - \dot{\hat{\alpha}}), \alpha - \hat{\alpha}\rangle. \tag{4.76}$$

Since $H\dot{s} = H\ddot{q}_r - H\ddot{q}$, hence by (4.74) yields

$$H\dot{s} = (H - \hat{H})\ddot{q}_r + (C - \hat{C})\dot{q}_r + (D - \hat{D}) - K_p(q_r - q) - K_D s - Cs, \tag{4.77}$$

from which follows

$$\frac{dV}{dt} = \langle \underbrace{(H - \hat{H})\ddot{q}_r + (C - \hat{C})\dot{q}_r + (D - \hat{D})}_{Y(\alpha - \hat{\alpha})} - K_p(q_r - q) - K_D s - Cs, s\rangle$$

$$+ \frac{1}{2}\langle \dot{H}s, s\rangle + \langle K_p(\dot{q}_r - \dot{q}), q_r - q\rangle + \langle \Gamma(\dot{\alpha} - \dot{\hat{\alpha}}), \alpha - \hat{\alpha}\rangle$$

$$= -\langle K_D s, s\rangle + \frac{1}{2}\underbrace{\langle (\dot{H} - 2C)s, s\rangle}_{0} + \langle \alpha - \hat{\alpha}, Y^T s + \Gamma(\dot{\alpha} - \dot{\hat{\alpha}})\rangle. \tag{4.78}$$

Here, the skew symmetry of $\dot{H} - 2C$ was exploited. The quadratic form of $\dot{H} - 2C$ is identically zero for any variable, thus also for s. Therefore, $Y^T s + \Gamma(\dot{\alpha} - \dot{\hat{\alpha}}) = 0$ can be chosen, since for $\alpha = const$ real robot parameter vector yields $\dot{\alpha} = 0$.

Adaptation law:

$$\frac{d\hat{\alpha}}{dt} = \Gamma^{-1} Y^T(q, \dot{q}, \dot{q}_r, \ddot{q}_r) s. \tag{4.79}$$

Since in this case

$$\frac{dV}{dt} = -\langle K_D s, s \rangle \tag{4.80}$$

is negative if $s \neq 0$, hence V is decreasing until $s = 0$ is reached, but in this case the error goes exponentially to zero, and the stability is proven.

4.5.3 Simulation Results for 2-DOF Robot

The self-tuning robot control algorithm was tested on a 2-DOF RR type manipulator. The simulations were performed in MATLAB/Simulink using the `ode45` (Dormand Prince) solver, with 10^{-6} relative tolerance.

The robot model, the control algorithm (4.73) and the parameter tuning algorithm (4.79) were implemented as three separate S-functions. The unknown parameters are given by the relations (4.60). During simulations, same robot parameters were chosen as in Sect. 4.6.2.

The controller parameters were chosen as follows: $K_p = \text{diag}([10, 50])$, $K_D = \text{diag}([500, 500])$, $\Lambda = \text{diag}([10, 10])$, $\Gamma^{-1} = \text{diag}([10, 1, 1, 100, 50])$.

During adaptive control, the reference trajectory was chosen to have acceleration, deceleration and constant velocity regimes both in positive and negative velocity domains with ± 1 rad/s velocity limits for both joints. The real joint trajectories are presented in Figs. 4.10 and 4.11. The convergence of the estimated parameters is shown in Fig. 4.12. All the robot parameters show fast convergence to their steady state values. Due to parameter adaptation, the position tracking error decreases in time. When the estimated parameters reached their steady state, 5×10^{-4} rad (0.0286 deg) position tracking precision was achieved for both joints, see Fig. 4.13.

4.5.4 Identification Strategy

It is useful to divide the identification of the parameters of the robot dynamic model into two phases: (i) Preidentification under laboratory circumstances. (ii) Online identification in case of varying load. In the two phases, different rules can be suggested for the proportional gain K_p in the self-tuning adaptive control algorithm.

Fig. 4.10 Motion of joint 1 during control

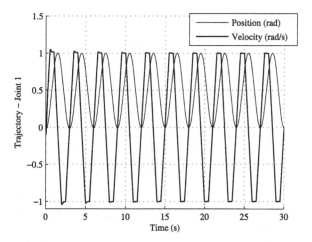

Fig. 4.11 Motion of joint 2 during control

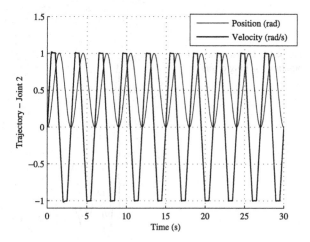

(i) Every time if the robot gripes an object (CLOSE), or releases an object (OPEN), the real α parameter vector assumes a new constant value for the next motion period and the self-tuning adaptive control has to find a good estimation of $\hat{\alpha}$ within a short transient time.

(ii) Although K_p has no influence on the stability (it is canceled in dV/dt), it may have influence on the transient of $\hat{\alpha}$. If there is no information about the robot parameters, then the error may be large so that the rotor current limits can be reached. Hence, it is suggested to switch off K_p in order the decrease the effect of the error in the controller output τ, that is, preidentify the unknown inertia parameters of the robot links using $K_p = 0$ (in case of zero or nominal load).

(iii) Later, as the control should react only to the varying load and therefore the error is smaller, K_p can be switched on in order to decrease the transient time of parameter estimation. Remember there is always an identification dilemma in adaptive control. We want to estimate the parameters, however we need to

Fig. 4.12 Convergence of estimated parameters

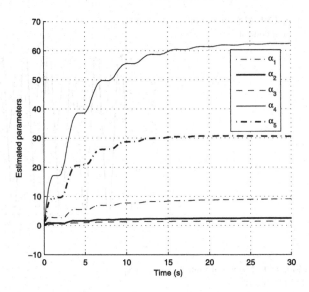

Fig. 4.13 Convergence of position tracking errors

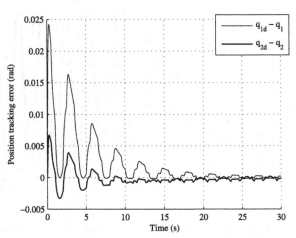

this the presence of the error in the adaptation law (see $s = \dot{q}_r - \dot{q}$), while the main goal of the control is to eliminate the error as soon as possible. If we want to find the new parameters, then we can do it only within a short time while the error is present.

4.6 Robust Backstepping Control in Case of Nonsmooth Path

Backstepping control was introduced in Sect. 2.6. A more robust version allowing limited disturbances is discussed in [130]. The method can be applied for the precise realizations of robot paths with sharp corners. The traveling period with large velocity will be as long as possible while in the neighborhood of sharp corners the velocity is decreased for a short time interval and continued after it again with the large traveling speed. For this purpose, the path and the velocity as a function of a scalar path parameter can be prescribed and backstepping control together with filtered-gradient technique can be used for the realization.

4.6.1 Gradient Update Laws for Speed Error

It has been shown in Sect. 2.6 that in case of backstepping and the notations there the Lyapunov function after the last n-th level is $V_n = z^T P z$ and its time deriva-tive is $\dot{V}_n = -z^T Q z + \tau_n \omega_s$ where $P > 0$ and $Q > 0$ (positive definite). The closed loop with backstepping control satisfies the state equation $\dot{z} = A_z(\bar{x}_n, \theta, t)z + b(\bar{x}_{n-1}, \theta, t)\omega_s$. The desired path is $y_d(\theta)$ where θ is the scalar path parameter, its time derivative is $\dot{\theta} = v_s(\theta, t) - \omega_s(\dot{\theta}, \theta, t)$ and $v_s(\theta, t)$ is the desired parameter speed. On the other hand, it follows from the derivations in Sect. 2.6 that

$$\tau_n = 2z_1^T P_1 y_d^\theta + 2z_2^T P_2 \alpha_1^\theta + \cdots + 2z_n^T P_{n-1}\alpha_{n-1}^\theta = 2z^T P b = 2b^T P z. \quad (4.81)$$

The system with backstepping control will be closed by an external tuning law for the speed error ω_s such that ω_s goes asymptotically to zero and the system will be input-to-state stable (ISS) with respect to a closed 0-invariant set.

In output maneuvering problem, the dynamic task may be *speed assignment* forc-ing the path speed $\dot{\theta}$ to the desired speed $v_s(\theta, t)$,

$$\lim_{t \to \infty} \left| \dot{\theta}(t) - v_s\big(\theta(t), t\big) \right| = 0. \quad (4.82)$$

Gradient update law
Setting $\omega_s = -\mu \tau_n(\bar{x}_n, \theta, t)$, $\mu \geq 0$ satisfies the speed assignment (4.82) asymptot-ically because $\dot{V}_n = -z^T Q z - \mu \tau_n^2 < 0$ and thus $z \to 0$ and $\tau \to 0$. We can call it gradient update law because $\tau_n(\bar{x}_n, \theta, t) = -V_n^\theta(\bar{x}_n, \theta, t)$ and thus ω_s is the negative gradient multiplied by the step length μ. The dynamic part becomes

$$\dot{\theta} = v_s(\theta, t) - \omega_s = v_s(\theta, t) + \mu \tau_n(\bar{x}_n, \theta, t) = v_s(\theta, t) + 2\mu b^T P z. \quad (4.83)$$

The closed loop with backstepping control and gradient update law satisfies

$$\begin{aligned} \dot{z} &= A_z z - 2\mu b b^T P z, \\ \dot{\theta} &= v_s(\theta, t) + 2\mu b^T P z. \end{aligned} \quad (4.84)$$

Assuming the path $y_d(\theta)$ and its n partial derivatives are uniformly bounded, and the speed assignment $v_s(\theta, t)$ and its $n - 1$ partial derivatives are uniformly bounded in θ and t, then the closed-loop system solves the maneuvering problem, that is, the system is ISS with respect to the closed and 0-invariant set $M = \{(z, \theta, t) : z = 0\}$. Choosing $x_1 := z$ and $x_2 = (\theta, t)^T$ the proof follows from Lemma A.1 in [130].

Filtered-gradient update law
Let the new Lyapunov function in the n-th step of the backstepping algorithm be modified to

$$V = V_n + \frac{1}{2\lambda\mu}\omega_s^2, \quad \lambda, \mu > 0, \tag{4.85}$$

then its time derivative is

$$\dot{V} = \dot{V}_n + \frac{1}{\lambda\mu}\omega_s\dot{\omega}_s = -z^T Qz + \left(\tau_n + \frac{1}{\lambda\mu}\dot{\omega}_s\right)\omega_s. \tag{4.86}$$

To make the second term negative, we can choose the update law

$$\dot{\omega}_s = -\lambda(\omega_s + \mu\tau_n), \tag{4.87}$$

which has the Laplace transform

$$\omega_s = -\frac{\lambda\mu}{s+\lambda}\tau_n = \frac{\lambda\mu}{s+\lambda}V_n^\theta. \tag{4.88}$$

Hence, the update law can be called filtered-gradient update. The dynamic part yields

$$\dot{\theta} = v_s(\theta, t) - \omega_s,$$
$$\dot{\omega}_s = -\lambda\omega_s + \lambda\mu\tau_n = -\lambda\omega_s + \lambda\mu V_n^\theta(\bar{x}, \theta, t), \tag{4.89}$$

and the cut-off frequency λ can support noise filtering and limitation of the control signals.

The closed loop with backstepping control and filtered-gradient update satisfies

$$\dot{z} = A_z z + b\omega_s,$$
$$\dot{\theta} = v_s(\theta, t) - \omega_s, \tag{4.90}$$
$$\dot{\omega}_s = -2\lambda\mu b^T Pz - \lambda\omega_s.$$

Under the same assumption as above, the closed-loop system solves the maneuvering problem, that is, the system is ISS with respect to the closed and 0-invariant set $M = \{(z, \omega_s, \theta, t) : (z^T, \omega_s)^T = 0\}$. Choosing $x_1 := (z^T, \omega_s)^T$ and $x_2 = (\theta, t)^T$, the proof follows from Lemma A.1 in [130].

4.6.2 Control of 2-DOF Robot Arm Along Rectangle Path

The backstepping control with filtered gradient update will be illustrated for 2-DOF RR-type robot arm moving in vertical plane.

Suppose the following geometrical and inertia data of the robot arm, all in SI units: $l_1 = l_2 = 1$ m, $l_{c1} = l_{c2} = 0.5$ m, $m_1 = m_2 = 5$ kg, $I_1 = I_2 = 1$ kg m^2, $g = 9.81$ m/s^2. The origin of the base frame is in $x = y = 0$, the tool is at the end of the second link.

Fig. 4.14 Rectangle-form path in vertical space

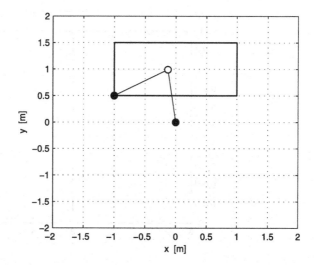

4.6.2.1 Path and Speed Design

The desired path is

$$y_d(\theta) = \begin{cases} (\theta - 1, 0.5)^T, & \text{if } \theta \in [0, 2), \\ (1, \theta - 1.5)^T, & \text{if } \theta \in [2, 3), \\ (4 - \theta, 1.5)^T, & \text{if } \theta \in [3, 5), \\ (-1, 6.5 - \theta)^T, & \text{if } \theta \in [5, 6). \end{cases} \tag{4.91}$$

The rectangle-form path $y_d(\theta)$ in vertical space is shown in Fig. 4.14. We assume upper right-hand configuration for $\theta = 0$ which is saved during motion.

The speed function $v_s(\theta, t)$ will be chosen such that the traveling velocity is large enough, however near to the corner the speed will be decreased almost to zero, then it returns again to the traveling speed. The continuously differentiable velocity function is defined between the k-th and $(k + 1)$-th corner points as follows:

$$v_s(\theta) = \begin{cases} \frac{m_s}{\pi |y_d^\theta|} \arctan(\frac{\theta - \theta_k - a_1}{a_2}) + \frac{m_s}{2|y_d^\theta|} & \text{if } \theta \in [\theta_k, \theta_k + \frac{\theta_{k+1} - \theta_k}{2}), \\ \frac{m_s}{\pi |y_d^\theta|} \arctan(\frac{\theta_{k+1} - a_1 - \theta}{a_2}) + \frac{m_s}{2|y_d^\theta|} & \text{if } \theta \in [\theta_k + \frac{\theta_{k+1} - \theta_k}{2}), \theta_{k+1}, \end{cases} \tag{4.92}$$

where $k = 1, 2, 3, 4$, $\theta_1 = 0$, $\theta_2 = 2$, $\theta_3 = 3$, $\theta_4 = 5$, and $\theta_6 = 6$. The maximal speed is $m_s = 0.1$. The parameters $a_1 = 0.005$ and $a_2 = 0.001$ take place in smoothing the square-wave form velocity profile. The path velocity $v_s(\theta)$ is shown in Fig. 4.15.

In the backstepping control, we need also the derivative of v_s by θ:

$$v_s^\theta(\theta) = \begin{cases} \frac{m_s}{\pi |y_d^\theta|} \frac{a_2}{a_2^2 + (\theta - \theta_k - a_1)^2} & \text{if } \theta \in [\theta_k, \theta_k + \frac{\theta_{k+1} - \theta_k}{2}), \\ \frac{m_s}{\pi |y_d^\theta|} \frac{a_2}{a_2^2 + (\theta_{k+1} - a_1 - \theta)^2} & \text{if } \theta \in [\theta_k + \frac{\theta_{k+1} - \theta_k}{2}), \theta_{k+1}. \end{cases} \tag{4.93}$$

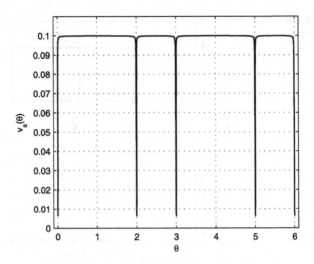

Fig. 4.15 Desired path velocity

4.6.2.2 Inverse Kinematics

First, we have to solve the inverse kinematic problem, that is, $P_2 = (x, y)$ is the end effector position along the path, what is the value of the joint variables q_1 and q_2 belonging to it. It is clear that if the connecting point of the two links is $P_1(x_1, y_1)$ then

$$q_1 = \arctan(y_1/x_1), \qquad q_1 + q_2 = \arctan(y/x),$$
$$q_2 = \arctan(y/x) - q_1. \tag{4.94}$$

Since P_1 is the intersection of two circles, the first has center $(0, 0)$ and radius l_1, the second has center (x, y) and radius l_2, hence $P_1 = (x_1, y_1)$ is the solution of the following equations:

$$x_1^2 + y_1^2 = l_1^2,$$
$$(x_1 - x)^2 + (y_1 - y)^2 = l_2^2 \quad \Rightarrow \quad x_1 x + y_1 y = \frac{l_1^2 + (x^2 + y^2) - l_2^2}{2} =: A.$$

From the second equation, we can express y_1 and substitute it into the first one resulting in a second order polynomial equation which can easily be solved:

$$y_1 = \frac{A - x_1 x}{y} \quad \Rightarrow \quad (x^2 + y^2)x_1^2 - 2Ax x_1 + (A^2 - y^2 l_1^2) = 0,$$
$$x_1 = \frac{2Ax \pm \sqrt{4A^2 x^2 - 4(x^2 + y^2)(A^2 - y^2 l_1^2)}}{2(x^2 + y^2)}.$$

Using the notations

$$d := x^2 + y^2, \qquad a := \frac{l_1^2 + d^2 - l_2^2}{2d}, \qquad A := ad, \qquad h := \sqrt{l_1^2 - a^2} \quad (4.95)$$

the following solutions are obtained:

$$x_1 = \frac{ax \pm hy}{d}, \qquad y_1 = \frac{ay \mp hx}{d},$$

$$P_{11} = \left(\frac{ax + hy}{d}, \frac{ay - hx}{d} \right), \qquad P_{12} = \left(\frac{ax - hy}{d}, \frac{ay + hx}{d} \right).$$

From the two solutions, we shall use in the sequel $P_1 := P_{11}$ (upper right-hand configuration) and the solution (4.94) belonging to.

4.6.2.3 Robot Dynamic Model for Backstepping

The sate variables for backstepping are $x_1 = (x, y)^T$ and $x_2 = (\dot{x}, \dot{y})^T$ where, together with the derivative of x_2, yields:

$$\begin{pmatrix} x \\ y \end{pmatrix} = \begin{pmatrix} l_1 C_1 + l_2 C_{12} \\ l_1 S_1 + l_2 S_{12} \end{pmatrix},$$

$$\begin{pmatrix} \dot{x} \\ \dot{y} \end{pmatrix} = \begin{bmatrix} -l_1 S_1 - l_2 S_{12} & -l_2 S_{12} \\ l_1 C_1 + l_2 C_{12} & l_2 C_{12} \end{bmatrix} \begin{pmatrix} \dot{q}_1 \\ \dot{q}_2 \end{pmatrix} = J(q) \begin{pmatrix} \dot{q}_1 \\ \dot{q}_2 \end{pmatrix}, \qquad (4.96)$$

$$\begin{pmatrix} \ddot{x} \\ \ddot{y} \end{pmatrix} = J(q) \begin{pmatrix} \ddot{q}_1 \\ \ddot{q}_2 \end{pmatrix} + \underbrace{\begin{bmatrix} -l_1 C_1 \dot{q}_1 - l_2 S_{12}(\dot{q}_1 + \dot{q}_2) & -l_2 C_{12}(\dot{q}_1 + \dot{q}_2) \\ -l_1 S_1 \dot{q}_1 - l_2 S_{12}(\dot{q}_1 + \dot{q}_2) & l_2 S_{12}(\dot{q}_1 + \dot{q}_2) \end{bmatrix}}_{\alpha(q, \dot{q})} \begin{pmatrix} \dot{q}_1 \\ \dot{q}_2 \end{pmatrix}.$$

The dynamic model in joint space is $H(q)\ddot{q} + h(q, \dot{q}) = \tau$ where

$$H(q) = \begin{bmatrix} D_{11} & D_{12} \\ D_{12} & D_{22} \end{bmatrix}, \qquad h(q, \dot{q}) = \begin{pmatrix} 2D_{121}\dot{q}_1 \dot{q}_2 + D_{122}\dot{q}_2^2 + D_1 \\ -D_{112}\dot{q}_1^2 + D_2 \end{pmatrix}, \quad (4.97)$$

and the functions $D_{11}, \ldots, D_2$ and their parameters were already given in (4.61) and (4.60). Exchange the variables $q, \dot{q}, \ddot{q}$ to the variables $x_1, \dot{x}_1, x_2$ in the functions of the dynamic model in joint space, and introduce the notations $H(x_1)$, $J(x_1)$, $h(x_1, x_2)$ and $\alpha(x_1, x_2)$ where (as usual in control engineering) the variables in the arguments denote also the modified function relations. Substituting the functions into the dynamic model, we obtain

$$H(x_1)J^{-1}(x_1)\big(\dot{x}_2 - \alpha(x_1, x_2)\big) + h(x_1, x_2) = \tau$$

$$\Rightarrow \quad \dot{x}_2 = J(x_1)H^{-1}(x_1)\tau - J(x_1)H^{-1}(x_1)h(x_1) + \alpha(x_1, x_2). \quad (4.98)$$

Now we are able to write down the robot dynamic model in the form suitable for backstepping control:

$$\dot{x}_1 = G_1(\bar{x}_1)x_2 + f_1(\bar{x}_1) = x_2 \quad \Rightarrow \quad G_1 = I_2, \qquad f_1 = 0,$$
$$\dot{x}_2 = G_2(\bar{x}_2)u + f_2(\bar{x}_2)$$
$$= J(x_1)H^{-1}(x_1)\tau - J(x_1)H^{-1}(x_1)h(x_1) + \alpha(x_1, x_2) \qquad (4.99)$$
$$\Rightarrow \quad G_2 = JH^{-1}, \qquad f_2 = -JH^{-1}h + \alpha, \qquad u = \tau,$$
$$y = h(x_1) = x_1 \quad \Rightarrow \quad h = x_1.$$

4.6.2.4 Backstepping Control with Filtered-Gradient Update Law

From stability consideration, we have to assure $P_i A_i + A_i^T P_i = -Q_i$ where $Q_i, P_i > 0$ and A_i is Hurwitz, $i = 1, 2$. Our choice is $P_1 = p_1 I_2$ and $P_2 = p_2 I_2$. On the other hand, we can choose $A_1 = -k_p I_2$ and $A_2 = -k_d I_2$ where $k_p, k_d > 0$, from which follows $Q_1 = 2p_1 k_p I_2 > 0$ and $Q_2 = p_2 k_d I_2 > 0$.

For the 2-DOF robot arm, the backstepping algorithm has the following form:

$$z_1 = y - y_d(\theta) = x_1 - y_d(\theta),$$
$$z_2 = x_2 - \alpha_1(x_1, \theta, t),$$
$$\alpha_1 = A_1 z_1 + y_d^\theta(\theta)v_s(\theta),$$
$$\alpha_1^\theta = -A_1 y_d^\theta(\theta) + y_d^{\theta^2}(\theta)v_s(\theta) + y_d^\theta(\theta)v_s^\theta(\theta), \qquad (4.100)$$
$$\sigma_1 = A_1 x_2,$$
$$\tau = G_2^{-1}\left[A_2 z_2 - P_2^{-1}G_1^T P_1 z_1 - f_2 + \sigma_1 + \alpha_1^\theta v_s\right] \quad \Rightarrow$$
$$\tau = HJ^{-1}\left[A_2 z_2 - \frac{p_1}{p_2}z_1 + JH^{-1}h - \alpha + \sigma_1 + \alpha_1\theta v_s\right].$$

The filtered-gradient law was chosen to update the speed error:

$$\dot{\theta} = v_s(\theta, t) - \omega_s,$$
$$\dot{\omega}_s = -2\lambda\mu\left(p_1 z_1^T y_d^\theta + p_2 z_2^T \alpha_1^\theta\right) - \lambda\omega_s. \qquad (4.101)$$

4.6.2.5 Simulation Experiment

For the 2-DOF robot arm, simulation experiments were performed with backstepping control and filtered-gradient speed error update law. The parameters were chosen $p_1 = 5$, $p_2 = 1$, $k_p = 50$, $k_d = 50$, $\mu = 1$ and $\lambda = 40$. The simulation system

Fig. 4.16 Realized robot configurations along the path

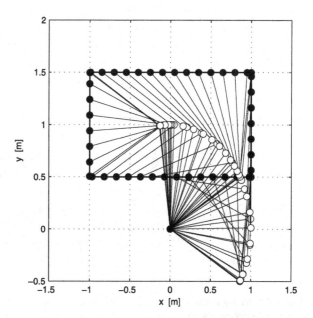

consists of three parts realized as S-functions in MATLAB/Simulink environment. The three parts are the controller, the plant (robot arm), and the guidance.

The *controller* is a continuous time system having state $x_c = (\theta, \omega_s)^T$, input $u_c = (x_1^T, x_2^T, y_d^T, (y_d^\theta)^T, (y_d^{\theta^2})^T)^T$ and output $y_c = (\tau^T, \theta)^T$. The computations are based on (4.100) and (4.101). The initial state is zero, and the driving torque is initially set to $\tau = (D_1, D_2)^T$ in order to perform gravity compensation. A separate function *dynamics* determines the dynamic model according to (4.99). Notice that y_d^θ is the short form for $\frac{dy_d}{d\theta}$, and similarly $y_d^{\theta^2}$ is the short form for $\frac{d^2 y_d}{d\theta^2}$.

The *plant* is a continuous time system with state $x_p = (x_1^T, x_2^T)^T = (x, y, \dot{x}, \dot{y})^T$, input $u_p = \tau$ and output $y_p = x_p$. From the input the derivative part (flag = 1) of the S-function solves the inverse kinematic problem (computation of q_1, q_2)), then it determines the Jacobian and computes the derivatives $\dot{q}_1, \dot{q}_2$. Based on them the function *dynamics* determines the dynamic model, that is, $H(x_1)$, $h(x_1, x_2)$ and $\alpha(x_1, x_2)$, followed by the computation of the right side of the state equation of the robot arm. The output is equal to the state. The initial state is set to $x_1(0) = (-1, 0.5)^T$ (position) and $x_2(0) = (0, 0)^T$ (velocity).

The *guidance* part is a system without state. Its input is $u_g = \theta$ and the output is $y_g = (y_d^T, (y_d^\theta)^T, (y_d^{\theta^2})^T, v_s, v_s^\theta)^T$.

The realized robot configurations along the path are presented in Fig. 4.16. The path error $z_1 = y - y_d$ can be seen in Fig. 4.17. The controller outputs (driving torques) are shown in Fig. 4.18. The speed is drawn in Fig. 4.19.

The results demonstrate the efficiency of the backstepping control with filtered-gradient update law.

Fig. 4.17 Path errors along the path

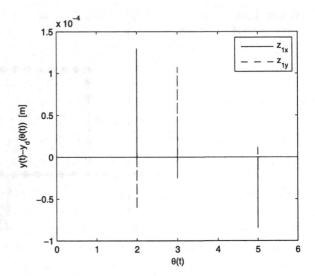

Fig. 4.18 Driving torques using backstepping control with filtered-gradient update

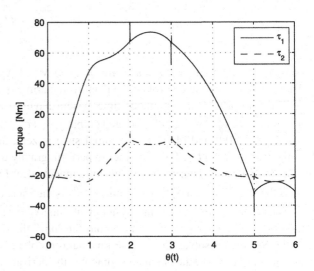

4.7 Closing Remarks

Industrial robots often use well proven decentralized cascade control based on linear control theory. On the other hand, advanced control methods based on the nonlinear dynamic model of the robot will be preferred for high precision applications, increased speed and compliance problems.

Decentralized three loop cascade control is a simple technique well known from the early period of numerical control of machines. First, the dynamic model of DC motors was considered then the design of the three-loop cascade control was presented. The internal PI current controller was designed by using a very simple ap-

Fig. 4.19 Realized speed
along the path

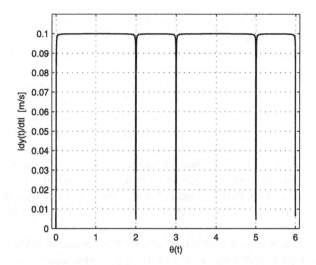

proximating model in which the back emf was considered as "disturbance" and the problem was enormously simplified. Then it was shown that the controller behaves also well without approximation. The intermediate PI velocity controller was designed to assure a second order system for velocity output in the asymptotic limit case (two equal poles). The external controller was assumed to be a PID controller, hence two integrators were present in the open loop for position output making the compensation harder. A method was presented for the design of an analog PID controller taking into consideration also the effect of the DAC converter to the open loop gain if the controller will be converted to DDC PID controller. The behavior of the control system was illustrated by root locus technique.

The applicability of decentralized cascade control is limited because the errors are increasing with the speed and the load. A simple extension was shown how can the average effective inertia contribute to the improvement of the control properties in case of repeated path.

From advanced control methods the computed torque technique, the nonlinear decoupling of the free motion in Cartesian space and the hybrid position/force control were considered.

Computed torque technique is nonlinear decoupling in joint space. Since industrial robots are typically full-actuated hence there is no zero dynamics and stability is assured. By using additional PID decentralized controllers, beside the centralized nonlinear decoupling, the control properties can be improved. For the choice of the controller parameters, a simple technique was presented.

In order to control the free and compliance motions in Cartesian space, first the conversion of sensory information between frames and the equivalent (generalized) torque for prescribed (generalized) force are discussed and the necessary computation rules were derived. Then the dynamic model of the robot in joint space was converted to the dynamic model in Cartesian space because the philosophy of control laws can easier be found in Cartesian space if the goal is hybrid position/force control.

The nonlinear decoupling of the free motion in Cartesian space was solved based on the transformed dynamic model. The method is very similar to nonlinear decoupling in joint space. It was assumed that the robot is 6-DOF one and the singular configurations are avoided.

From the large class of hybrid position/force control approaches the well-known operation space method was suggested. The space is divided in complementary position and force subspaces, and orientation and torque subspaces, respectively, which can easily be managed by the generalized task specification matrices. The (generalized) force was divided into four components in the centralized controller, while the decentralized controllers remained linear ones. The error supervision by the generalized task specification matrices are active only at the centralized level, the decentralized controllers need not deal with it. The control law has (generalized) force output and can be converted to (generalized) torque for the implementation at joint level.

It is a hard problem of finding the numerical values of the parameters in the nonlinear dynamic model of the robot, especially if the load is varying. The well-known method of self-tuning adaptive control was suggested for online parameter identification. Only the independent robot parameters can be identified because at joint level only one of the six components of the force and torque vectors between neighboring links has influence to the dynamic model. Fortunately, these independent parameters appear in linear form in the dynamic model as was illustrated for a 2-DOF arm. The self-tuning adaptive control needs the robot model in three different forms. The control law integrates the concept of the computed torque technique, sliding control, and PID control. The I-part appears in the definition of the reference signal and the errors are considered relative to it. It was shown by using Lyapunov theory that the self-tuning adaptive control is stable, moreover from the stability results the online parameter adaptation law was derived. The applicability of the method was demonstrated for the 2-DOF robot arm during simulation.

In some robotic application, high precision is necessary also if the path contains sharp corners. To solve the problem, the method of backstepping control with filtered gradient rule was suggested. The desired path has to be parametrized by a single scalar parameter and the path velocity should also be defined as the function of the path parameter. It was shown that the backstepping control with filtered gradient parameter update low solves the maneuvering problem. The effectiveness of the method was demonstrated for the 2-DOF robot arm moving in vertical plane along a path having sharp corners.

Chapter 5
Nonlinear Control of Cars

Overview This chapter focuses on the high level nonlinear control methods for cars (ground vehicles). The main tasks are path design, control algorithm development and control realization. These steps are demonstrated in the frame of the development of a collision avoidance system (CAS). From the large set of path design methods the principle of elastic band is chosen. The car can be modeled by full (nonaffine) or simplified (input affine) nonlinear models. Two nonlinear control methods are presented, the differential geometric approach (DGA), known as input-output linearization, and the nonlinear predictive control. For state estimation, Kalman filters and measurements of two antenna GPS and Inertial Measurement Unit (IMU) are used. The methods are illustrated through simulation results. Two software approaches are presented for the realization of the control algorithms. The first uses the MATLAB Compiler converting the whole control system to standalone program running under Windows or Linux. The second uses MATLAB/Simulink and Real-Time Workshop (RTW) to convert the control system for target processors.

5.1 Control Concept of Collision Avoidance System (CAS)

In a Collision Avoidance System (CAS), the path design has to be performed online in the presence of static and dynamic obstacles. The path information can be converted to reference signals for the control subsystems. Examples for static and dynamic obstacles may be the lost payload of a truck and an oncoming vehicle in the other lane. To solve the full obstacle identification problem, sensor fusion of environmental information produced by radar, laser and computer vision systems is preferred, but this is not part of the chapter.

Typical approaches for path design are constrained optimization using splines [71] and the elastic band originated from Quinlan and Khatib [118]. From engineering point of view, path design methods resulting in aesthetic paths, that is, paths with smooth limited curvatures, may be suggested.

Because of real-time considerations and the presence of moving obstacle, we prefer path design using elastic band with some modifications of the results in [22].

B. Lantos, L. Márton, *Nonlinear Control of Vehicles and Robots*,
Advances in Industrial Control,
DOI 10.1007/978-1-84996-122-6_5, © Springer-Verlag London Limited 2011

An improved method will be shown for reaction forces for road borders and static obstacles allowing quick computation of the force equilibrium [75]. The sensory information can be used to initiate the path design method. The CAS path (reference signal) has to be smoothed for control purposes.

The stability properties of the path can be divided into 5 categories called Characteristic Velocity Stability Indicator (CVSI) which may be interpreted as understeering, neutral steering, oversteering, high oversteering and braking away [19]. If the generation of the drivable trajectory cannot be finished within a given time limit $T_{\max}$, or if the estimated CVSI cannot be accepted because wrong stability properties, then emergency braking follows without steering. Otherwise the developed path will be performed by using an appropriate control method.

The speed of the computation of path design depends on the complexity of the situation and the number of iterations needed to find the force-equilibrium of the elastic band.

The control of cars is usually based on classical control methods (PID etc.), or advanced methods using Lie-algebra results of nonlinear control [56], optimal control [73] or predictive control [25, 73, 87]. Nonlinear control assumes known dynamic model of the car. Based on a simplified (input affine) nonlinear model, Freud and Mayr [43] derived a control method for cars using differential geometric approach (DGA). Although we shall concentrate on receding horizon control (RHC), the DGA solution of Freud and Mayr will be useful for initialization of the control sequence in the very first horizon and for comparison.

For state estimation of cars, Kalman filters and measurements of two antenna GPS and Inertial Navigation System (INS) are typical approaches [125].

An important question is the quick development and test of the full software. Two control softwares are presented for realization of the control algorithms. The first uses the C Compiler of MATLAB converting the whole control system running under Windows or Linux. In this case, the use of MATLAB toolboxes is allowed. The second uses MATLAB/Simulink and Real-Time Workshop (RTW) to convert the control system for target processors (MPC555, AutoBox etc.). In this second case, the use of MATLAB toolboxes is not possible, however we can use S-functions in C or MATLAB Embedded Functions.

5.2 Path Design Using Elastic Band

It is assumed that satisfactory sensory information is present about the path section and the static and dynamic obstacles that can be used to initiate an elastic band which is the basis for CAS path design.

In the actual state of the vehicle, an elastic band will be started from the origin r_0 of the vehicle coordinate system. The elastic band consists of N springs, the connection points r_i of the springs are the nodes. The spring constant and the initial length of spring are k_i and $l_{0,i}$, respectively. The goal is to find the force equilibrium. In every iteration step, the reaching time t_i will also be determined based on spline technique and the longitudinal velocity of the vehicle. If the path belonging to the force equilibrium is stabilized, then the vehicle can move along the path taking into consideration the reaching time. The elastic band is characterized by the information:

$$(r_0, t_0) \xrightarrow{\cdots} (r_i, t_i) \xrightarrow{k_i, l_{0,i}} (r_{i+1}, t_{i+1}) \xrightarrow{\cdots} (r_N, t_N).$$

Obstacles are modeled by safety circles. The center and diameter of the safety circle O_j are r^{O_j} and d_j, respectively. If the elastic band of the vehicle reaches the safety circle of a node, then new elastic bands will be generated from the existing ones so that typically more than one elastic bands are iterated in order to find force equilibrium. After reaching the force equilibrium, for each equilibrium the time-parametrized trajectory, curvature and lateral acceleration are determined based on spline technique. For the vehicle, that path will be chosen for which the lateral acceleration is the smallest.

The path design is based on the internal potentials V_i^{int} of the springs (forces at each node are given by the directional derivatives of the corresponding potential fields), the external forces $F_i^{B_q}$ of the left (B_l) and right (B_r) borders of the road, $q \in \{l, r\}$, and the external forces $F_{i,\text{stat}}^{O_j}$ and $F_{i,\text{mov}}^{O_j}$ of obstacles. For every node r_i, the nearest points $r_i^{B_l}$ and $r_i^{B_r}$ along the borders will be chosen.

The forces $F_i^{B_q}$ suggested in [22] were not able to place back the elastic band in the middle of the lane without obstacles and acted in different way at left and right road borders. Similarly, $F_{i,\text{stat}}^{O_j}$ in [22] were not able to guarantee satisfactory path curvature and centripetal acceleration in case of force equilibrium. Hence, our choice was:

$$V_i^{\text{int}} = \frac{1}{2} k_i \left(|r_{i+1} - r_i| - l_{0i} \right)^2, \tag{5.1}$$

$$V^{\text{int}} = \sum_{i=0}^{N-1} V_i^{\text{int}} = \sum_{i=0}^{N-1} \frac{1}{2} k_i \left(|r_{i+1} - r_i| - l_{0i} \right)^2, \tag{5.2}$$

$$F_i^{B_q} = M^B \exp\left[-\frac{1}{2} \left(|r_i - r_i^{B_q}| / \sigma^{B_q} \right)^2 \right] \frac{r_i - r_i^{B_q}}{|r_i - r_i^{B_q}|}, \tag{5.3}$$

$$F_{i,\text{stat}}^{O_j} = k^{O_j} \frac{d^{O_j}/2}{|r_i - r^{O_j}|} \cdot \frac{r_i - r^{O_j}}{|r_i - r^{O_j}|}, \tag{5.4}$$

$$F_{i,\text{mov}}^{O_j} = k^{O_j} \exp\left(-\left(|r_i - r^{O_j}(t_i)| - \frac{d_j}{2} \right)^2 \right) \frac{r_i - r^{O_j}(t_i)}{|r_i - r^{O_j}(t_i)|}. \tag{5.5}$$

In case of parallel borders of the road and no obstacles, it may be expected that the elastic band is positioned to the middle of the right lane. Hence, assuming road width b and 2 lanes (1 lane in each direction) and force equilibrium $F_i^{B_l} = F_i^{B_r}$, it should be satisfied

$$\frac{k_{B_l}}{k^{B_r}} = \frac{0.75b}{0.25b} = 3$$

for the friction constants of the road borders. For the forces belonging to the potential of the road borders, the other parameters were chosen according to

$$\sigma^{B_q} = k^{B_q}/\sqrt{2\ln(M^B/m^B)}, \quad M^B = 2, m^B = 0.05.$$

For static and dynamic obstacles, $k^{O_j} = 3$ was chosen.

Let M be the number of static and moving obstacles. Then the force equilibrium condition is:

$$F_i^{sum} = F_i^{int} + F_i^{B_l} + F_i^{B_r} + \sum_{j=1}^{M} F_{i*}^{O_j} = 0. \tag{5.6}$$

The equilibrium should be satisfied for each node. Introducing the notations $x = (r_1^T, r_2^T, \dots, r_N^T)^T$ and

$$f(x) = \left((F_1^{sum})^T, (F_2^{sum})^T, \dots, (F_N^{sum})^T \right)^T, \tag{5.7}$$

the numerical problem is to solve the nonlinear system of equations $f(x) = 0$. We used fsolve of the MATLAB Optimization Toolbox supplementing to it beside $f(x)$ also the Jacobian (derivative) of $f(x)$ in order to decrease computation time.

The path design was tested for a problem similar to that in [22]. The input parameters for path design are as follows:

- 41 nodes.
- Static obstacle of diameter 2.5 m and distance 40 m.
- Moving obstacle of diameter 3.5 m, initial distance 120 m, and assumed velocity 15 m/s.
- Average velocity of own car is 20 m/s.

The computation of the force equilibrium using fsolve needed approximately 1 s. The results are shown in Fig. 5.1.

5.3 Reference Signal Design for Control

In the developed program, first the elastic band was approximated by linear sections which is a data sequence of (x, y). The time instants belonging to the nodes were determined assuming known average velocity of the own car resulting in the sequences (t, x) and (t, y). The concept of the remaining part of reference signal design can be applied for any other path planning method, too.

Fig. 5.1 Result of path design using elastic band in case of 1 static obstacle (40 m) and 1 moving obstacle (initially 120 m)

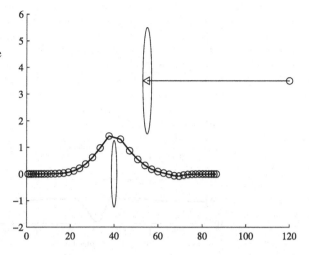

Based on the time distribution, the kinematic variables have been determined by using third order polynomial approximations between the nodes. The approximation was performed using the spline technique of MATLAB (spline, ppval, unmkpp, mkpp). However in case of third order polynomial approximation, the third derivative is stepwise constant, thus it is not continuous. Hence, in order to smooth $\dddot{x}$, $\dddot{y}$ the first derivatives $\dot{x}$, $\dot{y}$ were again approximated by third order polynomials resulting in continuous third order derivatives $\dddot{x}$, $\dddot{y}$. Based on the so smoothed derivatives, the kinematic variables v (velocity), ψ (orientation) and its derivatives, and κ (curvature) can easily be determined for the CAS reference path:

$$v = \left(\dot{x}^2 + \dot{y}^2\right)^{1/2},$$

$$\dot{v} = \frac{\dot{x}\ddot{x} + \dot{y}\ddot{y}}{(\dot{x}^2 + \dot{y}^2)^{1/2}},$$

$$\psi = \arctan(\dot{y}/\dot{x}),$$

$$\dot{\psi} = \frac{\ddot{y}\dot{x} - \dot{y}\ddot{x}}{\dot{x}^2 + \dot{y}^2},$$

$$\ddot{\psi} = -2\frac{(\dot{x}\ddot{x} + \dot{y}\ddot{y})(\ddot{y}\dot{x} - \dot{y}\ddot{x})}{(\dot{x}^2 + \dot{y}^2)^2} + \frac{\dddot{y}\dot{x} - \dot{y}\dddot{x}}{\dot{x}^2 + \dot{y}^2},$$

$$\kappa = \frac{\dot{x}\ddot{y} - \dot{y}\ddot{x}}{(\dot{x}^2 + \dot{y}^2)^{3/2}}.$$

For the reference signal, the side slip angle is assumed to be $\beta(t) = 0$, $\forall t$. The positions $x(t)$, $y(t)$ and the curvature $\kappa(t)$ are shown in Fig. 5.2. The signals are to be understood in SI units.

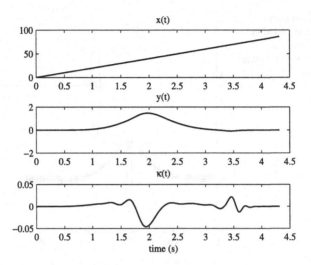

Fig. 5.2 Reference path position and curvature with 1 static obstacle and 1 moving obstacle

5.4 Nonlinear Dynamic Model

In Chap. 3, we have derived for cars a (full) nonlinear dynamic model based on the bicycle model of vehicles and two input affine approximations of it differing in the approximations of trigonometric functions. For convenience, we repeat here parts of them.

Denote β the side slip angle, v_G the velocity, X, Y the position, and δ_w the steering angle, respectively. Front and rear longitudinal forces are denoted by F_{lF} and F_{lR}, cornering stiffnesses by c_F and c_R. It is useful to introduce front and rear side forces S_v and S_h according to

$$S_h = c_R(-\beta + l_R \dot{\psi}/v_G) = c_R(-x_1 + l_R x_3/x_4), \tag{5.8}$$

$$S_v = c_F(\delta_w - \beta - l_F \dot{\psi}/v_G). \tag{5.9}$$

The front longitudinal force is assumed to be $F_{lF} = 0$. The air resistance disturbance $T(x)$ is not considered during control. Control input and state are $u = (S_v, F_{lR})^T$ and $x = (\beta, \psi, \dot{\psi}, v_G, X, Y)^T$, respectively. It is clear that steering angle δ_w can be determined for control implementation by

$$\delta_w = \frac{1}{c_F} S_v + \beta + l_F \dot{\psi}/v_G. \tag{5.10}$$

Full nonlinear dynamic model

$$\dot{\beta} = -\dot{\psi} + \frac{1}{m_v v_G}\left[-F_{lR}\sin(\beta) + S_v\cos(\delta_W - \beta) + S_h\cos(\beta)\right],$$

$$\dot{\psi} = \dot{\psi},$$

$$\ddot{\psi} = \frac{1}{I_{zz}}\left[l_F S_v\cos(\delta_W) - l_R S_h\right],$$

$$\dot{v}_G = \frac{1}{m_v}\left[F_{lR}\cos(\beta) - S_v\sin(\delta_W - \beta) + S_h\sin(\beta)\right],$$

$$\dot{X} = v_G\cos(\psi + \beta),$$

$$\dot{Y} = v_G\sin(\psi + \beta).$$

(5.11)

Approximated input affine nonlinear model

$$\dot{x} = \begin{pmatrix} -x_3 + S_h/(m_v x_4) \\ x_3 \\ -S_h l_R/I_{zz} \\ 0 \\ x_4 C_{12} \\ x_4 S_{12} \end{pmatrix} + \begin{bmatrix} 1/(m_v x_4) & -x_1/(m_v x_4) \\ 0 & 0 \\ l_F/I_{zz} & 0 \\ 0 & 1/m_v \\ 0 & 0 \\ 0 & 0 \end{bmatrix} u,$$

(5.12)

$$\dot{x} = A(x) + B(x)u, \qquad y = \begin{pmatrix} x_5 \\ x_6 \end{pmatrix} = C(x).$$

(5.13)

Here, we used the notation $C_{12} = \cos(x_1 + x_2)$ and $S_{12} = \sin(x_1 + x_2)$.

5.5 Differential Geometry Based Control Algorithm

The following algorithm is based on input-output linearization discussed in Appendix B and the results in [43]. First, we show that the approximated dynamic model has vector relative degrees. The proof is as follows:

$$y_1 = X = x_5,$$

$$\dot{y}_1 = \dot{X} = \dot{x}_5 = x_4 C_{12},$$

$$\ddot{y}_1 = \dot{x}_4 C_{12} + x_4(-S_{12})(\dot{x}_1 + \dot{x}_2)$$

$$= \frac{1}{m_v}u_2 C_{12} - x_4 S_{12}\left[-x_3 + \frac{S_h}{m_v x_4} + \frac{1}{m_v x_4}u_1 - \frac{x_1}{m_v x_4}u_2 + x_3\right]$$

$$= -\frac{1}{m_v}S_{12}S_h - \frac{1}{m_v}S_{12}u_1 + \frac{1}{m_v}(C_{12} + x_1 S_{12})u_2,$$

$$y_2 = Y = x_6,$$

$$\dot{y}_2 = x_4 S_{12},$$

$$\ddot{y}_2 = \dot{x}_4 S_{12} + x_4 C_{12}(\dot{x}_1 + \dot{x}_2)$$

$$= \frac{1}{m_v} u_2 S_{12} + x_4 C_{12}\left[-x_3 + \frac{S_h}{m_v x_4} + \frac{1}{m_v x_4}u_1 - \frac{x_1}{m_v x_4}u_2 + x_3\right]$$

$$= \frac{1}{m_v} C_{12} S_h + \frac{1}{m_v} C_{12} u_1 + \frac{1}{m_v}(S_{12} - x_1 C_{12})u_2.$$

Using the notations of Appendix B, we obtain:

$$S(x) = \frac{1}{m_v}\begin{bmatrix} -S_{12} & C_{12} + x_1 S_{12} \\ C_{12} & S_{12} - x_1 C_{12} \end{bmatrix}, \tag{5.14}$$

$$\det = (-S_{12})(S_{12} - x_1 C_{12}) - C_{12}(C_{12} + x_1 S_{12}) = -1,$$

$$S^{-1} = m_v \begin{bmatrix} x_1 C_{12} - S_{12} & C_{12} + x_1 S_{12} \\ C_{12} & S_{12} \end{bmatrix}, \tag{5.15}$$

$$q(x) = \frac{1}{m_v}\begin{pmatrix} -S_{12} \\ C_{12} \end{pmatrix} S_h. \tag{5.16}$$

Since $S(x)$ is nonsingular, therefore the vector relative degrees r_1, r_2 exist and satisfy $r_1 = r_2 = 2$. Thus, (the observable subsystem) has the form

$$\begin{pmatrix} \ddot{y}_1 \\ \ddot{y}_2 \end{pmatrix} = q(x) + S(x)u. \tag{5.17}$$

Hence, the system can be input-output linearized by the internal nonlinear feedback

$$u := S^{-1}[v - q(x)], \tag{5.18}$$

and the resulting system consists of two double integrators:

$$\ddot{y}_i = v_i, \quad i = 1, 2. \tag{5.19}$$

The stability of the zero dynamics will be discussed later.

5.5.1 External State Feedback Design

First, we assume a prescribed stable and sufficiently quick error dynamics

$$(\ddot{y}_{di} - \ddot{y}_i) + \alpha_{1i}(\dot{y}_{di} - \dot{y}_i) + \lambda_i(y_{di} - y_i) = 0 \tag{5.20}$$

where y_{di} is the reference signal for output y_i and $e_i = y_{di} - y_i$ is the error. Then it follows

$$\ddot{y}_i = \ddot{y}_{di} + \alpha_{1i}(\dot{y}_{di} - \dot{y}_i) + \lambda_i(y_{di} - y_i) = v_i, \tag{5.21}$$

$$\ddot{y}_i + \alpha_{1i}\dot{y}_i + \lambda_i y_i = \lambda_i \underbrace{\left[y_{di} + \frac{1}{\lambda_i}(\alpha_{1i}\dot{y}_{di} + \ddot{y}_{di}) \right]}_{w_i} = \ddot{y}_{di} + \alpha_{1i}\dot{y}_{di} + \lambda_i y_{di}, \quad (5.22)$$

$$\ddot{y}_i = v_i = \lambda_i w_i - \alpha_{1i}\dot{y}_i - \lambda_i y_i. \quad (5.23)$$

Observe that $\lambda_i w_i$ depends only on the reference signal and its derivatives (feed forward from the reference signal) and $-\alpha_{1i}\dot{y}_i - \lambda_i y_i$ is the state feedback stabilizing the system where α_{1i} and λ_i are positive for stable error dynamics. Let $\lambda_1 = \lambda_2 := \lambda$, $\alpha_{11} = \alpha_{12} = 2\sqrt{\lambda}$ where $\lambda > 0$, then two decoupled linear systems are arising whose characteristic equation and differential equation are, respectively,

$$s^2 + 2\sqrt{\lambda}s + \lambda = 0 \quad \Rightarrow \quad s_{1,2} = -\sqrt{\lambda}, \quad (5.24)$$

$$\ddot{y}_i + \alpha_{1i}\dot{y}_i + \lambda y_i = \lambda_i w_i. \quad (5.25)$$

The remaining part is to find a simple form for the composite (internal and external) feedback. Using (5.15), (5.18), (5.23) and the approximated model (5.12), we can write

$$\begin{pmatrix} u_1 \\ u_2 \end{pmatrix} = m_v \begin{bmatrix} x_1 C_{12} - S_{12} & C_{12} + x_1 S_{12} \\ C_{12} & S_{12} \end{bmatrix}$$

$$\times \begin{pmatrix} \overbrace{\lambda_1 w_1 - \alpha_{11}(x_4 C_{12}) - \lambda_1 x_5 + \frac{1}{m_v} S_{12} S_h}^{\bar{y}_1} \\ \underbrace{\lambda_2 w_2 - \alpha_{12}(x_4 S_{12}) - \lambda_2 x_6 - \frac{1}{m_v} C_{12} S_h}_{\bar{y}_2} \end{pmatrix}.$$

Let us consider the two controller outputs separately:

$$u_1 = m_v \big[(x_1 C_{12} - S_{12})(\bar{y}_1 + S_{12} S_h/m_v) + (C_{12} + x_1 S_{12})(\bar{y}_2 - C_{12} S_h/m_v) \big],$$

$$u_2 = m_v \big[C_{12}(\bar{y}_1 + S_{12} S_h/m_v) + S_{12}(\bar{y}_2 - C_{12} S_h/m_v) \big].$$

Taking into consideration that

$$(x_1 C_{12} - S_{12})S_{12} S_h - (C_{12} + x_1 S_{12})C_{12} S_h = -S_h \quad (5.26)$$

and omitting the canceling terms in u_2 we obtain the resulting composite (internal and external) feedback control algorithm in the following simplified form:

DGA Control Algorithm

Path:

$$y_{d1}(t) = X_d(t), \qquad \dot{y}_{d1}(t) = \dot{X}_d(t), \qquad \ddot{y}_{d1}(t) = \ddot{X}_d(t),$$

$$y_{d2}(t) = Y_d(t), \qquad \dot{y}_{d2}(t) = \dot{Y}_d(t), \qquad \ddot{y}_{d2}(t) = \ddot{Y}_d(t),$$

State:

$$x = \big(\beta, \psi, \dot{\psi}, v_G, X, Y\big)^T,$$

Control:

$$w_i := y_{di} + \frac{1}{\lambda_i}(\alpha_{1i} y_{di} + \ddot{y}_{di}), \quad i = 1, 2,$$

$$\bar{y}_1 := \lambda_1 w_1 - \alpha_{11}(x_4 C_{12}) - \lambda_1 x_5,$$

$$\bar{y}_2 := \lambda_2 w_2 - \alpha_{12}(x_4 S_{12}) - \lambda_2 x_6,$$

$$S_h := c_R\big[-x_1 + (l_R x_3/x_4)\big],$$

$$u_1 := -S_h + m_v\big[(x_1 C_{12} - S_{12})\bar{y}_1 + (x_1 S_{12} + C_{12})\bar{y}_2\big],$$

$$u_2 := m_v(C_{12}\bar{y}_1 + S_{12}\bar{y}_2),$$

$$\delta_w := (u_1/c_F) + x_1 + (l_F x_3/x_4),$$

$$F_{lR} := u_2.$$

Notice that because not all state variables can be measured a state estimator has to be implemented in order to supply the necessary state information for the controller. This problem will be discussed later.

5.5.2 Stability Proof of Zero Dynamics

We have already proven that the position errors $e_i = y_{di} - y_i$ are exponentially decaying by using the developed control algorithm. On the other hand, the car is underactuated therefore error can arise in the orientation which depends on the side slip angle $\beta = x_1$ and the angular rate $\dot{\psi} = r = x_3$. Making them small then the orientation error will also be small.

Remember that along the path the angle between the path velocity vector and the x-direction of the (quasi) inertial system is $\psi_d = \arctan(\dot{y}_d/\dot{x}_d)$ and the side slip angle is zero, while for the car the angle between the car velocity vector and the x-direction yields $\psi + \beta = \arctan(\dot{y}/\dot{x})$ and β may be nonzero. Since the position error is exponentially decaying, therefore the orientation error depends essentially on β. Hence, we introduce the nonlinear state transformation $T(x)$ and determine the Jacobian matrix and its determinant to check nonsingularity:

$$z = T(x) = \begin{pmatrix} x_5 \\ x_4 C_{12} \\ x_6 \\ x_6 S_{12} \\ x_1 \\ x_3 \end{pmatrix}, \tag{5.27}$$

$$\frac{dT}{dx} = \begin{bmatrix} 0 & 0 & 0 & 0 & 1 & 0 \\ -S_{12}x_4 & -S_{12}x_4 & 0 & C_{12} & 0 & 0 \\ 0 & 0 & 0 & 0 & 0 & 1 \\ C_{12}x_4 & C_{12}x_4 & 0 & S_{12} & 0 & 0 \\ 1 & 0 & 0 & 0 & 0 & 0 \\ 0 & 0 & 1 & 0 & 0 & 0 \end{bmatrix}, \tag{5.28}$$

$$\left| \frac{dT}{dx} \right| = 1 \cdot (-1) \cdot 1 \cdot (-1)\left(-S_{12}^2 x_4 - C_{12}^2 x_4\right) = -x_4. \tag{5.29}$$

The determinant was developed by the sixth, fifth and third columns, and the fifth row (according to the original numbering).

We shall assume that the following conditions are satisfied:

$$0 < v_{G,\min} \le v_G \le v_{G,\max}, \tag{5.30}$$

$$-\pi < \beta < \pi. \tag{5.31}$$

The second condition can be made stronger if the car moves in the path direction because in this case $\beta \in (-\pi/2, \pi/2)$. By the assumptions, it follows that $v_G = x_4$ is nonzero, assuring that the coordinate transformation is nonsingular, and x_4 and β are bounded.

After the coordinate transformation, the new variables are

$$z_o = (z_{11}, z_{12}, z_{21}, z_{22})^T, \tag{5.32a}$$

$$z_u = (x_1, x_3)^T = (\beta, r)^T. \tag{5.32b}$$

The stability of z_o is evident by the controller construction. The stability of the zero dynamics z_u has to be proven. We shall show that the zero dynamics is dominantly linear but the sate matrix contains the velocity $x_4 = v_G$ and the state equation contains additional bounded "disturbances". However, the stable system maps this disturbances into bounded signals.

Let us consider the state equations for $x_1 = \beta$ and $x_3 = \dot{\psi} = r$ under the developed control u_1 and u_2. Using earlier results, it follows

$$\bar{y}_1 = \lambda_1 w_1 - \alpha_{11}(x_4 C_{12}) - \lambda_1 x_5$$

$$= \lambda_1(w_1 - x_5) - \alpha_{11}(x_4 C_{12})$$

$$= \lambda_1(y_{d1} - x_5) - \alpha_{11}(x_4 C_{12}) + \alpha_{11}\dot{y}_{d1} + \ddot{y}_{d1},$$

$$\bar{y}_2 = \lambda_2 w_2 - \alpha_{12}(x_4 S_{12}) - \lambda_2 x_6$$

$$= \lambda_2(y_{d2} - x_6) - \alpha_{12}(x_4 S_{12}) + \alpha_{12}\dot{y}_{d2} + \ddot{y}_{d2},$$

which are bounded signals. Therefore, the developed control can be written as

$$u_1 = -S_h + m_v \overbrace{\left[(x_1 C_{12} - S_{12})\bar{y}_1 + (x_1 S_{12} + C_{12})\bar{y}_2\right]}^{d_{u1}(t)},$$

$$u_2 = m_v \underbrace{(C_{12}\bar{y}_1 + S_{12}\bar{y}_2)}_{d_{u2}(t)},$$

where $d_{u1}(t)$ and $d_{u2}(t)$ are also bounded signals. Using these results, we obtain

$$\dot{x}_1 = -x_3 + \frac{1}{m_v x_4} S_h + \frac{1}{m_v x_4} u_1 - \frac{x_1}{m_v x_4} u_2$$

$$= -x_3 + \frac{1}{m_v x_4} S_h + \underbrace{\frac{1}{m_v x_4}[-S_h + m_v d_{u1}(t)] - \frac{x_1}{m_v x_4} m_v d_{u2}(t)}_{d_1(t)}$$

$$= -x_3 + \frac{1}{x_4}[d_{u1}(t) - x_1 d_{u2}(t)],$$

$$\dot{x}_3 = -\frac{l_R}{I_{zz}} S_h + \frac{l_F}{I_{zz}} u_1$$

$$= -\frac{l_R}{I_{zz}} S_h + \frac{l_F}{I_{zz}}[-S_h + m_v d_{u1}(t)] = -\frac{l_R + l_F}{I_{zz}} S_h + \underbrace{\frac{l_F m_v}{I_{zz}} d_{u1}(t)}_{d_2(t)}$$

$$= -\frac{l_R + l_F}{I_{zz}} c_R\left[-x_1 + \frac{l_R}{v_G} x_3\right] + d_2(t)$$

$$= \frac{(l_R + l_F)c_R}{I_{zz}} x_1 - \frac{(l_R + l_F)c_R l_R}{v_G} x_3 + d_2(t).$$

By introducing the notations

$$A = \frac{(l_R + l_F)c_R}{I_{zz}} > 0, \qquad b = (l_R + l_F)c_R l_R > 0, \qquad (5.33)$$

it follows

$$\begin{pmatrix} \dot{x}_1 \\ \dot{x}_3 \end{pmatrix} = \begin{bmatrix} 0 & -1 \\ A & -\frac{b}{v_G} \end{bmatrix} \begin{pmatrix} x_1 \\ x_3 \end{pmatrix} + \begin{pmatrix} d_1(t) \\ d_2(t) \end{pmatrix}. \qquad (5.34)$$

For constant velocity $v_G = x_4$, the characteristic equation is

$$\begin{vmatrix} s & 1 \\ -A & s + \frac{b}{v_G} \end{vmatrix} = s\left(s + \frac{b}{v_G}\right) + A = s^2 + \frac{b}{v_G} s + A = 0, \qquad (5.35)$$

which proofs the stability of the zero dynamics because $A, b > 0$. Notice that the convolution of the decaying exponential matrix with the bounded disturbances resulults in bounded signal.

For nonconstant $v_G = x_4$, we can choose the Lyapunov function

$$V = \frac{1}{2}(Ax_1^2 + x_3^2) = \frac{1}{2}(A\beta^2 + r^2) \qquad (5.36)$$

and apply LaSalle's theorem. Since

$$\dot{V} = Ax_1\dot{x}_1 + x_3\dot{x}_3 = Ax_1(-x_3) + x_3\left(Ax_1 - \frac{b}{v_G}x_3\right) = -\frac{b}{v_G}x_3^2 \leq 0, \quad (5.37)$$

therefore $E = \{(x_1, x_3)^T : \dot{V} = 0\} = \{(x_1, 0)^T\}$. Its maximal invariant set $M \subset E$ satisfies $x_3 = 0$. Since trajectories starting in M remain also in M, therefore $x_3 \equiv 0$ and $\dot{x}_3 \equiv 0$, from which follows $0 = Ax_1$ and $x_1 = 0$. Hence, the maximal invariant set is the single point $M = \{(0, 0)^T\}$ and by LaSalle's theorem the system is asymptotically stable for $v_G \neq 0$.

Another possibility would be to choose a quadratic Lyapunov function $z_u^T P z_u$ with constant positive definite matrix $P = P^T > 0$ for all $1/v_G$ satisfying

$$\frac{1}{v_{G,max}} \leq \frac{1}{v_G} \leq \frac{1}{v_{G,min}}, \quad (5.38)$$

which is an LMI problem solvable with MATLAB Robust Control Toolbox, and apply quadratic stability theorems.

Remark The concept based on the properties of z_o, z_u proofs also the stability of the tracking problem under the developed control in the following interpretation. Consider the reference robot defined by the desired path and the real car. The path and the car's state equation are given in (quasi) inertial system hence we can define the error between reference car and real car through the "distance" between them. We can introduce new variables

$$e_1 = X - X_d,$$
$$e_2 = Y - Y_d,$$
$$e_3 = \psi + \beta - \psi_d,$$
$$e_4 = v_G - v_{G,d},$$
$$e_5 = \beta,$$
$$e_6 = \dot{\psi} = r,$$

where e_1, e_2, e_4 go to zero and e_3 goes to β which is bounded. Thus, the real car follows the reference car with limited error caused by β and r which cannot identically be zero because the car is underactuated. The stability of the zero dynamics assures the existence of the so called inverse dynamical model, too.

5.5.3 Simulation Results Using DGA Method

Based on the DGA method (internal input/output linearization and external linear state feedback) simulation experiments were performed from which we show the

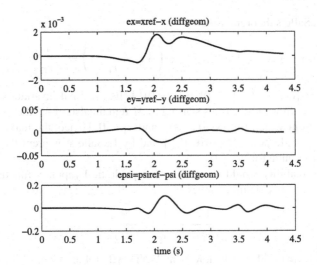

Fig. 5.3 Error signals using DGA method with measurable states and input affine model

results using approximated input affine model and measurable states. All signals are in SI units (m, rad, m/s, rad/s, N, N m etc.). Parameters of the dynamic model are $c_F = 100000$, $l_F = 1.203$, $c_R = 100000$, $l_R = 1.217$, $m_v = 1280$, and $I_{zz} = 2500$. For the external feedback, $\lambda = 10$ was chosen. The controller was discretized in time, that is, the continuous time DGA Control Algorithm was evaluated in integer multiples of the sampling time $T = 0.01$ s and held fixed for the sampling period.

Figure 5.3 shows the position and orientation errors. The control signals are drawn in Fig. 5.4(a). The states were assumed measurable, the state variables during DGA control are shown in Fig. 5.4(b).

The simulation results show that the errors and control signals are in acceptable order if we take into consideration the discrete time control realization and the underactuated property of the car.

The developed software makes it also possible to use state estimation and the full nonlinear model acting to the control, but these extensions were not used in the simulation example.

5.6 Receding Horizon Control

Receding horizon control [25, 73, 87] optimizes a cost function in open loop using the prediction of the system future behavior based on the dynamic model of the system, determines the future optimal control sequence within the horizon, applies the first element of the control sequence in closed loop to the real system, and repeats these steps for the new horizon which is the previous one shifted by the sampling time T. An open question is the stability of the closed loop nonlinear system, but the chance for stability is increasing with increasing the horizon length N (time NT).

Fig. 5.4 Control signals and measurable states using DGA method and input affine model

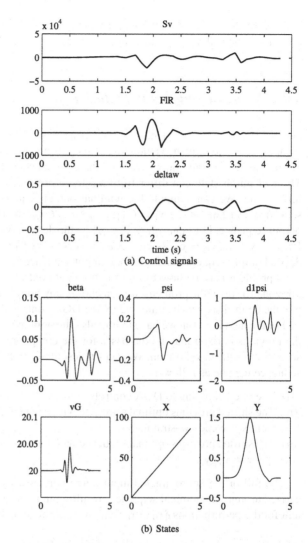

(a) Control signals

(b) States

If the nonlinear dynamic model is used for prediction, then the optimization is usually a nonlinear optimization problem in real-time which is time-critical for fast systems. Hence, linearization around the prescribed nominal trajectory may be suggested and optimization of the perturbations using quadratic cost and analytically manageable end-constraint seem to be a good compromise. This is also our concept [73]. However, it is a serious problem that the desired CAS emergency path can not easily be converted to the desired control which makes the linearization in the horizon more difficult (and hence nominal is not equal desired). The reason is that cars are underactuated hence not all paths can be (exactly) realized by control. However, if a control has already been determined in the previous horizon and the errors are small relative to the prescribed path, then this control can be used as nominal control for the actual horizon, for which we want to find optimal perturbations decreasing the path error in the actual horizon.

5.6.1 Nominal Values and Perturbations

Denote $\{u_0, u_1, \ldots, u_{N-1}\}$ the nominal control sequence within the horizon and compute the state sequence belonging to it by using the nonlinear state space model of the system. Denote $\{x_0, x_1, \ldots, x_N\}$ the nominal state determined in such a way. The real initial state can differ from x_0, therefore let $\hat{x}_0$ be the estimation of the initial state. We can compute the nominal output sequence belonging to the state sequence. Let $\{y_0 = Cx_0, y_1 = Cx_1, \ldots, y_N = Cx_N\}$ be the so computed nominal output sequence. Let $\{y_{d0}, y_{d1}, \ldots, y_{dN}\}$ be the desired output sequence and denote $\{e_0 = y_{d0} - y_0, e_1 = y_{d1} - y_1, \ldots, e_N = y_{dN} - y_N\}$ the error sequence between desired output sequence and nominal output sequence.

A problem may be how to find the nominal control sequence for the very first horizon. To solve this problem, we can choose any nonpredictive method, for example, the control sequence computed by the DGA control algorithm.

For the next horizon, we have to find also the nominal control sequence, however the previous horizon may be the bases for the computation of the nominal control sequence for the next horizon. We can suggest three strategies. The nominal control sequence may be as follows:

(i) The one belonging to DGA control.
(ii) The shifted previous optimal sequence with one new element at the end derived from some end-constraint.
(iii) The shifted previous optimal sequence with the last element repeated at the end (poor strategy).

The full or the approximated input affine nonlinear dynamic model can be linearized around the nominal sequences resulting in a linear time-varying (LTV) system for the perturbations $\delta x_0 = \hat{x}_0 - x_0, \delta x_1, \ldots, \delta x_N$ and $\delta u_0, \ldots, \delta u_{N-1}$:

$$\delta x_{i+1} = A_i \delta x_i + B_i \delta u_i, \tag{5.39}$$

$$\delta y_i = C \delta x_i. \tag{5.40}$$

The output errors are

$$y_{di} - C(x_i + \delta x_i) = e_i - \delta y_i. \tag{5.41}$$

Notice that the indexing starts from zero in each horizon.

The transients of the perturbed LTV system can be formulated based on the generalization of the following rules:

$$\delta x_1 = A_0 \delta x_0 + B_0 \delta u_0,$$

$$\delta x_2 = A_1(A_0 \delta x_0 + B_0 \delta u_0) + B_1 \delta u_1 = A_1 A_0 \delta x_0 + A_1 B_0 \delta u_0 + B_1 \delta u_1,$$

$$\delta x_3 = A_2(A_1 A_0 \delta x_0 + A_1 B_0 \delta u_0 + B_1 \delta u_1) + B_2 \delta u_2$$
$$= A_2 A_1 A_0 \delta x_0 + A_2 A_1 B_0 \delta u_0 + A_2 B_1 \delta u_1 + B_2 \delta u_2.$$

The transients in the horizon can be computed as follows:

$$
\begin{pmatrix} \delta x_1 \\ \delta x_2 \\ \vdots \\ \delta x_{N-1} \\ \cdots \\ \delta x_N \end{pmatrix}
=
\begin{bmatrix} A_0 \\ A_1 A_0 \\ \vdots \\ A_{N-2} \cdots A_1 A_0 \\ \cdots\cdots\cdots \\ A_{N-1} \cdots A_1 A_0 \end{bmatrix} \delta x_0
$$

$$
+
\begin{bmatrix}
B_0 & 0 & \cdots & 0 & 0 \\
A_1 B_0 & B_0 & \vdots & 0 & 0 \\
\vdots & \vdots & \ddots & \vdots & \vdots \\
A_{N-2} \cdots A_1 B_0 & A_{N-2} \cdots A_2 B_1 & \cdots & B_{N-2} & 0 \\
\cdots\cdots\cdots\cdots\cdots\cdots\cdots\cdots\cdots\cdots\cdots\cdots \\
A_{N-1} \cdots A_1 B_0 & A_{N-1} \cdots A_2 B_1 & \cdots & A_{N-1} B_{N-2} & B_{N-1}
\end{bmatrix}
$$

$$
\times
\begin{pmatrix} \delta u_0 \\ \delta u_1 \\ \vdots \\ \delta u_{N-2} \\ \delta u_{N-1} \end{pmatrix}.
$$

Now we are able to predict the output perturbations as a function of the perturbation of the initial state and the perturbations of the control inputs in the horizon:

$$
\begin{pmatrix} \delta y_1 \\ \delta y_2 \\ \vdots \\ \delta y_{N-1} \\ \cdots \\ \delta y_N \end{pmatrix}
=
\begin{bmatrix} C A_0 \\ C A_1 A_0 \\ \vdots \\ C A_{N-2} \cdots A_1 A_0 \\ \cdots\cdots\cdots \\ C A_{N-1} \cdots A_1 A_0 \end{bmatrix} \delta x_0
$$

$$
+
\begin{bmatrix}
C B_0 & 0 & \cdots & 0 & 0 \\
C A_1 B_0 & C B_0 & \vdots & 0 & 0 \\
\vdots & \vdots & \ddots & \vdots & \vdots \\
C A_{N-2} \cdots A_1 B_0 & C A_{N-2} \cdots A_2 B_1 & \cdots & C B_{N-2} & 0 \\
\cdots\cdots\cdots\cdots\cdots\cdots\cdots\cdots\cdots\cdots\cdots\cdots \\
C A_{N-1} \cdots A_1 B_0 & C A_{N-1} \cdots A_2 B_1 & \cdots & C A_{N-1} B_{N-2} & C B_{N-1}
\end{bmatrix}
$$

$$\times \begin{pmatrix} \delta u_0 \\ \delta u_1 \\ \vdots \\ \delta u_{N-2} \\ \delta u_{N-1} \end{pmatrix}. \tag{5.42}$$

In order to make the remaining discussion easier to understand, the following notations will be introduced in accordance with the structure of (5.42):

$$\delta U = \left(\delta u_0^T, \delta u_1^T, \ldots, \delta u_{N-1}^T \right)^T, \tag{5.43}$$

$$\begin{pmatrix} \delta y_1 \\ \delta y_2 \\ \vdots \\ \delta y_{N-1} \end{pmatrix} = P_1 \delta x_0 + H_1 \delta U, \tag{5.44}$$

$$\delta y_N = P_2 \delta x_0 + H_2 \delta U. \tag{5.45}$$

Starting from the continuous time nonlinear system, the state matrices of the LTV system for perturbations can easily be found using Euler formula for continuous-to-discrete time conversion:

$$\dot{x} = f_c(x, u), \tag{5.46}$$

$$x_{i+1} = x_i + T f_c(x_i, u_i), \tag{5.47}$$

$$A_i = I + T \frac{\partial f_c}{\partial x} \bigg|_{(x_i, u_i)}, \qquad B_i = T \frac{\partial f_c}{\partial u} \bigg|_{(x_i, u_i)}. \tag{5.48}$$

Notice that we have assumed that the output matrix C is constant which is evidently satisfied for cars because $y_1 = X = x_5$ and $y_2 = Y = x_6$.

5.6.2 RHC Optimization Using End Constraint

The cost function J is chosen to be a quadratic function penalizing both output errors and large deviations from the nominal control, furthermore we prescribe the constraint that the output error should be zero at the end of the horizon:

$$J = \frac{1}{2} \sum_{i=1}^{N-1} \| e_i - \delta y_i \|^2 + \frac{1}{2} \lambda \sum_{i=0}^{N-1} \| \delta u_i \|^2, \tag{5.49}$$

$$e_N - \delta y_N = e_N - (P_2 \delta x_0 + H_2 \delta U) = 0. \tag{5.50}$$

The problem can analytically be solved by using Lagrange multiplier rule. It is a compromise because for small sampling time the constraint optimization with bounds for δu and/or u is rarely possible in real-time. The order of input perturbation can be influenced with the choice of the value of λ. Larger value of λ penalizes more the δu perturbations. Because λ is reserved for weighting factor of the control perturbations, the vector Lagrange multiplier will be denoted by μ.

The concept of RHC optimization is simple. First, the cost function is augmented with the constraint multiplied by the vector Lagrange multiplier, then it will be differentiated and the derivative will be made to zero. The solution $\delta U(\mu)$ contains the Lagrange multiplier. Hence, the solution is substituted into the constraint equation which contains μ in a linear way. From the resulting linear equation, μ can be expressed and put into $\delta U(\mu)$. The result is the optimal control perturbation sequence δU. The details are as follows:

$$E = \left(e_1^T, e_2^T, \ldots, e_{N-1}^T\right)^T,$$

$$J = \frac{1}{2}\langle E - (P_1\delta x_0 + H_1\delta U), E - (P_1\delta x_0 + H_1\delta U)\rangle + \frac{1}{2}\lambda\langle \delta U, \delta U\rangle,$$

$$L = \frac{1}{2}\langle E - (P_1\delta x_0 + H_1\delta U), E - (P_1\delta x_0 + H_1\delta U)\rangle + \frac{1}{2}\langle \delta U, \Delta U\rangle$$
$$+ \langle \mu, e_N - P_2\delta x_0 - H_2\delta U\rangle$$

$$= \frac{1}{2}\big[\langle E, E\rangle + \langle P_1\delta x_0, P_1\delta x_0\rangle + \langle H_1\delta U, H_1\delta U\rangle + \lambda\langle \delta U, \delta U\rangle\big]$$
$$- \langle E, P_1\delta x_0\rangle - \langle E, H_1\delta U\rangle + \langle P_1\delta x_0, H_1\delta U\rangle$$
$$+ \langle \mu, e_N\rangle - \langle \mu, P_2\delta x_0\rangle - \langle \mu, H_2\delta U\rangle$$

$$= \frac{1}{2}\big[\langle E, E\rangle + \langle P_1\delta x_0, P_1\delta x_0\rangle + \langle H_1^T H_1\delta U, \delta U\rangle + \lambda\langle \delta U, \delta U\rangle\big]$$
$$- \langle E, P_1\delta x_0\rangle - \langle H_1^T E, \delta U\rangle + \langle H_1^T P_1\delta x_0, \delta U\rangle$$
$$+ \langle \mu, e_N\rangle - \langle \mu, P_2\delta x_0\rangle - \langle H_2^T \mu, \delta U\rangle,$$

$$\frac{dL}{d\delta U} = \overbrace{\left(H_1^T H_1 + \lambda I\right)}^{L_1}\delta U - H_1^T E + H_1^T P_1\delta x_0 - H_2^T \mu = 0,$$

$$\delta U(\mu) = L_1^{-1}\big[H_1^T E - H_1^T P_1\delta x_0 + H_2^T \mu\big],$$

$$0 = e_N - P_2\delta x_0 - H_2 L_1^{-1}\big[H_1^T E - H_1^T P_1\delta x_0 + H_2^T \mu\big]$$
$$= e_N - H_2 L_1^{-1} H_1^T E - \left(P_2 - H_2 L_1^{-1} H_1^T P_1\right)\delta x_0 - \underbrace{\left(H_2 L_1^{-1} H_2^T\right)}_{L_\mu}\mu,$$

$$\mu = L_\mu^{-1}\big[e_N - H_2 L_1^{-1} H_1^T E - \left(P_2 - H_2 L_1^{-1} H_1^T P_1\right)\delta x_0\big].$$

The final result can be summarized in the following form:

$$L_1 := H_1^T H_1 + \lambda I, \qquad L_\mu := H_2 L_1^{-1} H_2^T,$$

$$\delta U = L_1^{-1} \{ H_2^T L_\mu^{-1} e_N + (I - H_2^T L_\mu^{-1} H_2 L_1^{-1}) H_1^T E \qquad (5.51)$$

$$- [H_1^T P_1 + H_2^T L_\mu^{-1} (P_2 - H_2 L_1^{-1} H_1^T P_1)] \delta x_0 \}.$$

The closed loop control is $u_0 + \delta u_0$ where u_0 is the nominal control and δu_0 is the first element of the open loop optimal sequence δU.

RHC Control Algorithm
In every horizon, the following steps are repeated:

Step 1. From the initial state x_0 and the nominal control $\{u_0, u_1, \ldots, u_{N-1}\}$, the nominal state sequence $\{x_0, x_1, \ldots, x_N\}$ in the horizon is determined by using the discrete time nonlinear dynamic model of the car. Here, x_0 is coming from the shifted previous horizon and can differ from the estimated state $\hat{x}_0$. The desired state sequence is the one computed from the CAS emergency path with zero side slip angle. The output is assumed to be $y = (X, Y)^T$, hence the desired output sequence can easily be computed from the desired state sequence in the horizon using $C = [e_5 e_6]^T$ (here, e_5, e_6 are standard unit vectors). The nominal output sequence belonging to the nominal state sequence can also be computed using C. The nominal error sequence is the difference between desired output and nominal output sequences. In case of the very first horizon, the nominal control sequence is computed by the DGA method and x_0 is initialized from the desired CAS path with zero side slip angle.

Step 2. The discrete time LTV model $\delta x_{i+1} = A_i \delta x_i + B_i \delta u_i$ is determined from the approximated nonlinear model $\dot{x} = f_c(x, u)$ using Euler formula:

$$A_i := I + T \, df_c/dx|_{(x_i, u_i)}, \qquad B_i := T \, df_c/du|_{(x_i, u_i)}.$$

Step 3. The optimal change δU of the control sequence is computed by (5.51) using $\delta x_0 = \hat{x}_0 - x_0$ where $\hat{x}_0$ is the estimated state. The optimal control sequence is $U := U + \delta U$. Its first element u_0 will be applied in closed loop.

Step 4. In order to initialize the control sequence for the next horizon, the optimal state sequence belonging to the initial state x_0 and the optimal control sequence $\{u_0, u_1, \ldots, u_{N-1}\}$ are determined using the approximated nonlinear dynamic model. The result at the end of the transients is the state x_N. The unknown new u_N can be determined in three ways:

  (i) u_N is determined by using x_N and the DGA method.

  (ii) u_N is computed in such a way that the difference between x_{N+1} computed from the discrete time nonlinear input affine model $x_{i+1} = x_i + f(x_i) + G(x_i)u_i$ and $x_{d,N+1}$ according to the CAS path is minimized in LS-sense. The goal and the solution are, respectively,

$$x_{d,N+1} - x_N - f(x_N) = G(x_N)u_N,$$

$$u_N = G(x_N)^+ \{x_{d,N+1} - x_N - f(x_N)\},$$

where G^+ is LS pseudoinverse.

(iii) The last control signal is simply repeated (poor strategy): $u_N := u_{N-1}$.

Step 5. The nominal control sequence for the next horizon is $\{u_1, u_2, \ldots, u_N\}$ which is the augmented optimal control sequence $\{u_0, u_1, \ldots, u_N\}$ shifted by 1.

Remark Linearizing the discrete time nonlinear model in u_{-1} and $\hat{x}_0$ yields LTI system for the actual horizon. RHC based on LTI model is simpler but less precise for time varying paths. Hence, we preferred the technique based on LTV model.

Integral control
It is possible to put integrator into the controller using augmented state $\delta x_i :=$ $(\delta x_i^T, \delta u_{i-1}^T)^T$ and $\delta u_i = \delta u_{i-1} + \delta r_i$ where the change of the control δr_i has to be optimized. Substituting

$$A_i := \begin{bmatrix} A_i & B_i \\ 0 & I \end{bmatrix} \quad \text{and} \quad B_i := \begin{bmatrix} B_i \\ I \end{bmatrix},$$

the earlier results remain valid in the new variables. However, in this case, δR is the optimal change of the control differences and the optimal δU is the cumulative sum of δR.

5.7 State Estimation Using GPS and IMU

It is assumed that the measured signals are supplied by a two antenna GPS system and an IMU (Inertial Measurement Unit) containing accelerometers and gyroscopes.

The evaluation of GPS/IMU signals is based on the results of Ryu and Gerdes [125]. It is assumed that the GPS/IMU system software package provides high level information in form of the following signals:

V_m^{GPS}: measured velocity of the car in the GPS coordinate system
ψ_m^{GPS}: the orientation of the car in the GPS coordinate system
$a_{x,m}$: longitudinal acceleration of the car in the car fixed coordinate system
$a_{y,m}$: transversal acceleration of the car in the car fixed coordinate system
r_m: the angular velocity of the car in z (yaw) direction in the car fixed coordinate system

It is assumed that the software package expresses the GPS information in the NED frame belonging to the ground vehicle moving in the horizontal plane. It is assumed that the first antenna is immediately above the IMU system which is located in the COG (center of gravity), otherwise small modifications are needed. The sensors have biases which have also to be estimated. The accuracy of sensors can be characterized by the σ value of the noise and the bias. The additive Gaussian noise is concentrated into the domain $[-3\sigma, 3\sigma]$ to which comes yet the value of the bias.

For a vehicle moving in the horizontal plane, $V_m^{GPS} = (V_1^{GPS}, V_2^{GPS}, 0)^T$ is satisfied, from which the measured value of the side slip angle can be determined and from it the components of the car velocity u can be computed:

$$\gamma = \text{atan2}(V_2^{GPS}, V_1^{GPS}) \quad \Rightarrow \quad \beta^{GPS} = \gamma - \psi_m^{GPS}, \tag{5.52}$$

$$u_{x,m}^{GPS} = \|V_m^{GPS}\| \cos(\beta^{GPS}) + noise, \tag{5.53}$$

$$u_{y,m}^{GPS} = \|V_m^{GPS}\| \sin(\beta^{GPS}) + noise. \tag{5.54}$$

The state estimation can be based on a two stage Kalman filter. The first stage estimates the angular velocity $\dot{\psi} = r$. For this purpose, two methods can be suggested, the second method estimate the sensitivity s_r of the sensor too:

$$\begin{pmatrix} \dot{\psi} \\ \dot{r}_{bias} \end{pmatrix} = \begin{bmatrix} 0 & -1 \\ 0 & 0 \end{bmatrix} \begin{pmatrix} \psi \\ r_{bias} \end{pmatrix} + \begin{pmatrix} 1 \\ 0 \end{pmatrix} r_m + noise, \tag{5.55a}$$

$$\psi_m^{GPS} = \begin{bmatrix} 1 & 0 \end{bmatrix} \begin{pmatrix} \psi \\ r_{bias} \end{pmatrix} + noise, \tag{5.55b}$$

$$\frac{d}{dt} \begin{pmatrix} \psi \\ 1/s_r \\ r_{bias}/s_r \end{pmatrix} = \begin{bmatrix} 0 & r_m & -1 \\ 0 & 0 & 0 \\ 0 & 0 & 0 \end{bmatrix} \begin{pmatrix} \psi \\ 1/s_r \\ r_{bias}/s_r \end{pmatrix} + noise, \tag{5.56a}$$

$$\psi_m^{GPS} = \begin{bmatrix} 1 & 0 & 0 \end{bmatrix} \begin{pmatrix} \psi \\ 1/s_r \\ r_{bias}/s_r \end{pmatrix}^T + noise. \tag{5.56b}$$

The state estimation is performed by Kalman filter, thus the continuous time models will be converted to discrete time ones:

$$A_{d1} = \begin{bmatrix} 1 & -T \\ 0 & 1 \end{bmatrix}, \qquad B_{d1} = \begin{bmatrix} T \\ 0 \end{bmatrix}, \qquad C_{d1} = \begin{bmatrix} 1 & 0 \end{bmatrix}, \tag{5.57a}$$

$$A_{d1} = I_3 + A_{c1}T, \qquad B_{d1} = 0_{3\times 1}, \qquad C_{d1} = \begin{bmatrix} 1 & 0 & 0 \end{bmatrix}. \tag{5.57b}$$

From the estimated values, the angular velocity can be computed:

$$r = \hat{\dot{\psi}} := -\hat{r}_{bias} + r_m, \tag{5.58a}$$

$$r = \hat{\dot{\psi}} := r_m(1/\hat{s}_r) - (\hat{r}_{bias}/\hat{s}_r). \tag{5.58b}$$

This value r of the angular velocity can be applied in the second Kalman filter which is based on the relation $a = \dot{u} + \omega \times u$ from which the following continuous time model arises for the estimation of the velocities:

$$\frac{d}{dt} \begin{pmatrix} u_x \\ a_{x,bias} \\ u_y \\ a_{y,bias} \end{pmatrix} = \begin{bmatrix} 0 & -1 & r & 0 \\ 0 & 0 & 0 & 0 \\ -r & 0 & 0 & -1 \\ 0 & 0 & 0 & 0 \end{bmatrix} \begin{pmatrix} u_x \\ a_{x,bias} \\ u_y \\ a_{y,bias} \end{pmatrix}$$

$$+ \begin{bmatrix} 1 & 0 \\ 0 & 0 \\ 0 & 1 \\ 0 & 0 \end{bmatrix} \begin{pmatrix} a_{x,m} \\ a_{y,m} \end{pmatrix} + noise. \tag{5.59}$$

The model contains r which is varying in real-time. Hence, it is useful that the discrete time form of the model can analytically be found:

$$A_{d2} = \begin{bmatrix} C(rT) & \frac{-S(rT)}{r}/r & S(rT) & \frac{-[1-C(rT)]}{r} \\ 0 & 1 & 0 & 0 \\ -S(rT) & \frac{[1-C(rT)]}{r} & C(rT) & \frac{S(rT)}{r} \\ 0 & 0 & 0 & 1 \end{bmatrix}, \tag{5.60}$$

$$B_{d2} = \begin{bmatrix} S(rT)/r & [1-C(rT)]/r \\ 0 & 0 \\ -[1-C(rT)]/r & S(rT)/r \\ 0 & 0 \end{bmatrix}, \tag{5.61}$$

where in the system matrices C and S stand for cos and sin, respectively.

The estimation is performed by the second Kalman filter. From the estimated values $\hat{u}_x$, $\hat{u}_y$ of the second Kalman filter together with $\hat{\psi}$ of the first Kalman filter the remaining state variables can be estimated:

$$\hat{v}_G = \sqrt{\hat{u}_x^2 + \hat{u}_y^2}, \tag{5.62}$$

$$\hat{\beta} = \text{atan} \, 2(\hat{u}_y, \hat{u}_x). \tag{5.63}$$

Since the state variables X, Y of the position are not supported by (accurate enough DGPS) absolute measurements, hence their values were determined by numerical integration:

$$\hat{X} = \hat{X} + T\hat{v}_G \cos(\hat{\psi} + \hat{\beta}), \tag{5.64}$$

$$\hat{Y} = \hat{Y} + T\hat{v}_G \sin(\hat{\psi} + \hat{\beta}). \tag{5.65}$$

Kalman filters were implemented in the following form (Q and R are, respectively, the covariances of the system and measurement noises, and $x_+ = \hat{x}$ is the estimated state):

Time update:

$$x_-(t+1) = A_d x_+(t) + B_d u(t),$$
$$P_-(t+1) = A_d P_+(t) A_d^T + Q. \tag{5.66}$$

Measurement update:

$$x_+(t) = x_-(t) + K\big[y(t) - Cx_-(t)\big],$$

$$K = P_-(t)C^T\big[CP_-(t)C^T + R\big]^{-1}, \tag{5.67}$$

$$P_+(t) = [I - KC]P_-(t).$$

The measurement sampling times (frequencies) were $T_{\text{IMU}} = T = 0.01$ s (100 Hz), $T_{\text{GPS,vel}} = 0.1$ s (10 Hz), $T_{\text{GPS,att}} = 0.2$ s (5 Hz).

5.8 Simulation Results with RHC Control and State Estimation

Based on the developed theory, simulation results will be presented for the realization of the CAS reference path using predictive control and state estimation. The dimensions of all signals are to be understood in SI units (m, rad, m/s, rad/s, N, N m etc.). Parameters of the dynamic model are $c_F = 100000$, $l_F = 1.203$, $c_R = 100000$, $l_R = 1.217$, $m_v = 1280$, and $I_{zz} = 2500$. The sampling time was chosen $T = 0.01$ s for RHC Control Algorithm. The sampling times (frequencies) of sensor signals have already been given above.

During the simulation example, integrating RHC controller was used. The horizon length is $N = 10$, that is, $NT = 0.1$ s. The initialization of the very first horizon and all other ones uses DGA technique.

The "real" sensor signals are computed as the system answer for the control signal in the actual sampling interval. To the simulated states, Gaussian noises with realistic sigma are added and the covariance matrices are computed from the sigma values. The noisy signals contain also biases (offsets) which are also estimated in the two level Kalman filters.

Figure 5.5(a) shows the position and orientation errors. The control signals are drawn in Fig. 5.5(b). The states and estimated states are shown in Fig. 5.6(a), especially the side slip angle and its estimated value are shown in Fig. 5.6(b). Real and estimated states are almost coinciding in the figures. The side slip angle illustrates the order of the difference between real (simulated) and estimated state.

The simulation results show that the errors and control signals are in acceptable order if we take into consideration the underactuated property of the car.

5.9 Software Implementations

For advanced controller realizations, we usually leave the MATLAB environment. Two control softwares were elaborated for the realization of the control algorithms. During the development, MATLAB R2006a was available for us thus the discussion is based on this MATLAB version.

Fig. 5.5 Error and control signals using RHC method and state estimation

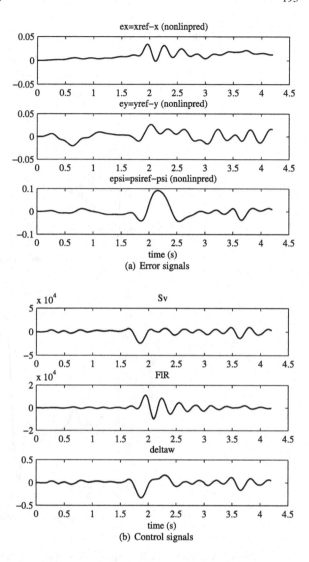

(a) Error signals

(b) Control signals

5.9.1 Standalone Programs

The first approach uses the MATLAB Compiler for converting the whole control system into a separate program running under Windows or Linux without the need of MATLAB license (standalone program). It is useful to know that MATLAB is an interpreted language containing 3 interpreters (MATLAB, Simulink, Java). On the other hand, standalone programs do not need the MATLAB license from which follows that they do not contain the interpreters. Another problem is that leaving the MATLAB environment cannot be made independent of the C/C++ development environment (including libraries) that contains the MATLAB Compiler and

Fig. 5.6 Real and estimated
state variables using RHC
method

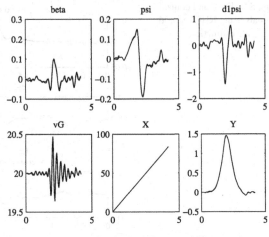

(a) Real and estimated states variables

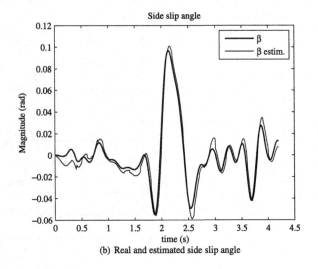

(b) Real and estimated side slip angle

the operating system under which the standalone program will run. We decided for
Microsoft Visual C++ 6.0 and Windows operating system.

Although MATLAB R2006a and its toolboxes are object oriented, because of
the differences between the development and the executive systems not all services of MATLAB toolboxes are permitted for use in standalone programs. Fortunately, some toolboxes are exceptions and amongst them is the Optimization Toolbox whose almost all functions can be used in standalone programs. This is important because the CAS path design uses the function `fsolve` of the Optimization
Toolbox and the spline technique of MATLAB.

The MATLAB Compiler converts the application program into a special CTF
(Component Technology File) format depending on the executing system. Unfortunately CTF is not a C/C++ code (except two components for start). Special docu-

mentation exists describing the installation of standalone programs for the running environment and how to start running. If the standalone program is started, then first the CTF code is interpreted followed by the program execution. Some constraints regarding input/output should be kept, but fortunately reading control parameters from prepared files and storing figures are possible. The use of shared libraries (DLL etc.) is allowed hence this technique can be used for calling drivers for sensors and actuators, and scheduling execution.

5.9.2 *Quick Prototype Design for Target Processors*

The second approach uses MATLAB/Simulink and Real-Time Workshop (RTW) to convert the control system for target processors (MPC555, AutoBox etc.). Unfortunately, this approach does not allow the use of toolboxes, hence path design (using `fsolve` and spline technique) cannot be part of the Simulink model to be converted. It means that a separate path design program has to be developed for the target system (using its standard development tools). Communication between the two parts can be realized through memory or file transfers.

The Simulink model of the controller can be based on Simulink S-functions in C/C++ or the so called Embedded MATLAB Function technology allowed in Simulink. We decided mainly for this latter one because it allows the use of an important subset of linear algebraic functions (qr, svd etc.) which are used in controller implementation. Fusion of the two techniques is also possible, furthermore standard (amplifier, integrator, multiplexer, demultiplexer etc.) and/or special (CAN communication) Simulink blocks can be used, too.

The development of the controller written in Simulink should always contain the test of the controller together with the system model using fixed step size and simple solution technique (Euler method), because this is the only allowed method for the conversion in the sequence Simulink $\rightarrow$ RTW $\rightarrow$ Target Processor. It is a hard constraint that the conversion does not allow the use of variable dimension in data specifications (arrays etc.) whose memory size cannot be identified in the compilation phase, that is, the use of concepts needing dynamic memory reservation is forbidden.

The result of the conversion is a source file in C which can be compiled and loaded with the development tools of the target system. By using this approach and taking into consideration the limits, we have successfully developed controllers realizing DGA and RHC control methods for dSPACE AutoBox, DS1103 board (running under ControlDesk) and MPC555 target processor (using CodeWarrior development environment).

5.10 Closing Remarks

High level nonlinear control of cars was considered in the chapter. The problem was divided into path design, control development and test, and software realization based on advanced control platforms. The steps were illustrated in the frame of the design of a Collision Avoidance System (CAS).

The CAS path design was solved by using the principle of elastic band amongst static and dynamic obstacles. Typical obstacles may be the lost load in the lane of the own car and an incoming car in the other lane of the road. The elastic band reduces the problem of finding an aesthetic path with acceptable curvature to the solution of a large set of nonlinear equations. This task was solved by using the function fsolve of MATLAB Optimization Toolbox. Supporting the optimization with the derivatives (Jacobian) of the vector-vector functions it was possible to find the solution within 1 s. The own car was sent through the elastic band with its known average velocity in order to find initial time functions. The smoothed time functions of the path were determined by using the spline technique of MATLAB.

The full (non-affine) and the approximated (simplified input affine) dynamic models of the car are used in the design of high level control algorithms. For input-output linearization, a method was presented based on differential geometric principles of nonlinear control theory. This method, called DGA algorithm, simultaneously solves the internal input-output linearization and external linear state feedback problem assuring stable second order error dynamics. It was proven that the zero dynamics is stable. The DGA method can be used independently for control or may be a tool of finding the initial control sequence for the very first horizon in predictive control. Simulation example illustrates the efficiency of the control for measurable states, although state estimation is also possible in the developed system.

Because of high computational costs of nonlinear RHC (Receding Horizon Control) with constraints a compromise solution was suggested. In our approach, the solution of the shifted optimal control sequence in the previous horizon is considered as nominal control in the actual horizon and the perturbations around it are optimized regarding a quadratic cost function containing the errors with respect to the reference path and the weighted control perturbations. The only constraint is that the error at the end of the horizon should be zero. The nonlinear system in each horizon is approximated by LTV model. Neglecting other constraints the optimization problem can be analytically solved using Lagrange multiplier rule. Integral control is also possible. For finding the missing element of the control sequence for the next horizon, three methods are suggested. This approach takes into consideration that the system is underactuated, the desired paths assumes zero side slip angle, hence usually no control exists exactly realizing the desired path. RHC control may be the tool to find the realizable suboptimal control that minimizes the path errors. Since this control is not known a priori therefore the linearization in the actual horizon is performed based on the nominal control derived from the previous horizon. The estimated state belonging to the begin of the horizon can be taken into consideration in the optimization. This version of RHC gives a good chance for real-time realization because the computational costs are low.

State estimation is an important subtask of control system design. It was assumed that the measured signals are supplied by a two antenna GPS system and an IMU containing accelerometers and gyroscopes. For state estimation, a two-level approach is suggested based on Kalman filters. The first Kalman filter estimates the angular velocity and its bias. The such found estimated angular velocity is used in the second Kalman filter to find the estimations of the velocities and the biases of the accelerations.

The efficiency of RHC and state estimation was illustrated in a simulation example. The controller was an RHC one containing integrators, the sampling time was 0.01 s and the horizon length was 10. During the simulation, the sensory system was also emulated. The noisy signals contained biases, too, the estimators worked from the noisy signals. The estimated states well cover the real states, that is, the simulated noiseless states. The errors are in acceptable order considering the underactuated property of the car, and the control signals are realistic regarding the mass of the car.

Two software approaches are presented for realization of the control algorithms. The first uses the MATLAB Compiler converting the whole control system into a standalone program running under Windows or Linux. The services of the Optimization Toolbox are usable in the standalone program, but sensor drivers and time scheduling has to be solved using shared libraries. The second approach uses MATLAB/Simulink and Real-Time Workshop (RTW) to convert the control system for target processors (MPC555, AutoBox etc.). This version can contain high level tools (CAN bus etc.) for data communication between controller and sensors, but the functions of the Optimization Toolbox cannot be used. The conversion needs fixed step size and simple Euler method in the Simulink model of the controller, hence the system has also to be tested under these conditions before conversion.

Chapter 6
Nonlinear Control of Airplanes and Helicopters

Overview This chapter deals with the predictive control of airplanes and the back-stepping control of quadrotor helicopters. The airplane is an LPV system which is internally stabilized by disturbance observer and externally controlled by high level RHC. The control design is illustrated for the pitch rate control of an aircraft. The internal system is linearized at the begin of every horizon and RHC is applied for the resulting LTI model. The RHC controller contains integrators. The management of hard and soft constraints and the use of blocking parameters are elaborated. Simulation examples show the disturbance suppressing property of disturbance observer and the efficiency of high level RHC. For the indoor quadrotor helicopter, two (precise and simplified) dynamic models are derived. The sensory system contains onboard IMU and on-ground vision subsystems. A detailed calibration technique is presented for IMU. The vision system is based on motion-stereo and a special virtual camera approach. For state estimation, two level EKF is used. The quadrotor helicopter is controlled using backstepping in hierarchical structure. The embedded control system contains separate motor and sensor processors communicating through CAN bus with the main processor MPC555 realizing control and state estimation. Simulation results are shown first with MATLAB/Simulink then during the hardware-in-the-loop test of the embedded control system where the helicopter is emulated on dSPACE subsystem.

6.1 Receding Horizon Control of the Longitudinal Motion of an Airplane

Airplanes are large scale nonlinear dynamic systems having time varying parameters. Their control system usually contains many levels. The control problem is often divided into the control of the longitudinal and lateral motion for different flight conditions. High level control integrates these subsystems into the complex control system of the airplane using gain scheduling technique or similar methods. Modern directions are LPV (Linear Parameter-Varying) and MPC (Model Predictive) control systems.

B. Lantos, L. Márton, *Nonlinear Control of Vehicles and Robots*,
Advances in Industrial Control,
DOI 10.1007/978-1-84996-122-6_6, © Springer-Verlag London Limited 2011

In this subsection, we consider the pitch rate control subsystem of an F-16 VISTA (Variable Stability In-Flight Simulator Test Aircraft) based on internal observer-based control and external RHC (Receding Horizon) control. The LPV longitudinal model for pitch control was published in [16]. The model consists of the short periodic equations and the actuator model:

$$\frac{d}{dt}\begin{pmatrix} \alpha \\ q \end{pmatrix} = \begin{bmatrix} Z_\alpha & 1 \\ M_\alpha & M_q \end{bmatrix}\begin{pmatrix} \alpha \\ q \end{pmatrix} + \begin{bmatrix} Z_{\delta_e} \\ M_{\delta_e} \end{bmatrix}\delta_{ed} = A_{\text{LPV}}\begin{pmatrix} \alpha \\ q \end{pmatrix} + B_{\text{LPV}}\delta_{ed}, \quad (6.1)$$

$$q = \begin{bmatrix} 0 & 1 \end{bmatrix}\begin{pmatrix} \alpha \\ q \end{pmatrix} = C_{\text{LPV}}\begin{pmatrix} \alpha \\ q \end{pmatrix} + D_{\text{LPV}}\delta_{ed}, \quad (6.2)$$

$$\delta_{ed} = P_{\text{act}}\delta_{ec} = -\frac{20.2}{s+20.2}\delta_{ec}, \quad (6.3)$$

where α is the angle of attack (deg), q is the pitch rate (deg/s), δ_{ed} is the elevator angle (deg) and δ_{ec} is the command signal (deg). The negative sign in the actuator model is a convention. The dimensional coefficients Z_α, M_α, M_q, M_{δ_e} and Z_{δ_e} depend especially on the hight h (ft) and the Mach number M. Notice that for F-16 VISTA hight is given in ft, velocity is given in ft/s and 1 ft $= 0.3126$ m.

The LPV version of F-16 VISTA is based on the high precision model of Air Force Research Laboratory linearized in the operation points

$$M = \begin{bmatrix} 0.35 & 0.45 & 0.55 & 0.65 & 0.75 & 0.85 \end{bmatrix}, \quad (6.4)$$

$$h = \begin{bmatrix} 1000 & 5000 & 15000 & 25000 \end{bmatrix}. \quad (6.5)$$

After linearizing a polynomial model in h, M was fitted to the models resulting in

$$Z_\alpha = 0.22 - 4.1 \times 10^{-7}h - 2.6M + 5.15 \times 10^{-5}Mh,$$

$$M_\alpha = 17.1 - 8.07 \times 10^{-4}h - 68.4M + 3.31 \times 10^{-3}Mh$$
$$+ 56.2M^2 - 2.92 \times 10^{-3}M^2h,$$

$$M_q = -0.228 + 7.06 \times 10^{-6}h - 2.12M + 4.86 \times 10^{-5}Mh, \quad (6.6)$$

$$Z_{\delta_e} = -1.38 \times 10^{-3} + 8.75 \times 10^{-8}h - 0.34M + 7.98 \times 10^{-6}Mh,$$

$$M_{\delta_e} = -8.16 + 1.73 \times 10^{-4}h + 40.6M - 8.96 \times 10^{-4}Mh$$
$$- 99.3M^2 + 2.42 \times 10^{-3}M^2h.$$

For the airplane with actuator, we shall apply the notation

$$P(h, M) = P_{\text{LPV}}(h, M)P_{\text{act}}. \quad (6.7)$$

Aircraft control should satisfy different level specification standards. The Level 1 specifications for the pitch rate are defined as follows. The step response has decaying oscillation character since it is a second order system with damping. The tangent

Fig. 6.1 Control system with disturbance observer

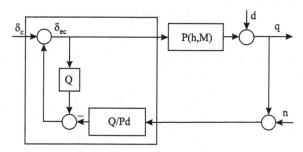

of the step response with maximal slope (taken in the inflexion point) intersects the time axis at t_1 and the line belonging to the steady state value at t_2. Their difference is $\Delta_t = t_2 - t_1$. The difference between the first maximum and the steady state value and the steady state value and the first minimum of the step response are denoted by Δq_1 and Δq_2, respectively. If the true speed is V_T, then the Level 1 specifications are $t_{1,max} = 0.12$ s, $\Delta q_2/\Delta q_1 = 0.30$, $\Delta_{t,max} = (500/V_T)$ s, $\Delta_{t,min} = (9/V_T)$ s. A desired system satisfying Level 1 conditions may be

$$P_d = \frac{4^2}{s^2 + 2 \cdot 0.5 \cdot 4s + 4^2}. \tag{6.8}$$

The main goal is RHC control of the pitch rate q. However, stable RHC is a hard problem if the LPV system to be controlled can become unstable. Hence, first an internal control based on disturbance observer architecture will be designed, see Fig. 6.1, in which the filter $Q(s)$ has low order (order 2) and assures stability and robustness for the closed loop system within the envelop $(h, M) \in$ [5000 ft, 25000 ft] × [0.4, 0.8]. Then the external RHC control has to produce optimal command signal (reference signal δ_c) for the internal stabilizing controller such that the composite system is optimal in RHC sense.

6.1.1 Robust Internal Stabilization Using Disturbance Observer

The transfer functions of the disturbance observer structure can be determined from Fig. 6.1 in the following steps:

$$\delta_{ec} = \delta_c + Q\delta_{ec} - \frac{Q}{P_d}(q + n),$$

$$\delta_{ec} = \frac{1}{1 - Q}\left[\delta_c - \frac{Q}{P_d}(q + n)\right],$$

$$q = P(h, M)\delta_{ec} + d$$

$$= P(h, M)\frac{1}{1 - Q}\left[\delta_c - \frac{Q}{P_d}(q + n)\right] + d,$$

$$\frac{P_d(1-Q)+P(h,M)Q}{(1-Q)P_d}q = P(h,M)\frac{1}{1-Q}\left[\delta_c - \frac{Q}{P_d}n\right]+d.$$

The resulting transfer functions for pitch rate output and the different (control, disturbance, noise) inputs yield

$$\frac{q(s)}{\delta_c(s)} = \frac{P(h,M)P_d}{P_d(1-Q)+P(h,M)Q} =: H(h,M), \tag{6.9}$$

$$\frac{q(s)}{d(s)} = \frac{P_d(1-Q)}{P_d(1-Q)+P(h,M)Q} =: S(h,M), \tag{6.10}$$

$$\frac{-q(s)}{n(s)} = \frac{P(h,M)Q}{P_d(1-Q)+P(h,M)Q} =: T(h,M). \tag{6.11}$$

If $Q(s)$ is a low pass filter of order 2 having the form

$$Q(s) = \frac{\omega_Q^2}{s^2+2\xi_Q\omega_Q s+\omega_Q^2}, \tag{6.12}$$

then for low frequencies $Q \approx 1$ and therefore

$$H \approx P_d, \qquad S \approx 0, \qquad T \approx 1. \tag{6.13}$$

$S \approx 0$ gives ground for calling the structure disturbance observer. However, the steady state behavior needs guarantee for stability.

On the other hand, assuming the system can well be approximated by the fixed $P_d(s)$ of order 2, we can characterize the performance of the system by the transfer function

$$H_p(s) = P_d(s) - H(h,M)(s). \tag{6.14}$$

In order to make the system stable for varying parameters satisfying robust stability and performance, we prescribe the following conditions to be satisfied for every h, M belonging to the flight envelope:

$$\text{roots}\big(P_d(1-Q)+P(h,M)Q\big) \quad \Rightarrow \quad s_i, \text{Re}(s_i) < 0 \quad \text{Hurwitz stability}, \tag{6.15}$$

$$\|W_u T\|_\infty \le 1 \quad \text{robust stability}, \tag{6.16}$$

$$\|W_p H_p\|_\infty \le 1 \quad \text{nominal performance}. \tag{6.17}$$

The filters $W_u(s)$ and $W_p(s)$ were chosen according to the suggestion in [136]:

$$W_u(s) = 2\frac{s+0.2\cdot 1256}{s+2\cdot 1256}, \tag{6.18}$$

$$W_p(s) = 0.4\frac{s+100}{s+4}. \tag{6.19}$$

Notice that $W_u(0) = 0.2$, $W_u(\infty) = 2$ and $W_p(0) = 10$, $W_p(\infty) = 0.1$ which are appropriate choices for uncertainty weights.

The remaining part is to find the feasibility region of ω_Q and ξ_Q. A simplified technique is to consider the four corners of the flight envelop, then to determine the feasibility regions belonging to each corners and taking the intersections of the four feasibility regions. For the computations of the H_∞ norm specifications, the function hinfnorm of the Robust Control Toolbox of MATLAB can be used for stable systems. A more exact method is to choose a fine raster in h, M of the flight envelope, then to determine the feasibility regions satisfying the three specifications for each h, M of the raster, and taking the intersections of all feasibility regions. We have found that $\omega_Q = 37$ and $\xi = 0.7$ are in the resulting feasibility region therefore we have chosen

$$Q(s) = \frac{1369}{s^2 + 51.8s + 1369}. \tag{6.20}$$

6.1.2 High Level Receding Horizon Control

RHC was suggested by Keviczky and Balas in [61] to compute optimized control reference signals for the internal stabilizing system according to the architecture in Fig. 6.2.

The closed loop with internal stabilizing control is augmented with integrator (for each control command). The internal controller may be any robust stabilizing controller (H_∞, disturbance observer based etc.).

The high level RHC control problem can be formulated in a general fashion. The initial time of the actual horizon will be denoted by t, however we use relative time within the actual horizon, i.e. $0, 1, 2, \ldots$ instead of $t, t+1, t+2, \ldots$. Let the internal closed loop system linearized at the beginning of each horizon (without the augmented integrators) be given by

$$x_{i+1} = Ax_i + Br_i, \tag{6.21}$$

$$\begin{pmatrix} y_i \\ z_i \\ u_i \end{pmatrix} = Cx_i + Dr_i, \tag{6.22}$$

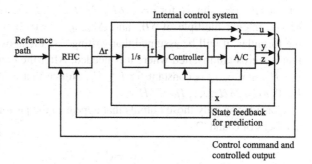

Fig. 6.2 Control system with high level RHC and low level stabilizing controller

where x is the state, r is the command input for the internal control, and y, z, u are distinguished measurable outputs. For example y should be regulated, z is part of IMU measurements (vertical acceleration etc.) and u are actuator command signals and their trends. For y a cost function will be defined while for z and u constraints can optionally be prescribed. Here $A := A(\rho(t)), B := B(\rho(t)), C := C(\rho(t)), D := D(\rho(t))$ belong to the actual parameter value $\rho(t)$ valid at the time moment of linearization, that is, at the beginning of the actual horizon. These matrices are frozen within the actual horizon, that is, the system is assumed to be LTI within the horizon. Augmented integrators will be added to the system satisfying $r_i = r_{i-1} + \Delta r_i$. Constant disturbance $d_{i+1} = d_i$ is assumed at the output having influence only on y and z. For predictive control, a prediction model is needed which can be assumed in the form

$$
\underbrace{\begin{pmatrix} \hat{x}_{i+1} \\ \hat{d}_{i+1} \\ r_i \end{pmatrix}}_{\hat{\xi}_{i+1}} = \overbrace{\begin{bmatrix} A & 0 & B \\ 0 & I & 0 \\ 0 & 0 & I \end{bmatrix}}^{\mathcal{A}} \underbrace{\begin{pmatrix} \hat{x}_i \\ \hat{d}_i \\ r_{i-1} \end{pmatrix}}_{\hat{\xi}_i} + \overbrace{\begin{bmatrix} B \\ 0 \\ I \end{bmatrix}}^{\mathcal{B}} \Delta r_i,
\tag{6.23}
$$

$$
\underbrace{\begin{pmatrix} \hat{y}_i \\ \hat{z}_i \\ \hat{u}_i \end{pmatrix}}_{\hat{w}_i} = \underbrace{\begin{bmatrix} C & \begin{vmatrix} I & 0 \\ 0 & I \\ 0 & 0 \end{vmatrix} & D \end{bmatrix}}_{\mathcal{C}} \underbrace{\begin{pmatrix} \hat{x}_i \\ \hat{d}_i \\ r_{i-1} \end{pmatrix}}_{\hat{\xi}_i} + \underbrace{D}_{\mathcal{D}} \Delta r_i.
\tag{6.24}
$$

A quadratic cost function for the actual horizon to be minimized is assumed in the form

$$
J = \sum_{i=1}^{H_p} \| \hat{y}_i - y_{\text{ref},i} \|_Q^2 + \sum_{i=0(\delta H_c)}^{H_c-1} \| \Delta r_i \|_R^2 + \rho \varepsilon,
\tag{6.25}
$$

where H_p is the number of steps in the prediction horizon, H_c denotes the control horizon and $Q \geq 0$, $R > 0$ are suitably chosen weighting matrices. The slack variable $\varepsilon \geq 0$ and its weight ρ is used for softening some constraints.

The control signal is allowed to change only at integer multiples of the "blocking parameter" δH_c, that is, for $i = 0, \delta H_c, 2\delta H_c, \ldots, H_c - \delta H_c$. Therefore, $H_c = N_c \cdot \delta H_c$ is integer multiple of δH_c and $\Delta r_{k \cdot \delta H_c + i} = 0$ for $i = 1, 2, \ldots, \delta H_c - 1$. In the formulas, it will be assumed that writing Δr_i the above convention is satisfied, that is, zeros are inserted where it is needed. After the control horizon $\Delta r_i = 0$ for $i \geq H_c$, that is, r_i is constant for $i \geq H_c$. Notice that r_i is changing also only for $i = 0, \delta H_c, 2\delta H_c, \ldots, H_c - \delta H_c$.

Considering only those outputs that appear in the performance index, that is, the y-blocks of $\mathcal{C}$ and $\mathcal{D}$,

$$
\hat{y}_i = \mathcal{C}_y \hat{\xi}_i + \mathcal{D}_y \Delta r_i,
\tag{6.26}
$$

the predicted output sequence can be written as

$$
\overbrace{\begin{pmatrix} \hat{y}_{i+1} \\ \hat{y}_{i+2} \\ \vdots \\ \hat{y}_{i+H_c} \\ \hat{y}_{i+H_c+1} \\ \vdots \\ \hat{y}_{i+H_p} \end{pmatrix}}^{y} = \overbrace{\begin{bmatrix} C_y \mathcal{A} \\ C_y \mathcal{A}^2 \\ \vdots \\ C_y \mathcal{A}^{H_c} \\ C_y \mathcal{A}^{H_c+1} \\ \vdots \\ C_y \mathcal{A}^{H_p} \end{bmatrix}}^{\Psi_y} \hat{\xi}_0
$$

$$
+ \overbrace{\begin{bmatrix} C_y \mathcal{B} & D_y & \cdots & & 0 \\ C_y \mathcal{A}\mathcal{B} & C_y \mathcal{B} & D_y & & \vdots \\ \vdots & \vdots & & \ddots & \\ C_y \mathcal{A}^{H_c-1}\mathcal{B} & C_y \mathcal{A}^{H_c-2}\mathcal{B} & \cdots & & C_y \mathcal{B} \\ C_y \mathcal{A}^{H_c}\mathcal{B} & C_y \mathcal{A}^{H_c-1}\mathcal{B} & \cdots & & C_y \mathcal{A}\mathcal{B} \\ \vdots & \vdots & & \ddots & \\ C_y \mathcal{A}^{H_p-1}\mathcal{B} & C_y \mathcal{A}^{H_p-2}\mathcal{B} & \cdots & & C_y \mathcal{A}^{H_p-H_c}\mathcal{B} \end{bmatrix}}^{\Theta_y} \overbrace{\begin{pmatrix} \Delta r_0 \\ \Delta r_1 \\ \vdots \\ \Delta r_{H_c-1} \end{pmatrix}}^{\Delta \mathcal{R}},
$$

$$(6.27)$$

$$
\mathcal{Y} = \Psi_y \hat{\xi}_0 + \Theta_y \Delta \mathcal{R}. \tag{6.28}
$$

Let Q_e and R_e be block diagonal matrices of appropriate dimension, repeating Q matrix H_p times and R matrix H_c times in the block diagonal, respectively. Then the cost function without the slack variable term can be written in the form

$$
\begin{aligned}
J &= \langle Q_e(\mathcal{Y}_{\text{ref}} - \mathcal{Y}), (\mathcal{Y}_{\text{ref}} - \mathcal{Y}) \rangle + \langle R_e \Delta \mathcal{R}, \Delta \mathcal{R} \rangle \\
&= \langle Q_e \underbrace{(\mathcal{Y}_{\text{ref}} - \Psi_y \hat{\xi}_0 - \Theta_y \Delta \mathcal{R})}_{\mathcal{E}}, \underbrace{(\mathcal{Y}_{\text{ref}} - \Psi_y \hat{\xi}_0 - \Theta_y \Delta \mathcal{R})}_{\mathcal{E}} \rangle + \langle R_e \Delta \mathcal{R}, \Delta \mathcal{R} \rangle \\
&= \langle Q_e(\mathcal{E} - \Theta_y \Delta \mathcal{R}), \mathcal{E} - \Theta_y \Delta \mathcal{R} \rangle + \langle R_e \Delta \mathcal{R}, \Delta \mathcal{R} \rangle \\
&= \langle Q_e \mathcal{E}, \mathcal{E} \rangle - 2 \langle \Delta \mathcal{R}, \Theta_y^T Q_e \mathcal{E} \rangle + \langle (\Theta_y^T Q_e \Theta_y + R_e) \Delta \mathcal{R}, \Delta \mathcal{R} \rangle \\
&= \Delta \mathcal{R}^T (\Theta_y^T Q_e \Theta_y + R_e) \Delta \mathcal{R} + \Delta \mathcal{R}^T (-2\Theta_y^T Q_e \mathcal{E}) + \mathcal{E}^T Q_e \mathcal{E}, \tag{6.29}
\end{aligned}
$$

which is a standard quadratic cost function. The term $\mathcal{E}^T Q_e \mathcal{E} = \text{const}$ can be neglected because constant term does not influence the argument of the minimum.

However, programs supporting quadratic optimization need the problem to be solved without the inserted zero terms in $\Delta \mathcal{R}$. We can introduce the matrix $\mathcal{S}$ which computes the terms containing the inserted zeros from the terms not containing the inserted zeros such that

$$\Delta \mathcal{R} = S \delta \mathcal{R}. \tag{6.30}$$

For example if $\delta H_c = 4$ and $H_c = 5 \cdot \delta H_c$ then (in MATLAB convention)

$$S1 = \begin{bmatrix} I \\ 0 \\ 0 \\ 0 \end{bmatrix} \quad \Rightarrow \quad S = \texttt{blockdiag(S1,S1,S1,S1,S1)}.$$

Substituting (6.30) into (6.29), we obtain

$$J = \delta \mathcal{R}^T S^T \left(\Theta_y^T Q_e \Theta_y + R_e \right) S \delta \mathcal{R} + \delta \mathcal{R}^T \left(-2 S^T \Theta_y^T Q_e \mathcal{E} \right) + \mathcal{E}^T Q_e \mathcal{E}$$
$$= 0.5 x^T H x + f^T x + \text{const},$$

where

$$x = \delta \mathcal{R}, \qquad H = 2 S^T \left(\Theta_y^T Q_e \Theta_y + R_e \right) S, \qquad f = -2 S^T \Theta_y^T Q_e \mathcal{E}.$$

Notice that $\texttt{x=quadprog(H,f,A,b,Aeq,beq)}$ is the function in MATLAB Optimization Toolbox for solving QP (Quadratic Programming) problem $\min 0.5 x^T H x + f^T x$ under the constraints $Ax \le b$ and $Aeqx = beq$. It is clear that the original problem containing the term with the slack variable can also be brought to this standard form and can be solved in the new variables $(x^T, \varepsilon)^T$.

Constraints for the change of the command signal
Let us consider first the constraints

$$\Delta r_L \le \delta r_i \le \Delta r_H, \tag{6.31}$$

where only that i should be considered at which Δr_i is changing. This is marked by δr_i. We shall use the following notations:

$$1_{N,n\times n} = \begin{bmatrix} I_{n\times n} \\ I_{n\times n} \\ \vdots \\ I_{n\times n} \end{bmatrix}_{Nn\times n}, \qquad T_{N,n\times n} = \begin{bmatrix} I_{n\times n} & 0 & \cdots & 0 \\ I_{n\times n} & I_{n\times n} & \cdots & 0 \\ \vdots & \cdots & \ddots & \vdots \\ I_{n\times n} & I_{n\times n} & \cdots & I_{n\times n} \end{bmatrix}_{Nn\times Nn}. \tag{6.32}$$

Then the above constraints can be written as

$$\begin{bmatrix} I_{N_c \times N_c} \\ -I_{N_c \times N_c} \end{bmatrix} \delta \mathcal{R} \le \begin{pmatrix} 1_{N_c, n_r \times n_r} \Delta r_H \\ -1_{N_c, n_r \times n_r} \Delta r_L \end{pmatrix}. \tag{6.33}$$

Constraints for the command signal
A bit more complicated is the handling of the constraint

$$r_L \le r_i \le r_H. \tag{6.34}$$

However, taking into consideration

$$r_0 = r_{-1} + \delta r_0,$$

$$r_1 = r_0 + \delta r_1 = r_{-1} + \delta r_0 + \delta r_1,$$

$$\vdots$$

$$r_{N_c-1} = r_{-1} + \delta r_0 + \delta r_1 + \cdots + \delta r_{N_c-1},$$

we can write

$$\mathcal{R} = \mathbf{T}_{N_c,n_r \times n_r} \delta \mathcal{R} + \mathbf{1}_{N_c,n_r \times n_r} r_{-1}, \tag{6.35}$$

from which it follows

$$\begin{bmatrix} \mathbf{T}_{N_c,n_r \times n_r} \\ -\mathbf{T}_{N_c,n_r \times n_r} \end{bmatrix} \delta \mathcal{R} \leq \begin{pmatrix} \mathbf{1}_{N_c,n_r \times n_r} r_H - \mathbf{1}_{N_c,n_r \times n_r} r_{-1} \\ -\mathbf{1}_{N_c,n_r \times n_r} r_L + \mathbf{1}_{N_c,n_r \times n_r} r_{-1} \end{pmatrix}. \tag{6.36}$$

Constraints for optional output signals
Similarly to $\hat{y}$ the optional output signals $\hat{z}$ and $\hat{u}$ can also be written in the form

$$\mathcal{Z} = \Psi_z \hat{\xi}_0 + \Theta_z \mathcal{S} \delta \mathcal{R}, \tag{6.37}$$

$$\mathcal{U} = \Psi_u \hat{\xi}_0 + \Theta_u \mathcal{S} \delta \mathcal{R}. \tag{6.38}$$

Constraints for them are assumed in the form

$$z_L \leq \hat{z}_i \leq z_H, \quad i = 1, \ldots, H_p, \tag{6.39}$$

$$u_L \leq \hat{u}_i \leq u_H, \quad i = 1, \ldots, H_p, \tag{6.40}$$

which can be converted to

$$\begin{bmatrix} \Theta_z \mathcal{S} \\ -\Theta_z \mathcal{S} \end{bmatrix} \delta \mathcal{R} \leq \begin{pmatrix} \mathbf{1}_{H_p,n_z \times n_z} z_H - \Psi_z \hat{\xi}_0 \\ -\mathbf{1}_{H_p,n_z \times n_z} z_L + \Psi_z \hat{\xi}_0 \end{pmatrix}, \tag{6.41}$$

$$\begin{bmatrix} \Theta_u \mathcal{S} \\ -\Theta_u \mathcal{S} \end{bmatrix} \delta \mathcal{R} \leq \begin{pmatrix} \mathbf{1}_{H_p,n_u \times n_u} u_H - \Psi_u \hat{\xi}_0 \\ -\mathbf{1}_{H_p,n_u \times n_u} u_L + \Psi_u \hat{\xi}_0 \end{pmatrix}. \tag{6.42}$$

Constraints can easily be softened if no feasible solution exists. For example, taking $z_L - \varepsilon$ and $z_H + \varepsilon$ the constraints for $\hat{z}$ are weakened making them easier to satisfy. As a general rule, a constraint $\Omega \delta \mathcal{R} \leq \omega$ can be softened if it is substituted by $\Omega \delta \mathcal{R} \leq \omega + \varepsilon$, where ε is a scalar slack variable and $\varepsilon \geq 0$. In case of softening, some constraints the term $\rho \varepsilon$ has to be added to the cost function. (Notice that in MATLAB the sum of vector and scalar is allowed and the scalar is added to all components of the vector.)

6.1.3 Simulation Results with External RHC and Internal Disturbance Observer

The discussion will be divided into two steps. First, we present the results with disturbance observer alone. Then the results with augmented RHC will be shown and the two results will be compared demonstrating the efficiency of RHC.

In the simulation examples, we assumed the height and Mach number trajectories shown in Fig. 6.3.

6.1.3.1 Control with Disturbance Observer Alone

First, the simulation results with the internal control loop without predictive control will be discussed. The noise n will be not considered during the experiments.

The actuator has the state equation $\dot{\delta}_{ed} = -20.2\delta_{ed} + \delta_{ec}$, with input δ_{ec} from the controller and output δ_{ed} for the LTV system. The state equation of the LTV system with augmented actuator has the form

$$\begin{pmatrix} \dot{\alpha} \\ \dot{q} \\ \dot{\delta}_{ed} \end{pmatrix} = \begin{bmatrix} Z_\alpha & 1 & Z_{\delta_e} \\ M_\alpha & M_q & M_{\delta_e} \\ 0 & 0 & -20.2 \end{bmatrix} \begin{pmatrix} \alpha \\ q \\ \delta_{ed} \end{pmatrix} + \begin{bmatrix} 0 \\ 0 \\ 20.2 \end{bmatrix} \delta_{ec}, \tag{6.43}$$

$$q = \begin{bmatrix} 0 & 1 & 0 \end{bmatrix} \begin{pmatrix} \alpha \\ q \\ \delta_{ed} \end{pmatrix}. \tag{6.44}$$

For the internal loop, the controller is a disturbance observer with inputs δ_c and q, and output δ_{ec}. The controller components and the resulting controller are, re-

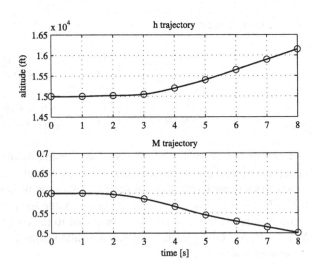

Fig. 6.3 Height and Mach number trajectories during simulation

spectively,

$$Q = \frac{\omega_Q^2}{s^2 + 2\xi_Q\omega_Q s + \omega_Q^2}, \quad \omega_Q = 37, \quad \xi_Q = 0.7,$$

$$P_d = \frac{16}{s^2 + 4s + 16},$$

$$\frac{1}{1-Q} = \frac{s^2 + 2\xi_Q\omega_Q s + \omega_Q^2}{s^2 + 2\xi_Q\omega_Q s},$$

$$\frac{1}{P_d}\frac{Q}{1-Q} = \frac{\frac{\omega_Q^2}{16}(s^2 + 4s + 16)}{s^2 + 2\xi_Q\omega_Q s},$$

$$\delta_{ec}(s) = \frac{(s^2 + 2\xi_Q\omega_Q s + \omega_Q^2)\delta_c(s) - \frac{\omega_Q^2}{16}(s^2 + 4s + 16)q(s)}{s^2 + 2\xi_Q\omega_Q s}.$$

The controller has 2 states and its state equation has the form

$$\dot{x}_c = A_c x_c + B_{c1}\delta_c + B_{c2}q,$$

$$\delta_{ec} = C_c x_c + D_{c1}\delta_c + D_{c2}q.$$

The closed loop system consists of LTV system, actuator and disturbance observer, the total number of states is 5. The state equation can be determined as follows:

$$\dot{x}_p = A_p x_p + B_p(C_c x_c + D_{c1}\delta_c + D_{c2}x_p)$$

$$= (A_p + B_p D_{c2})x_p + B_p C_c x_c + B_p D_{c2}\delta_c,$$

$$\dot{x}_c = A_c x_c + B_{c1}\delta_c + B_{c2}C_p x_p,$$

which can be collected in the resulting state equation of the internal closed loop system

$$\begin{pmatrix} \dot{x}_p \\ \dot{x}_c \end{pmatrix} = \begin{bmatrix} A_P + B_p D_{c2} & B_p C_c \\ B_{c2}C_p & A_c \end{bmatrix} \begin{pmatrix} x_p \\ x_c \end{pmatrix} + \begin{bmatrix} B_p D_{c1} \\ B_{c1} \end{bmatrix}\delta_c, \qquad (6.45)$$

$$q = \begin{bmatrix} C_p & 0 \end{bmatrix}\begin{pmatrix} x_p \\ x_c \end{pmatrix}, \qquad (6.46)$$

where δ_c is the reference input for the internal control and the pitch rate q is the system output.

Two simulation examples were considered under MATLAB/Simulink, one without disturbance and one with output disturbance.

Figure 6.4 shows the pith rate output (q) and the actuator signal (δ_{ed}) for 10 deg/s reference signal input and no disturbance. The transients satisfy Level 1 specifications.

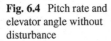

Fig. 6.4 Pitch rate and elevator angle without disturbance

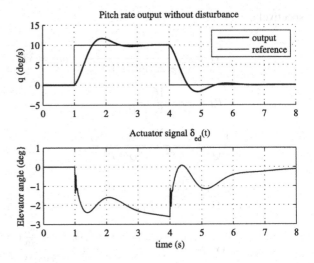

Fig. 6.5 Pitch rate and elevator angle in case of disturbance

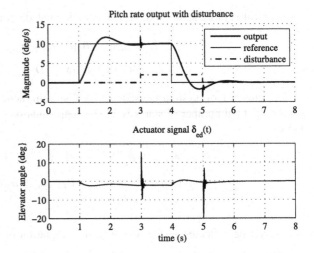

Simulation results in case of output disturbance are shown in Fig. 6.5 containing the pith rate output (q) and the actuator signal (δ_{ed}) for 10 deg/s reference signal input and 2 deg/s disturbance. It can be seen that the disturbance observer really suppresses the effect of the output disturbance after a very short transient. The price is the enlarged control signal at the rising and falling edges of the disturbance.

6.1.3.2 Control with High Level RHC and Low Level Disturbance Observer

Notice that the block scheme of the complex control system is shown in Fig. 6.2. During the simulation with external RHC and internal disturbance observer zero disturbance ($d = 0$) was assumed. The sampling time was $T = 0.05$ s, the predic-

Fig. 6.6 Pitch rate and
actuator signal using RHC
and disturbance observer
(command change limits are
inactive)

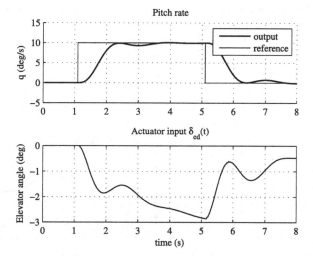

Fig. 6.7 RHC controller's
command change and
command outputs (command
change limits are inactive)

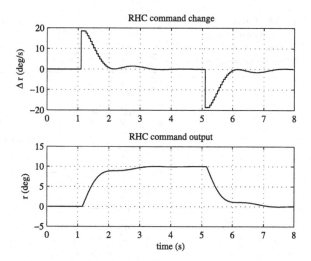

tion horizon $H_p = 10$ (0.5 s), $\delta H_c = 1$ (0.05 s) and the control horizon $H_c = 5$
(0.25 s). The weighting matrices in the cost function were chosen to be $Q_e = I$
and $R_e = 0.8I$. The prescribed limits for the change of the control command were
$\Delta r_H = 60$ deg/s and $\Delta r_L = -60$ deg/s while for r no limits were given. The optimization was performed using quadprog of MATLAB Optimization Toolbox.

In the first experiment, the reference signal order is small (10 deg/s), hence the
command change limits are inactive. Figure 6.6 shows the reference signal input
and the pitch rate output (q), and the actuator signal (elevator angle δ_{ed}). Comparing
with the results in Fig. 6.4 it can be seen that the actuator signal is smoother with
RHC and disturbance observer than with disturbance observer alone. In Fig. 6.7, the
RHC controller's command input change and command signal can be observed, the
latter deals as reference signal for the disturbance observer.

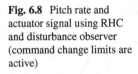

Fig. 6.8 Pitch rate and actuator signal using RHC and disturbance observer (command change limits are active)

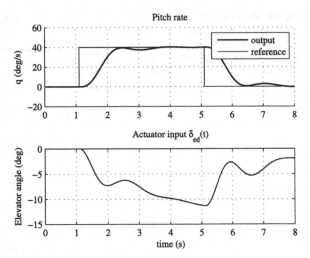

Fig. 6.9 RHC controller's command change and command outputs (command change limits are active)

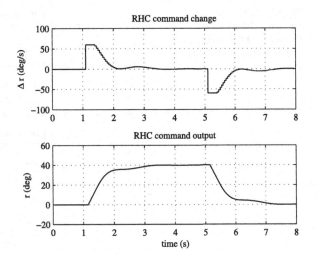

In the second experiment, the reference signal order is larger (40 deg/s), hence the command change limits ± 60 deg/s are active. Figure 6.8 shows the reference signal input and the pitch rate output (q), and the actuator input (elevator angle δ_{ed}). In Fig. 6.9, the RHC controller's command input change and command signal can be observed, the latter deals as reference signal for the disturbance observer. It can be seen that Δr is saturated at ± 60 deg/s. The actuator signal is smooth with RHC together with disturbance observer. The smooth actuator signals and the possibility of actuator limit specifications illustrate the preference of high level RHC.

In both experiments, it can be observed that the pitch rate is not monotone increasing. Notice that as a possible extension constraints can be formulated for the predicted output signals enforcing monotone transient for the output [25, 73].

6.2 Backstepping Control of an Indoor Quadrotor Helicopter

Unmanned aerial vehicles (UAVs) increasingly attract the attention of potential appliers, vehicle professionals and researchers. UAV field seems to step out of the exclusivity of the military applications, a lot of potential civil applications have emerged, research and development in this field has gained increasing significance.

Research teams are interested in many respects in this field, many efforts are spent in several research areas in connection with the individual and cooperative control of aerial and land vehicles including the unmanned ones. Developing an unmanned outdoor mini quadrotor helicopter that is able to execute autonomously a mission, for example, performing a series of measurements in predefined positions, or completing a surveillance task above a given territory is one of the goals formulated for the near future. Indoor mini quadrotor helicopters can constitute a research field to gain experiences both for single UAV and formation control of multi agent UAVs.

The results of this section are related to a project initiated in the spring of 2006 as a cooperation between Budapest University of Technology and Economics Dept. Control Engineering and Information Technology (BME IIT) and Computer and Automation Research Institute of the Hungarian Academy of Sciences (MTA SZTA-KI). The first version of the helicopter body was reported in [134, 135].

The development at BME IIT started in five areas: sensor fusion, control system algorithms, system architecture, electronic components and vision system. In the current phase, the primary goal of this research is to build an autonomous indoor quadrotor helicopter, which will serve later as a research test bed for advanced algorithms in the areas of control, path planning, maneuvering in formation, position estimation, sensor fusion, embedded real-time (RT) vision system, spatial map building, robust and efficient mechanical and aerodynamic construction. In 2008 the hardware-in-the-loop tests were successfully performed, which verified that the on-board CPU can be programmed via Simulink to perform the backstepping control and extended Kalman filter algorithms together with CAN bus I/O in real time [68].

This section mainly focuses on the development of the sensory and control system of the indoor quadrotor helicopter. Because of the indoor character GPS signals cannot be received hence the sensory system is based on image processing, IMU (Inertial Measurement Unit) and extended Kalman filters (EKFs). The control system uses backstepping based control algorithm integrating state estimation and path tracking. A detailed treatment of the control system can be found in [121].

The concept of the quadrotor helicopter can be seen in Fig. 6.10. The helicopter has four actuators (four brushless DC motors) which exert lift forces f_i proportional to the square of the angular velocities Ω_i of the actuators. The rotational directions of the propellers are shown in the sketch.

The structure of the system is drawn in Fig. 6.11. The control loop of the helicopter requires accurate position and orientation information. The primary sensor for this is a Crossbow MNAV100CA inertial measurement unit (IMU), which provides acceleration and angular velocity measurements. Based on the physical prop-

Fig. 6.10 Concept of the
quadrotor helicopter

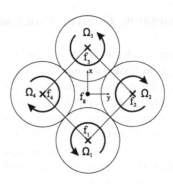

Fig. 6.11 System
architecture of the quadrotor
helicopter

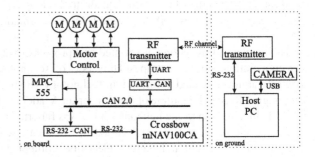

erties of the sensors [66], the raw sensor output is loaded with noise and bias errors, which can build up increasing orientation and position error during integration. In order to reduce the effect of noise and bias, extended Kalman filters are used instead of integration. For 6-degrees of freedom mobile robots, one needs absolute measurements performed frequently. For outdoor autonomous vehicles, GPS or differential GPS is often used. For indoor vehicles, vision based measurements are widely used to compensate integration errors.

The quadrotor helicopter has an on-board CPU. The rotors are driven by brushless DC (BLDC) motors. The motor controllers and the IMU are connected to the CPU via CAN bus. Also, a spread spectrum code division multiplexing (CDM) radio link is connected to the CPU, providing bidirectional communication between the quadrotor and the ground station. The ground station sends commands and reference path to the CPU, together with the absolute position measurements. The helicopter sends status information to the ground. The ground station is connected to the computer performing vision algorithms via Ethernet. We use commercial, high resolution web cameras for imaging. More details on the sensor and actuator system can be found in [66].

The on-board computer is a phyCORE-MPC555, since it is lightweight, equipped with a floating point unit and can be programmed in MATLAB/Simulink. An alternative goal of the project is to compare high level model based programming of embedded systems with traditional methods.

6.2.1 Dynamic Model of the Quadrotor Helicopter

Let us assume that a frame (coordinate system) K_E fixed to the Earth can be considered as an inertial frame of reference. In our case, it may be the NED frame considered to be fixed.

Kinematic model

The frame fixed to the center of gravity of the helicopter K_H can be described by its position $\xi = (x, y, z)^T$ and orientation (RPY angles) $\eta = (\Phi, \Theta, \Psi)^T$ relative to K_E. Translational and angular velocities v and ω of the helicopter are given in K_H.

The orientation can be described by the (orthonormal) matrix R_t in the following way:

$$R_t = \begin{bmatrix} C_\Theta C_\Psi & S_\Phi S_\Theta C_\Psi - C_\Phi S_\Psi & C_\Phi S_\Theta C_\Psi + S_\Phi S_\Psi \\ C_\Theta S_\Psi & S_\Phi S_\Theta S_\Psi + C_\Phi C_\Psi & C_\Phi S_\Theta S_\Psi - S_\Phi C_\Psi \\ -S_\Theta & S_\Phi C_\Theta & C_\Phi C_\Theta \end{bmatrix}. \tag{6.47}$$

The relation between $\dot{\xi}$ and $\dot{\eta}$ and translational and angular velocities v and ω of the helicopter take the form

$$\dot{\xi} = R_t v, \tag{6.48}$$
$$\omega = R_r \dot{\eta},$$

where time derivative is denoted by dot and the (nonorthonormal) matrix R_r has the form

$$R_r = \begin{bmatrix} 1 & 0 & -S_\Theta \\ 0 & C_\Phi & S_\Phi C_\Theta \\ 0 & -S_\Phi & C_\Phi C_\Theta \end{bmatrix}. \tag{6.49}$$

It is worth mentioning that the inverse of R_r can be computed as

$$R_r^{-1} = \begin{bmatrix} 1 & S_\Phi T_\Theta & S_\Phi T_\Theta \\ 0 & C_\Phi & -S_\Phi \\ 0 & S_\Phi/C_\Theta & C_\Phi/C_\Theta \end{bmatrix}, \tag{6.50}$$

and the derivative of ω can be written as

$$\varepsilon = \dot{\omega} = R_r \ddot{\eta} + \dot{R}_r \dot{\eta}. \tag{6.51}$$

Dynamic model

Applying Newton's laws, the translational and rotational motions of the helicopter in K_H are described by

$$\sum F_{\text{ext}} = m\dot{v} + \omega \times (mv),$$
$$\sum T_{\text{ext}} = I_c \dot{\omega} + \omega \times (I_c \omega), \tag{6.52}$$

where I_c is the inertia matrix of the helicopter and it is supposed that it can be described by a diagonal matrix $I_c = \text{diag}(I_x, I_y, I_z)$. $\sum F_{\text{ext}}$ and $\sum T_{\text{ext}}$ represent the forces and torques respectively applied to the quadrotor helicopter expressed in K_H. These forces and torques are partly caused by the rotation of the rotors (F and T), the aerodynamic friction (F_a and T_a), the gravitational effect (F_g) in the translational motion and the gyroscopic effect (T_g) in the rotational motion:

$$\sum F_{\text{ext}} = F + F_a + F_g,$$
$$\sum T_{\text{ext}} = T + T_a + T_g. \tag{6.53}$$

The helicopter has four actuators (four brushless DC motors) which exert a lift force proportional to the square of the angular velocities Ω_i of the actuators ($f_i = b\Omega_i^2$). The BLDC motors' reference signals can be programmed in Ω_i. The resulting torque and lift force are

$$T = \begin{pmatrix} lb(\Omega_4^2 - \Omega_2^2) \\ lb(\Omega_3^2 - \Omega_1^2) \\ d(\Omega_2^2 + \Omega_4^2 - \Omega_1^2 - \Omega_3^2) \end{pmatrix},$$
$$f = f_1 + f_2 + f_3 + f_4 = b \sum_{i=1}^{4} \Omega_i^2, \tag{6.54}$$

where l, b, d are helicopter and rotor constants. The force F can then be rewritten as $F = (0, 0, f)^T$.

The gravitational force points to the negative z-axis, hence $F_g = -mR_t^T (0, 0, g)^T = -mR_t^T G$. The gyroscopic effect can be modeled as

$$T_g = -(\omega \times k)I_r(\Omega_2 + \Omega_4 - \Omega_1 - \Omega_3) = -\omega \times (I_r \Omega_r), \tag{6.55}$$

where I_r is the rotor inertia and k is the third unit vector.

The aerodynamic friction at low speeds can well be approximated by the linear formulas $F_a = -K_t v$ and $T_a = -K_r \omega$.

Using the equations above, we can derive the motion equations of the helicopter:

$$F = mR_t^T \ddot{\xi} - K_t R_t^T \dot{\xi} - mR_t^T G,$$
$$T = I_c R_r \ddot{\eta} + I_c \left(\frac{\partial R_r}{\partial \Phi} \dot{\Phi} + \frac{\partial R_r}{\partial \Theta} \dot{\Theta} \right) \dot{\eta} \tag{6.56}$$
$$+ K_r R_r \dot{\eta} + (R_r \dot{\eta}) \times (I_c R_r \dot{\eta} + I_r \Omega_r).$$

Simplified dynamic model

A simplified model of the quadrotor helicopter can be obtained by neglecting certain effects and applying reasonable approximations. The purpose of the construction of such a model is to reduce the complexity of the controller while keeping its performance.

Since the helicopter's motion is planned to be relatively slow, it is reasonable to neglect all the aerodynamic effects, namely, K_t and K_r are approximately zero matrices. The other simplification is also related to the low speeds. Slow motion in lateral directions means little roll and pitch angle changes, therefore R_r can be approximated by a 3-by-3 unit matrix. Such simplification cannot be applied to R_t.

Consequently, the dynamic equations in (6.56) become

$$F \approx mR_t^T \ddot{\xi} - mR_t^T G,$$
$$T \approx I\ddot{\eta} + \dot{\eta} \times (I\dot{\eta} + I_r \Omega_r). \tag{6.57}$$

The six equations in detail are the ones that can be found in [21] and [68].

$$m\ddot{x} \approx (C_\Phi S_\Theta C_\Psi + S_\Phi S_\Psi)f,$$
$$m\ddot{y} \approx (C_\Phi S_\Theta S_\Psi - S_\Phi C_\Psi)f,$$
$$m\ddot{z} \approx C_\Phi C_\Theta f - mg,$$
$$I_x \ddot{\Phi} \approx \dot{\Theta}\dot{\Psi}(I_y - I_z) - I_r \dot{\Theta}\Omega_r + T_1,$$
$$I_y \ddot{\Theta} \approx \dot{\Psi}\dot{\Phi}(I_z - I_x) - I_r \dot{\Phi}\Omega_r + T_2,$$
$$I_z \ddot{\Psi} \approx \dot{\Phi}\dot{\Theta}(I_x - I_y) + T_3. \tag{6.58}$$

Rotor dynamics
The four BLDC motors' dynamics can be described as ($k = 1, \ldots, 4$):

$$L\frac{di_k}{dt} = u_{m,k} - Ri_k - k_e \Omega_k,$$
$$I_r \dot{\Omega}_k = k_m i_k - k_r \Omega_k^2 - k_s, \tag{6.59}$$

where k_e, k_m and k_s represent the back emf constant, the motor torque constant and the friction constant, respectively. If the motors' inductance is negligible, (6.59) can be rewritten to

$$\dot{\Omega}_k = -k_{\Omega,0} - k_{\Omega,1}\Omega_k - k_{\Omega,2}\Omega_k^2 + k_u u_{m,k}. \tag{6.60}$$

The mechanical parameters of the helicopter and the BLDC motors with the rotors are based on the planned dimensions, the masses of purchased elements. These values are summarized in Table 6.1.

6.2.2 Sensor System of the Helicopter

A Crossbow MNAV100CA inertial measurement unit (IMU) is used for inertial sensing. It has three dimensional (3D) accelerometers and 3D angular velocity sen-

Table 6.1 The physical parameters of the helicopter and the motors

Parameter	Value
l	0.23 m
b	1.759×10^{-5} kg m
d	1.759×10^{-6} kg m^2
I_x, I_y	1.32×10^{-2} kg m^2
I_z	2.33×10^{-2} kg m^2
I_r	5.626×10^{-5} kg m^2
m	1.4 kg
K_t	diag($[0.1, 0.1, 0.15]$) N s/m
K_r	diag($[0.1, 0.1, 0.15]$) N s m
$k_{\Omega,0}$	94.37 s^{-2}
$k_{\Omega,1}$	3.02 s^{-1}
$k_{\Omega,2}$	0.005
k_u	139.44 V/s^2

sors. There are other sensors on this unit as well, but the GPS, magnetometers, the pressure sensor and the air-flow meter cannot be used in indoor circumstances. There are also three thermometers built in the angular velocity sensors. The IMU is a commercial one and it was implemented on an Atmel ATMega128 microcontroller, which has no CAN interface. The RS-232 to CAN transmission is solved by an Atmel AT90CAN microcontroller.

The position and orientation sensing is based on a separated vision system using one camera. The algorithms run over a host PC. This PC implements also a user interface, where one can describe the path, send start/stop commands or do logging. The whole module sends the information to the helicopter via radio frequency (RF) channel based on MaxStream XBee Pro units. Each RF transmitter has serial interface. On the board of the helicopter the translation into CAN packages is done by a similar Atmel AT90CAN as in the case of the IMU.

Six frames (coordinate systems) can be defined according to the sensory system. The graph of the frames can be seen in Fig. 6.12. K_W is the world frame. Its origin and orientation is defined later. K_C is the frame of the camera and K_{virt} is a virtual coordinate system. K_H is the frame of the helicopter, this moves together with the helicopter's body. K_S is the desired frame of the IMU and K_{S_0} is the original frame of the IMU, defined by the manufacturer. The homogeneous transformations $T_{C,W}$ and $T_{\text{virt},C}$ define the relations between them. T_S is originated from the mechanical fixation of the IMU on the body of the helicopter. The fact of the equality of the transformations between K_{virt}, K_H and between K_C, K_W, is explained later.

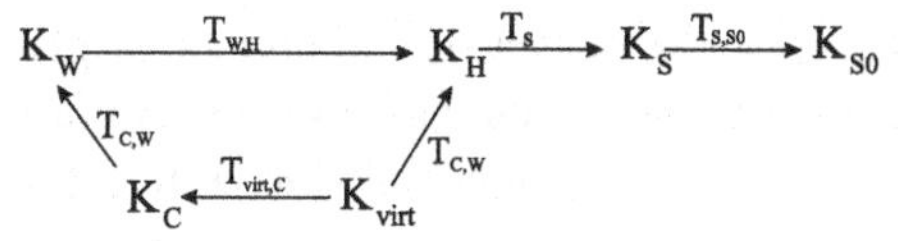

Fig. 6.12 Graph of the frames

6.2.2.1 Inertial Measurement Unit

There are accelerometers and angular velocity sensors on the board of the helicopter. To get acceptable measures from the IMU, few types of errors should be noted.

Primary information is listed on the datasheets of the sensors. To estimate the values of these errors, calibrations can be done. There are three different phases when error values can be calculated.

The first phase is before assembling the sensor to the board of the helicopter, called off-line calibration. Here there are a lot of opportunities, because time efficiency is not so important. In this case, time independent errors can be measured. These errors are the nonperpendicularity of the axis of the sensors, the gain errors and the gravity effect on the angular velocity sensor. The gain error of the acceleration is also temperature dependent and the temperature can change from flight to flight, but this dependency is linear and can be calculated off-line.

The second phase is just before the flight. In this case, there are only few seconds for calibration and usually measurements can be done only in one orientation. During this method, errors can be calculated which are supposed to be constant during the whole flight, and measurement values can be used to initiate the third phase. The values to be measured are the temperature and the initial biases.

The third phase is even more an estimation than a calibration. Based on the before start-up measurements, the extended Kalman filters can be initiated to estimate the biases of the acceleration or the angular velocity.

For each calibration step, 3D acceleration and angular velocity are measured. These measurements are always mean values of a few second long data acquisition. By this way, noise effect is reduced considerably.

Acceleration calibration
Offline calibration should be general and should not need any special equipment. The concept is the following. If there is not any error in the acceleration in stationary position in different orientations, then the measured gravity vector should be on the surface of a sphere with radius of 1 (the sensor measures the acceleration in G units) and with a center in the origin. Because of the composite error sources like gain, bias, and nonperpendicularity errors this surface is in reality an ellipsoid with unknown center, that is, a quadratic surface. We want to find a calibration rule which is a linear transformation of the measured value a_{measured} into the calibrated value a_{cal}. The transformation rule is sought in the form

$$a_{\text{cal}} = N a_{\text{measured}} - N p_e = N(a_{\text{measured}} - p_e), \tag{6.61}$$

where N is 3×3 matrix and the 3D vector p_e can be interpreted as the bias. For simplicity, denote a_c and a_m the calibrated and the measured acceleration, respectively.

If the sensor is placed on the horizontal plane in K_{NED} and it is not moving (stationary situation), then the correctly calibrated value a_c should satisfy $|a_c| = 1$ because the measurement is scaled in G. If Q parametrize a quadratic surface, then it yields:

$$|a_c|^2 = 1 = \langle Na_m - Np_e, Na_m - Np_e \rangle$$

$$= \langle Na_m, Na_m \rangle - 2\langle N^T Np_e, a_m \rangle + \langle Np_e, Np_e \rangle \quad \Rightarrow$$

$$0 = \underbrace{\langle N^T N}_{M} a_m, a_m \rangle + 2\underbrace{\langle -N^T Np_e, a_m \rangle}_{v} + \overbrace{\underbrace{\langle -Np_e, -Np_e \rangle}_{v}}^{c} - 1, \quad (6.62)$$

$$0 = \langle Ma_m, a_m \rangle + 2\langle v, a_m \rangle + c,$$

$$0 = \begin{pmatrix} a_m^T & 1 \end{pmatrix} \underbrace{\begin{bmatrix} M & v \\ v^T & c \end{bmatrix}}_{Q} \begin{pmatrix} a_m \\ 1 \end{pmatrix},$$

where a_m is the measured 3D acceleration vector in the frame of K_{S_0}, and Q is a 4×4 symmetric, positive definite matrix with 10 free parameters.

We can perform measurements in different orientations of the sensor in stationary situation and put the $a_{m,i}$ measurements in appropriate way in a vector f_i^T and the parameters of Q in a vector p:

$$f_i^T = [a_x^2 \; a_y^2 \; a_z^2 \; 2a_x a_y \; 2a_x a_z \; 2a_y a_z \; 2a_x \; 2a_y \; 2a_z \; 1],$$

$$p = (m_{11} \; m_{22} \; m_{33} \; m_{12} \; m_{13} \; m_{23} \; v_1 \; v_2 \; v_3 \; c)^T.$$

Let F be the matrix containing the "measured" f_i^T in its rows, then $Fp = 0$ should be satisfied, which can be solved under constraint $p^T p = 1$ in order to eliminate trivial solution.

Using Lagrange multiplier rule, the derivative of $L = p^T F^T F p - \lambda p^T p$ should be zero, that is, $F^T F p = \lambda p$, which is an eigenvalue, eigenvector problem for the matrix $F^T F$. To find the optimal λ, we can take into consideration that

$$|Fp|^2 = \langle Fp, Fp \rangle = \langle F^T F p, p \rangle = \lambda \langle p, p \rangle = \lambda \to \min,$$

which means that λ is the minimal eigenvalue.

On the other hand, we can consider the singular value decomposition of F which is $F = U \Sigma V^T$ and use the fact that the square of singular value σ_i of F is λ_i, i.e. $\lambda_i(F^T F) = \sigma_i^2(F)$. Hence, we have to chose the minimal singular value of F. Direct computation shows that the eigenvector of $F^T F$ belonging to λ is the column vector of V belonging to σ. Then p will be the column vector of V belonging to the least singular value in the SVD of F. Now it follows from (6.62)

$$M = N^T N \quad \text{and} \quad p_e = -M^{-1} v. \quad (6.63)$$

However, the constant term c should satisfy $\langle -Np_e, -Np_e \rangle - 1 = \langle v, M^{-1}v \rangle - 1 = c$ which is not necessarily the case. Hence, a scaling factor $r = 1/(\langle v, M^{-1}v \rangle - c)$ can be chosen and the problem is solved by substituting $M := rM$, $v := rv$ and recomputing $p_e := -M^{-1}v$.

The matrix $M = N^T N$ is symmetric and positive definite and has singular value decomposition $M = U_M S_M U_M^T$, hence $M = U_M S_M^{1/2} S_M^{1/2} U_M^T$ and N can be chosen as $N = S_M^{1/2} U_M^T$. We can apply QR-factorization to the matrix N such that $N = QR$ where Q is orthonormal, that is, $Q^T = Q^{-1}$, and S is upper triangular, therefore choosing $\tilde{N} = Q$ and $S = R$ we can write

$$N = S_M^{1/2} U_M^T = \tilde{N} \begin{bmatrix} s_{11} & s_{12} & s_{13} \\ 0 & s_{22} & s_{23} \\ 0 & 0 & s_{33} \end{bmatrix}, \tag{6.64}$$

$$a_{\text{cal},S_0} = \tilde{N} S a_{\text{measured}} - \tilde{N} S p_e. \tag{6.65}$$

If not all elements in the main diagonal of S are positive, then a diagonal matrix D with $D^2 = I$ can be chosen such that DS already has positive elements in the main diagonal. Substituting $\tilde{N} := \tilde{N} D$ and $S := DS$ the above calibration rule still remain valid. The elements of S in the main diagonal are the sensor gains while the remaining elements are responsible for the nonorthogonality of the sensors for different directions.

The next step is to define the frame K_S which is preferred as sensor frame for the application. Notice that the initial sensor frame is the result of internal mounting and may have nonorthogonal axes. At this point, the desired frame of the IMU (K_S) can be defined. The sensor (having frame K_{S_0}) must be rotated to the orientation where the desired z_S-axis is equal to z_{S_0} and a measurement a_{cal,S_0} has to be taken whose components are the components of the unit vector in z_S-direction expressed in the basis of K_{S_0}. Then the sensor has to be rotated again to a new orientation where the x_S'-direction is approximately equal to z_{S_0} and a measurement a_{cal,S_0} has to be taken whose components are the components of the unit vector in x_S'-direction expressed in the basis of K_{S_0}. Since there is no guarantee that x_S' is orthogonal to z_S, hence we consider the plane defined by z_S and x_S', take the vector product between z_S and x_S' defining y_S in the basis of K_{S_0}. Then we perform the vector product of the vectors belonging to y_S and z_S (in this order) resulting in x_S. After normalizing, we put the unit vectors into the columns of the orthonormal matrix $A_{S_0,S}$ in x_S, y_S, z_S order and take the inverse (which is the transpose) resulting in A_{S,S_0} and describing the unit vectors of K_{S_0} in the basis of K_S, that is, the coordinate transformation from K_{S_0} into K_S. Then we can determine the calibrated measurement in the sensor frame K_S according to the following formula:

$$a_{\text{cal}} = A_{S,S_0} a_{\text{cal},S_0} = A_{S,S_0} \tilde{N} S a_{\text{measured}} - A_{S,S_0} \tilde{N} S p_e. \tag{6.66}$$

In the sequel, we shall simply call this calibrated value the acceleration a_s. Notice that this algorithm calculates also the biases that are valid during the calibration.

The calibrated measurements will be used later during the off-line calibration of the angular velocity sensor.

The gain error of the accelerometer is *temperature dependent*. This is a linear dependency so at least two measurements in different temperature conditions are needed to find the approximation $s_{ii} = m_{ii}T_{\text{act}} + b_{ii}$ where T_{act} is the actual temperature.

During *calibration before start-up*, we calculate the bias immediately before start-up. It is assumed that in the starting position the orientation of the helicopter is known. For example, the helicopter should start from an orientation where the $[x, y]$ plane of the sensor frame is horizontal. In this case, the measured gravity vector should be $[0\ 0\ -1]$. Acceleration information should be collected and using the measured temperature the calibrated value can be computed. Then the bias will be the deviation from the expected value.

During flight, state estimation is applied determining also the estimation of the bias. The necessity of bias estimation during flight is caused by the experimental observation that within 10 min the bias changes approximately 0.2 m/ s^2.

For the start of the filters, the correlation matrices of the noises are needed. It is always difficult to give a starting estimation for the correlation of the state noises, because these are related to the internal properties of the system. In the case of position and velocity estimation, we have to characterize the acceleration noise, the acceleration bias noise the and velocity noise. To calculate an initial value, a few seconds long acquisition in stationary position is taken and the mean of the data is subtracted. The first two noises are measured by their sum, so it is difficult to separate them, the third noise cannot be measured. One solution is to separate the measured acceleration noise in frequency, the low frequency part is the bias noise and the high frequency part is the other. The covariance matrices can be approximated with diagonal ones, because the order of the off-diagonal elements are smaller than the diagonal elements. Initial estimates can be found in [68] for the covariance matrices $R_{w,\text{acc}_{\text{noise}}0}$, $R_{w,\text{acc}_{\text{bias}}0}$, and $R_{w,\text{vel}0}$. It can be assumed that there is no correlation between different state noises, so the covariance matrix of the full state noise vector can be approximated by $R_{w,0} = \text{diag}(R_{w,\text{acc}_{\text{noise}}0}, R_{w,\text{acc}_{\text{bias}}0}, R_{w,\text{vel}0})$.

Angular velocity calibration

During *offline calibration*, measurements are taken mainly in stationary position, where the expected angular velocity is $0°/$s. Since there is also an acceleration effect in the measured angular velocity, first the acceleration effect and the biases are reduced, then the gain error is calculated. The remaining error is the nonperpendicularity of the axes. A similar process as for the accelerometer can be used if a known constant angular velocity is valid during calibration.

The type of the angular velocity sensor is an Analog Devices ADXRS150. According to the datasheet, the acceleration effect is in a linear connection with the acceleration vector. So in stationary position, where the gain error is negligible, the measured angular velocity in K_{S_0} is

$$\omega_{\text{measured}} = \omega_{\text{real}} + \omega_{\text{bias}} + K^T a_{\text{cal}}, \tag{6.67}$$

where K is the acceleration compensation matrix and a_{cal} is the three dimensional calibrated acceleration vector, measured at the same time as the angular velocity. Note that a_{cal} is in the desired sensor frame (K_S) and $\omega_{measured}$ is in the original one (K_{S_0}) of the IMU, so in K matrix there is a rotation part as well. If ω_{real} is expected to 0°/s because of stationary position, then the following equation has to be solved:

$$\omega_{measured}^T = \begin{bmatrix} a_{cal}^T & 1 \end{bmatrix} \begin{bmatrix} K \\ \omega_{bias}^T \end{bmatrix}. \tag{6.68}$$

If the measured angular velocities belonging to different orientations of K_{S_0} in stationary position are put into the rows of a matrix Ω and the calibrated acceleration vectors with augmented 1 into the rows of the matrix A, then the problem can be solved with LS method:

$$\begin{bmatrix} K \\ \omega_{bias}^T \end{bmatrix} = \left(A^T A\right)^{-1} A^T \Omega, \tag{6.69}$$

from which K and ω_{bias}^T can easily be separated. The calibrated value can be transformed from K_{S_0} into K_S:

$$\omega_s = A_{S,S_0}\left(\omega_{measured} - K^T a_{cal} - \omega_{bias}\right). \tag{6.70}$$

For *gain calibration*, we refer to [66] where a method is described to find $\text{diag}(s_\Phi, s_\Theta, s_\Psi)$ by which

$$\omega_{cal_S} = \text{diag}(s_\Phi, s_\Theta, s_\Psi)\omega_s. \tag{6.71}$$

During *calibration before start-up*, the fact can be used that according to the sensor's datasheet only the bias error is temperature dependent. This can be compensated before start-up. If the starting position is stationary, the expected value is 0°/s, so the measured, calibrated angular velocity will be the bias error. It can be used as the initial value for the extended Kalman filter.

During *flight*, the initial value of the covariance matrix $R_{w,ori}$ has to be determined for the extended Kalman filter (EKF$_1$). This can be measured from a few second long angular velocity data acquisition. The results of these data show that the covariance matrix can be described by a diagonal one, as in the case of acceleration. One result of the covariance matrices $R_{w,ori_{noise}0}$ and $R_{w,ori_{bias}0}$ can be found in [66], where also a detailed Calibration Algorithm is formulated.

6.2.2.2 Vision System

The basic concept of the vision system is originated from motion-stereo approach, but only one camera is used. This camera is attached to the ceiling and the working space of the helicopter is in front of the camera. The helicopter has at least 7 active markers. Then the algorithm is the following:

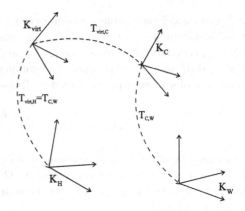

Fig. 6.13 Graphs of the vision system

Denote K_C the frame of the camera. In the start-up situation, the world frame (K_W) is defined by the frame of the helicopter (K_H). In this situation, K_W and K_H are equal. An image should be taken with the camera. Let this image be I_1 and store it. Denote $T_{C,W}$ the unknown transformation between K_C and K_W. Once the spatial positions of the markers are known, from I_1 the transformation $T_{C,W}$ can be specified.

During flight, the frame of the helicopter will be in a different position. This situation is shown in Fig. 6.13. Let a virtual frame (K_{virt}) be defined as a frame linked to K_H with the transformation $T_{\text{virt},H} = T_{C,W}$. A virtual camera is represented by the attached frame K_{virt}. A virtual image (I_{virt}) should be taken with this virtual camera. Because of the fact that the transformation between K_{virt}, K_H and between K_C, K_W is defined to be same, therefore in the points of the markers I_1 and I_{virt} are same.

Let the live image of the real camera during flight be denoted by I_{live}. Using the 7-point algorithm, described in [46, 47], the transformation ($T_{\text{virt},C}$) between K_{virt} and K_C can be calculated. The positions of the markers in the first, stationary situation can also be produced in K_C by this algorithm. Thus, $T_{C,W}$ and $T_{\text{virt},C}$ can be calculated. Since the 7 point algorithm is not robust enough, hence we applied modifications that take into consideration the marker structure as well.

Then the position and orientation of the helicopter in K_W is described by

$$T_{W,H} = (T_{C,W})^{-1} \cdot (T_{\text{virt},C})^{-1} \cdot T_{C,W}. \tag{6.72}$$

The operation of the 7 point algorithm is the following. Reference [46] describes the method how to calculate the essential matrix (Λ) which gives the relationship between the pairs of markers on image I_{live} and I_1. Denote $p_{i,1}$ the two dimensional (2D) coordinate of a marker on I_1 and $p_{i,\text{live}}$ the 2D coordinate of the same marker on I_{live}. Then the essential matrix $\Lambda_{1,\text{live}}$ satisfies

$$p_{i,1}^T \Lambda_{1,\text{live}} p_{i,\text{live}} = 0 \tag{6.73}$$

for each marker. In ideal case

$$\Lambda_{1,\text{live}} = R \cdot \begin{bmatrix} 0 & -t_z & t_y \\ t_z & 0 & -t_x \\ -t_y & t_x & 0 \end{bmatrix}, \tag{6.74}$$

where t is the vector from the origin of K_{virt} to the origin of K_C in K_C and t_x, t_y, t_z are its components. R describes the orientation between these frames.

A solution to produce an R matrix and a t vector are given in [46] having the following form:

$$\Lambda_{1,\text{live}} = USV^T, \tag{6.75}$$

$$R = U \begin{bmatrix} 0 & \pm 1 & 0 \\ \mp 1 & 0 & 0 \\ 0 & 0 & 1 \end{bmatrix} V^T, \tag{6.76}$$

$$t = V \cdot [001]^T, \tag{6.77}$$

where (6.75) is the SVD decomposition of $\Lambda_{1,\text{live}}$. The right R matrix can be determined from the fact that the markers should be in front of the camera. Hence $T_{\text{virt},C}$ can be assembled.

Light emitting spheres are used as markers, because it is necessary for image preprocessing that the size of the markers on the images be at least a few pixel large in all situations. Due to the spheres, the color intensity of the markers will not be uniform on the images.

Because of the nonuniform intensity of the markers, their central points cannot be measured accurately. Therefore, the essential matrix differs from the real one. Hence, R and t may have errors. The effect of these errors can be seen on a plot in Fig. 6.14. Let l_i be the projection line from the principal point of the real camera

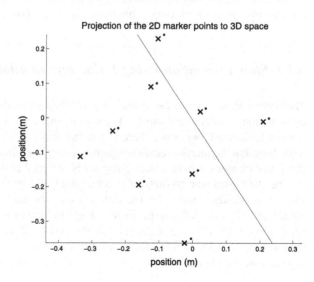

Fig. 6.14 Position error of the reprojected markers

through the central of the ith marker on I_{live}. Let r_i be a similar projection line from the principal point of the virtual camera through the corresponding marker's central on $I_{\text{virt}} = I_1$. Then the crosses and dots in Fig. 6.14 are the nearest points of l_i and r_i for each marker. Finally, the line in Fig. 6.14 is the epipolar axis (also referred as baseline) which can also be determined by the method in [46].

According to lots of measurements, it can be said that the lines through the corresponding dots and crosses are almost perpendicular to the epipolar axis. Hence, this misalignment error would be smaller if R would be corrected with a small rotation around the epipolar axis. Starting from the Rodriguez formula an optimal rotation angle can be calculated. One solution is described in [67].

Connection between the vision and state estimation system

The vision system produces measured output (y) information for the state estimation, in the case of the position and orientation as well. The extended Kalman filter deals with measurement noises (z). Their effect appear in the covariance matrices ($R_{z,k}$).

As in the case of the IMU, some kind of approximations should be given as initial values. Because of the fact that the helicopter starts from a stationary position, a few second long measurements can be taken. The initial value can be the mean value of the position and orientation components of the state vectors. The remaining parts are the noises with 0 expected values and their correlation matrices can be calculated. Initial estimations of the covariance matrices R_{z,ori_0} and R_{z,pos_0} for position and orientation, respectively, can be found in [67].

Finalizing the state estimation

The remaining matrix, which should be initiated, is the covariance matrix Σ_0 of the initial state. These Σ_0 can be formed from the existing information in [67]. Finally, the time aspect of the sensor fusion problem should be solved. There are two different sampling times (0.01 s for IMU and 0.04 s for vision). The problem can be solved in the way that if there is new information from the vision system then the state update $\hat{x}$ is performed. Otherwise, $\hat{x}_k := \bar{x}_k$ should be taken.

6.2.3 State Estimation Using Vision and Inertial Measurements

The control algorithm discussed later requires the signals shown in Table 6.2 for the computation of the control inputs. The signals measured by the sensor system are marked by asterisk ($*$). Since there is a signal that is not measured and long time tests show that the inertial sensor's signals not only contain noises, but also an offset (bias) that changes slowly, a state estimator is included in the control algorithm.

The state estimator consists of two hierarchically structured extended Kalman filters that are responsible for the estimation of the attitude and position related signals (EKF_1 and EKF_2, respectively). The structure of the state estimator can be seen in Fig. 6.15. The role of the block in the middle of the diagram is to transform the measured acceleration from the sensor frame to the helicopter frame. This block separates the two EKFs.

Table 6.2 The signals required by the control algorithm

Signal	Meaning
ξ	Position*
$\dot{\xi}$	Velocity
η	Attitude*
$\dot{\eta}$	Angular velocity*
Ω	Angular velocities of the rotors*

Fig. 6.15 The structure of the state estimator

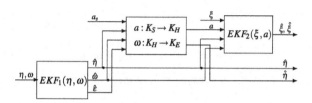

Before the description of the state estimation, it is necessary to consider that the inertial sensor's frame is not identical to that of the helicopter's. Therefore, the measured acceleration and angular velocity have to be transformed. In what follows, A_s (orthonormed) and p_s represent the angular and spatial displacements of the two frames.

Estimating the attitude and angular velocity of the helicopter
As indicated before, the IMU's outputs consist of three components: the real values and an additional bias and noise with zero mean. In the sensor frame, it can be written as

$$\omega_m = \omega_s = \omega_{s,0} + \omega_{s,b} + \omega_{s,n}. \tag{6.78}$$

The real value of the angular velocity in the helicopter's frame is

$$\omega = A_s \omega_{s,0} = A_s(\omega_s - \omega_{s,b} - \omega_{s,n}), \tag{6.79}$$

which can be transformed to the reference frame K_E as

$$\dot{\eta} = R_r^{-1}\omega = R_r^{-1}A_s(\omega_s - \omega_{s,b} - \omega_{s,n}). \tag{6.80}$$

Since the bias changes slowly, it can be assumed that its time derivative is close to zero. This can be formulated by the aid of small virtual noise that affects the change of the bias's value:

$$\dot{\omega}_{s,b} = \omega_{s,b,n}. \tag{6.81}$$

As the measurements of the positioning system do not contain offset, the third equation that can be used for state estimation along with (6.80) and (6.81) is the following:

$$\eta_m = \eta + \eta_n. \tag{6.82}$$

If T_s denotes the sampling time, these three equations can be transformed to discrete time using the Euler's formula as

$$x_{1,k+1} = x_{1,k} + T_s R_{r,k}^{-1} A_s (-x_{2,k} + u_k + w_{1,k}),$$

$$x_{2,k+1} = x_{2,k} + T_s w_{2,k}, \qquad\qquad (6.83)$$

$$y_k = x_{1,k} + z_k,$$

with the following notations:

$$x_1 = \eta, \qquad x_2 = \omega_{s,b}, \qquad x = \left(x_1^T, x_2^T \right)^T,$$

$$w_1 = -\omega_{s,n}, \qquad w_2 = \omega_{s,b,n}, \qquad w = \left(w_1^T, w_2^T \right)^T, \qquad (6.84)$$

$$u = \omega_m, \qquad y = \eta_m, \qquad z = \eta_n.$$

The equations (6.83) can be rewritten to the form of

$$x_{k+1} = f(x_k, u_k, w_k),$$

$$y_k = g(x_k, z_k).$$

Assuming w and z are not correlated, the EKF algorithm can be performed by introducing the following notations:

$$R_{w,k-1} = E\left[w_{k-1} w_{k-1}^T \right], \qquad R_{z,k} = E\left[z_k z_k^T \right],$$

$$A_{k-1} = \frac{\partial f(\hat{x}_{k-1}, u_{k-1}, 0)}{\partial x}, \qquad B_{w,k-1} = \frac{\partial f(\hat{x}_{k-1}, u_{k-1}, 0)}{\partial w}, \qquad (6.85)$$

$$C_k = \frac{\partial g(\bar{x}_k, 0)}{\partial x}, \qquad C_{z,k} = \frac{\partial g(\bar{x}_k, 0)}{\partial z},$$

where $\hat{x}_k$ is the estimated value of x. The well-known steps of the extended Kalman filter algorithm are:

1. Prediction:

$$\bar{x}_k = f(\hat{x}_{k-1}, u_{k-1}, 0),$$

$$M_k = A_{k-1} \Sigma_{k-1} A_{k-1}^T + B_{w,k-1} R_{w,k-1} B_{w,k-1}^T. \qquad (6.86)$$

2. Time update:

$$S_k = C_k M_k C_k^T + C_{z,k} R_{z,k} C_{z,k}^T,$$

$$G_k = M_k C_k^T S_k^{-1},$$

$$\Sigma_k = M_k - G_k S_k G_k^T, \qquad\qquad (6.87)$$

$$\hat{x}_k = \bar{x}_k + G_k \left(y_k - g(\bar{x}_k, 0) \right).$$

Estimating the position and velocity of the helicopter

The estimation method is similar to that in the previous part. First, it has to be considered what an accelerometer senses. Its output not only contains the three components of the acceleration, but also the effect of the gravity. Using the same notations as before, it can be formulated as

$$a_s = a_{s,0} + a_{s,b} + a_{s,n} - A_s^T R_t^T g. \tag{6.88}$$

As the quadrotor's frame is not an inertial frame, the connection between the acceleration in K_S and K_H is the following:

$$a_{s,0} = A_s^T \left(a + \varepsilon \times p_s + \omega \times (\omega \times p_s) \right). \tag{6.89}$$

From this equation, the acceleration in K_H (a) can be obtained as

$$a = A_s a_s - \varepsilon \times p_s - \omega \times (\omega \times p_s) + R_t^T g - A_s a_{s,b} - A_s a_{s,n}. \tag{6.90}$$

The first part of (6.90) can be interpreted as a transformed value of the accelerometer's output:

$$a_t = A_s a_s - \varepsilon \times p_s - \omega \times (\omega \times p_s) + R_t^T g. \tag{6.91}$$

This transformation is performed by the central block in Fig. 6.15. The equation is also an explanation why the two EKFs need to be arranged hierarchically.

Applying the differentiation rule in a moving frame ($\dot{\xi} = \dot{v} + \omega \times v$) once again and making the same assumptions about the bias and the positioning system yields the equations that can be used for the state estimation:

$$\begin{aligned}
\dot{v} &= -\omega \times v - A_s a_{s,b} + a_t + A_s a_{s,n}, \\
\dot{a}_{s,b} &= a_{s,b,n}, \\
\dot{\xi} &= R_t v + v_{\xi,n}, \\
\xi_m &= \xi + \xi_n.
\end{aligned} \tag{6.92}$$

Following the same steps as previously, the discrete time equations of the system above are

$$\begin{aligned}
x_{1,k+1} &= \left(I_3 - T_s [\omega \times]_k \right) x_{1,k} - T_s A_s x_{2,k} + T_s u_k + T_s A_s w_{1,k}, \\
x_{2,k+1} &= x_{2,k} + T_s w_{2,k}, \\
x_{3,k+1} &= x_{3,k} + T_s R_{t,k} x_{1,k} + T_s w_{3,k}, \\
y_k &= x_{3,k} + z_k,
\end{aligned} \tag{6.93}$$

with the notations

$$\begin{aligned}
x_1 &= v, & x_2 &= a_{s,b}, & x_3 &= \xi, \\
w_1 &= -a_{s,n}, & w_2 &= a_{s,b,n}, & w_3 &= v_{\xi,n}, \\
u &= a_t, & y &= \xi_m, & z &= \xi_n.
\end{aligned} \tag{6.94}$$

Here $[\omega\times]$ represents the matrix of the cross product. From here, the EKF can easily be formed. In order to find $\varepsilon = \dot{\omega}$ needed in a_t, numerical differentiation can be used.

6.2.4 Backstepping Control Algorithm

There are numerous control algorithms that can be applied to a quadrotor helicopter including linear [20], nonlinear and even soft computing techniques [32]. Among the nonlinear control algorithms, the backstepping approach has gained the most attention, although several other methods are elaborated including sliding mode [21] and feedback linearization control algorithms [33].

These pieces of research not only differ from each other on the control algorithm, but also on the types of simplification of the dynamic model of the helicopter. Several methods exist for dynamic models that retain the basic behavior of the vehicle. Some neglect the rotor dynamics assuming the transients of the rotors are fast compared to those of the helicopter, some others do not consider the aerodynamics or the gyroscopic effect.

A full state backstepping algorithm is presented in [88], where the control law is obtained step by step through three virtual subsystems' stabilization. The quadrotor dynamic model described in Sect. 6.2.1 is similar to that in this work. In [21, 135] and [68], a backstepping algorithm is applied to simplified helicopter dynamic model. These are the base of the algorithm that is presented in this section.

The following parts focus on the construction of such an algorithm that is capable of explicitly handling all the effects appearing in (6.56), while being ignorant to realistic measurement noises.

Applying backstepping algorithm to the helicopter
The control algorithm has evolved from the results of [21] and our research [68]. The algorithm presented in this part intends to exploit the advantages of two approaches, that are the ability to control a dynamic model with the least possible simplification and the good handling of measurement noises experienced in the case of our earlier algorithm based on [68].

First, we have to reformulate the equations (6.56) and (6.59).

$$\ddot{\xi} = f_\xi + g_\xi u_\xi,$$
$$\ddot{\eta} = f_\eta + g_\eta u_\eta,$$
$$\dot{\Omega}_k = f_{\Omega,k} + g_{\Omega,k} u_{\Omega,k},$$

(6.95)

where f_ξ, g_ξ and u_ξ are

$$f_\xi = -G - \frac{1}{m} R_t K_t R_t^T \dot{\xi},$$
$$g_\xi = \frac{1}{m} \operatorname{diag}(r_{t,3}),$$
$$u_\xi = (f, f, f)^T,$$

(6.96)

while f_η, g_η and u_η stand for

$$f_\eta = (I_c R_r)^{-1}\left[-I_c\left(\frac{\partial R_r}{\partial \Phi}\dot{\Phi} + \frac{\partial R_r}{\partial \Theta}\dot{\Theta}\right)\dot{\eta}\right.$$

$$\left. - K_r R_r \dot{\eta} - (R_r \dot{\eta}) \times (I_c R_r \dot{\eta} + I_r \Omega_r)\right],\qquad(6.97)$$

$$g_\eta = (I_c R_r)^{-1},$$

$$u_\eta = T,$$

and $f_{\Omega,k}$, $g_{\Omega,k}$ and $u_{\Omega,k}$ yield

$$f_{\Omega,k} = -k_{\Omega,0} - k_{\Omega,1}\Omega_k - k_{\Omega,2}\Omega_k^2,$$

$$g_{\Omega,k} = k_u,\qquad(6.98)$$

$$u_{\Omega,k} = u_{m,k}.$$

Since the vector F contains only one nonzero element, hence

$$g_\xi u_\xi = \frac{1}{m}r_{t,3}f = \frac{1}{m}\,\mathrm{diag}(r_{t,3})u_\xi,\qquad(6.99)$$

where $r_{t,3}$ is the third column of R_t.

The construction of the control law is similar to that presented in [21]. Since the helicopter is underactuated, the concept is that the helicopter is required to track a path defined by its (x_d, y_d, z_d, Ψ_d) coordinates. The helicopter's roll and pitch angles are stabilized to 0 internally. The control algorithm can be divided into three main parts. At first, the translational part of the vehicle dynamics is controlled, which then produces the two missing reference signals to the attitude control system. The third part is responsible for generating the input signals of the BLDC motors. The hierarchical structure of the controller is shown in Fig. 6.16, where indexes d and m denote desired and measured values, respectively. The speed ratio of the three parts of the hierarchical structure depends on the physical properties of the components, especially on the measurement frequency of the sensors. The ideal values of the sampling times for position and orientation control are between 10–30 ms. Kalman filters can tolerate the difference of measurement frequencies of the

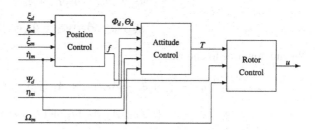

Fig. 6.16 The hierarchical structure of the controller

position and orientation (vision system) and acceleration and velocity (IMU). The sampling time of the motor control is set to 10 ms.

Position control
The concept of backstepping control will be explained for position control. First, let us define the path tracking error

$$q_{\xi_1} = \xi_d - \xi. \tag{6.100}$$

Applying Lyapunov's theorem we construct a positive definite function whose time derivative is negative definite:

$$V(q_{\xi_1}) = \frac{1}{2} q_{\xi_1}^T q_{\xi_1}, \tag{6.101}$$

$$\dot{V}(q_{\xi_1}) = q_{\xi_1}^T \dot{q}_{\xi_1} = q_{\xi_1}^T (\dot{\xi}_d - \dot{\xi}). \tag{6.102}$$

If we were free to choose

$$\dot{\xi} = \dot{\xi}_d + A_{\xi_1} q_{\xi_1} \tag{6.103}$$

then the time derivative of the Lyapunov function would be

$$\dot{V}(q_{\xi_1}) = -q_{\xi_1}^T A_{\xi_1} q_{\xi_1} < 0 \tag{6.104}$$

if the matrix A_{ξ_1} was positive definite. Therefore, introducing a virtual tracking error

$$q_{\xi_2} := \dot{\xi} - \dot{\xi}_d - A_{\xi_1} q_{\xi_1} = -\dot{q}_{\xi_1} - A_{\xi_1} q_{\xi_1} \tag{6.105}$$

and an augmented Lyapunov function

$$V(q_{\xi_1}, q_{\xi_2}) = \frac{1}{2} \left(q_{\xi_1}^T q_{\xi_1} + q_{\xi_2}^T q_{\xi_2} \right), \tag{6.106}$$

we are able to find the actual lift force f needed. The augmented Lyapunov function's time derivative is

$$\begin{aligned}
\dot{V}(q_{\xi_1}, q_{\xi_2}) &= q_{\xi_1}^T \dot{q}_{\xi_1} + q_{\xi_2}^T \dot{q}_{\xi_2} = q_{\xi_1}^T (-\dot{q}_{\xi_1} - A_{\xi_1} q_{\xi_1}) \\
&\quad + q_{\xi_2}^T (\ddot{\xi} - \ddot{\xi}_d - A_{\xi_1} (-\dot{q}_{\xi_1} - A_{\xi_1} q_{\xi_1})) \\
&= -q_{\xi_1}^T A_{\xi_1} q_{\xi_1} - q_{\xi_2}^T q_{\xi_1} + q_{\xi_2}^T (f_\xi + g_\xi u_\xi) \\
&\quad - q_{\xi_2}^T [\ddot{\xi}_d - A_{\xi_1} (q_{\xi_2} + A_{\xi_1} q_{\xi_1})].
\end{aligned} \tag{6.107}$$

We are now free to choose

$$\begin{aligned}
u_\xi &= g_\xi^{-1} [q_{\xi_1} - f_\xi + \ddot{\xi}_d - A_{\xi_1} (q_{\xi_2} + A_{\xi_1} q_{\xi_1}) - A_{\xi_2} q_{\xi_2}] \\
&= g_\xi^{-1} [\ddot{\xi}_d - f_\xi + (I_3 + A_{\xi_2} A_{\xi_1}) q_{\xi_1} + (A_{\xi_2} + A_{\xi_1}) \dot{q}_{\xi_1}],
\end{aligned} \tag{6.108}$$

where I_3 is a unit matrix. It could be assumed here that $\ddot{\xi}_d$ is negligible as in [21]. However, the further goal is that the helicopter tracks certain waypoints, which means that it is in continuous motion. Therefore, $\ddot{\xi}_d$ does have an important role in the control. If A_{ξ_2} is positive definite, the time derivative of the Lyapunov function is

$$\dot{V}(q_{\xi_1}, q_{\xi_2}) = -q_{\xi_1}^T A_{\xi_1} q_{\xi_1} - q_{\xi_2}^T A_{\xi_2} q_{\xi_2} < 0. \tag{6.109}$$

Applying the control law (6.108) to (6.95), it results in

$$\ddot{\xi} = \ddot{\xi}_d + (I_3 + A_{\xi_2} A_{\xi_1}) q_{\xi_1} + (A_{\xi_2} + A_{\xi_1}) \dot{q}_{\xi_1}, \tag{6.110}$$

which is equivalent to

$$0 = (I_3 + A_{\xi_2} A_{\xi_1}) q_{\xi_1} + (A_{\xi_2} + A_{\xi_1}) \dot{q}_{\xi_1} + \ddot{q}_{\xi_1}. \tag{6.111}$$

Assuming positive definite and diagonal A_{ξ_1}, A_{ξ_2} matrices with diagonal elements $a_{\xi_1,i}, a_{\xi_2,i}$, the characteristic equations have the form

$$s^2 + (a_{\xi_2,i} + a_{\xi_1,i})s + (1 + a_{\xi_2,i} a_{\xi_1,i}) = 0, \tag{6.112}$$

which guarantees stability.

This means that the errors exponentially converge to zero if the calculated values of f_ξ and g_ξ are close to the real ones.

Algebraic manipulations can be performed in $g_\xi u_\xi$. The third component of u_ξ is the lift force f. The other two components are for different purposes. Multiplying the formula of (6.108) in brackets by $\frac{m}{f}$ instead of the reciprocal of the appropriate element of g_ξ yields an expression for the third column of R_t. Since this change has no effect on the stability, then if the entire controlled system is stable, g_ξ has to be convergent and its limit is $(0, 0, 1)^T$. Therefore, the reference signals Φ_d and Θ_d can be obtained as follows. First, we modify g_ξ and u_ξ:

$$\tilde{g}_\xi = \frac{1}{m} \begin{bmatrix} f & 0 & 0 \\ 0 & f & 0 \\ 0 & 0 & C_\Phi C_\Theta \end{bmatrix}, \tag{6.113}$$

$$\tilde{u}_\xi = \begin{pmatrix} C_\Phi S_\Theta C_\Psi + S_\Phi S_\Psi \\ C_\Phi S_\Theta S_\Psi - S_\Phi C_\Psi \\ f \end{pmatrix} = \begin{pmatrix} u_{\xi_x} \\ u_{\xi_y} \\ f \end{pmatrix}. \tag{6.114}$$

We can extract f as before and then we obtain u_{ξ_x} and u_{ξ_y} using the elements of $\tilde{g}_\xi$ and

$$S_{\Phi_d} = S_\Psi u_{\xi_x} - C_\Psi u_{\xi_y},$$
$$S_{\Theta_d} = \frac{C_\Psi u_{\xi_x} + S_\Psi u_{\xi_y}}{C_\Phi} \tag{6.115}$$

yield Φ_d and Θ_d. The reason why these signals can be considered as reference signals is that as the helicopter approaches the desired coordinates, they converge to zero. Conversely, if the helicopter follows the appropriate attitude and lift force, it will get to the desired position and orientation.

Attitude control
The design is similar to that described in the previous part. Again, let us define the attitude error

$$q_{\eta_1} = \eta_d - \eta, \tag{6.116}$$

and introduce a virtual tracking error

$$q_{\eta_2} = \dot{\eta} - \dot{\eta}_d - A_{\eta_1} q_{\eta_1} = -\dot{q}_{\eta_1} - A_{\eta_1} q_{\eta_1}. \tag{6.117}$$

Following the same steps, the result is $T = u_\eta$ where

$$u_\eta = g_\eta^{-1}\left[\ddot{\eta}_d - f_\eta + (I_3 + A_{\eta_2} A_{\eta_1})q_{\eta_1} + (A_{\eta_2} + A_{\eta_1})\dot{q}_{\eta_1}\right]. \tag{6.118}$$

A simplified control of the position and attitude
According to (6.56) and (6.57), a change in f_ξ, f_η and g_η results in simpler controller equations that are formally identical to (6.108) and (6.118):

$$f_\xi = -G,$$
$$f_\eta = -I_c^{-1}\left[\dot{\eta} \times (I_c\dot{\eta} + I_r\Omega_r)\right], \tag{6.119}$$
$$g_\eta = I_c^{-1}.$$

A further simplification is to assume that the second derivatives of the reference signals are zero ($\ddot{\xi}_d = 0$ and $\ddot{\eta}_d = 0$), thus these terms disappear from the control laws. In the following sections, it will be shown that these simplifications do not deteriorate the overall performance of the control system.

Rotor control
There is a slight difference in the calculation of u_m compared to the other inputs since in the third equation of (6.95) only the first derivative of Ω_k appears. This means that there is no need for the virtual error q_{m_2} and the Lyapunov function remains $V(q_{m_1}) = \frac{1}{2}q_{m_1}^T q_{m_1}$. However, it is worth including the derivative of q_{m_1} similarly as in the previous sections because of the error dynamics. Then

$$u_m = g_m^{-1}\left[\dot{\Omega}_d - f_m + (I_4 + A_{m_2} A_{m_1})q_{m_1} + (A_{m_2} + A_{m_1})\dot{q}_{m_1}\right] \tag{6.120}$$

with q_{m_1} and f_m being

$$q_{m_1} = \begin{pmatrix} \Omega_{1_d} - \Omega_1 \\ \Omega_{2_d} - \Omega_2 \\ \Omega_{3_d} - \Omega_3 \\ \Omega_{4_d} - \Omega_4 \end{pmatrix} \quad \text{and} \quad f_m = \begin{pmatrix} f_{m,1} \\ f_{m,2} \\ f_{m,3} \\ f_{m,4} \end{pmatrix}. \tag{6.121}$$

Since the four motors are considered to be identical, g_m can be any of $g_{m,k}$-s and therefore it is a scalar. It is worth noticing that since T and f are linear combinations of Ω_k^2, hence Ω_{k_d} are the element-wise square roots of

$$
\begin{pmatrix} \Omega_{1_d}^2 \\ \Omega_{2_d}^2 \\ \Omega_{3_d}^2 \\ \Omega_{4_d}^2 \end{pmatrix} = \begin{bmatrix} 0 & -(2lb)^{-1} & -(4d)^{-1} & (4d)^{-1} \\ -(2lb)^{-1} & 0 & (4d)^{-1} & (4d)^{-1} \\ 0 & (2lb)^{-1} & -(4d)^{-1} & (4d)^{-1} \\ (2lb)^{-1} & 0 & (4d)^{-1} & (4d)^{-1} \end{bmatrix} \begin{pmatrix} T \\ f \end{pmatrix}. \tag{6.122}
$$

The time derivative of the Lyapunov function becomes

$$
\begin{aligned}
\dot{V}(q_{m_1}) &= q_{m_1}^T \dot{q}_{m_1} \\
&= -q_{m_1}^T \big[(I + A_{m_2} A_{m_1}) q_{m_1} - (A_{m_2} + A_{m_1}) \dot{q}_{m_1} \big] \\
&= -q_{m_1}^T (I + A_{m_2} + A_{m_1})^{-1} (I + A_{m_2} A_{m_1}) q_{m_1} < 0 \tag{6.123}
\end{aligned}
$$

if A_{m_1} and A_{m_2} are positive definite matrices.

Tuning of controller parameters
The parameters of the controllers are related to coefficients of low order characteristic polynomials or elements of positive definite diagonal matrices, hence their choice is simple. The numerical values are immediately related to the speed of the control. It should also be taken into account that the increased speed of the control can cause saturation in the actuators. Simulation experiments can help in parameter tuning.

The tracking algorithm
The purpose of the control design is to track a predefined trajectory with the smallest possible tracking error. In practice a navigation point must be approximated with a predefined accuracy considering the positions and orientations. Apart from the tracking error it is also important to keep the helicopter in continuous motion. In other words, the helicopter should not slow down when it reaches a waypoint but move towards the next one.

These principles can be formulated by setting up certain conditions. If one of them is satisfied, the quadrotor helicopter may advance towards the next navigation point in the algorithm.

The conditions for position tracking are

$$
\sum_{j=1}^{3} \big(\xi_{d_j}^{(i+1)} - \xi_j \big)^2 < \Delta \xi_0,
$$

$$
\sum_{j=1}^{3} \big(\xi_{d_j}^{(i+1)} - \xi_j \big)^2 < \lambda \sum_{j=1}^{3} \big(\xi_{d_j}^{(i+1)} - \xi_{d_j}^{(i)} \big)^2, \tag{6.124}
$$

where ξ is the current position, $\Delta \xi_0$ is a predefined constant distance, $\xi_{d_j}^{(i)}$ and $\xi_{d_j}^{(i+1)}$ are the coordinates of two consecutive navigation points of the trajectory (the helicopter has already stepped towards $\xi_{d_j}^{(i+1)}$). The first condition ensures that the helicopter will remain in the proximity of the navigation points, while the other one is responsible for keeping the motion continuous. $\lambda = (3/4)^2$ is a suitable compromise if the navigation points are close to each other. In the proximity of obstacles, the navigation points should be chosen denser. Obstacle avoidance is a high level motion design problem, which is not part of this discussion.

The tracking of the yaw angle is somewhat different since it might be important how the attitude of the helicopter behaves during flight. Therefore, the only condition is that the absolute value of the yaw angle error has to be lower than a predefined limit ($\Delta\Psi_0$):

$$\left|\Psi_d^{(i+1)} - \Psi\right| < \Delta\Psi_0. \tag{6.125}$$

As a further refinement, $\Psi_d^{(i+1)}$ should be chosen such that the helicopter rotates in the desired direction. If Ψ_d values can be outside the interval $[0, 2\pi)$, this can also be taken into account.

Ensuring smooth motion between navigation points
Abrupt changes may occur between two navigation points during the maneuver. In order to guarantee the smooth motion, the predefined path must be refined by using a filtering procedure in order to avoid the risk of numerical problems for large tracking errors. A block with the following transfer function is able to perform this task:

$$W(s) = \frac{\gamma|p_s|^{n+1}}{(s + \gamma p_s)(s + p_s)^n} \tag{6.126}$$

with $0 < p_s$ and $0 < \gamma$. By setting $n = 5$ and $\gamma = 3$, the control inputs will still remain smooth and the transient will mostly be determined by p_s.

6.2.5 Embedded Control Realization

Flying systems are complex ones, hence thorough test of their control system before the first real flight is highly suggested. Before implementing the control algorithm on the embedded target, it was tested using hardware-in-the-loop method.

The mechanical parameters of the helicopter and the BLDC motors with the rotors are based on the planned dimensions, the masses of purchased elements. These values are summarized in Table 6.1.

Simulations performed using Simulink
The components are included in the control loop gradually from the simplest case to the most complex. The simplest simulation contained only the model of the quadrotor helicopter and the controller. No measurement noises were considered and the helicopter only had to reach a single spatial point with a defined yaw angle. The reference signals were smoothed using the same method presented in the previous section. The parameters of the controller could be tuned this way (see Table 6.3

Table 6.3 The physical parameters of the controller	Parameter	Value
	A_{ξ_1}	diag(2, 1.6, 1.6)
	A_{ξ_2}	diag(1.2, 1.2, 1.2)
	A_{η_1}	diag(12, 12, 12)
	A_{η_2}	diag(8, 8, 8)
	A_{m_1}	diag(0.04, 0.04, 0.04, 0.04)
	A_{m_2}	diag(0.016, 0.016, 0.016, 0.016)

for the exact values). The desired sample time of the control algorithm was set to $T_s = 10$ ms, since the maximum operating frequency of the IMU is 100 Hz. Therefore, this was the fixed step size during the simulations.

An example is shown below, including some signals of interest. Figure 6.17 shows the position and attitude of the quadrotor helicopter, while the control inputs of the motors and the actual angular velocities of the rotors appear in Fig. 6.18.

Control behaviors using the simplified and more complex dynamic models were compared. According to the simulation results, the simplified position and attitude control performed similarly to the more complex model, although the latter performed better. This can be explained by the low speeds of the helicopter and the small roll and pitch angles during flight. Therefore, it is reasonable to use it instead of the more complex algorithm if hardware resources are limited. Figure 6.19 shows the tracking errors of x and Ψ during the simple test (measurable states) presented above. The dotted lines correspond to the motion of the helicopter controlled by the simpler controller. As a final result of the integration of all the components including state estimation, a tracking of a complex trajectory is presented in Figs. 6.20 and 6.21. These simulations were performed by setting the two EKFs initial parameters as shown in Table 6.4 (I_3 is a 3-by-3 unit matrix). The parameters were set according to the results of the analysis of the sensor signals [66, 67].

Real-time tests
Real-time tests were performed using the hardware-in-the-loop method. The tests were aided by a dSPACE DS1103 board. First, the model of the helicopter, the sensor and the vision system's measurements were emulated on the board, while further experiments included the real IMU and vision system. The communication channel between the MPC555 and the DS1103 board was the CAN bus, as in the final version of the helicopter. The scheme of the tests can be seen in Fig. 6.22.

Software environment
The central unit is a Freescale MPC555 microcontroller mounted on a board produced by Phytec. The processor is equipped with a 64-bit floating point unit. The control algorithm is designed using MATLAB/Simulink environment with Real-Time Workshop, Real-Time Workshop Embedded Coder and Embedded Target for Motorola MPC555. The generated code is compiled by MetroWerks CodeWarrior cross-development tool.

Fig. 6.17 The position and
attitude of the helicopter
during flight

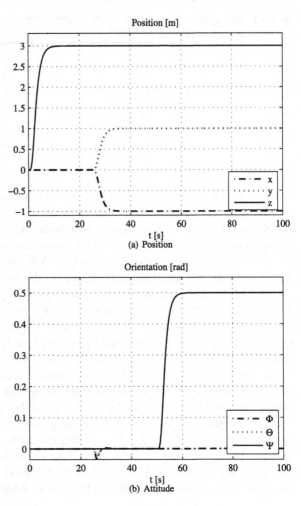

(a) Position

(b) Attitude

Hardware-in-the-loop tests are performed with the use of a dSPACE DS1103 board that is a powerful means of rapid prototype development. The software package includes a Simulink block library and ControlDesk, which provides a graphical user interface that controls program flow, data monitoring and collection. The collected data can easily be analyzed then in MATLAB. Result of the hardware-in-the-loop tests for a complex spiral motion is illustrated in Fig. 6.23.

Communication

The majority of Simulink's blocks are supported MATLAB's Target Language Compiler, while the Embedded Target for Motorola MPC555 includes blocks that can be used for handling peripherals such as communication via serial or CAN interface. However, experiments show that serial communication causes a significant delay (20–30 ms) in the signal propagation. Since CAN communication does not cause such delays, there is an extra component in the system which is responsible

Fig. 6.18 The control inputs and the angular velocities of the helicopter

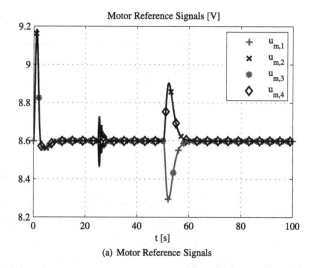

(a) Motor Reference Signals

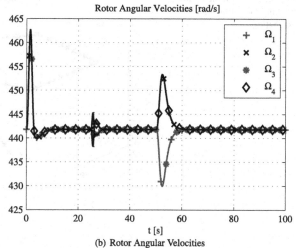

(b) Rotor Angular Velocities

for converting all the serial packets into CAN packets. Wireless communication and the IMU's output is affected by this problem.

The MPC555 contains 16 buffers that can be used for transmitting or receiving CAN packets, while the number of data to be transmitted in each cycle exceeds the buffer number. Since the packet size is limited to 8 bytes, groups of the measurement data need to be transmitted in each packet. It is also crucial to maintain data integrity during the hardware-in-the-loop test, since starting the calculation of the control inputs before receiving all measurement data may make the control loop unstable. To ensure that the time delay between receiving the sensor data and starting the calculation of the new control inputs is minimal, data acquisition (checking the buffers for new packets) on the target processor is performed at a higher frequency compared to that of the control algorithm.

Fig. 6.19 Comparison of the tracking errors using simplified (index 2) and more complex (index 1) dynamic models

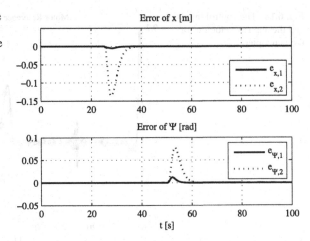

Fig. 6.20 A complex path tracking including state estimation

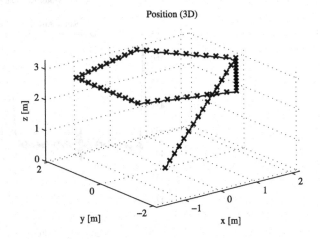

Reachable sampling time for the control algorithm

Preliminary calculations showed that the MPC555 is theoretically capable of performing the computation of the control inputs every 10 ms. However, Execution Profiling shows that calculating the control inputs using double precision floating point numbers takes slightly more than 20 ms. Using single precision numbers does not cause significant deterioration in the calculations, however, it saves about 5 ms.

Software related issues

We faced minor software problems during the development that can be avoided with little attention. If the Simulink model to be run on the dSPACE DS1103 board contains several S-functions written in C that contain global variables inside, then these variables should have different names in order to avoid unexpected behavior during execution. The reason for is that during the compilation procedure these variables are overwritten by each other.

Table 6.4 The initial parameters of the EKFs

Parameter	Value
$R_{\omega w,0}$	$\mathrm{diag}(\sigma_\omega I_3, \sigma_{\omega,b} I_3)$
$R_{\omega z,0}$	$\mathrm{diag}(\sigma_\eta I_3)$
$R_{aw,0}$	$\mathrm{diag}(\sigma_a I_3, \sigma_{a,b} I_3, \sigma_v I_3)$
$R_{az,0}$	$\mathrm{diag}(\sigma_\xi I_3)$
σ_ω	$(2 \times 10^{-2}\,\mathrm{s}^{-1})^2$
$\sigma_{\omega,b}$	$(5 \times 10^{-4}\,\mathrm{s}^{-1})^2$
σ_η	$(\pi/180\,\mathrm{rad})^2$
σ_a	$(10^{-2}\,\mathrm{m/s}^2)^2$
$\sigma_{a,b}$	$(2 \times 10^{-3}\,\mathrm{m/s}^2)^2$
σ_v	$(2.5 \times 10^{-3}\,\mathrm{m/s})^2$
σ_ξ	$(2 \times 10^{-2}\,\mathrm{m})^2$

Sampling times are also of high importance, especially when the model contains multiple sample times. It is not a good practice to set the sampling time property "inherited" of any Simulink block. For the same reason, the usage of discrete time blocks that do not have sampling time property (like discrete derivative blocks) is not recommended.

6.3 Closing Remarks

The control of airplanes and helicopters is a complex problem in which many research institutes and companies carry out researches all over the world. Classical approaches use LQ or robust (H_2/H_∞) control, newer ones apply parameter varying control or model predictive control. We considered only two subproblems from the large field of these control problems.

The first approach assumes a defined flight envelope and tries to stabilize and optimize the control behavior of an aircraft within the envelope. Two controllers are used to solve the problem.

The internal controller is a disturbance observer well known in motor control. The role of this controller is to stabilize the system without dealing with optimality. The aimed closed loop transfer function was derived from Level 1 flight conditions. Three criteria were formulated for the LPV system, such as stability, robust stability and nominal performance. A numerical technique was presented to find the free part of the disturbance observer satisfying the criteria.

The external control method is Receding Horizon Control that performs open loop optimization in the actual horizon and applies the first element of the optimal control sequence in closed loop, then repeats the process. The problem of RHC control of airplanes was considered in a general enough formulation assuring integral control and allowing constraints both for change of the control signal and for

Fig. 6.21 The position and
attitude of the helicopter
during flight including state
estimation

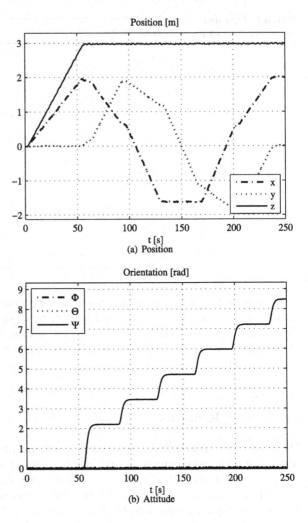

(a) Position

(b) Attitude

Fig. 6.22
Hardware-in-the-loop test
setup

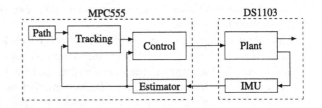

the control signal itself. The nonlinear system was substituted by the linearized LTI
system at the beginning of each horizon which dealt as system model for prediction during the optimization. The resulting optimum problem is standard Quadratic
Programming (QP).

Fig. 6.23 Real-time test results for a complex motion

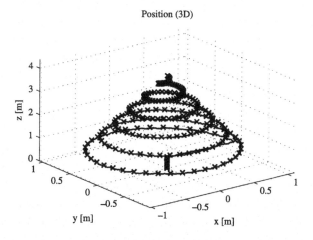

Position (3D)

The efficiency of the composite control law consisting of internal disturbance observer and external RHC control was illustrated in the example of pitch rate control of the longitudinal motion of an aircraft. It was demonstrated that the developed disturbance observer can stabilize the LPV system during measurable Mach number and height trajectories and it can really eliminate the effect of output disturbance after a short transient but the price is increased control signal magnitude. The external RHC control determines optimal command signal reference for the internal disturbance observer. The transients of the composite control system satisfy the flight specification without overshoot in the pitch rate while the control signal remains smooth and constraints can be taken into consideration.

The second approach dealt with the problem of controlling an indoor helicopter (UAV). In the investigations, the theoretical foundations and the real-time realization of the embedded control system of an indoor quadrotor helicopter were presented. The components of the controller were described in detail. A backstepping algorithm is responsible for stabilizing the quadrotor helicopter. From theoretical point of view, our results differ from earlier ones in that backstepping is integrated with state estimation based on advanced sensors. The state estimator consists of two extended Kalman filters.

The embedded controller was realized by using a Freescale MPC555 processor, a Crossbow MNAV100CA IMU and a marker-based vision system developed by BME IIT. Quick prototype design of the controller was performed based on MAT-LAB/Simulink, Real-Time Workshop and MPC555 Target Compiler. Hardware-in-the-loop real-time tests were performed using the DS1103 board of dSPACE which emulated the helicopter and the sensor system before the first flight with the real helicopter, while the state estimation and control algorithm were running on MPC555 processor. A CAN network served as communication channel between the components of the units of the distributed control system. Path tracking results show the effectiveness of the embedded control system under real-time conditions.

Chapter 7
Nonlinear Control of Surface Ships

Overview   This chapter deals with the dominant control methods of ships. First, the typical structure of the control system is shown. A short overview is given about the path design methods including Line-of-Sight guidance and wave filtering. The state estimation using GPS and IMU is discussed and observers are suggested to solve the problem. From the wide spread control methods the acceleration feedback with nonlinear PD, the nonlinear decoupling both in body and in NED frames, the adaptive feedback linearization and the MIMO backstepping in 6 DOF are presented including stability considerations. An overview is given about the most common actuators and the principles for control allocations. Simulation results are presented for the backstepping control of a multipurpose naval vessel including surge, sway, roll and yaw interactions in the nonlinear dynamic model.

7.1 Control System Structure

From the large field of guidance, navigation and control of ships, we concentrate here mainly on control and state estimation methods. The reader interested deeper with marine control systems can find further suggestions in the excellent book of Fossen [41].

For better readability, we repeat here the dynamic models of ships elaborated in Sects. 3.5.7 and 3.5.8:

$$\dot{\eta} = J(\eta)v,$$
$$M\dot{v} + C(v)v + D(v)v + g(\eta) = \tau + g_0 + w, \tag{7.1}$$

$$M^*(\eta)\ddot{\eta} + C^*(\eta, \dot{\eta})\dot{\eta} + D^*(\eta, \dot{\eta})\dot{\eta} + g^*(\eta) = J^{-T}(\eta)(\tau + g_0 + w), \tag{7.2}$$

B. Lantos, L. Márton, *Nonlinear Control of Vehicles and Robots*,
Advances in Industrial Control,
DOI 10.1007/978-1-84996-122-6_7, © Springer-Verlag London Limited 2011

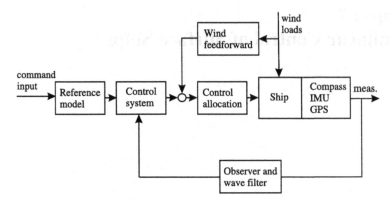

Fig. 7.1 Typical structure of the control system of ships

where the system matrices in the second equation are as follows:

$$M^*(\eta) = J^{-T}(\eta)MJ^{-1}(\eta),$$
$$C^*(\eta, v) = J^{-T}(\eta)\big[C(v) - MJ^{-1}(\eta)\dot{J}(\eta)\big]J^{-1}(\eta),$$
$$D^*(\eta, v) = J^{-T}(\eta)D(v)J^{-1}(\eta),$$
$$g^*(\eta) = J^{-T}(\eta)g(\eta).$$

(7.3)

Here, $\eta = (x, y, z, \phi, \theta, \psi)^T$ is the position (north, east, down) and orientation (Euler = RPY angles), $v = (u, v, w, p, q, r)^T$ is the velocity and angular velocity, $M = M_{RB} + M_A$ is the mass matrix, $C(v) = C_{RB}(v) + C_A(v)$ is the skew symmetric matrix of centripetal and Coriolis effects, $D(v) = D + D_n(v)$ is the symmetric and positive definite damping matrix consisting of linear and nonlinear parts, $g(\eta)$ is the vector of gravity and buoyancy effects, τ is the control force and torque, g_0 is the ballast force and torque, and finally w is the force and torque of environmental effects caused by wind, waves and current. As usual, $\eta \in K_{NED}$ and $v \in K_{BODY}$. It is assumed that K_{NED} can be considered as (quasi) inertia system and the Jacobian $J(\eta)$ is nonsingular during the motion.

In order to simplify the notation, vectors expressed in K_{NED} and K_{BODY} will be denoted by right superscript n and b, respectively. Coordinate transformation from K_{BODY} to K_{NED} will be denoted by R_b^n.

For the simpler 3D problems, $\eta = (x, y, \psi)^T$, $v = (u, v, r)^T$ and the Jacobian matrix has type 3×3. Notice that for surface ships moving with large speed and in waves $M_A \neq M_A^T > 0$ which can cause stability problems.

The typical structure of the control system of ships is shown in Fig. 7.1. It can be seen that, beside the control algorithms, reference path design and state estimation are important subtasks thus they will be considered first.

7.1.1 Reference Path Design

The path is usually formulated in a scalar path variable $\theta(t)$. The geometric task is to make the error between developed path and desired path asymptotically to zero. Different methods exist to solve the problem from which polynomial interpolation methods are dominating. The way-point data usually contains the set of the desired point $(x_k, y_k, z_k)^T$ and the speed U_k and the heading ψ_k.

We shall deal only with the planar case. The path can be divided into sections (θ_{k-1}, θ_k) in which the path is approximated by cubic polynomial in each component:

$$x_d(\theta) = a_{k,3}\theta^3 + a_{k,2}\theta^2 + a_{k,1}\theta + a_{k,0}, \tag{7.4}$$

$$y_d(\theta) = b_{k,3}\theta^3 + b_{k,2}\theta^2 + b_{k,1}\theta + b_{k,0}. \tag{7.5}$$

The speed can be computed as:

$$U_d(t) = \sqrt{x'_d(\theta)^2 + y'_d(\theta)^2}\dot{\theta}(t). \tag{7.6}$$

Two methods are available in MATLAB for polynomial interpolation. The function `pchip` performs piecewise cubic Hermite interpolation and it assures that the first derivatives $\dot{x}_d(t)$, $\dot{y}_d(t)$ are continuous. The function `spline` uses cubic spline interpolation and assures that the second derivatives $\ddot{x}_d(t)$, $\ddot{y}_d(t)$ at the endpoints are equal. This gives a smooth path. Hermite interpolation has less oscillations if the way-point data is non-smooth.

Leaving MATLAB, we can easily formulate the general cubic spline interpolation algorithm in each variable. At the boundary of two neighboring sections, we can prescribe that the position, velocity and acceleration are the same computed from the left by $(a_{k-1,3}, a_{k-1,2}, a_{k-1,1}, a_{k-1,0})^T$ and from the right by $(a_{k,3}, a_{k,2}, a_{k,1}, a_{k,0})^T$, respectively. At the end-points, we have to satisfy that the position and the velocity (or acceleration) are equal to the prescribed position and the prescribed velocity (or acceleration), respectively. From the velocity and the acceleration, only one can be prescribed for the endpoints. The path design problem results in a linear set of equations in the parameters $(a_{0,3}, \ldots, a_{n,0})^T$ for x and in a similar one for y. The problem can be rewritten to an optimization problem if constraints should be satisfied.

7.1.2 Line-of-Sight Guidance

The concept of Line-of-Sight (LOS) is shown in Fig. 7.2. Because of disturbances (wind, waves, ocean currents etc.) the cross-track error can increase hence a correction should be made to prescribe realistic new desired point and orientation along the path. Denote L_{pp} the length of the ship then line-of-sight vector

Fig. 7.2 The principle of
Line-of-Sight

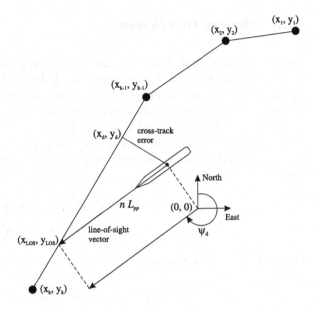

(x_{LOS}, y_{LOS}) can be determined based on elementary geometry and the desired yaw
angle $\psi_d = \operatorname{atan}2(y_{LOS} - y(t), x_{LOS} - x(t))$:

$$\big(y_{LOS} - y(t)\big)^2 + \big(x_{LOS} - y(t)\big)^2 = (nL_{pp})^2, \tag{7.7}$$

$$\frac{y_{LOS} - y_{k-1}}{x_{LOS} - x_{k-1}} = \frac{y_k - y_{k-1}}{x_k - x_{k-1}}. \tag{7.8}$$

The next way-point is chosen in a circle with acceptable radius R_0. If $|x_k -
x(t)|^2 + |y_k - y(t)|^2 \le R_0^2$ then $k := k + 1$ and a new way-point is selected. Typical
values in practice are $n = 5$ and $R_0 = 2L_{pp}$.

7.1.3 Filtering Wave Disturbances

Let us consider surface ships and in order to simplify the discussion let us collect
the centrifugal and damping effects in a common term Dv where D is constant
matrix. Only position measurements will be assumed. The wave effects in NED
axes directions are modeled by three separate second order systems in state space
(state variable $x_{w,i} \in R^2$, parameters are ω_0, λ, K_w) and three bias terms ($\dot{b}_i \in R^1$).
The orientation matrix $R(\psi) := R_b^n(\psi)$ is:

$$R(\psi) = \begin{bmatrix} C_\psi & -S_\psi & 0 \\ S_\psi & C_\psi & 0 \\ 0 & 0 & 1 \end{bmatrix}. \tag{7.9}$$

The resulting system can be described by the following state equations:

$$\dot{\xi} = A_w \xi + E_w w_1, \tag{7.10}$$

$$\dot{\eta} = R(\psi)v, \tag{7.11}$$

$$\dot{b} = w_2, \tag{7.12}$$

$$\dot{v} = M^{-1}\left[-Dv + R^T(\psi)b + \tau + w_3\right], \tag{7.13}$$

$$y = \eta + C_w \xi + z, \tag{7.14}$$

where system noises are denoted by w_i and measurement noise by z. The noises are zero-mean Gaussian ones.

The nonlinear system can be brought to the form:

$$\dot{x} = f(x) + Bu + Ew, \tag{7.15}$$

$$y = Cx + z, \tag{7.16}$$

where $x = (\xi^T, \eta^T, b^T, v^T)^T \in R^{15}$ is the state vector, $u = \tau \in R^3$ is the control vector, $w = (w_1^T, w_2^T, w_3^T)^T$ is the system noise vector and $z \in R^3$ is the measurement noise, furthermore

$$f(x) = \begin{bmatrix} A_w \xi \\ R(\psi)v \\ 0 \\ M^{-1}(-Dv + R^T(\psi)b) \end{bmatrix}, \quad B = \begin{bmatrix} 0 \\ 0 \\ 0 \\ M^{-1} \end{bmatrix}, \tag{7.17}$$

$$C = \begin{bmatrix} C_w & I & 0 & 0 \end{bmatrix}, \quad E = \begin{bmatrix} E_w & 0 & 0 \\ 0 & 0 & 0 \\ 0 & I & 0 \\ 0 & 0 & M^{-1} \end{bmatrix}. \tag{7.18}$$

The remaining part is standard. The continuous time system can be transformed to discrete time one using Euler approximation and extended Kalman filter (EKF) can be suggested to estimate the state and to filter the noises.

7.1.4 State Estimation Using IMU and GPS

Inertial Measurement Unit (IMU) measures the 3D specific force $f_{\text{IMU}} = a_{\text{IMU}} - g^b$ in body frame where g^b is the gravity acceleration. The other measurement is the 3D angular velocity also expressed in body frame. GPS measures the 3D position and optionally the 3D velocity of the vehicle (now the ship) in the NED frame which will be considered a (quasi) inertial frame.

The orientation can be described by Euler (RPY) angles $\Psi = (\phi, \theta, \psi)^T$ or the quaternion $q = (s, w^T)$ where $s \in R^1$ and $w \in R^3$. For better readability, we repeat here the orientation matrix transforming vectors from the BODY frame to the NED frame:

$$R_b^n(\Psi) = \begin{bmatrix} C_\psi C_\theta & C_\psi S_\theta S_\phi - S_\psi C_\phi & C_\psi S_\theta C_\phi + S_\psi S_\phi \\ S_\psi C_\theta & S_\psi S_\theta S_\phi + C_\psi C_\phi & S_\psi S_\theta C_\phi - C_\psi S_\phi \\ -S_\theta & C_\theta S_\phi & C_\theta C_\phi \end{bmatrix}. \tag{7.19}$$

7.1.4.1 Integration of IMU and GPS Position

Denoting the position in the NED frame by $p^n = (n, e, d)^T$ and the velocity in the NED frame by v^n, the bias of the acceleration measurement by $b_{acc} \in R^3$ and the GPS position measurement by $y_1 = (n_{GPS}, e_{GPS}, d_{GPS})^T$, then the following kinematic equations can be stated:

$$\dot{p}^n = v^n, \tag{7.20}$$

$$\dot{v}^n = R_b^n(\Psi)(f_{IMU} + b_{acc} + w_1) + g^n, \tag{7.21}$$

$$\dot{b}_{acc} = w_3, \tag{7.22}$$

$$y_1 = p^n, \tag{7.23}$$

where w_1, w_3 are zero mean Gaussian noises. The following Luenberger type observer can be proposed:

$$\dot{\hat{p}}^n = \hat{v}^n + K_1 \tilde{y}_1, \tag{7.24}$$

$$\dot{\hat{v}}^n = R_b^n(\Psi)\left(f_{IMU} + \hat{b}_{acc}\right) + g^n + K_2 \tilde{y}_1, \tag{7.25}$$

$$\dot{\hat{b}}_{acc} = K_3\left[R_b^n(\Psi)\right]^T \tilde{y}_1, \tag{7.26}$$

$$\hat{y}_1 = \hat{p}^n, \tag{7.27}$$

where $\tilde{y}_1 = y_1 - \hat{y}_1 = p^n - \hat{p}^n$. The body frame velocity estimation is

$$\hat{v}^b = \left[R_b^n(\Psi)\right]^T \hat{v}^n. \tag{7.28}$$

Denote the estimation errors $\tilde{p}^n = p^n - \hat{p}^n$, $\tilde{v}^n = v^n - \hat{v}^n$ and $\tilde{b}_{acc} = b_{acc} - \hat{b}_{acc}$, respectively, then the observer error dynamics becomes:

$$\begin{pmatrix} \dot{\tilde{p}}^n \\ \dot{\tilde{v}}^n \\ \dot{\tilde{b}}_{acc} \end{pmatrix} = \begin{bmatrix} -K_1 & I & 0 \\ -K_2 & 0 & R_b^n(\Psi) \\ -K_3[R_b^n(\Psi)]^T & 0 & 0 \end{bmatrix} \begin{pmatrix} \tilde{p}^n \\ \tilde{v}^n \\ \tilde{b}_{acc} \end{pmatrix} + \begin{bmatrix} 0 & 0 \\ R_b^n(\Psi) & 0 \\ 0 & I \end{bmatrix} \begin{pmatrix} w_1 \\ w_2 \end{pmatrix}, \tag{7.29}$$

$$\Updownarrow$$

$$\dot{\tilde{x}} = A(\Psi)\tilde{x} + Ew. \tag{7.30}$$

We have to assure that $\tilde{x}(t) \to 0$. The main problem is that $A(\Psi)$ is not constant. To solve the problem, the coordinate transformation $\bar{x} = T\tilde{x}$ will be applied where, using simplified notation R for $R_b^n(\Psi)$, $T = \text{diag}(R^T, R^T, I)$ and $T^{-1} = T^T$:

$$\bar{A} = TAT^T = \begin{bmatrix} -R^T K_1 R & I & 0 \\ -R^T K_2 R & 0 & I \\ -K_3 & 0 & 0 \end{bmatrix} =: \begin{bmatrix} -\bar{K}_1 & I & 0 \\ -\bar{K}_2 & 0 & I \\ -K_3 & 0 & 0 \end{bmatrix}. \tag{7.31}$$

Notice that $\|\tilde{x}\| = \|\bar{x}\|$ hence for stability it is sufficient to make $\bar{A}$ constant and Hurwitz.

Consider the fictitious observable system defined by A_0 and C_0:

$$A_0 = \begin{bmatrix} 0 & I & 0 \\ 0 & 0 & I \\ 0 & 0 & 0 \end{bmatrix}, \quad C_0 = \begin{bmatrix} I & 0 & 0 \end{bmatrix}, \tag{7.32}$$

$$A_0 - GC_0 = \begin{bmatrix} 0 & I & 0 \\ 0 & 0 & I \\ 0 & 0 & 0 \end{bmatrix} - \begin{bmatrix} \bar{K}_1 \\ \bar{K}_2 \\ K_3 \end{bmatrix} \begin{bmatrix} I & 0 & 0 \end{bmatrix} = \begin{bmatrix} -\bar{K}_1 & I & 0 \\ -\bar{K}_2 & 0 & I \\ -K_3 & 0 & 0 \end{bmatrix} = \bar{A}. \tag{7.33}$$

Hence, we can prescribe the stable characteristic equation $\phi_0(s) = \det(sI - (A_0 - GC_0))$ and solve the pole placement problem:

$$\left(A_0^T, C_0^T\right) \xrightarrow{\phi_0(s)} G^T \to G = \begin{bmatrix} \bar{K}_1 \\ \bar{K}_2 \\ K_3 \end{bmatrix}. \tag{7.34}$$

The problem can be solved with the function `place` in MATLAB Control System Toolbox. From G^T, we can obtain G by taking the transpose. From G, we can determine K_1, K_2 and K_3 according to

$$\bar{K}_1 = R^T K_1 R \to K_1 = R(\Psi)\bar{K}_1 R^T(\Psi), \tag{7.35}$$

$$\bar{K}_2 = R^T K_2 R \to K_2 = R(\Psi)\bar{K}_2 R^T(\Psi), \tag{7.36}$$

$$K_3 \checkmark \tag{7.37}$$

Remark Notice that in the above problem we neglected wave effects. The problem in (7.10)–(7.14) including the wave effects is more complicated to discuss. The main idea is the use of commuting matrices $(AR(\alpha) = R(\alpha)A)$ where $R(\alpha)$ is rotation. To find the observer with stability guarantee, an LMI feasibility problem has to be solved with structural constraints, see [84].

7.1.4.2 Attitude Observer Using Quaternion

First, we shall assume that the ship does *not accelerate*, hence $f_{\text{IMU}} = -g^b = -[R_b^n(\Psi)]^T g^n$, that is,

$$\begin{pmatrix} f_x \\ f_y \\ f_z \end{pmatrix} = -\begin{pmatrix} -gS_\theta \\ gC_\theta S_\phi \\ gC_\theta C_\phi \end{pmatrix} \quad \Rightarrow \quad \phi = \text{atan2}(f_y, f_z), \quad \theta = \text{atan2}(-C_\phi f_x, f_z).$$

$$(7.38)$$

When IMU is combined with compass, then the heading (yaw angle ψ) is measured and the orientation $\Phi = (\phi, \theta, \psi)^T$ is theoretically known. Unfortunately, the measurements contain noises and biases, hence the result should be filtered taking into consideration the angular velocity measurements.

The orientation can also be represented by quaternion $q = (s, w^T)^T$. We have already determined the relation between the angular velocity and the quaternion:

$$\dot{q} = \frac{1}{2}\begin{bmatrix} -w^T \\ sI + [w\times] \end{bmatrix} \omega^b =: T(q)\omega^b. \tag{7.39}$$

The differential equation of the attitude in quaternion form can be assumed as:

$$\dot{q} = T(q)[\omega_{\text{IMU}} + b_{\text{gyro}} + w_2], \tag{7.40}$$

$$\dot{b}_{\text{gyro}} = w_4, \tag{7.41}$$

where b_{gyro} is the bias expressed in body frame, w_2 and w_4 are zero mean Gaussian noises.

The nonlinear observer is assumed in Luenberger form, see [153]:

$$\dot{\hat{q}} = T(\hat{q})\tilde{R}_b^n(\hat{q})\big[\omega_{\text{IMU}} + \hat{b}_{\text{gyro}} + K_1 \tilde{w}\, \text{sgn}(\tilde{s})\big], \tag{7.42}$$

$$\dot{\hat{b}}_{\text{gyro}} = \frac{1}{2}K_2 \tilde{w}\, \text{sgn}(\tilde{s}). \tag{7.43}$$

The factor $\tilde{R}_b^n(\hat{q})$ in the differential equation of $\hat{q}$ is used in order to simplify later the differential equations for the estimation errors. Notice that by (A.25), if R is an orientation matrix and q is the quaternion identifying R, then it yields $q \star (0, r) \star q^* = (0, Rr)$. By evaluating the transformation for $\hat{q} = (\hat{s}, \hat{w})$ and $\tilde{R}_b^n(\hat{q})$, it follows:

$$\tilde{R}_b^n(\hat{q}) = I + 2\hat{s}[\hat{w}\times] + 2[\hat{w}\times][\hat{w}\times]. \tag{7.44}$$

The quaternion estimation error is defined by using quaternion product (see Appendix A) as follows:

$$\tilde{q} = \hat{q}^* \star q = \begin{pmatrix} \hat{s}s + \hat{w}^T w \\ -s\hat{w} + \hat{s}w - [\hat{w}\times]w \end{pmatrix}, \tag{7.45}$$

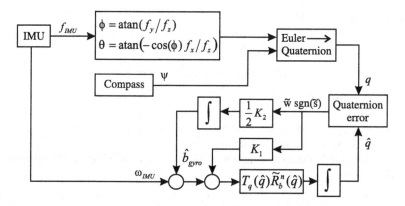

Fig. 7.3 Block scheme of attitude observer using quaternion

where $\hat{q} = (\hat{s}, \hat{w}^T)^T$ and its conjugate is $\hat{q}^* = (\hat{s}, -\hat{w}^T)^T$. The observer error dynamics can be brought to the form

$$\dot{\tilde{q}} = T(\tilde{q})[\tilde{b}_{gyro} - K_1\tilde{w}\,\mathrm{sgn}(\tilde{s})] = \frac{1}{2}\begin{bmatrix} -\tilde{w}^T \\ \tilde{s}I + [\tilde{w}\times] \end{bmatrix}[\tilde{b}_{gyro} - K_1\tilde{w}\,\mathrm{sgn}(\tilde{s})],$$
$$\dot{\tilde{b}}_{gyro} = -\frac{1}{2}K_2\tilde{w}\,\mathrm{sgn}(\tilde{s}),$$
(7.46)

where $K_i = K_i^T > 0$, $i = 1, 2$.

For stability examination (neglecting noises since they have zero mean), we choose the following Lyapunov function:

$$V = \langle K_2^{-1}\tilde{b}_{gyro}, \tilde{b}_{gyro}\rangle + (1 - |\tilde{s}|),$$
(7.47)

and determine its time derivative:

$$\dot{V} = 2\langle K_2^{-1}\dot{\tilde{b}}_{gyro}, \tilde{b}_{gyro}\rangle - \mathrm{sgn}(\tilde{s})\dot{\tilde{s}}$$
$$= -2\langle K_2^{-1}\frac{1}{2}K_2\tilde{w}\,\mathrm{sgn}(\tilde{s}), \tilde{b}_{gyro}\rangle + \mathrm{sgn}(\tilde{s})\frac{1}{2}\tilde{w}^T[\tilde{b}_{gyro} - K_1\tilde{w}\,\mathrm{sgn}(\tilde{s})]$$
$$= -\frac{1}{2}\langle K_1\tilde{w}, \tilde{w}\rangle \leq 0.$$
(7.48)

Hence, V is decreasing until $\tilde{s} = \pm 1$. Using LaSalle's theorem, it follows that $\tilde{s} = \pm 1$ and $\tilde{w} = 0$ are stable equilibrium points. Notice that $\tilde{s} = \pm 1$ is necessary in equilibrium point because $|\tilde{q}| = 1$.

The resulting structure of the attitude observer using quaternion is shown in Fig. 7.3.

Remark In the above discussions, we have assumed that the ship is not accelerating. If the ship *does accelerate*, but the estimation of the acceleration $a^b = \dot{v}^b + \omega^b \times v^b$

is available, then it yields:

$$\hat{a}^n \approx R_b^n(\Psi)\big[f_{\text{IMU}} + \hat{b}_{\text{acc}}\big] + g^n + K_2\tilde{y}_1$$

$$\Rightarrow \quad \hat{a}^b \approx f_{\text{IMU}} + \hat{b}_{\text{acc}} + g^b + \big(R_b^n\big)^T K_2\tilde{y}_1,$$

$$-g^b \approx f_{\text{IMU}} - \hat{a}^b + \hat{b}_{\text{acc}},$$

and with this correction we have an approximating method for accelerating ships. For initialization in stationary situation, the original method can be applied. Notice that more general (but time consuming) methods exist for GPS and IMU integration, see [38].

7.2 Acceleration Feedback and Nonlinear PD

First, we consider the case $M = M^T > 0 = $ const. The controller consists of acceleration feedback and nonlinear PD compensation, the desired position is constant, that is, $\eta_d = $ const $\Rightarrow \dot{\eta}_d = 0$, and the error is defined as $\tilde{\eta} = \eta_d - \eta \Rightarrow \dot{\tilde{\eta}} = -\dot{\eta}$. The controller output is the generalized force τ:

$$\tau := g(\eta) - K_m\dot{v} + J^T(K_p\tilde{\eta} - K_d\dot{\eta}), \qquad (7.49)$$

where $-K_m\dot{v}$ is the "acceleration feedback" and the PD compensation is defined in K_{NED}. Because of the nonlinear weighting matrix $J^T(\eta)$, the PD controller is nonlinear in this sense. Notice that $\dot{v}$ is not equal to the acceleration defined in inertia system, it is simply the formal derivative of the velocity in K_{BODY}. It is assumed that the matrices in the control law satisfy $K_m = K_m^T > 0$, $K_p = K_p^T > 0$ and $K_d = K_d^T > 0$.

Neglecting the disturbance, the closed loop system satisfies

$$M\dot{v} + (C + D)v + g = g - K_m\dot{v} + J^T(K_p\tilde{\eta} - K_d\dot{\eta})$$

$$\Rightarrow \quad \underbrace{(M + K_m)}_{H}\dot{v} = -\big(C + D + J^T K_d J\big)v + J^T K_p\tilde{\eta}.$$

Let us consider the following Lyapunov function and its derivative by the time:

$$V := \frac{1}{2}\langle Hv, v\rangle + \frac{1}{2}\langle K_p\tilde{\eta}, \tilde{\eta}\rangle, \qquad (7.50)$$

$$\dot{V} = \langle H\dot{v}, v\rangle + \langle K_p\dot{\tilde{\eta}}, \tilde{\eta}\rangle$$

$$= \langle -(C + D + J^T K_d J)v, v\rangle + \langle J^T K_p\tilde{\eta}, v\rangle - \langle K_p Jv, \tilde{\eta}\rangle, \qquad (7.51)$$

$$\dot{V} = -\langle (D + J^T K_p J)v, v\rangle \le 0.$$

Here it was exploited that C is skew symmetric hence its quadratic form is identically zero in any variable.

In order to prove stability, LaSalle's theorem can be applied. Denote E the set where $\dot{V}$ becomes zero: $E = \{(\tilde{\eta}, \nu) : \nu = 0\}$ and let M be the maximal invariant set contained in E. Trajectories starting in M remain in M, hence $\dot{\nu} = 0 = H^{-1} J^T K_p \tilde{\eta}$, from which follows $\tilde{\eta} = 0$. Hence, $M = \{(0,0)\}$ and the system is globally asymptotically stable (GAS).

Let us consider now the more general case $M_A \neq M_A^T > 0$. It is clear that M can be written in the form

$$M = M_{RB} + \frac{1}{2}\left(M_A + M_A^T\right) + \frac{1}{2}\left(M_A - M_A^T\right). \tag{7.52}$$

Then

$$K_m := -\frac{1}{2}\left(M_A - M_A^T\right) + \Delta K, \quad \Delta K = \Delta K^T > 0, \tag{7.53}$$

$$H = (M + K_m) = M_{RB} + \frac{1}{2}\left(M_A + M_A^T\right) + \Delta K, \quad H = H^T > 0, \tag{7.54}$$

and the new closed loop system with the modified controller is GAS.

7.3 Nonlinear Decoupling

It is useful to write the dynamic model in the compact form

$$M\dot{\nu} + n(\eta, \nu) = \tau. \tag{7.55}$$

Similarly to robots, nonlinear decoupling can be reached in body frame and in NED frame.

7.3.1 Nonlinear Decoupling in Body Frame

Centralized nonlinear control law:

$$\tau := Mu^b + n(\eta, \nu). \tag{7.56}$$

Closed loop:

$$M\dot{\nu} + n(\eta, \nu) = Mu^b + n(\eta, \nu) \quad \Rightarrow \quad \dot{\nu} = u^b \quad \Rightarrow \quad \dot{\nu}_i = u_i^b. \tag{7.57}$$

Decentralized linear controllers:

$$u_i^b := \dot{\nu}_{i,d} + k_{p,i} \underbrace{(\nu_{d,i} - \nu_i)}_{\tilde{\nu}_i} + k_{I,i} \int_0^t (\nu_{d,i} - \nu_i)\, d\vartheta, \tag{7.58}$$

$$\ddot{\tilde{v}} + k_{p,i}\dot{\tilde{v}}_i + k_{I,i}\tilde{v}_i = 0 \quad \Rightarrow \quad (s + \lambda_i)^2 = 0 \quad \Rightarrow \quad k_{p,i} = 2\lambda_i > 0, k_{I,i} = \lambda_i^2.$$

(7.59)

As can be seen, the velocities and angular velocities in the directions of the body frame are decoupled.

7.3.2 Nonlinear Decoupling in NED Frame

It follows from $\dot{\eta} = J(\eta)v$ that the acceleration in NED frame is $\ddot{\eta} = J\dot{v} + \dot{J}v$, hence the dynamic model can be written as

$$M\dot{v} + n(\eta, v) = \tau \quad \Rightarrow \quad MJ^{-1}(\ddot{\eta} - \dot{J}v) + n(\eta, v) = J^T F,$$

$$\underbrace{J^{-T}MJ^{-1}}_{M^*(\eta)}\ddot{\eta} + \underbrace{\left[J^{-T}n - J^{-T}MJ^{-1}\dot{J}v\right]}_{n^*(\eta, v)} = F.$$

(7.60)

The nonlinear decoupling in NED frame can be performed similarly to the approach in body frame.

System:

$$M^*(\eta)\ddot{\eta} + n^*(\eta, v) = F.$$

(7.61)

Centralized nonlinear control law:

$$F := M^* u^n + n^*.$$

(7.62)

Closed loop:

$$M^*\ddot{\eta} + n^* = M^* u^n + n^* \quad \Rightarrow \quad \ddot{\eta} = u^n \quad \Rightarrow \quad \ddot{\eta}_i = u_i^n.$$

(7.63)

Decentralized linear controllers:

$$u^n := \ddot{\eta}_d + K_p(\eta_d - \eta) + K_I \int_0^t (\eta_d - \eta)\, d\vartheta + K_d(\dot{\eta}_d - \dot{\eta}).$$

(7.64)

Here $\ddot{\eta}_d$ is the desired acceleration along the path in NED frame. Denote $\tilde{\eta} = \eta_d - \eta$ the error in the closed loop, then the closed loop error satisfies the following differential equation:

$$\dddot{\tilde{\eta}} + K_d\ddot{\tilde{\eta}} + K_p\dot{\tilde{\eta}} + K_I\tilde{\eta} = 0.$$

(7.65)

Based on $(s + \lambda_i)^3 = s^3 + 3s^2\lambda_i + 3s\lambda_i^2 + \lambda_i^3$, $\lambda_i > 0$, the following choice stabilizes the closed loop system with centralized nonlinear and decentralized linear controllers:

$$\Lambda = \mathrm{diag}(\lambda_1, \lambda_2, \ldots, \lambda_n), \quad K_d = 3\Lambda, \quad K_p = 3\Lambda^2, \quad K_I = \Lambda^3.$$

(7.66)

The centralized nonlinear controller can be implemented as $\tau = J^T F$:

$$\tau := MJ^{-1}(\eta)\left[u^n - \dot{J}(\eta)v\right] + n(\eta, v).$$

(7.67)

7.4 Adaptive Feedback Linearization

The parameters of the ship's dynamic model may be varying during long time motion. Adaptive control tries the find the new parameters from the signals available in the control system. We shall assume that the parameters to be identified appear in the dynamic model in linear way multiplied with signals of the system:

$$Mv + n(\eta, v) = \tau \quad \Leftrightarrow \quad Y(\eta, v, \dot{v})\vartheta = \tau, \tag{7.68}$$

where ϑ is the vector containing the parameters to be identified and $Y(\eta, v, \dot{v})$ is the so called regressor matrix containing only the signals in the system. Evidently if the control law is based on the dynamic model then the real parameters of the system and the available estimated parameters for the control may be different. If the control law is based on the dynamic model, it uses nonlinear decoupling in body frame and decentralized PD controllers, then

$$\tau := \hat{M}a^b + \hat{n}(\eta, v) = Y(\eta, v, a^b)\hat{\vartheta}, \tag{7.69}$$

and the closed loop satisfies:

$$M\dot{v} + n(\eta, v) = \hat{M}a^b + \hat{n}(\eta, v),$$
$$M(\dot{v} - a^b) + n(\eta, v) = \hat{M}a^b + \hat{n}(\eta, v) - Ma^b, \tag{7.70}$$
$$M(\dot{v} - a^b) = \hat{M}a^b + \hat{n}(\eta, v) - Ma^b - n(\eta, v) = Y(\eta, v, a^b)(\hat{\vartheta} - \vartheta).$$

On the other hand, the term $\dot{v} - a^b$ can be expressed by an appropriately chosen acceleration in NED frame:

$$\ddot{\eta} = J(\eta)\dot{v} + \dot{J}(\eta)v,$$
$$a^n = J(\eta)a^b + \dot{J}(\eta)v, \tag{7.71}$$
$$\ddot{\eta} - a^n = J(\dot{v} - a^b) \quad \Rightarrow \quad \dot{v} - a^b = J^{-1}(\ddot{\eta} - a^n).$$

Hence, the closed loop system can be considered also in NED frame:

$$MJ^{-1}(\ddot{\eta} - a^n) = Y(\eta, v, a^b)(\hat{\vartheta} - \vartheta),$$
$$M^*(\ddot{\eta} - a^n) = J^{-T}(\eta, v, a^b)(\hat{\vartheta} - \vartheta). \tag{7.72}$$

We have to choose the control law and adaptation law, and prove the stability. We introduce the notation $\tilde{\eta} = \eta_d - \eta$ for the path error and $\tilde{\vartheta} = \vartheta - \hat{\vartheta}$ for the parameter estimation error. Let the part of the *control law* be

$$a^n := \ddot{\eta}_d + K_p(\eta_d - \eta) + K_d(\dot{\eta}_d - \dot{\eta}), \tag{7.73}$$

then the closed loop will be

$$M^*[\ddot{\eta} - \ddot{\eta}_d - K_p(\eta_d - \eta) - K_d(\dot{\eta}_d - \dot{\eta})] = -J^{-T}Y\tilde{\vartheta},$$

$$M^* \left[\ddot{\tilde{\eta}} + K_d \dot{\tilde{\eta}} + K_p \tilde{\eta} \right] = J^{-T} Y \tilde{\vartheta},$$

and assuming $\exists [M^*(\eta)]^{-1}$ it follows

$$\ddot{\tilde{\eta}} + K_d \dot{\tilde{\eta}} + K_p \tilde{\eta} = (M^*)^{-1} J^{-T} Y \tilde{\vartheta}. \tag{7.74}$$

By introducing the notations

$$x = \begin{pmatrix} \tilde{\eta} \\ \dot{\tilde{\eta}} \end{pmatrix}, \qquad A = \begin{bmatrix} 0 & I \\ -K_p & -K_d \end{bmatrix}, \qquad B = \begin{bmatrix} 0 \\ (M^*)^{-1} \end{bmatrix},$$

the closed loop system can be described as follows:

$$\dot{x} = Ax + B J^{-T} Y \tilde{\vartheta}. \tag{7.75}$$

Assume the Lyapunov function has the form

$$V = \langle P(t)x, x \rangle + \langle \Gamma \tilde{\vartheta}, \tilde{\vartheta} \rangle, \tag{7.76}$$

where $P(t) = P(t)^T > 0$ has to be chosen later and $\Gamma = \Gamma^T > 0$. Determine the derivative of V by the time:

$$\dot{V} = \langle \dot{P}x, x \rangle + \langle P\dot{x}, x \rangle + \langle Px, \dot{x} \rangle + 2 \langle \Gamma \dot{\tilde{\vartheta}}, \tilde{\vartheta} \rangle$$

$$= \langle (\dot{P} + PA + A^T P)x, x \rangle + 2 \langle P B J^{-T} Y \tilde{\vartheta}, x \rangle + 2 \langle \Gamma \dot{\tilde{\vartheta}}, \tilde{\vartheta} \rangle$$

$$= \langle (\dot{P} + PA + A^T P)x, x \rangle + 2 \langle Y^T J^{-1} B^T Px + \Gamma \dot{\tilde{\vartheta}}, \tilde{\vartheta} \rangle.$$

First, the adaptation law will be chosen. It can be assumed that the real parameter is actually not varying, that is, $\theta = \text{const} \Rightarrow \dot{\tilde{\vartheta}} = \dot{\vartheta}$. Hence, we can choose the following *adaptation law* that makes the second term in $\dot{V}$ to zero:

$$\dot{\tilde{\vartheta}} := -\Gamma^{-1} Y^T \left(\eta, v, a^b \right) J^{-1}(\eta) B^T(\eta) Px. \tag{7.77}$$

The remaining part of the stability proof is based on the results of Asare and Wilson [9, 41]. First, the notations $y = Cx$, $C := B^T P$, $C = [c_0 I c_1 I]$ are introduced where $c_0 > 0$ and $c_1 > 0$ are scalars. Then P and Q are chosen according to

$$PA + A^T P = -Q, \qquad Q = Q^T > 0,$$

$$P = \begin{bmatrix} c_0 M^* K_d + c_1 M^* K_p & c_0 M^* \\ c_0 M^* & c_1 M^* \end{bmatrix},$$

$$Q = \begin{bmatrix} 2 c_0 M^* K_p & 0 \\ 0 & 2(c_1 M^* K_d - c_0 M^*) \end{bmatrix}.$$

Observe that $M^*(\eta)$ is not constant hence $\dot{P}$ is not zero:

$$\dot{P} = \begin{bmatrix} c_0 \dot{M}^* K_d + c_1 \dot{M}^* K_p & c_0 \dot{M}^* \\ c_0 \dot{M}^* & c_1 \dot{M}^* \end{bmatrix}.$$

If there exists $\beta > 0$ scalar such that

$$x^T \dot{P} x \leq \beta x^T \begin{bmatrix} M^* & 0 \\ 0 & M^* \end{bmatrix} x,$$

then the choice

$$(c_0 K_d + c_1 K_p) c_1 > c_0^2 I, \qquad 2 c_0 K_p > \beta I, \qquad 2(c_1 K_d - c_0 I) > \beta I$$

assures that

$$\dot{V} = x^T (\dot{P} - Q) x \leq 0. \tag{7.78}$$

The stability follows from Barbalat's lemma by which $\lim_{t \to \infty} \dot{V}(x(t)) = 0$ and therefore $\tilde{\eta}$ and $\dot{\tilde{\eta}}$ go to zero. It can be shown that the parameter error $\tilde{\vartheta}$ remains bounded but not necessarily convergent. Good choices are small β, and diagonal matrices $K_p > 0$ and $K_d > 0$.

7.5 MIMO Backstepping in 6 DOF

The dynamic model of 6-DOF ship in the NED frame is given by (7.2) and (7.3). Notice that M^*, C^*, D^* and g^* have arguments from the original state variables η and v. Assume that the reference path η_D and its derivatives $\dot{\eta}_d$, $\ddot{\eta}_d$ and $\dddot{\eta}_d$ are available as a result of path design.

The following signals will be intensively used later:

$$\begin{aligned}
\tilde{\eta} &:= \eta_d - \eta, \\
\dot{\eta}_r &:= \dot{\eta}_d + \Lambda(\eta_d - \eta), \quad \Lambda > 0 \text{ is diagonal}, \\
v_r &:= J^{-1}(\eta) \dot{\eta}_r, \\
v_d &:= J^{-1}(\eta) \dot{\eta}_d, \\
s &:= \dot{\eta}_r - \dot{\eta} = \dot{\tilde{\eta}} + \Lambda \tilde{\eta}, \\
\dot{s} &= \ddot{\eta}_r - \ddot{\eta}, \\
\dot{\eta} &= \dot{\eta}_r - s, \\
\ddot{\eta}_r &= J(\eta) \dot{v}_r + \dot{J}(\eta) v_r.
\end{aligned} \tag{7.79}$$

Notice that similarly to robots η_r is the reference signal and s can be interpreted as the sliding variable. The position and orientation error is $\tilde{\eta}$. For the derivation of the backstepping control algorithm, we need two important relations.

Firstly, we need the detailed form of $M^* \dot{s}$. With the new variables it yields:

$$M^*(\eta)\dot{s} = M^*(\eta)\ddot{\eta}_r - M^*(\eta)\ddot{\eta}$$
$$= M^*(\eta)\ddot{\eta}_r + C^*(\eta, v)\dot{\eta} + D^*(\eta, v)\dot{\eta} + g^*(\eta) - J^{-T}(\eta)\tau$$
$$= M^*(\eta)\ddot{\eta}_r + C^*(\eta, v)(\dot{\eta}_r - s) + D^*(\eta, v)(\dot{\eta}_r - s) + g^*(\eta) - J^{-T}(\eta)\tau$$
$$= -C^*(\eta, v)s - D^*(\eta, v)s$$
$$+ M^*(\eta)\ddot{\eta}_r + C^*(\eta, v)\dot{\eta}_r + D^*(\eta, v)\dot{\eta}_r + g^*(\eta) - J^{-T}(\eta)\tau.$$

Taking into consideration the definitions in (7.3) and (7.79), it follows:

$$M^*\ddot{\eta}_r + C^*\dot{\eta}_r = J^{-T}MJ^{-1}(J\dot{v}_r + \dot{J}v_r) + J^{-T}\big[C - MJ^{-1}\dot{J}\big]\underbrace{J^{-1}\dot{\eta}_r}_{v_r}$$

$$= J^{-T}[M\dot{v}_r + Cv_r], \tag{7.80}$$

$$M^*\dot{s} = -C^*s - D^*s + J^{-T}[M\dot{v}_r + Cv_r + Dv_r + g - \tau].$$

Secondly, we have to show that $\dot{M}^* - 2C^*$ is skew symmetric, that is, its quadratic form in any variable is identically zero, if $M = M^T > 0$ and $\dot{M} = 0$. Notice that although C is skew symmetric, C^* is not. Hence, consider the definitions in (7.3) and apply the rule for the derivative of the inverse matrix by a scalar variable:

$$\dot{M}^* - 2C^*$$

$$= -\big(J^{-1}\dot{J}J^{-1}\big)^T MJ^{-1} - J^{-T}MJ^{-1}\dot{J}J^{-1} - 2J^{-T}\big[C - MJ^{-1}\dot{J}\big]J^{-1}$$

$$= -\big(J^{-1}\dot{J}J^{-1}\big)^T MJ^{-1} + J^{-T}MJ^{-1}\dot{J}J^{-1} - 2J^{-T}CJ^{-1}.$$

The quadratic form of the third matrix is zero, since $\langle T^T ATx, x\rangle = \langle ATx, Tx\rangle = \langle Ay, y\rangle$ for invertible T and now $J(\eta)$ is assumed to be nonsingular and $A = C$ is skew symmetric. On the other hand, the transpose of $-(J^{-1}\dot{J}J^{-1})^T MJ^{-1}$ is the second matrix with negative sign therefore their quadratic forms are canceling.

Backstepping control will be developed in two steps.

Step 1:

Let the virtual control signal v and the auxiliary signal α_1 be defined as follows:

$$v := -s + \alpha_1, \tag{7.81}$$

$$\alpha_1 := v_r = J^{-1}(\eta)(\dot{\eta}_d + \Lambda\tilde{\eta}). \tag{7.82}$$

Then the error dynamics satisfies:

$$\dot{\tilde{\eta}} = \dot{\eta}_d - \dot{\eta} = J(\eta)(v_d - v) = J(\eta)(v_d + s - \alpha_1)$$

$$= J(\eta)\big[v_d + s - \underbrace{J^{-1}(\eta)\dot{\eta}_d - J^{-1}\Lambda\tilde{\eta}}_{v_d}\big]$$

$$= J(\eta)s - \Lambda\tilde{\eta}. \tag{7.83}$$

Define the Lyapunov function and take its time derivative:

$$V_1 := \frac{1}{2}\langle K_p \tilde{\eta}, \tilde{\eta} \rangle, \quad K_p = K_p^T \rangle 0, \tag{7.84}$$

$$\dot{V}_1 = \langle K_p \dot{\tilde{\eta}}, \tilde{\eta} \rangle = \langle K_p [J(\eta)s - \Lambda \tilde{\eta}], \tilde{\eta} \rangle$$

$$= -\langle K_p \Lambda \tilde{\eta}, \tilde{\eta} \rangle + \langle J^T(\eta) K_p \tilde{\eta}, s \rangle. \tag{7.85}$$

Observe that the first term is negative but the second is not.

Step 2:
Define the Lyapunov function for the closed loop as the sum of V_1 and the "kinetic energy of the error", and determine its time derivative:

$$V_2 := V_1 + \frac{1}{2}\langle M^* s, s \rangle. \tag{7.86}$$

$$\dot{V}_2 = \dot{V}_1 + \frac{1}{2}\langle \dot{M}^* s, s \rangle + \langle M^* \dot{s}, s \rangle$$

$$= \dot{V}_1 + \frac{1}{2}\langle \dot{M}^* s, s \rangle - \langle C^* s, s \rangle - \langle D^* s, s \rangle$$

$$+ \langle J^{-T}[Mv_r + Cv_r + Dv_r + g - \tau], s \rangle$$

$$= -\langle K_p \Lambda \tilde{\eta}, \tilde{\eta} \rangle + \langle J^T K_p \tilde{\eta}, s \rangle + \frac{1}{2}\langle (\dot{M}^* - 2C^*)s, s \rangle - \langle D^* s, s \rangle$$

$$+ \langle J^{-T}[M\dot{v}_r + Cv_r + Dv_r + g - \tau], s \rangle. \tag{7.87}$$

Hence, the control law can be chosen as:

$$\tau = M\dot{v}_r + Cv_r + Dv_r + g + \left[J^T(\eta)\right]^2 K_p \tilde{\eta} + J^T(\eta) K_d s, \quad K_d = K_d^T > 0. \tag{7.88}$$

With this control law, the time derivative of the resulting Lyapunov function will be negative:

$$\dot{V}_2 = -\langle K_p \Lambda \tilde{\eta}, \tilde{\eta} \rangle - \langle [D^*(\eta, v) + K_d]s, s \rangle < 0. \tag{7.89}$$

Hence, the equilibrium point $(\tilde{\eta}, s)$ is GAS. Since $s \to 0$ and $\tilde{\eta} \to 0$ hence it yields also $\dot{\tilde{\eta}} \to 0$.

For the implementation of τ, we need $\dot{v}_r$ which can be determined either by numerical differentiation of $v_r = J^{-1}\dot{\eta}_r$, see [41], or by performing the computation $\dot{v}_r = -J^{-1}\dot{J}J^{-1}\eta_r + J^{-1}\ddot{\eta}_r$ where $\ddot{\eta}_r = \ddot{\eta}_d + \Lambda[\dot{\eta}_d - J(\eta)v]$. Here, η_d, $\dot{\eta}_d$, $\ddot{\eta}_d$ is the result of the path design and η and v are state measurements or their estimated values.

If $\tau = Bu$ is the actuator model and the system is fullactuated or overactuated, moreover BB^T is nonsingular, then the actuator signal is $u = B^+\tau$ where B^+ is LS or Moore–Penrose pseudoinverse.

7.6 Constrained Control Allocation

For ships in n-DOF, it is necessary to distribute the generalized control forces $\tau \in R^n$ to the actuators in terms of control inputs $u \in R^r$. The system is overactuated if $r > n$, otherwise it is underactuated for $n < r$ or fullactuated if $n = r$. Computation of u from τ is typically a model based optimum problem while physical limitations have to be taken into consideration.

The most common actuators are:

- *Main propellers* are mounted aft of the hull usually in conjunction with rudders. They produce the necessary force in x-direction of the ship.
- *Tunnel thrusters* for low speed maneuvering go through the hull of the ship. The propeller unit is mounted inside a transverse tube and produces force in y-direction of the ship.
- *Azimuth thruster* units that can rotate an angle α about the z-axis and produce two force components in the horizontal plane. They are usually mounted under the hull of the ship.
- *Aft rudders* are the primary steering device located aft of the shift and the rudder force in y-direction of the ship will produce a yaw moment for steering.
- *Stabilizing fins* are used for damping the vertical vibrations and roll motions. They produces a force in z-direction of the ship which is a function of fin deflection.
- *Control surfaces* can be mounted in different directions to produce lift and drag forces. For underwater vehicles these can be fins for diving, rolling, pitching etc.
- *Water jets* is an alternative to main propellers aft of the ship used for high-speed motion.

A unified representation of control forces and moments is

$$\tau = T(\alpha) \underbrace{Ku}_{f}, \tag{7.90}$$

where $\alpha = (\alpha_1, \ldots, \alpha_p)^T$ and $u = (u_1, \ldots, u_r)^T$ are control inputs, and $f = Ku$ is a vector of control surfaces [41].

Unconstrained control allocation problems of the form $J = \langle Wf, f \rangle \to$ min subject to $\tau - Tf = 0$ can be solved using Lagrange multiplier rule. Here, W is a weighting matrix and "unconstrained" means that there are no bounds on f_i, α_i and u_i.

Constrained allocation problems can be formulated as $\langle Wf, f \rangle + \langle Qs, s \rangle + \beta \max_i |f_i| \to$ min subject to $Tf = \tau + s$, $f_{min} \leq f \leq f_{max}$ etc. where s is a vector of slack variables and β is a weighting factor. The QP problem can be solved by using MATLAB Optimization Toolbox.

For azimuthing thrusters, the optimum problem is more complicated because the pseudo-inverse can be singular for certain α values. The cost function is more general resulting in a nonconvex NP problem which requires a large amount of computations at each sampling instant [41].

7.7 Simulation Results

The following results were obtained during backstepping control of a multipurpose naval vessel controlled in surge, sway, roll and yaw variables (4 DOF). The dynamic model of the ship was taken over from the GNC Toolbox [40] and used in our simulation software.

The important ship variables are $\eta = (x, y, \phi, \psi)^T$ for position and orientation, and $v = (u, v, p, r)^T$ for velocity and angular rate.

Wave effects were added to the ship model describing environmental effects. The desired path was designed using spline technique and the path derivatives were determined assuming roll angle $p = 0$ and considering elementary geometrical relations. The Jacobian matrix was approximated by

$$J = \begin{bmatrix} C_\psi & -S_\psi & 0 & 0 \\ S_\psi & C_\psi & 0 & 0 \\ 0 & 0 & 1 & 0 \\ 0 & 0 & 0 & 1 \end{bmatrix}. \tag{7.91}$$

The control error was defined by $\tilde{\eta} = \eta_d - \eta$. The control law was implemented based on (7.88). For state estimation in the presence of wave effects, extended Kalman filter (EKF) was applied.

First, we show the simulation results and later we give some comments to the implementation.

Figure 7.4 shows the realized, desired and measured paths. Since the wave effects are in the order of ± 0.5 m in position and ± 0.01 rad in orientation, respectively, hence the realized, desired and measured paths are well covered in the resolution of the path dimensions.

Real and estimated speeds u and v are shown in Fig. 7.5. Similarly, real and estimated roll rate p and yaw rate r are shown in Fig. 7.6.

The generalized forces are $\tau = (X, Y, K, N)^T$, where X and Y are the force control signals, while K and N are the torque control ones. For illustration, the control forces were shown in Fig. 7.7.

It can be seen that the control signals are realistic if we take into consideration the main parameters of the ship, i.e. effective length $L_{pp} = 51.5$ m and displacement disp $= 357$ m^3. The control allocation problem was not considered in the simulation example.

Let us consider now the implementation details. First of all, the naval model in GNC Toolbox is a MATLAB function having the following function prototype:

$$[out] = nv\_nlin\_model(in),$$

where $in = [u, v, p, r, \phi, \psi, Xe, Ye, Ke, Ne]$ and $out = M^{-1}[X, Y, K, N, p, r]$. The last four input variables are the external forces to be determined by the control law. Notice that out support the computation of $\dot{v} = M^{-1}[-Dv - Cv - g + \tau_{external}]$ in state space assumption, hence the matrices D, C should be multiplied by -1.

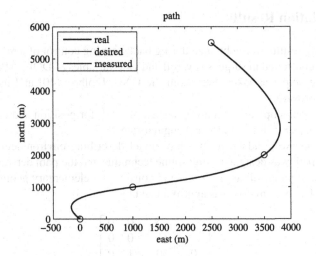

Fig. 7.4 Desired and realized paths under wave effects

Fig. 7.5 Real and estimated
velocities

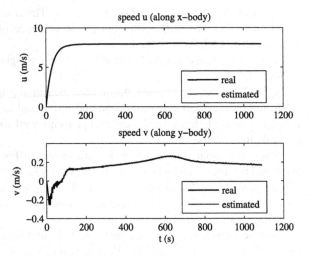

The mass matrix contains also the added mass and has the form:

$$
M = \begin{bmatrix}
m - X_{\dot{u}} & 0 & 0 & 0 & 0 & 0 \\
0 & m - Y_{\dot{v}} & -(mz_G + Y_{\dot{p}}) & mx_G - Y_{\dot{r}} & 0 & 0 \\
0 & -(mz_G + K_{\dot{v}}) & I_{xx} - K_{\dot{p}} & -K_{\dot{r}} & 0 & 0 \\
0 & mx_G - N_{\dot{v}} & -N_{\dot{p}} & I_{zz} - N_{\dot{r}} & 0 & 0 \\
0 & 0 & 0 & 0 & 1 & 0 \\
0 & 0 & 0 & 0 & 0 & 1
\end{bmatrix}.
$$

Since $Y_{\dot{r}} \neq N_{\dot{v}}$ hence $M \neq M^T$. Perhaps there is an error in the data set of the function because $K_{\dot{v}} = -Y_{\dot{p}}$ (while for symmetry $K_{\dot{v}} = Y_{\dot{p}}$ is necessary). Therefore the

Fig. 7.6 Real and estimated angular rates

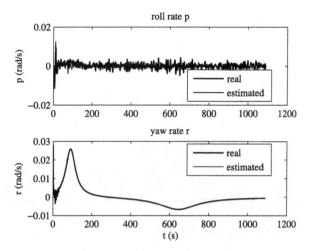

Fig. 7.7 Surge and sway forces during control

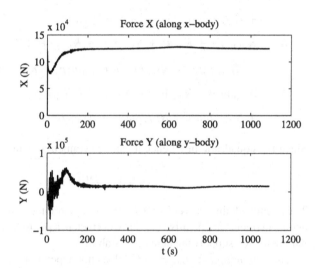

positive definiteness of M is uncertain although fortunately the diagonal elements are positive. This can cause stability problems since in Step 2 of the proof of the backstepping method $V_2 > 0$ is not guaranteed.

The centripetal and Coriolis effects are given in the naval model as follows:

$$\begin{aligned}
X_c &= m\left(rv + x_G r^2 - z_G pr\right), \\
Y_c &= -mur, \\
K_c &= mz_G ur, \\
N_c &= -mx_G ur,
\end{aligned} \tag{7.92}$$

which can be rewritten as

$$
\begin{pmatrix} X_c \\ Y_c \\ K_c \\ N_c \end{pmatrix} = \begin{bmatrix} 0 & 0 & 0 & m(v + x_G r - z_G p) \\ 0 & 0 & 0 & -mu \\ 0 & 0 & 0 & m z_G u \\ -m(v + x_G r - z_G p) & mu & -m z_G u & 0 \end{bmatrix} \begin{pmatrix} u \\ v \\ p \\ r \end{pmatrix}
$$

$$
= -C(v)v,
$$

where the matrix C is skew symmetric.

The hydrodynamic forces (without the added mass which already was considered in M) were modeled as $\tau_h + \tau_{h0}$. For simplifying the MATLAB source of the function, ϕ was denoted by b. The first term τ_h has the following components:

$$
X_h = X_{u|u|}u|u| + X_{vr}vr,
$$

$$
Y_h = Y_{|u|v}|u|v + Y_{ur}ur + Y_{v|v|}v|v| + Y_{v|r|}v|r| + Y_{r|v|}r|v|
$$
$$
+ Y_{b|uv|}b|uv| + Y_{b|ur|}b|ur| + Y_{buu}bu^2,
$$

$$
K_h = K_{|u|v}|u|v + K_{ur}ur + K_{v|v|}v|v| + K_{v|r|}v|r| + K_{r|v|}r|v| + K_{b|uv|}b|uv|
$$
$$
+ K_{b|ur|}b|ur| + K_{buu}bu^2 + K_{|u|p}|u|p + \big[K_{p|p|}|p| + K_p\big]p,
$$

$$
N_h = N_{|u|v}|u|v + N_{|u|r}|u|r + N_{r|r|}r|r| + N_{r|v|}r|v|
$$
$$
+ N_{b|uv|}b|uv| + N_{bu|r|}bu|r| + N_{bu|u|}bu|u|.
$$

Since τ_h is quadratic in v hence it can be symmetrized and rewritten as

$$
\tau_h = -D(b, v)v.
$$

The positive definiteness of $-D(b, v)$ is not guaranteed although fortunately the diagonal elements are positive (we have corrected $K_{v|v|}$ to be negative). The controller parameter K_d support the chance of stability.

The remaining part $\tau_{h0} = -g(b)$ does not depend on v:

$$
X_{h0} = 0,
$$

$$
Y_{h0} = 0,
$$

$$
K_{h0} = K_{bbb}b^3 - (\rho_w g g_m \cdot \text{disp})b,
$$

$$
N_{h0} = 0.
$$

Notice that $b = \phi$ is small in case of efficient control. In the control law, we used the term $D(b, v)v_r + g(b)$.

Do not forget that for the EKF, we have to differentiate the functions of the state equation which is not an easy task in 4 DOF. Symbolic computation tools can support the computation of the derivatives.

Table 7.1 The parameters of the wave models

Direction	ω_0	λ_0	K_w
x	1	0.15	0.2
y	1	0.15	0.2
ϕ	1	0.15	0.001
ψ	1	0.15	0.001

The wave model in each direction $i \in \{x, y, \phi, \psi\}$ was chosen to be

$$
\dot{x}_{w,i} = \begin{bmatrix} 0 & 1 \\ -\omega_{0,i}^2 & -2\lambda_{0,i}\omega_{0,i} \end{bmatrix} x_{w,i} + \begin{bmatrix} 0 \\ K_{w,i} \end{bmatrix} w_i,
$$
$$
y_i = \begin{bmatrix} 0 & 1 \end{bmatrix} x_{w,i}.
$$

(7.93)

The parameters of the wave models are shown in Table 7.1.

The controller parameters were chosen as follows:

$$
K_p = \mathrm{diag}(1e4, 1e4, 1e6, 1e8),
$$
$$
K_d = \mathrm{diag}(1e2, 1e2, 1e5, 1e7),
$$
$$
\Lambda = \mathrm{diag}(1, 1, 1, 4).
$$

The control signal τ was determined in every sampling instant $T_s = 0.1$ s and held fixed in the sampling interval during simulation.

Summarizing the simulation results it can be stated that backstepping control was a successful method in the tested order of wave effects although not all assumptions of the theory were completely satisfied.

7.8 Closing Remarks

The nonlinear control of surface vessels is an important field of control theory and applications. The control is usually model based.

The nonlinear dynamic model of ships was summarized both in body frame and NED frame, and the typical structure of the control system was shown. For reference path design, a short survey was given and the LOS concept was presented.

An important part of navigation is the state estimation in the presence of noisy measurements and wave effects. Wave effects in different directions were modeled in state space. Luenberger type observer was presented which integrates IMU and GPS position. For attitude observer, a quaternion based technique was shown using IMU and compass measurements.

For constant goal position, acceleration feedback with additional nonlinear PD was suggested. Stability was proven using LaSalle's theorem. Corrections are suggested if the mass matrix M is nonsymmetrical.

Nonlinear decoupling methods both in body frame and in NED frame were presented by showing the analogy with similar control methods of robots. The controller in both cases consists of a centralized nonlinear controller and the set of decentralized linear controllers.

Adaptive feedback linearization was suggested for the case if some parameters of the nonlinear dynamic model are unknown or slowly varying during the application. The method is similar to the selftuning adaptive control of robots. In the nonlinear model the parameters should be separated in linear vector form weighted by known signals. Some constraints should be satisfied for the controller parameters in which case stability follows from Lyapunov theory and Barbalat's lemma.

MIMO backstepping control was shown in 6 DOF based on the dynamic model in NED frame. The control law integrates many concepts like computed torque, sliding variable and PID control. Stability was proven if the mass matrix M is symmetrical and constant, the matrix C describing centripetal and Coriolis effects is skew symmetric, and the damping matrix D is symmetrical and positive definite. The controller implementation needs only the dynamic model in body frame and the Jacobian J beside the controller parameters, the path error, the reference velocity and the reference acceleration.

Simulation results were presented for a multipurpose naval vessel using backstepping control in 4 DOF. Path design was performed by spline technique. Wave models were applied in all four directions whose effects appear in the measured outputs (position and orientation of the ship under wave effects). For state estimation, EKF was implemented. Although the theoretical conditions for backstepping of ships were not exactly satisfied, the control behavior was acceptable by the simulation results.

Chapter 8
Formation Control of Vehicles

Overview This chapter deals with the formation control of unmanned ground and marine vehicles moving in horizontal plane. For stabilization of ground vehicles in formation, the potential field method is applied. The controller consists of three levels, the low level linearizing controller, the high level formation controller and the medium level stabilizing controller. The chosen potential function is structured such that each vehicle asymptotically approaches its prescribed position in the formation while a minimal distance to other vehicles is guaranteed. Simulation results illustrate the theory for UGVs. For stabilization of marine vehicles in formation, the passivity theory is applied. The control structure can be divided into synchronization level and the level of control subsystems stabilizing the different vehicles. The communication topology between the members of the formation is described by a graph. Only vehicles connected in the graph exchange their synchronization parameters. The stability of the formation is proven based on passivity theory and the Nested Matrosov Theorem. Simulation results illustrate the theory for UMVs.

8.1 Selected Approaches in Formation Control of Vehicles

In the last years, the increased capabilities of processors, communication technologies and sensor systems made possible that highly automated unmanned vehicles are able to cooperate each other and can perform complex tasks beyond the abilities of individual vehicles. Although the application fields are different, in control design several common points can be found. From the large set of control methods for stabilizing vehicles in formation, we select two popular methods, formation control based on potential function and formation control using passivity theory.

The first approach will be applied to unmanned ground vehicles (UGVs), the second to unmanned marine vehicles (UMVs). The discussion is based on the works of Peni and Bokor [112, 113] in case of UGVs, and that of Ihle, Arcak and Fossen [53] for UMVs.

Interested readers can find further methods in [114] and [122].

B. Lantos, L. Márton, *Nonlinear Control of Vehicles and Robots*,
Advances in Industrial Control,
DOI 10.1007/978-1-84996-122-6_8, © Springer-Verlag London Limited 2011

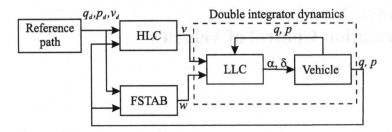

Fig. 8.1 Multilevel structure for formation control using potential field method

8.2 Stabilization of Ground Vehicles Using Potential Field Method

We present a multilevel structure for formation control consisting of the low level linearizing controller (LLC), the high level formation controller (HFC) and the intermediate level stabilizing controller (FSTAB), see Fig. 8.1.

8.2.1 Low Level Linearizing Controller

Let us start with the control of a single vehicle. We shall use the second input affine model of ground vehicles moving in horizontal plane which will be repeated here for better readability:

$$
\dot{x} = \begin{pmatrix} x_4 C_3 \\ x_4 S_3 \\ (a_{11}/x_4)x_5 + (a_{12}/x_4^2)x_6 \\ 0 \\ (a_{11}/x_4)x_5 + ((a_{12}/x_4^2) - 1)x_6 \\ a_{21}x_5 + (a_{22}/x_4)x_6 \end{pmatrix} + \begin{bmatrix} 0 & 0 \\ 0 & 0 \\ b_1/x_4 & 0 \\ 0 & 1 \\ b_1/x_4 & 0 \\ b_2 & 0 \end{bmatrix} u,
$$

$$
\dot{x} = A(x) + B(x)u, \qquad y = (x_1, x_2)^T = C(x),
$$

$$
x = (X, Y, \beta + \psi, v, \beta, r)^T, \qquad u = (\delta_w, \alpha)^T, \qquad y = (X, Y)^T.
$$

(8.1)

The state variables $x = (X, Y, \beta + \psi, v, \beta, r)^T$ have to be understood in the quasi inertial frame K_0 where X, Y are the position, $\beta + \psi$ is the angle between the velocity vector and the x_0-axis, ψ is the yaw angle (orientation) of the car, that is, the angle between the x-axis of the car's own frame K and the x_0-axis of the quasi inertial frame, β is the side slip angle, v is the absolute value of the velocity, and $r = \dot{\psi}$ is the angular velocity. The control signal is $u = (\delta_w, \alpha)$ where δ_w is the steering angle and α is the longitudinal acceleration. The output is the position $y = (X, Y)^T$.

The parameters in the dynamic model are

$$a_{11} = -\frac{c_F + c_R}{m_v}, \qquad a_{12} = \frac{c_R l_R - c_F l_F}{m_v}, \qquad b_1 = \frac{c_F}{m_v},$$

$$a_{21} = \frac{c_R l_R - c_F l_F}{I_z}, \qquad a_{22} = -\frac{c_R l_R^2 - c_F l_F^2}{I_z}, \qquad b_2 = \frac{c_F l_F}{I_z}, \tag{8.2}$$

where front and rear wheels are denoted by F and R, c_F, c_R are the cornering stiffnesses, l_f, l_R are the distances of the wheels from COG (which is the origin of the car's own frame K), and m_v is the mass and I_{zz} is the inertia moment of the car.

The state variables can be grouped as

$$x_1 = (X, Y)^T, \qquad x_2 = (\beta + \psi, v)^T, \qquad x_3 = (\beta, r)^T, \tag{8.3}$$

where x_1, x_2 can be measured. The state equation can be assumed as an LPV system with the parameter vector $\rho = (v, v^2)^T$ which can be measured as well. It is useful to introduce the following factorization:

$$\dot{x}_1 = \begin{pmatrix} v \cos(\beta + \psi) \\ v \sin(\beta + \psi) \end{pmatrix} =: h(x_2), \tag{8.4}$$

$$\dot{x}_2 = \begin{bmatrix} \frac{a_{11}}{v} & \frac{a_{12}}{v^2} \\ 0 & 0 \end{bmatrix} x_3 + \begin{bmatrix} \frac{b_1}{v} & 0 \\ 0 & 1 \end{bmatrix} u =: A_2(v)x_3 + B_2(v)u, \tag{8.5}$$

$$\dot{x}_3 = \begin{bmatrix} \frac{a_{11}}{v} & \frac{a_{12}}{v^2} - 1 \\ a_{21} & \frac{a_{22}}{v} \end{bmatrix} x_3 + \begin{bmatrix} \frac{b_1}{v} & 0 \\ b_2 & 0 \end{bmatrix} u =: A_3(v)x_3 + B_3(v)u. \tag{8.6}$$

The Jacobian of $h(x_2)$ will be used later:

$$J = \frac{dh}{dx_2} = \begin{bmatrix} -v \sin(\beta + \psi) & \cos(\beta + \psi) \\ v \cos(\beta + \psi) & \sin(\beta + \psi) \end{bmatrix}, \tag{8.7}$$

$$\det J = -v \neq 0. \tag{8.8}$$

The state variables will be changed in order to make the stabilization easier:

$$z_1 := x_1 = y, \qquad z_2 := \dot{x}_1 = h(x_2), \qquad z_3 := x_3. \tag{8.9}$$

The new state equations are as follows:

$$\dot{z}_1 = z_2,$$

$$\dot{z}_2 = \frac{dh}{dx_2} \frac{dx_2}{dt} = J \big[A_2(\rho)x_3 + B_2 u \big] = J A_2 z_3 + J B_2 u, \tag{8.10}$$

$$\dot{z}_3 = A_3 z_3 + B_3 u.$$

Observe that $y = x_1$ was twice differentiated and now u has appeared in $\ddot{y}$ and $J B_2$ is invertible. It follows that input-output linearization can be performed, but since z_3 is not measured, therefore the controller can only approximate z_3 by its own state z_{3c}:

$$u_c = -B_2^{-1}J^{-1}[JA_2z_{3c} + v] = -B_2^{-1}A_2z_{3c} - B_2^{-1}J^{-1}v, \tag{8.11}$$

$$\dot{z}_{3c} = A_3z_{3c} + B_3u_c - w = A_3z_{3c} - B_3B_2^{-1}A_2z_{3c} - B_3B_2^{-1}J^{-1}v - w, \tag{8.12}$$

where w is the input of the controller beside v. Substituting u_c into the state equations it follows:

$$\dot{z}_2 = JA_2z_3 - JB_2B_2^{-1}J^{-1}\left[JA_2z_{3c} + v + \overbrace{JA_2z_3 - JA_2z_3}^{0}\right]$$

$$= JA_2(z_3 - z_{3c}) + v, \tag{8.13}$$

$$\dot{z}_3 = A_3z_3 - B_3B_2^{-1}A_2z_{3c} + B_3B_2^{-1}J^{-1}v, \tag{8.14}$$

$$\dot{z}_{3c} = A_3z_{3c} - B_3B_2^{-1}A_2z_{3c} - B_3B_2^{-1}J^{-1}v - w, \tag{8.15}$$

$$\dot{z}_3 - \dot{z}_{3c} = A_3(z_3 - z_{3c}) + w. \tag{8.16}$$

The last equation is the stable zero dynamics (see earlier results for CAS), hence the dynamic inversion controller is applicable for each separate vehicle.

Observe that $q := x_1 = (X, Y)^T$ is the position while $p := \dot{x}_1 = (\dot{X}, \dot{Y})^T$ is the velocity of the vehicle. As can be seen the vehicle behaves as double integrator (one for X and one for Y) if $z_3 - z_{3c}$ goes to zero. Hence, $e := z_3 - z_{3c}$ can be considered as error.

Introducing index i for the ith vehicle then the state equations of the ith vehicle can be brought to the following final form:

$$\dot{q}_i = p_i, \tag{8.17}$$

$$\dot{p}_i = v_i + J_iA_{2,i}e_i, \tag{8.18}$$

$$\dot{e}_i = A_{3,i}e_i + w_i. \tag{8.19}$$

Here v_i and w_i are high level control signals which have to be designed for the formation.

8.2.2 High Level Formation Controller

We shall assume that a master vehicle is defined and the positions of the other vehicles in the formation are prescribed in comparison with the master. In this case, it is enough to define the path $q_D(t)$ for the master and the relative position $r_i = q_i - q_M$ between the ith vehicle and the master. The task can be formulated as

$$\lim_{t\to\infty} \|q_M(t) - q_D(t)\| = 0 \quad \text{and} \quad \lim_{t\to\infty}[y_i(t) - y_M(t)] = r_i. \tag{8.20}$$

Further constraint is that the distance between the vehicles must be larger than a minimal distance d during the transients.

It is clear that this specification can be easier checked in a moving coordinate system relative to the path. Hence, let us consider the moving frame K_D with the origin $q_D(t)$ and x_D-axis tangential to the path and y_D-axis to the left of x_D. The z_D axis is vertical and points upwards in this case. Denote ψ_D the angle between x_D and x_0, then the transformation matrix R_{ψ_D} in 2D from K_D into K_0 and its derivatives by the time are as follows:

$$R_{\psi_D} = \begin{bmatrix} C_{\psi_D} & S_{\psi_D} \\ -S_{\psi_D} & C_{\psi_D} \end{bmatrix}, \tag{8.21}$$

$$\dot{R}_{\psi_D} = \begin{bmatrix} -S_{\psi_D} & C_{\psi_D} \\ -C_{\psi_D} & -S_{\psi_D} \end{bmatrix} \dot{\psi}_D, \tag{8.22}$$

$$\ddot{R}_{\psi_D} = \begin{bmatrix} -C_{\psi_D} & -S_{\psi_D} \\ S_{\psi_D} & -C_{\psi_D} \end{bmatrix} \dot{\psi}_D^2 + \begin{bmatrix} -S_{\psi_D} & C_{\psi_D} \\ -C_{\psi_D} & -S_{\psi_D} \end{bmatrix} \ddot{\psi}_D. \tag{8.23}$$

If the vehicle is in q_i then there is an error in K_0 which can be transformed in K_D and can be differentiated further:

$$\tilde{q}_i = R_{\psi_D}(q_i - q_D), \tag{8.24}$$

$$\dot{\tilde{q}}_i = \dot{R}_{\psi_D}(q_i - q_D) + R_{\psi_D}(\dot{q}_i - \dot{q}_D) =: \tilde{p}_i, \tag{8.25}$$

$$\dot{\tilde{p}}_i = \ddot{R}_{\psi_D}(q_i - q_D) + 2\dot{R}_{\psi_D}(\dot{q}_i - \dot{q}_D) + R_{\psi_D}(v_i - \ddot{q}_D) =: \tilde{v}_i. \tag{8.26}$$

From the last equation, we can express v_i:

$$v_i = \ddot{q}_D + R_{\psi_D}^T\left[-\ddot{R}_{\psi_D}(q_i - q_D) - 2\dot{R}_{\psi_D}(\dot{q}_i - \dot{q}_D) + \tilde{v}_i\right]. \tag{8.27}$$

Hence, at high level it is sufficient to determine $\tilde{v}_i$, because v_i can be determined from it by the last equation.

Let us introduce the following notations where r is the number of vehicles:

$$\tilde{q} = \left(\tilde{q}_1^T, \ldots, \tilde{q}_r^T\right)^T,$$

$$\tilde{p} = \left(\tilde{p}_1^T, \ldots, \tilde{p}_r^T\right)^T,$$

$$\tilde{e} = \left(\tilde{e}_1^T, \ldots, \tilde{e}_r^T\right)^T,$$

$$\tilde{v} = \left(\tilde{v}_1^T, \ldots, \tilde{v}_r^T\right)^T,$$

$$\tilde{w} = \left(\tilde{w}_1^T, \ldots, \tilde{w}_r^T\right)^T,$$

$$\tilde{A}_2 = \text{diag}[R_{\psi_D} J_i A_{2,i}],$$

$$\tilde{A}_3 = \text{diag}[A_{3,i}].$$

Using these notations, the composite system of vehicles can be written in the following form:

$$\dot{\tilde{q}} = \tilde{p}, \tag{8.28}$$

$$\dot{\tilde{p}} = \tilde{v}(\tilde{q}, \tilde{p}) + \tilde{A}_2 e, \tag{8.29}$$

$$\dot{e} = \tilde{A}_3 e + w. \tag{8.30}$$

An energy function $\mathcal{V}(\tilde{q}, \tilde{p})$ will be introduced whose first term is the potential function $\mathcal{V}(\tilde{q})$ while its second term is the normalized kinetic energy:

$$\mathcal{V}(\tilde{q}, \tilde{p}) = \mathcal{V}(\tilde{q}) + \frac{1}{2}\|\tilde{p}\|, \tag{8.31}$$

$$\mathcal{V}(\tilde{q}) = \sum_{i=1}^{r}\left[\mu\big(\delta(\tilde{q}_i)\big) + \sum_{j\neq i}\mu\big(d - \|\tilde{q}_j - \tilde{q}_i\|\big)\right]. \tag{8.32}$$

The functions $\mu : R \rightarrow R^+$ and $\delta : R \rightarrow R^+$ should be chosen such that $\mathcal{V}(\tilde{q})$ becomes zero if the vehicles are in the prescribed formation. A possible choice is the following [113]:

$$\delta(\tilde{q}_i) = \begin{cases} \|\tilde{q}_i - (0, \mathrm{sgn}(\tilde{q}_{i,y}) \cdot 2d)^T\| & \text{if } |\tilde{q}_{i,y}| > 2d, \\ \tilde{q}_{i,x} & \text{if } |\tilde{q}_{i,y}| \leq 2d, \end{cases} \tag{8.33}$$

$$\mu(x) = \begin{cases} 0 & \text{if } x < 0, \\ \frac{1}{2}mx^2 & \text{if } 0 \leq x \leq \frac{M}{m}, \\ Mx - \frac{1}{2}\frac{M^2}{m} & \text{if } \frac{M}{m} < x. \end{cases} \tag{8.34}$$

Notice that $\delta(\tilde{q}_i)$ is the distance of vehicle i from the formation. The scaling function $\mu(x)$ consists of a quadratic part followed be a linear part, M and m are its parameters.

The control input is chosen according to

$$\tilde{v} = -\frac{\partial \mathcal{V}(\tilde{q})}{\partial \tilde{q}} - k\tilde{p}, \tag{8.35}$$

where $k > 0$. If $\mathcal{V}(\tilde{q}, \tilde{p})$ is considered as Lyapunov function then, with this control law, the time derivative of the Lyapunov function becomes negative semidefinite:

$$\dot{\mathcal{V}}(\tilde{q}, \tilde{p}) = \left\langle \frac{\partial \mathcal{V}(\tilde{q})}{\partial \tilde{q}}, \dot{\tilde{q}} \right\rangle + \langle \tilde{p}, \dot{\tilde{p}} \rangle = \left\langle \frac{\partial \mathcal{V}(\tilde{q})}{\partial \tilde{q}}, \tilde{p} \right\rangle + \langle \tilde{p}, \tilde{v} \rangle$$

$$= \left\langle \frac{\partial \mathcal{V}(\tilde{q})}{\partial \tilde{q}}, \tilde{p} \right\rangle + \left\langle \tilde{p}, \left[-\frac{\partial V(\tilde{q})}{\partial \tilde{q}} - k\tilde{p} \right] \right\rangle$$

$$= -k\|\tilde{p}\|^2 \leq 0. \tag{8.36}$$

8.2.3 Passivity Based Formation Stabilization

It still remains to develop the high level control signal w such that the formation becomes stable.

First of all, we shall assume that the system $\dot{e} = \tilde{A}_3 e$ is quadratically stable, that is, there exist Lyapunov functions $\mathcal{W}_i(e_i) = \frac{1}{2}\langle W_i e_i, e_i \rangle$, $W_i = W_i^T > 0$, such that $\dot{\mathcal{W}}_i = \frac{1}{2}[\langle W_i A_{3,i} e_i, e_i \rangle + \langle W_i e_i, A_{3,i} e_i \rangle] < 0$, $\forall \rho$ in the limited domain of nonzero velocities. To find the appropriate W_i, a small number of LMIs of the form $W_i A_{3,i} + A_{3,i}^T W_i < 0$ have to be solved if the velocity belongs to an interval. To find the feasible W_i, MATLAB Robust Control Toolbox can be used. Then $W = \text{diag}(W_i)$ and $\mathcal{W}(e) = \frac{1}{2}\langle We, e \rangle$ can be chosen to assure quadratic stability, that is, $\dot{\mathcal{W}} < 0$, $\forall \rho$.

It will be shown that the formation described by (8.28)–(8.30) can be assumed as two interconnected dissipative systems.

Let the two subsystems be given by

$$\text{S1:} \quad \dot{e} = \tilde{A}_3 e + w, \tag{8.37}$$

$$\text{S2:} \quad \dot{\tilde{q}} = \tilde{p}, \quad \dot{\tilde{p}} = \tilde{v} + \tilde{A}_2 e. \tag{8.38}$$

Consider first S2 for which yields:

$$\dot{V} = \underbrace{\frac{\partial V(\tilde{q}, \tilde{p})}{\partial \tilde{q}} \tilde{p} + \frac{\partial V(\tilde{q}, \tilde{p})}{\partial \tilde{p}} \tilde{v}}_{\leq 0} + \underbrace{\frac{\partial V(\tilde{q}, \tilde{p})}{\partial \tilde{p}} \tilde{A}_2}_{y_2^T} \underbrace{e}_{u_2} \leq y_2^T u_2, \tag{8.39}$$

that is, S2 is strictly output passive with storage function V for input u_2 and output y_2.

Consider now S1 which is quadratically stable with Lyapunov function $\mathcal{W}$, hence it satisfies:

$$\dot{\mathcal{W}} = \underbrace{e^T W \tilde{A}_3 e}_{<0} + \underbrace{e^T}_{y_1} \underbrace{W w}_{u_1} \leq y_1^T u_1, \tag{8.40}$$

that is, S1 is strictly output passive with storage function $\mathcal{W}(e)$ and input u_1 and output y_1.

The real formation is the interconnection of the two subsystems satisfying $u_1 = -y_2$, that is, $Ww = -\tilde{A}_2^T \frac{\partial V}{\partial \tilde{p}}$, from which follows the control law for FSTAB:

$$w := -W^{-1} \tilde{A}_2^T \frac{\partial V(\tilde{q}, \tilde{p})}{\partial \tilde{p}} = -W^{-1} \tilde{A}_2^T \tilde{p}, \tag{8.41}$$

$$\Downarrow$$

$$w_i := -W_i^{-1} (R_{\psi_D} J_i A_{2,i})^T \tilde{p}_i. \tag{8.42}$$

The interconnection is globally asymptotically stable (GAS) according to Theorem 2.14.

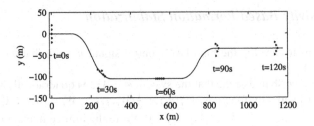

Fig. 8.2 Prescribed path having continuous curvature

8.3 Simulation Results for UGVs

The formation consists of five ($r = 5$) heavy-duty vehicles [123]. All vehicles have the same parameters (in SI units):

$$a_{11} = -147.1481, \qquad a_{12} = 0.0645, \qquad b_1 = 66.2026,$$

$$a_{21} = 0.0123, \qquad a_{22} = -147.1494, \qquad b_2 = 31.9835.$$

The assumed velocity domain is $1 \leq v \leq 25$, in which case the following choice is sufficient for quadratic stability:

$$W_i = \begin{bmatrix} 246.7608 & -4.7350 \\ -4.7350 & 247.7231 \end{bmatrix}, \quad \forall i.$$

The controller parameters were chosen to be $M = 6$, $m = 2$, $d = 4$, and $k = 2$. The continuous curvature path consists of clothoid, circle, clothoid, straight line sequences with extra straight lines at the beginning and at the end of the path. The corner points (P_i) of the straight lines are as follows:

$$[P_1, P_2, P_3, P_4, P_5, P_6, P_7, P_8, P_9]$$

$$= \begin{bmatrix} 0 & 100 & 185.1 & 220.5 & 305.6 & 655.6 & 740.8 & 825.9 & 1175.9 \\ 0 & 0 & -35.3 & -70.6 & -105.9 & -105.9 & -70.6 & -35.4 & -35.4 \end{bmatrix}.$$

Figure 8.2 shows the prescribed path. The vehicles start in orthogonal formation, then at $t = 30$ s the formation begins to change to longitudinal formation, while at 90 s command is given to go over V-shape formation.

Figure 8.3 shows the control signals v and w determined by HLC and FSTAB, respectively.

The steering angle and the acceleration actuator signals of the vehicles are shown in Fig. 8.4. The magnitude of the control signals is realistic regarding the mass of the vehicles.

The minimal distance amongst the vehicles was between 7.5 m and 10 m during

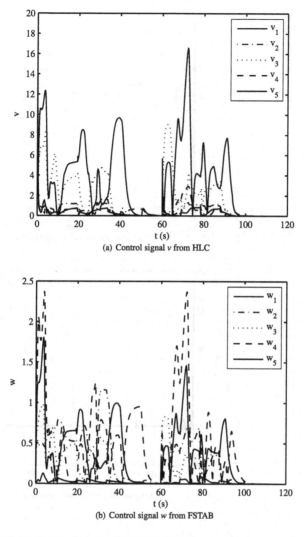

(a) Control signal v from HLC

(b) Control signal w from FSTAB

Fig. 8.3 Control signals determined by HLC and FSTAB

the transients, hence collision avoidance was guaranteed during the motion of the formation.

8.4 Stabilization of Marine Vehicles Using Passivity Theory

We present a formation control method for surface vessels with limited communication properties based on the results of Ihle, Arcak and Fossen [53]. The approach uses synchronized path following and stabilization based on passivity theory.

Fig. 8.4 Steering angle and acceleration actuator signals determined by LLC

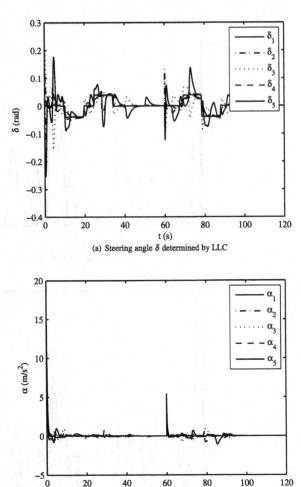

(a) Steering angle δ determined by LLC

(b) Longitudinal acceleration α determined by LLC

8.4.1 Problem Formulation for Synchronized Path Following

For each component of the formation, say i, there is a path $y_{di}(\theta_i)$ defined in the scalar path parameter θ_i, and the aim is to develop control laws ω_i such that the path velocity $\dot{\theta}(t) = v(t) - \omega_i(t)$ goes to $v(t)$ if $t \to \infty$. It is assumed that $v(t)$ is identical for each vehicle. The definition of the prescribed paths of the vehicles according to the aimed formation and the choice of the common path velocity is the responsibility of the high level guidance system and it will not be discussed here. For simple formations (orthogonal, longitudinal or V-shape formations, circle-shaped formations with different radii etc.), the path design problem can easily be solved. Common path velocity simplifies the stability proof.

The communication topology between the members of the formation is described by a graph $\mathcal{G}$. Two vehicles, i and j are neighbors, if they access the synchronization error $\theta_i - \theta_j$. In this case, the vertexes i and j are connected by an edge in the graph. The information flow is bidirectional, but for simplifying the discussion, an orientation is assigned to the graph by considering one of the nodes to be the positive end of the edge. For a group of r members with p edges, the $r \times p$ incidence matrix $D(\mathcal{G})$ is defined, for the edge k:

$$d_{ik} = \begin{cases} +1 & \text{if } i\text{-th vertex is the positive end of } k, \\ -1 & \text{if } i\text{-th vertex is the negative end of } k, \\ 0 & \text{otherwise.} \end{cases} \tag{8.43}$$

It will be assumed that $\mathcal{G}$ is a connected graph, that is, a path exists between any two distinct vertexes. By the definition, the sum of the elements in each column of D (the sum of the elements in each row of D^T) is zero. For example, if there are $r = 3$ vehicles, only 1 and 2, and 2 and 3 are exchanging their path parameters, then the communication topology is given by the incidence matrix

$$D(\mathcal{G}) = \begin{bmatrix} 1 & 0 \\ -1 & 1 \\ 0 & -1 \end{bmatrix}. \tag{8.44}$$

8.4.2 Control Structure

The block scheme of the control system for synchronized path following is shown in Fig. 8.5. The system Σ is the diagonal interconnection of the subsystems Σ_i,

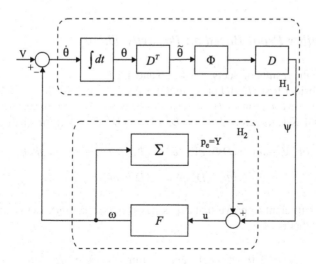

Fig. 8.5 Block scheme of formation control using synchronized path following

and similarly F is the diagonal interconnection of the subsystems F_i. (Σ_i, F_i) is the control system of vehicle i.

The subsystem Σ_i is part of the control system of vehicle i. It is assumed that the vehicle's dynamic model satisfies the condition for backstepping, and the backstepping control approach is applied for the control. The closed loop system of the vehicle can be brought to the form (see Sect. 2.6)

$$\dot{z}_i = A_i(x_i)z_i - b_i(t, x_i, \theta_i)\omega_i, \tag{8.45}$$

$$\dot{\theta}_i = v(t) - \omega_i. \tag{8.46}$$

The output is the path error $p_{ei} =: y_i$. According to the backstepping control the path error is $p_{ei} = 2z_i^T P_i b_i$ where $P_i A_i + A_i^T P_i = -Q_i$, $P_i > 0$, $Q_i > 0$ and A_i is Hurwitz.

The subsystem F_i computes the synchronization signal ω_i based on the available synchronization error ψ_i and the position error p_{ei}. This subsystem is a low order dynamic system, for example one realizing filtered-gradient update (see Sect. 4.6). The input $u_i = -p_{ei} + \psi_i$ reacts to the path error and the synchronization error. Its output ω_i is used also by the system H_1.

The system H_1 determines the path synchronization errors based on the communication topology given in the incidence matrix D, the synchronization vector ω and the specified path velocity vector V (with equal components v). The mapping $\Phi(\cdot)$ is the diagonal interconnection of static nonlinearities whose ith component $\phi_i(\cdot)$ is an odd static nonlinearity, typically linear or cubic function. The output $\psi_i(\theta) = \sum_{k=1}^{P} d_{ik}\phi_k(\theta_i - \theta_k)$ depends only on the path errors of the vehicles communicating with vehicle i.

The composite system consisting of H_1 and $H_2 = (\Sigma, F)$ should be stable and should guarantee synchronized path following for all vehicles. The aim is $\omega_i \to 0$ as $t \to \infty$, $\forall i$.

8.4.3 Stability Proof Based on Passivity Theory

We shall use the results of Sects. 2.3.4, 2.6 and 4.6.1.

Let us consider first H_1 and prove that it is passive from $-\omega$ to Ψ. Since the sum of the elements in each row of D^T is zero, v is constant, $V = v\bar{1}$ and each component of the vector $\bar{1}$ is 1, therefore $D^T V = D^T v\bar{1} = v(D^T \bar{1}) = 0$, from which it follows:

$$\langle \Psi, V - \omega \rangle = \langle D\Phi, v\bar{1} - \omega \rangle = \langle \Phi, D^T (v\bar{1} - \omega) \rangle = \langle \Phi, \dot{\theta} \rangle$$

$$= \langle \Phi, -D^T \omega \rangle = -\langle D\Phi, \omega \rangle = -\Psi^T \omega. \tag{8.47}$$

Hence, it is sufficient to prove that H_1 is passive from $\dot{\theta}$ to Φ. Let ϕ_k be a sector nonlinearity, that is

$$x\phi_k(x) > 0 \quad \forall x \neq 0 \quad \text{and} \quad \lim_{|x| \to \infty} \int_0^x \phi_k(\sigma)\, d\sigma = \infty. \tag{8.48}$$

Define the storage function and take its time derivative:

$$V_\psi(\Psi, t) := \sum_{k=1}^{p} \int_0^{\tilde{\theta}_k(t)} \phi_k(\sigma) \, d\sigma, \qquad (8.49)$$

$$\dot{V}_\psi(\Psi, t) = \sum_{k=1}^{p} \phi_k \big(\theta_k(t) - \theta_{k+1}(t) \big) \big(\dot{\theta}_k(t) - \dot{\theta}_{k+1}(t) \big) = \big\langle \Phi(t), \dot{\tilde{\theta}}(t) \big\rangle, \qquad (8.50)$$

which proves that H_1 is passive from $\dot{\tilde{\theta}}$ to Φ, hence it is also passive from $-\omega$ to Ψ.

Let us consider now the second subsystem H_2. It will be assumed that the control law is backstepping control (see Sect. 2.6) with filtered-gradient update (see Sect. 4.6).

Let $V_\Sigma = \sum_1^r z_i^T P_i z_i$ be the storage function for the Σ block using backstepping control then its time derivative is

$$\dot{V}_\Sigma = - \sum_{i=1}^{r} \big[z_i^T Q_i z_i + \underbrace{2 b_i^T P_i z_i}_{p_{ei} = y_i} \omega_i \big] = \underbrace{- \sum_{i=1}^{r} z_i^T Q_i z_i + \omega^T Y}_{<0}, \qquad (8.51)$$

which proves that Σ is strictly passive from ω to Y.

Let consider now the subsystem F in H_2. By (4.87), it yields for each component subsystem:

$$\dot{\omega}_i = -\lambda \omega_i - \lambda \mu u_i. \qquad (8.52)$$

Hence, the storage function $V_F = \sum_{i=1}^{r} \frac{1}{2\lambda\mu} \omega^2$ can be chosen for F whose time derivative is

$$\dot{V}_F = \frac{1}{\lambda\mu} \langle \omega, \dot{\omega} \rangle = \frac{1}{\lambda\mu} \langle \omega, -\lambda(\omega + \mu u) \rangle = \underbrace{-\frac{1}{\mu} \|\omega\|^2}_{<0} + \langle \omega, u \rangle, \qquad (8.53)$$

which proves that F is strictly passive from u to ω.

Let consider now the composite system $H_2 = (\Sigma, F)$ with input Ψ and output ω. Let the storage function be defined by $V_{H_2} = V_\Sigma + V_F$ and substitute $u = \Psi - Y$, then the time derivative of V_{H_2} results in

$$\dot{V}_{H_2} = - \sum_{i=1}^{n} z_i^T P_i z_i + \omega^T Y - \frac{1}{\mu} \|\omega\|^2 + \omega^T (\Psi - Y)$$

$$= \underbrace{- \sum_{i=1}^{n} z_i^T P_i z_i - \frac{1}{\mu} \|\omega\|^2}_{<0} + \omega^T \Psi, \qquad (8.54)$$

which means that H_2 is strictly passive from Ψ to ω.

To prove the stability of $(\tilde{\theta}, z, \omega) = 0$ we use the Lyapunov function $V(\tilde{\theta}, z, \omega) = V_\psi(\tilde{\theta}) + V_{H_2}(z, \omega)$ and determine its time derivative:

$$\dot{V} = -\Psi^T \omega - \sum_{i=1}^{n} z_i^T P_i z_i - \frac{1}{\mu}\|\omega\|^2 + \Psi^T \omega$$

$$= -\sum_{i=1}^{n} z_i^T P_i z_i - \frac{1}{\mu}\|\omega\|^2 \leq 0. \tag{8.55}$$

Since $\dot{V}$ is only negative semidefinite, hence it follows only the stability, however we need global absolute stability.

First of all, we can conclude that the trajectories $(z(t), \omega(t), \tilde{\theta}(t))^T$ are uniformly bounded on every finite time interval $[t_0, t_0 + T]$. Since the speed assignment $v(t)$ is bounded, hence $(\theta(t), x(t))$ is bounded by a continuous function of T which implies that $\tilde{\theta}(t)$ and $z(t)$ are well defined for all $t \geq t_0$ and the equilibrium $(z(t), \omega(t), \tilde{\theta}(t))^T = 0$ is uniformly globally stable (UGS). The proof of absolute stability is based on the Nested Matrosov theorem [85].

Theorem 8.1 (Nested Matrosov theorem) *Under the following assumptions, the $x = 0$ equilibrium point of the system $\dot{x} = F(t, x)$ is uniformly globally asymptotically stable (UGAS):*

1. *The origin is UGS.*
2. *There exist integers $j, m > 0$ and for each $r > 0$ there exist*
   (a) *A number $\mu > 0$.*
   (b) *Locally Lipschitz continuous functions $V_i : R^+ \times R^n \to R, i = 1, \ldots, j$.*
   (c) *A function $\phi : R^+ \times R^n \to R^m$.*
   (d) *Continuous functions $Y_i : R^n \times R^m \to R, i = 1, \ldots, j$, such that for almost all $(t, x) \in R^+ \times B(r)$ and all $i = 1, \ldots, j$*

$$\max\{|V_i(t, x)|, |\phi(t, x)|\} \leq \mu,$$

$$\dot{V}_i(t, x) \leq Y_i(x, \phi(t, x)).$$

3. *For each integer $k \in \{1, \ldots, j\}$, we have that*

$$Y_i(z, \psi) = 0 \quad \forall i \in \{1, \ldots, k-1\} \quad \text{and all } (z, \psi) \in B(r) \times B(\mu) \text{ implies that}$$

$$Y_k(z, \psi) \leq 0 \quad \forall(z, \psi) \in B(r) \times B(\mu).$$

4. *We have that the statement*

$$Y_i(z, \psi) = 0 \quad \forall i \in \{1, \ldots, j\} \text{ and all } (z, \psi) \in B(r) \times B(\mu) \text{ implies that}$$

$$z = 0.$$

Observe that an external input is present therefore we cannot use the theorem for interconnected dissipative systems to prove GAS. We shall use the Nested Matrosov

theorem to prove UGAS. In our case, $j = 2$ and we have only V_1, Y_1 and V_2, Y_2.
Notice that the Nested Matrosov theorem does not need the positivity of V_2.

Let $V_1 := V$ and denote Y_1 the right side of (8.55). The auxiliary function V_2 will
be chosen

$$V_2 = -\tilde{\theta}^T D^+ \Lambda \omega, \tag{8.56}$$

where D^+ denotes Moore–Penrose pseudo-inverse and Λ is a diagonal matrix sat-
isfying $\Lambda_{ii} = -\frac{1}{\lambda \mu}$, $i = 1, \dots, r$ (see filtered-gradient update). Since $\dot{\tilde{\theta}} = D^T \theta$,
therefore

$$Y_2 := \dot{V}_2 = -\dot{\tilde{\theta}}^T D^+ \Lambda \omega - \theta^T D D^+ \Lambda \dot{\omega}. \tag{8.57}$$

On the other hand, we know that for filtered-gradient update $\dot{\omega}_i = -\lambda \omega_i - \lambda \mu u_i$.
Hence, we can claim that

$$Y_1 = 0 \to z, \qquad \omega = 0, \tag{8.58}$$

$$Y_1 = 0 \to Y_2 = -\theta^T D D^+ \Lambda \dot{\omega} = -\theta^T D D^+ \Lambda \left(\Lambda^{-1} u \right) - \theta^T \underbrace{D D^+ D}_{D} \Phi \left(\tilde{\theta} \right)$$

$$= -\tilde{\theta}^T \Phi \left(\tilde{\theta} \right) \leq 0, \tag{8.59}$$

where we have also used (8.48) to show $Y_2 \leq 0$ if $Y_1 = 0$. Since $Y_1 = 0$, $Y_2 = 0$
together imply $(z, \omega, \tilde{\theta}) = 0$, we conclude from the Nested Matrosov theorem that
the equilibrium point $(z, \omega, \tilde{\theta}) = 0$ is UGAS.

8.5 Simulation Results for UMVs

For simulation test of the formation control algorithm, equal container ships were
chosen whose mass and length are 4000 t and 76.2 m, respectively. The ships are
moving in the horizontal plane. The block scheme and the signal flow of the control
system of each UMV is shown in Fig. 8.6. Path, formation and synchronization
specifications are variable.

The formation consists of five ships moving on circle-form paths of different
radii. First, the formation is divided into two groups, the first group consists of three
ships, and the second group contains the remaining two ships. At 1500 s, the formation is changing, all ships form a single group. The communication topology is
varying according with the division of the groups. The initial state (position and orientation) of the formation is randomly chosen but not far away from the prescribed
paths of the different ships.

Each ship has state variables $\eta(x, y, \psi)^T$ and $v = (u, v, r)$. The rotation matrix
that transforms vectors from the body frame in the NED frame is $R = Rot(z, \psi)$:

$$R(\psi) = \begin{bmatrix} C_\psi & -S_\psi & 0 \\ S_\psi & C_\psi & 0 \\ 0 & 0 & 1 \end{bmatrix}.$$

Fig. 8.6 Block scheme and signal flow in the control system of each ship of the formation

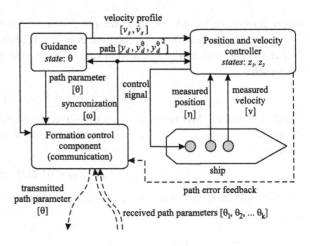

The dynamic model has the form

$$\dot{\eta} = R(\psi)v,$$

$$\dot{v} = M^{-1}\big[-\big(C(v) + D\big)v + \tau\big],$$

$$y = \eta,$$

where τ is the control signal and y is the output signal.

We use the model identified in [42], but in the simulation example D was modified to be symmetrical:

$$M = \begin{bmatrix} m - X_{\dot{u}} & 0 & 0 \\ 0 & m - Y_{\dot{v}} & mx_G - Y_{\dot{r}} \\ 0 & mx_G - Y_{\dot{r}} & I_z - N_{\dot{r}} \end{bmatrix},$$

$$D = \begin{bmatrix} -X_u & 0 & 0 \\ 0 & -Y_v & -Y_r \\ 0 & -Y_r & -N_r \end{bmatrix},$$

$$C(v) = \begin{bmatrix} 0 & 0 & -m_{22}v - m_{23}r \\ 0 & 0 & m_{11}u \\ m_{22}v + m_{23}r & -m_{11}u & 0 \end{bmatrix}.$$

The mechanical parameters of the ships are summarized in Table 8.1. The parameters were taken over from [23]. The final form of the matrices in the dynamic model of the ships are as follows:

$$M = 10^3 \cdot \begin{bmatrix} 4.5096 & 0 & 0 \\ 0 & 7.5608 & -22.68 \\ 0 & -22.68 & 2968.3 \end{bmatrix},$$

Table 8.1 The dimensional hydrodynamical parameters of the ships

Parameter	Value	Dimension
m	4000	t
$X_{\dot{u}}$	-0.5096×10^3	t
$Y_{\dot{v}}$	-3.5608×10^3	t
x_G	0	m
$Y_{\dot{r}}$	0.02268×10^6	t m
I_z	2.0903×10^6	$t\,m^2$
$N_{\dot{r}}$	-0.8780×10^6	$t\,m^2$
X_u	-0.05138×10^3	t/s
Y_v	-0.1698×10^3	t/s
Y_r	1.5081×10^3	t m/s
N_r	-0.2530×10^6	$t\,m^2/s$
m_{11}	4.5096×10^3	t
m_{22}	7.5608×10^3	t
m_{23}	-22.68×10^3	t m/s

Table 8.2 The correspondence between the notations in backstepping and the ship model

Backstepping	Ship model
x_1	η
x_2	ν
$G_1(x_1)$	$R(\psi)$
f_1	0
$G_2(\bar{x}_2)$	M^{-1}
f_2	$-M^{-1}[C(\nu) + D]\nu$
h	I
u	τ

$$D = 10^3 \cdot \begin{bmatrix} 0.05138 & 0 & 0 \\ 0 & 0.1698 & -1.5081 \\ 0 & -1.5081 & 253 \end{bmatrix},$$

$$C(\nu) = 10^3 \cdot \begin{bmatrix} 0 & 0 & -7.5608\nu + 22.68r \\ 0 & 0 & 4.5096u \\ 7.5608\nu - 22.68r & -4.5096u & 0 \end{bmatrix}.$$

The correspondence between the notations of backstepping control design in Sect. 2.6 and the ship's dynamic model can be seen in Table 8.2. For each ship, the backstepping control design consists of the following steps (notations as in Sect. 2.6, in the signals the index of the ship is omitted):

Step 1:

$$z_1 = y - y_d = \eta - y_d,$$
$$\alpha_1 = R^{-1}[A_1 z_1 + y_d^\theta v_s],$$
$$\dot{z}_1 = A_1 z_1 + R z_2 + y_d^\theta \omega.$$

Step 2:

$$\dot{\alpha}_1 = -r S R^{-1}[A_1 z_1 + y_d^\theta v_s] + R^{-1}[A_1(Rv - y_d^\theta \dot{\theta}) + y_d^{\theta^2} \dot{\theta} v_s + y_d^\theta \dot{v}_s],$$
$$\sigma_1 = -r S R^{-1}[A_1 z_1 + y_d^\theta v_s] + R^{-1}[A_1 R v + y_d^\theta \dot{v}_s],$$
$$\alpha_1^\theta = -R^{-1}[A_1 y_d^\theta - y_d^{\theta^2} v_s],$$
$$\dot{z}_2 = -P_2^{-1} R^{-1} P_1 z_1 + A_2 z_2 + y_d^\theta \omega,$$
$$u = \alpha_2 = M[A_2 z_2 - P_2^{-1} R^{-1} P_1 z_1 + \sigma_1 + \alpha_1^\theta v_s] + (C + D)v.$$

Notice that $R^{-1} = R^T$ and $S = [k \times]$. At start the initial values of z_1 and z_2 can be chosen according to

$$z_{10} = \eta_0 - y_d(\theta_0),$$
$$z_{20} = v_0 - R^{-1}(\psi_0)[A_1 z_{10} + y_d^\theta(\theta_0) v_s(0)].$$

Here, θ_0 can also be chosen. A possible choice is to take the nearest point of the path relative to the initial position of the ship.

The free parameters of the controller are A_1, Q_1, A_2, Q_2, from which A_1, A_2 are related to the length of transient times for the controlled ship. Taking into consideration the magnitude of the control signals, the following controller parameters were chosen after experiments:

$$A_1 = -10^{-2} \operatorname{diag}(1, 2, 2), \qquad Q_1 = 10^{-4} \operatorname{diag}(2, 2, 5),$$
$$A_2 = -10^{-2} \operatorname{diag}(1, 1, 1), \qquad Q_2 = \operatorname{diag}(10, 10, 100).$$

For filtered-gradient update, $\lambda = 1$ and $\mu = 0.1$ were used. For scaling of the synchronization errors by $\phi(x)$ in H_1, the function $\phi(x) = \alpha \tanh(0.005x)$ was chosen that has saturation (approximately) for $|x| > 600$. Avoiding large synchronization errors, $\alpha = 50$ was chosen after experiments.

Figure 8.7(a) shows the main signals of the control system, that is, the path error $P_e = Y$ of the vehicles, the synchronization control input ω for H_1 and the synchronization error $\tilde{\theta}$ inside H_1. Figure 8.7(b) shows the velocities of the ships. Notice that at $t = 1500$ s the group division is changed.

From the actuator inputs of the five ships the actuator inputs of ship 3 and ship 5 were selected and shown in Fig. 8.8. It can be seen that the magnitudes of the actuator inputs are realistic regarding the mass and the displacement of the ships.

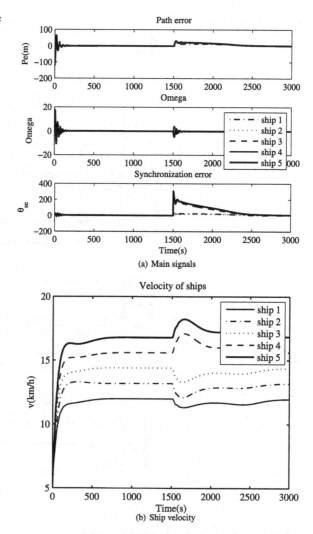

Fig. 8.7 Main signals of the control system and the velocity of the ships

Figure 8.9 illustrates that the control method based on backstepping, filtered-gradient update and path synchronization is working well also in case of varying communication topology. Notice that the formation was started from a randomly chosen initial state and it reached the prescribed first formation. After changing the communication topology at $t = 1500$ s, the UMVs reached the second prescribed formation.

8.6 Closing Remarks

The formation control of unmanned ground and marine vehicles moving in horizontal plane was considered based on potential function method and passivity theory.

Fig. 8.8 Actuator inputs of
ship 3 and ship 5

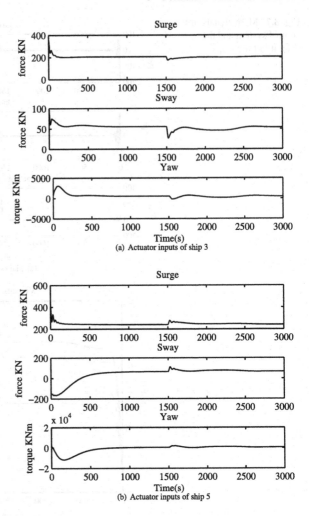

(a) Actuator inputs of ship 3

(b) Actuator inputs of ship 5

For the formation stabilization of ground vehicles, first a control structure was elaborated consisting of three control levels (LLC, HLC, FSTAB). The dynamic model was assumed to be input affine with control signals consisting of the steering angle and the longitudinal acceleration of the vehicles.

The low level stabilizing controller (LLC) applied input-output linearization for each vehicle and its state variables approximated the side slip angle and the yaw rate of the vehicle. If the approximation error goes to zero, then each vehicle in the formation can be described by its position and velocity alone. LLC has two input vectors, one from HLC and an other from FSTAB.

The high level formation controller (HLC) has the task of stabilizing the position of the vehicles in the formation. As Lyapunov function, an energy type function was chosen which is the sum of a potential function depending on the position and the square of the velocity. The potential function was chosen such that it becomes zero

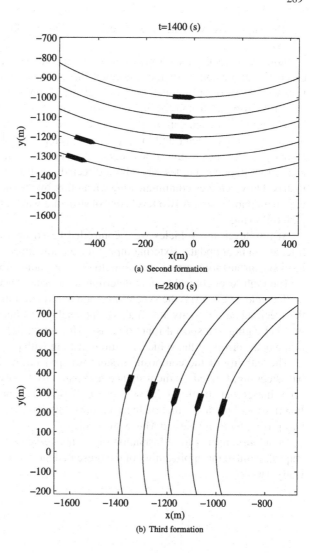

Fig. 8.9 Reached goal formations of the UMVs in case of varying communication topology

if the vehicles are in the prescribed formation. It consists of a distance and a scaling part. The output of HLC is the negative gradient of the Lyapunov function which assures the stability of the formation.

The intermediate level stabilization controller (FSTAB) was responsible for eliminating the error between the vehicle states and controller states regarding side slip angle and yaw rate. This system is parameter varying hence for stabilization quadratic stability approach was applied. In order to find a quadratic Lyapunov function for all velocity values in the assumed working domain, a small number of LMI problems were solved. The output of FSTAB was developed from the quadratic stability criterion. The composite system can be divided into two subsystems which are strictly output passive hence their interconnection is globally asymptotically stable (GAS).

Simulation results for a formation of five unmanned heavy-duty vehicles illustrated the applicability of the presented theory. The formation can dynamically varies in the set of orthogonal, longitudinal and V-shape formations.

For the formation stabilization of marine vehicles, the results of passivity theory were applied. For each component of the formation, separate path can be defined in its scalar path parameter and the aim is to develop control laws such that the path velocity approaches a prescribed identical value for each vehicle. The communication topology can be described by a directed graph represented by its incidence matrix. Only vehicles communicating each other by the incidence matrix exchange their path parameters. A two level control structure was suggested for synchronized path following.

At low level, the vehicles were controlled based on backstepping method and filtered-gradient update. External inputs for the low level controllers were the high level synchronization errors according to the communication topology.

The high level control system determines the path synchronization errors based on the incidence matrix, the synchronization control coming from the low level and the specified path velocity. It contains a diagonal interconnection of static nonlinearities belonging to a special function class. The output depends only on the integral of the paths errors of the vehicles communicating with each other.

The stability of the composite system was proven by integrating the results of backstepping, filtered-gradient update and passivity theory of dissipative systems. First, the global stability (GS) was proven based on appropriately chosen energy functions and passivity theory, then uniformly global asymptotic stability (UGAS) was proven by using Nested Matrosov theorem.

Simulation results for a formation of marine vessels containing heavy container ships illustrated the applicability of the presented theory. The formation was dynamically varying.

Chapter 9
Modeling Nonsmooth Nonlinearities in Mechanical Systems

Overview   The first part of this chapter presents the basic theoretical notions from the field of modeling and stability of nonsmooth systems. Afterward, it focuses on modeling of two nonsmooth nonlinearities that can influence the performances of mechanical control systems: friction and backlash. The most important static and dynamic friction models, that are applied in friction compensation algorithms, are discussed. After presenting the well known friction models from the literature, a novel, piecewise linearly parameterized model is introduced based on which the problem of friction compensation with unknown friction parameters can easily be solved. At the end of the chapter, it is presented, how the backlash type nonlinearity influences the motion of mechanical systems.

9.1 Modeling and Stability of Nonsmooth Systems

Common physical phenomena in mechanical control systems such as friction and backlash can be described by mathematical models with discontinuous, nonsmooth nonlinearities. The models of mechanical systems which incorporate such phenomena are generally given by a set of interconnected first order ordinary differential equations with discontinuous right-hand side.

Due to the discontinuous functions that appear in the system model, classical solution theories of ordinary differential equations are no longer applicable. The solution concept can be addressed using Filippov's solution theory. The stability concepts and the Lyapunov-like stability theorems have to be reformulated for these systems.

In this section, two approaches of modeling and stability of nonsmooth systems are treated. First, an overview of switched systems is given. Second, the solution and stability notions based on the theory of differential inclusions are presented.

B. Lantos, L. Márton, *Nonlinear Control of Vehicles and Robots*,
Advances in Industrial Control,
DOI 10.1007/978-1-84996-122-6_9, © Springer-Verlag London Limited 2011

9.1.1 Modeling and Stability of Switched Systems

A *switched system* is defined as

$$\dot{x} = f_\sigma(t, x), \tag{9.1}$$

where $x \in R^n$ and $f_\sigma : R^n \times R \to R^n$ can take its value from finite set F_N of smooth vector fields

$$F_N = \{f_1, f_2, \ldots f_i, \ldots f_N\}. \tag{9.2}$$

The *switching signal* σ determines which element of the set the function f_σ will take in a given time instant t. The switching signal is a piecewise continuous function of time with a finite number of discontinuities (finite switches in finite time)

$$\sigma : [0, \infty) \to I, \tag{9.3}$$

where $I = \{1, 2, \ldots i, \ldots N\}$ is the index set of F_N. σ may also depend on the state x. It can be autonomous or controlled.

A *piecewise affine system* is defined as

$$\dot{x} = A_\sigma x + B_\sigma u + a_\sigma, \tag{9.4}$$

where $x \in R^n$ and $u \in R^m$. The system matrices $A_\sigma (n \times n)$, $B_\sigma (n \times m)$ and the offset vector $a_\sigma (n \times 1)$ depend on the switching signal σ.

Consider that σ in (9.4) depends only on the state x, and the state space is partitioned into polyhedra type regions denoted by X_i.

$X \subseteq R^n$ is called *polyhedral partition* if it is a collection of n dimensional convex polyhedras X_i such that

- $\bigcup X_i = X$.
- The polyhedras have disjoint interior, $\text{int}(X_i) \cap \text{int}(X_j) = \emptyset$ for $i \neq j$.
- The polyhedras only share their common boundaries.

The model (9.4) is called *polyhedral piecewise affine system* if the space of the states is a polyhedral partition.

Since closed convex polyhedra can be represented as the intersection of finite number of closed halfspaces, there exists a matrix E_i and a vector e_i such that

$$X_i = \{E_i x + e_i \geq 0\}. \tag{9.5}$$

The vector inequality means that each element of the vector is nonnegative.
The boundary of two polyhedras can be represented by surfaces:

$$X_i \cap X_j = \{F_i x + f_i = 0\} = \{F_j x + f_j = 0\}. \tag{9.6}$$

The stability and asymptotic stability of switched systems (9.1) are defined as in the case of continuous systems (see Chap. 2) with the extension that the stability conditions must hold for every switching signal (uniform with respect to σ) [82].

There are two approaches for the stability analysis of switched systems, see [34, 83] and references therein. The first approach assumes that for all f_σ the switched system share a common Lyapunov function. The second approach allows different Lyapunov functions for different f_σ, but they should satisfy extra conditions.

Theorem 9.1 *Let a switched system be described by (9.1). Consider that for each smooth function f_i from the set F_N defined in (9.2) holds: $f_i(0) = 0$. Suppose for each f_i the systems $\dot{x} = f_i(x)$ share a common Lyapunov function candidate $V(x)$ with the following properties*

- *V is a continuous function with continuous first order derivate.*
- $x \neq 0 \Rightarrow V > 0; x = 0 \Rightarrow V = 0.$
- *V is radially unbounded ($\|x\| \to \infty \Rightarrow V \to \infty$).*

Then if $\frac{\partial V}{\partial x} f_i(x) < 0$, $\forall i$, the equilibrium point $x = 0$ of the system (9.1) is globally uniformly asymptotically stable.

Theorem 9.2 *Let a polyhedral piecewise affine autonomous system be described by*

$$\dot{x} = A_\sigma x. \tag{9.7}$$

A_σ *may take any value form the finite set of matrices $\{A_1, A_2, \ldots A_i, \ldots A_N\}$ in function of the switching signal σ. If there exists a positive definite symmetric matrix P such that*

$$A_i^T P + P A_i < 0, \quad \forall i = 1, \ldots, N \tag{9.8}$$

then the equilibrium point $x = 0$ is globally uniformly asymptotically stable.

If the matrices $\{A_1, A_2, \ldots A_i, \ldots A_N\}$ commute pairwise, that is, $A_i A_j = A_j A_i$, $\forall i, j = 1, \ldots, N$, and all are stable, then the matrix P in the common Lyapunov function $V = x^T P x$ can be found by solving the following Lyapunov equations successively

$$A_1^T P_1 + P_1 A_1 = -I, \tag{9.9}$$

$$A_2^T P_2 + P_2 A_2 = -P_1, \tag{9.10}$$

$$\vdots$$

$$A_N^T P_N + P_N A_N = -P_{N-1}, \tag{9.11}$$

$$P = P_N. \tag{9.12}$$

The previous theorems assume that the individual subsystems share a common Lyapunov function. The properties of the admissible switching signals can sometimes be used to prove the asymptotic stability of the switched system even in the absence of the common Lyapunov function.

Theorem 9.3 *Let a switched system be described by* (9.1). *Consider a switching signal* σ *with the switching times* $t_1 < t_2 < \cdots t_k < \cdots t_l < \cdots$ *and assume that the signals are continuous from the right, that is,* $\lim_{t \to t_{k+0}} \sigma(t) = \sigma(t_k)$, $\forall k$.

Assume that all the individual subsystems $\dot{x} = f_i(x)$ *are asymptotically stable, and accordingly they possess their own strictly decreasing Lyapunov function.*

Suppose that there exists a constant $\rho > 0$ *such that for any two switching times* t_k *and* $t_l > t_k$ *for which* $\sigma(t_k) = \sigma(t_l)$ *holds* $V_{\sigma(t_l)}(x(t_{l+1})) - V_{\sigma(t_k)}(x(t_{k+1})) \leq -\rho |x(t_{k+1})|^2$.

Then the switched system (9.1) *is globally asymptotically stable.*

In the case of polyhedral piecewise affine autonomous systems, the stability analysis reduces to LMI type conditions [60]. Consider the system $\dot{x} = A_\sigma x + a_\sigma$, the state space of which is given by the polyhedral partition X. If the state is in the partition X_i, we have $A_\sigma = A_i$ and $a_\sigma = a_i$. Let x be a piecewise continuous trajectory of the system. Assume that the trajectory cannot stay on the boundary of two polyhedras unless it satisfies the differential equation for both and simultaneously.

The index set of the polyhedras is denoted I. Let $I_0 \subseteq I$ be the set of indices for the polyhedras that contain origin and $I_1 \subseteq I$ be the set of indices for cells that do not contain the origin. It is assumed that $i \in I_0$ implies $a_i = 0$.

The polyhedras and their boundaries are defined as in (9.5) and (9.6). Let us introduce the following notations

$$\overline{A}_i = \begin{bmatrix} A_i & a_i \\ 0 & 0 \end{bmatrix}, \qquad \overline{E}_i = [E_i e_i], \qquad \overline{F}_i = [F_i f_i] \quad \text{if } i \in I_1. \qquad (9.13)$$

Theorem 9.4 *Consider the symmetric matrices* T, U_i *and* W_i, *such that* U_i *and* W_i *have nonnegative entries, while* $P_i = F_i^T T F_i$ *for* $i \in I_0$ *and* $\overline{P}_i = \overline{F}_i^T T \overline{F}_i$ *for* $i \in I_1$ *satisfy*

$$\begin{cases} A_i^T P_i + P_i A_i + E_i^T U_i E_i < 0, \\ P_i - E_i^T W_i E_i > 0, \end{cases} \quad i \in I_0, \qquad (9.14)$$

$$\begin{cases} \overline{A}_i^T \overline{P}_i + \overline{P}_i \overline{A}_i + \overline{E}_i^T U_i \overline{E}_i < 0, \\ \overline{P}_i - \overline{E}_i^T W_i \overline{E}_i > 0, \end{cases} \quad i \in I_1. \qquad (9.15)$$

Then every piecewise continuous trajectory $x \in X$ *tends to zero exponentially.*

In solving the inequalities of the theorem above, it is advisable to consider first only the Lyapunov inequality in each region and ignore the positivity condition. Once a solution to this reduced problem has been found, it remains to investigate whether or not the resulting piecewise quadratic function is nonnegative in the entire state space. This can also be done in each region separately.

9.1.2 Modeling, Solution and Stability of Differential Inclusions

Consider nonsmooth dynamic system described by the differential equation

$$\dot{x} = f(t, x), \quad x \in R^n \tag{9.16}$$

the right-hand side of which is a discontinuous function. Assume that the discontinuous function $f : R^n \times R \to R^n$ is measurable and essentially locally bounded.

Let us restrict ourself on differential equations which is discontinuous only on a single hyper surface in R^n: the state space R^n is splitted into two subspaces V_- and V_+ by a hyper-surface V_0 such that $R^n = V_- \cup V_0 \cup V_+$. The hyper-surface V_0 is defined by a scalar indicator function $h(x(t))$. The state $x(t)$ is in V_0 if $h(x(t)) = 0$.

The subspaces V_-, V_+ and the hyper-surface V_0 can be formulated as

$$V_- = \{x \in R^n | h(x(t)) < 0\}, \tag{9.17}$$

$$V_0 = \{x \in R^n | h(x(t)) = 0\}, \tag{9.18}$$

$$V_- = \{x \in R^n | h(x(t)) > 0\}. \tag{9.19}$$

Now consider the following differential equation

$$\dot{x} = f(t, x) = \begin{cases} f_+(t, x), & \text{if } x \in V_+, \\ f_-(t, x), & \text{if } x \in V_-. \end{cases} \tag{9.20}$$

Assume that $f_+(t, x)$ is continuous in $V_+ \cup V_0$, $f_-(t, x)$ is continuous in $V_- \cup V_0$ but $f_+(t, x) \neq f_-(t, x)$ for $h(x(t)) = 0$. Thus, the system (9.20) does not define $f(t, x)$ if $h(x(t)) = 0$.

The natural idea to analyze these systems is to replace the right-hand side by a set of continuous functions denoted by F. This representation is called *differential inclusion* [78, 79]

$$\dot{x} \in F(t, x(t)). \tag{9.21}$$

In order to define the *solution* for these type of systems, the notion of the convex hull has to be introduced.

A set $A \subset R^n$ is *closed* if it contains all of its limit points. Every limit point of a set is the limit of a sequence x_k where all the elements of the sequence are in A.

A set $A \subset R^n$ is *convex* if for each $x, y \in A$ also $qx + (1 - q)y \in A$ where $q \in [0, 1]$.

The *closed convex hull* of a set of points A_P is the smallest closed convex set containing all the elements of A_P.

The closed convex hull of $x, y \in R^n$ is the smallest closed convex set containing x and y, that is, the line segment between x and y

$$\overline{co}\{x, y\} = \{qx + (1 - q)y, q \in [0, 1]\}. \tag{9.22}$$

Based on the definition (9.22), the differential equation with discontinuous right-hand side (9.20) can be extended into a convex differential inclusion (Filippov convex method)

$$\dot{x} \in F(t, x) = \begin{cases} f_+(t, x), & \text{if } x \in V_+, \\ \overline{co}\{f_+, f_-\}, & \text{if } x \in V_0, \\ f_-(t, x), & \text{if } x \in V_-. \end{cases} \tag{9.23}$$

As example consider an autonomous mechanical system with Coulomb friction. Denote v the velocity of the system, m is the mass, F_C is the Coulomb friction coefficient. The value of the friction is not defined for $v = 0$. Using differential inclusions, the system can be modeled as

$$m\dot{v} \in -F_C \, \text{sgn}(v) = \begin{cases} -F_C, & \text{if } v > 0, \\ [-F_C, F_C], & \text{if } v = 0, \\ F_C, & \text{if } v < 0. \end{cases} \tag{9.24}$$

Filippov Solution An absolute continuous function $x(t) : [t_0, t_f] \rightarrow R^n$ is said to be a solution of (9.16) in the sense of Filippov if for almost everywhere $t \in [t_0, t_f]$ it holds $\dot{x} \in F(t, x)$, where $F(t, x)$ is the smallest closed convex set containing all the limit values of the vector function $f(t, x_B)$, where x_B tends to x

$$F(t, x) = \bigcap_{\delta > 0} \bigcap_{\lambda(N)=0} \overline{co}\{f(t, B(x, \delta) \setminus N)\}. \tag{9.25}$$

Here x_B ranges a ball B of center x and any radius $\delta > 0$, with t fixed, except for arbitrary sets (N) of Lebesque measure 0. Almost everywhere means that for all t, except for a set of Lebesque measure 0. λ is the Lebesgue measure over R^n.

Let $x(t)$ be a solution of the differential inclusion (9.21). Denote by S_{x_0} the set of solutions of (9.21) such that $x(t_0) = x_0$.

$x = 0$ is *equilibrium point* if $0 \in F(t, 0)$, $\forall t$.

Stability The differential inclusion (9.21) is stable at the equilibrium point $x = 0$ if for all $\varepsilon > 0$ there exists $\delta > 0$ such that for each initial condition x_0 and each solution $x(t) \in S_{x_0}$ holds

$$\|x_0\| < \delta \quad \Rightarrow \quad \|x(t)\| \leq \varepsilon, \quad \forall t \geq t_0. \tag{9.26}$$

Global Asymptotic Stability The differential inclusion (9.21) is globally asymptotically stable at the equilibrium point $x = 0$ if it is stable and each solution $x(t) \in S_{x_0}$ satisfies: $\lim_{t \to \infty} \|x(t)\| = 0$.

In order to formulate Lyapunov like stability theorems for differential inclusions, the notion of generalized derivate and generalized Clarke gradient has to be introduced [45, 129, 158].

Consider a scalar, continuous, piecewise differentiable function $f : R \rightarrow R$. Assume that the function is not differentiable in some points x, but possesses in each

x left and right-hand side derivatives

$$f'_- = \lim_{x_i \to x-0} \frac{f(x_i) - f(x)}{x_i - x}, \qquad f'_+ = \lim_{x_i \to x+0} \frac{f(x_i) - f(x)}{x_i - x}. \qquad (9.27)$$

The *generalized differential* of f at x is defined as

$$\partial f = \overline{co}\{f'_+, f'_-\}. \qquad (9.28)$$

For example, consider the function $\mathrm{abs}(x) = |x|$. The generalized differential of this function can be calculated as

$$\partial\, \mathrm{abs}(x) = \begin{cases} -1, & \text{if } x < 0, \\ [-1, 1], & \text{if } x = 0, \\ 1, & \text{if } x > 0. \end{cases} \qquad (9.29)$$

The generalized differential is the set of slopes of all lines included in the cone bounded by the left and right tangent lines.

Clarke Generalized Gradient For a locally Lipschitz function $V(t, x) : R^n \times R \to R$, the generalized gradient of V at (t, x) is defined as

$$\partial V = \overline{co}\left\{ \lim_{x_i \to x} \begin{pmatrix} \frac{\partial V(t,x_i)}{\partial x} \\ \frac{\partial V(t,x_i)}{\partial t} \end{pmatrix} \,|\, x_i \notin N_V \right\}. \qquad (9.30)$$

N_V is a set of measure 0 where the gradient of V is not defined.

Lemma 9.1 (Chain rule) *Let $x(t)$ be a Filippov solution of (9.16) defined in $[t_0, t_f]$ and $V(t, x) : R^n \times R \to R$ is a regular function. Then for $t \in [t_0, t_f]$*

$$\frac{dV}{dt} \in \bigcap_{\xi \in \partial V(t,x)} \xi^T \begin{pmatrix} F \\ 1 \end{pmatrix}. \qquad (9.31)$$

F is defined as in (9.25).

The $V(t, x) : R^n \times R \to R$ regular function is called *Lyapunov function candidate* if $V(t, 0) = 0$, $\forall t \geq 0$, and there exists a positive continuous nondecreasing function $\alpha : [0, \infty) \to [0, \infty)$ such that $V(t, x) > \alpha(\|x\|)$, $\forall x \neq 0$, where $\alpha(0) = 0$ and $\alpha(\|x\|) \to \infty$ as $\|x\| \to \infty$.

Note that a Lyapunov function candidate need not to be differentiable.

Theorem 9.5 (Lyapunov stability) *Let $x(t)$ be a Filippov solution defined for $t \geq 0$ of the system (9.16) with the equilibrium point $x = 0$. Let V be a Lyapunov function candidate. If*

$$\frac{dV}{dt} \leq 0, \quad \text{if } t \geq 0 \text{ almost everywhere} \qquad (9.32)$$

then the origin of (9.16) is stable.

Theorem 9.6 (Lyapunov global asymptotic stability) *Let $x(t)$ be a Filippov solution defined for $t \geq 0$ of the system (9.16) with the equilibrium point $x = 0$. Let V be a Lyapunov function candidate. Let $\Phi(t, V) : (0, \infty) \times [0, \infty) \to R$ be a continuous function with $\inf_{t \in [0, \infty)} \Phi(t, V) > 0$, $\forall V > 0$ and $\lim_{t \to \infty} \Phi(t, V)$ exists, $\forall V > 0$. If*

$$\frac{dV}{dt} \leq -\Phi(t, V), \quad \text{if } t \geq 0 \text{ almost everywhere} \tag{9.33}$$

then the origin of (9.16) is globally asymptotically stable.

9.2 Static Friction Models

Friction is a result of complex interactions between the surface and the near-surface regions of two moving bodies in contact. Due to the interaction of moving surfaces, a highly nonlinear and discontinuous friction force arises.

In many industrial mechatronic systems, the surfaces in contact are lubricated with oil or grease, which also contributes to nonlinear behavior of friction. Many models were developed to explain the friction phenomenon. These models are based on experimental results rather than analytical deductions. The classic model of friction—friction force that is proportional to normal force generated by load (N) and independent of contact area—was known to Leonardo Da Vinci, but remained hidden for centuries. His model was rediscovered after the enunciation of Newton's laws and modern conception of force by Amontons and written in a well-known form: $F_f = \mu N$, where μ is the coefficient of friction, which is an empirical property of the contacting materials.

The static friction models generally describe the friction force in function of velocity [7], see Fig. 9.1.

Coulomb Friction Model This classical friction model was developed by Coulomb, who discovered that the friction force depends on sign of velocity. Hence, the friction force can be written in the following form

$$F_f(v) = F_C \, \text{sgn}(v), \tag{9.34}$$

where F_C denotes the Coulomb friction coefficient that depends on the load (it is proportional with the normal force generated by load) and v denotes the relative tangential velocity between the two surfaces in contact. This model can be used in the case of dry contacts.

Coulomb + Viscous Friction Model The viscous term is the friction component that is proportional to velocity. This term has a dominant influence when the contact of the bodies in motion are lubricated with oil or grease (hydrodynamic lubrication). It was introduced by Reynolds who studied the friction occurring in fluids

$$F_f(v) = F_C \, \text{sgn}(v) + F_V v, \tag{9.35}$$

where F_V denotes the viscous friction coefficient.

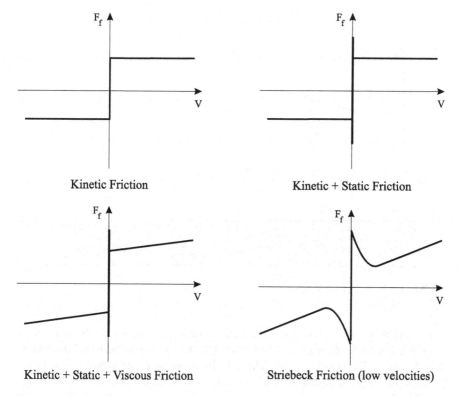

Kinetic Friction   Kinetic + Static Friction

Kinetic + Static + Viscous Friction   Striebeck Friction (low velocities)

Fig. 9.1 Static friction models

The static friction coefficient (F_S), introduced initially by Artur Morin, represents the force necessary to initiate motion from rest and in most of the cases its value is grater than the Coulomb friction coefficient (F_C). The motion does not starts until the tangent force (τ) that generates the motion does not reach the level of the static friction force (sticking). This phenomenon can be incorporated into static friction models as

$$F_f(v, \tau) = \begin{cases} \min(|\tau|, F_S)\, \mathrm{sgn}(\tau), & \text{if } v = 0, \\ F_f(v), & \text{otherwise.} \end{cases} \tag{9.36}$$

$F_f(v)$ represents the velocity dependent part of the friction that may be given by (9.34) or (9.35).

Striebeck Friction Models Tribological experiments showed that in the case of lubricated contacts the simple static + Coulomb + viscous model cannot explain some phenomena in low velocity regime, such as the Striebeck effect. This friction phenomenon arises from the use of fluid lubrication and gives rise to decreasing friction with increasing velocities.

To describe this low velocity friction phenomenon, four regimes of lubrications can be distinguished (see Fig. 9.2 [7]). *Static Friction*: (*I.*) the junctions deform elas-

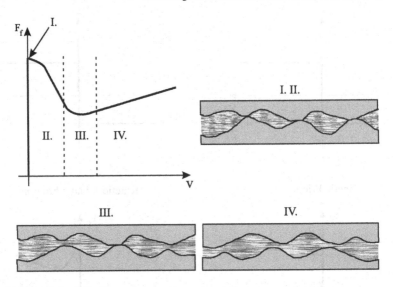

Fig. 9.2 The Stribeck friction regimes

tically and there is no excursion until the control force does not reach the level of static friction force. *Boundary Lubrication*: (*II.*) this is also solid to solid contact, the lubrication film is not yet built. The velocity is not adequate to build a solid film between the surfaces. A sliding of friction force occurs in this domain of low velocities. The friction force decreases with increasing velocity but generally is assumed that friction in boundary lubrication is higher than for fluid lubrication (regimes three and four). *Partial Fluid Lubrication*: (*III.*) the lubricant is drawn into the contact area through motion, either by sliding or rolling. The greater the viscosity or motion velocity, the thicker the fluid film will be. Until the fluid film is not thicker than the height of aspirates in the contact regime, some solid-to-solid contacts will also influence the motion. *Full Fluid Lubrication*: (*IV.*) When the lubricant film is sufficiently thick, separation is complete and the load is fully supported by fluids. The viscous term dominates the friction phenomenon, the solid-to-solid contact is eliminated and the friction force can be considered as proportional with the velocity.

From these domains results a highly nonlinear behavior of the friction force. Near zero velocities the friction force decreases in function of velocity and at higher velocities the viscous term will be dominant and the friction force increases with the velocity. Moreover, it also depends on the sign of velocity with an abrupt change when the velocity passes through zero.

For the moment, no theoretically founded model of the Stribeck effect is available. Several empirical models were introduced to explain the Stribeck phenomenon: Tustin model, exponential in velocity ($e^{-|v|/v_S}$), Gaussian model ($e^{-(v/v_S)^2}$) and Lorentzian model ($1/(1 + (v/v_S)^2)$). The constant value v_S is the Stribeck velocity which describes the shape of the Stribeck curve. These terms can be introduced in the models previously presented to obtain a more precise description of the frictional behavior. The different models are enumerated below

$$\text{Tustin model:} \quad F_f(v) = \left(F_C + (F_S - F_C)e^{-|v|/v_S}\right)\text{sgn}(v) + F_V v, \quad (9.37a)$$

$$\text{Gaussian model:} \quad F_f(v) = \left(F_C + (F_S - F_C)e^{-(v/v_S)^2}\right)\text{sgn}(v) + F_V v, \quad (9.37b)$$

$$\text{Lorentzian model:} \quad F_f(v) = \left(F_C + (F_S - F_C)\frac{1}{1 + (v/v_S)^2}\right)\text{sgn}(v) + F_V v. \tag{9.37c}$$

As other static friction models, the Stribeck models have to be extended to describe the fact, that the motion does not starts until the force that generates the motion does not reach the level of the static friction (see (9.36)). After that the motion starts the moving object slips. Hence, for example, using (9.37a), a static stick–slip friction model can be formulated as

$$F_f(v, \tau) = \begin{cases} \min(|\tau|, F_S)\,\text{sgn}(\tau), & \text{if } v = 0, \\ (F_C + (F_S - F_C)e^{-|v|/v_S})\,\text{sgn}(v) + F_V v, & \text{otherwise.} \end{cases} \tag{9.38}$$

For simulation purposes (9.38) can hardly be used since the stick state appears only in the moment when the velocity is exactly zero. For this reason, to describe the stick-slip phenomenon in simulation environments, an approximate model may be applied

$$F_f(v, \tau) = F_{\text{Stick}}(v, \tau)\mu(v) + F_{\text{Slip}}(v)\left(1 - \mu(v)\right), \tag{9.39}$$

where the function $\mu(v)$ is defined as

$$\mu(v) = \begin{cases} 0, & \text{if } |v| \leq v_\varepsilon, \\ 1, & \text{otherwise.} \end{cases} \tag{9.40}$$

The value of $v_\varepsilon > 0$ is a small limit value for the velocity that has to be set in function of simulation conditions (simulation step size, tolerance). F_{Slip} is given by (9.37a); $F_{\text{Stick}}(v, \tau)$ is defined as

$$F_{\text{Stick}}(v, \tau) = \begin{cases} F_S\,\text{sgn}(\tau), & \text{if } |\tau| \geq F_S, \\ \tau, & \text{otherwise.} \end{cases} \tag{9.41}$$

9.2.1 Stick–Slip Motion

The motion in the presence of stick–slip friction can be investigated by considering the mechanical system in Fig. 9.3. An object with mass m is connected to a spring with constant k. The spring is pulled with constant speed v_{ext}. Denote with l the elongation of the spring and with v the speed of the moving object. Let the friction between the object and the base surface be described with Coulomb + static friction model ((9.36) with (9.34)).

Fig. 9.3 Stick–slip
experiment

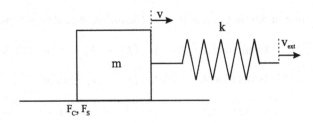

Fig. 9.4 Stick–slip motion

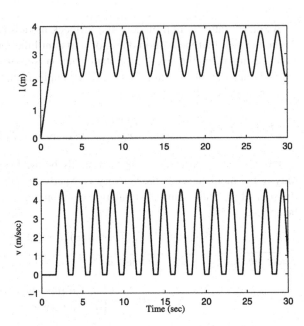

If $v = 0$, the system remains in stick mode until the force generated by the spring does not reach the level of static friction, namely $F_{spring} \leq k l_S$ where $l_S = F_S / k$. When the force generated by the spring exceeds the level of static friction, the object starts moving. The slip and stick conditions can be formulated as

$$\text{Slip mode:} \quad v \neq 0 \quad \text{or} \quad (v = 0 \text{ and } |l| > l_S),$$

$$\text{Stick mode:} \quad \text{otherwise.} \tag{9.42}$$

The dynamics of the motion can be described as a hybrid model. In stick mode, the equations of motion are given by

$$\dot{l} = v_{ext},$$

$$\dot{v} = 0. \tag{9.43}$$

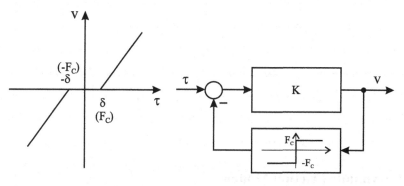

Fig. 9.5 Dead zone

In slip mode, the dynamics reads as

$$\dot{l} = v - v_{\text{ext}},$$
$$m\dot{v} = kl - F_C \, \text{sgn}(v). \tag{9.44}$$

The position and velocity profile of the stick-slip motion are presented in Fig. 9.4. The following system parameters were considered: $m = 0.1$ kg, $k = 1$ N/m, $v_{\text{ext}} = 2$ m/s, $F_C = 3$ N, $F_S = 3.5$ N. The responses of the system show that the stick-slip phenomenon can lead to hunting (limit cycling). The object stops periodically and the motion restarts when the slip condition is satisfied.

9.2.2 Friction-Induced Dead Zone

Dead zone is a static transfer which is characterized by the property that for a range of input (τ) the output (v) is zero [142]. Once the magnitude of input signal is greater than the breakpoint (δ) the output increases linearly in function of the input. As it is shown in Fig. 9.5, it can be described with a piecewise linear model

$$D(\tau) = \begin{cases} K(\tau - \delta), & \text{if } \tau > \delta, \\ 0, & \text{if } |\tau| \leq \delta, \\ K(\tau + \delta), & \text{if } \tau < -\delta, \end{cases} \tag{9.45}$$

where K denote the slope of the linearly increasing part.

Consider an electromechanical drive in which the dynamics of the transfer from the input force (τ) to the velocity (v) can be described with a stable dynamics with amplification K and the friction in the system is described by the Coulomb friction model (9.34) as it is shown in Fig. 9.5. The steady-state characteristic of this system is given by dead zone. If the input force τ is smaller than the value of the Coulomb friction coefficient F_C there is no motion, the output velocity is zero, hence $\delta = F_C$. When the value of τ becomes greater than F_C, the motion starts, in steady-state the amplification is given by K.

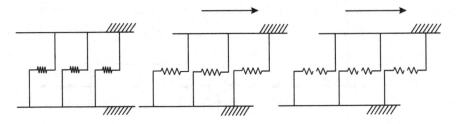

Fig. 9.6 Dahl model for frictional surfaces

9.3 Dynamic Friction Models

In order to give a more precise description for friction phenomenon in the low velocity regime and to assure a smooth transition at velocity reversal, dynamic terms can also be introduced in friction models [10, 26]. In the static models, it is considered that there is no motion until the applied control force for the machine does not reach the level of static friction. The experiments made for small displacements (µm domain) showed that even in this domain there is a slip of the machine, named *presliding displacement*. In this domain, the friction behaves like a spring. When force is applied, the asperities of the contact will deform, but recover when the force is removed. That is, the force is a linear function of the displacement, up to a critical force level at which the brake away occurs (see Fig. 9.6). When the force exceeds the level of static friction, the real slip occurs which phenomenon is described by the previously introduced Stribeck model. Moreover, it was observed that in this presliding domain at the change of the direction of motion (change in the sign of velocity) a *hysteresis loop* also appears in the motion (when the friction force is represented as a function of displacement).

A general form of a dynamic friction model can be written as

$$\dot{z} = z(v, z, x),$$
$$F_f = F_f(v, z, x),$$
$$(9.46)$$

where x denotes the relative displacement of the surfaces in contact and z denotes the internal state of the friction the dynamics of which is described by a nonlinear differential equation.

9.3.1 Classic Dynamic Friction Models

Dahl Model To explain the presliding behavior for solid to solid contacts Dahl proposed, the following dynamic model which captures the variation of friction force in function of displacement

$$\frac{dF_f}{dx} = \sigma_0 \left(1 - \frac{F_f}{F_C} \operatorname{sgn}(v) \right)^{\alpha},$$
$$(9.47)$$

Fig. 9.7 The bristles in
LuGre friction model

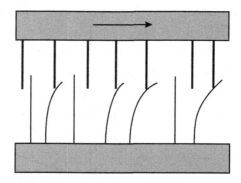

where F_C denotes the Coulomb friction term, σ_0 is the contact stiffness and α is the model parameter that should be chosen empirically in order to fit experimental data. Differentiating (9.47) with respect to time and taking $\alpha = 1$, we obtain

$$\frac{dF_f}{dt} = \frac{dF_f}{dx}\frac{dx}{dt} = \sigma_0\left(1 - \frac{F_f}{F_C}\,\mathrm{sgn}(v)\right)v. \tag{9.48}$$

By introducing the state variable z such that $F_f = \sigma_0 z$, we obtain

$$\frac{dz}{dt} = v - \frac{\sigma_0|v|}{F_C}z,$$
$$F_f = \sigma_0 z. \tag{9.49}$$

In steady-state ($dz/dt = 0$), the Dahl model is identical with the Coulomb friction model (9.34).

The presented model is developed for dry contacts and it does not capture the Stribeck phenomenon which occurs in the case of lubricated contacts. However, it explains the presliding displacement and hysteresis phenomena near zero velocities.

LuGre Model As it was mentioned before, the dynamic model introduced by Dahl neither explains the Stribeck phenomenon nor incorporates the viscous friction term. The LuGre model is the generalization of the Dahl model that can explain the presliding behavior and the slipping behavior simultaneously.

The contact of two rigid bodies is modeled as a set of elastic bristles (see Fig. 9.7). When a tangential force is applied, the bristles deflect like springs which gives rise to friction force. If the force is sufficiently large, some of the bristles deflect so much that they will slip. This phenomenon is highly random due to irregular forms of the surfaces. Let us denote the average deflection of the bristles with z and consider that it is modeled by

$$\frac{dz}{dt} = v - \sigma_0\frac{|v|}{g(v)}z, \tag{9.50a}$$
$$g(v) = \left(F_C + (F_S - F_C)e^{-|v|/v_S}\right). \tag{9.50b}$$

Fig. 9.8 Presliding
displacement

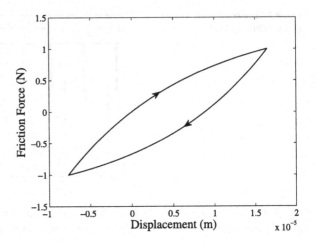

$g(v)$ is a positive continuous function which is meant to describe the Stribeck effect. It was chosen based on the Tustin model (9.37a). σ_0 is a positive constant parameter representing the stiffness.

The friction force generated from the bending of the bristles and from the viscous lubrication is described as

$$F_f = \sigma_0 z + \sigma_1 \frac{dz}{dt} + F_V v, \tag{9.51}$$

where σ_1 is a damping coefficient.

The LuGre model is defined by the relations (9.50a) and (9.51).

Figure 9.8 shows the presliding phenomenon predicted by the LuGre model for the case when the input torque of the mechanical systems is smaller than the static friction coefficient. Figure 9.9 shows the hysteresis-like behavior during direction change predicted by the LuGre model for the case when the input torque of the mechanical system is greater than the static friction coefficient. The simulation results were generated using the following friction parameters: $F_C = 1$ N, $F_S = 1.5$ N, $F_V = 1$ N s/m, $v_S = 0.01$ m/s, $\sigma_0 = 10^5$ N/m, $\sigma_1 = \sqrt{10^5}$ N s/m.

In steady-state ($dz/dt = 0$), the LuGre model is equivalent to the Tustin model (9.37a).

Beside the slip motion and the presliding displacement, this model can also explain another low velocity friction phenomenon, the *friction lag*, a hysteresis in the relation between friction and velocity during unidirectional motion. The friction force is lower for decreasing velocities than for increasing velocities. Moreover, the hysteresis loop becomes wider at higher rates of velocity changes.

In the special case of $g(v) = F_C$ and $\sigma_1 = F_V = 0$, the LuGre model reduces to the Dahl model.

The model has two important properties: state boundedness and input to state dissipativity [10, 26]. The first property can be exploited during robust control design. The verification of dissipativity is important because the nature of friction: it

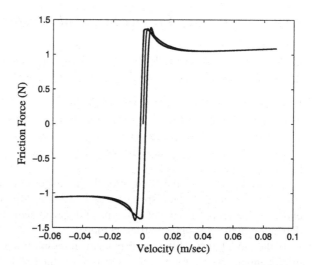

Fig. 9.9 Hysteresis-like behavior during direction change

always dissipates energy during motion if the mechanical system moves. To analyze this fundamental properties of the model, consider the following storage function

$$V = \frac{1}{2}z^2(t).$$ (9.52)

Assume that $z(0)$ is bounded as follows: $|z(0)| \leq F_S/\sigma_0$. The time derivative of the storage function (9.52) using (9.50a) is given by

$$\dot{V} = z\dot{z} = z\left(v - \sigma_0\frac{|v|}{g(v)}z\right) = -|v||z|\left(\sigma_0\frac{|z|}{g(v)} - \text{sgn}(v)\,\text{sgn}(z)\right).$$ (9.53)

The nonlinear function $g(v)$ is bounded ($F_C \leq g(v) \leq F_S$) if it is defined as in (9.50b). $\text{sgn}(v)\,\text{sgn}(z) = -1$ implies $\dot{V} < 0$ for $v, z \neq 0$. $\text{sgn}(v)\,\text{sgn}(z) = 1$ and $|z| \geq F_S/\sigma_0$ also implies $\dot{V} < 0$ for $v, z \neq 0$. Hence, $|z| \leq F_S/\sigma_0$, the internal state of the model is bounded.

By definition, the mapping from v to z is passive if the following inequality is satisfied

$$\int_0^t z(\tau)v(\tau)\,d\tau \geq V(t) - V(0).$$ (9.54)

By expressing v from (9.50a), we have

$$zv = z\dot{z} - \sigma_0\frac{|v|}{g(v)}z^2 \geq z\dot{z}.$$ (9.55)

Hence, it yields that the mapping from v to z is passive

$$\int_0^t z(\tau)v(\tau)\,d\tau \geq \int_0^t z(\tau)\dot{z}(\tau)\,d\tau \geq V(t) - V(0).$$ (9.56)

9.3.2 Modified and Advanced Dynamic Friction Models

Modified versions of the LuGre model were also proposed to increase its accuracy mostly in the presliding regime and during the presliding–sliding transition.

Elastoplastic Model The LuGre model shows a nonphysical drift for very small displacements (near 0.1 μm) which results from modeling of presliding motion as a combination of elastic and plastic displacement. To avoid this undesired phenomenon, the elastoplastic model clearly distinguishes the elastic and plastic motion [37].

For the LuGre (and the Dahl) model, the displacement can be modeled by the sum of elastic and plastic displacements $x = z + w$, where $\dot{x} = v$, z represents the elastic and w represents the plastic type of displacement. This relation appears after derivation in the elastoplastic friction model: $\dot{z} = v - \dot{w}$.

In the model, three different displacement regimes may be considered: I. elastic displacement ($v = \dot{z}$) for $z \leq z_{ba}$, where z_{ba} is the break away displacement under which the model behaves purely elastic. II. mixed elastic and plastic displacement ($v = \dot{z} + \dot{w}$) if $z_{ba} < z < |g(v)|/\sigma_0$ and III. plastic displacement ($v = \dot{w}$) for $z \geq |g(v)|/\sigma_0$. Here, $g(v)$ and σ_0 are defined as in the LuGre model. The parameter z_{ba} has to be chosen as $0 \leq z_{ba} < |g(v)|$.

This modification can be applied by defining a switching function $\alpha(v, z)$ which is introduced in the LuGre model as follows

$$\frac{dz}{dt} = v - \sigma_0 \frac{|v|}{g(v)} z,$$

$$F_f = \sigma_0 z + \alpha(z, v)\sigma_1 \frac{dz}{dt} + F_V v. \tag{9.57}$$

The function α is defined as

$$\alpha(z, v) = \begin{cases} 0, & \text{if } (|z| \leq z_{ba} \text{ and } \text{sgn}(v) = \text{sgn}(z)) \text{ or } \text{sgn}(v) \neq \text{sgn}(z), \\ \alpha_m(z), & \text{if } z_{ba} < |z| < \frac{|g(v)|}{\sigma_0} \text{ and } \text{sgn}(v) = \text{sgn}(z), \\ 1, & \text{if } |z| \geq \frac{|g(v)|}{\sigma_0} \text{ and } \text{sgn}(v) = \text{sgn}(z), \end{cases} \tag{9.58}$$

where $0 < \alpha_m(z) < 1$ is a continuous function of z that describes the mixed elastic and plastic behavior. For positive z values, it has to be chosen as an increasing function with $\alpha_m(z_{ba}) = 0$ and $\alpha_m(\frac{|g(v)|}{\sigma_0}) = 1$. For negative z values, it has to be a decreasing function with $\alpha_m(-z_{ba}) = 0$ and $\alpha_m(-\frac{|g(v)|}{\sigma_0}) = 1$.

For $|z| \geq \frac{|g(v)|}{\sigma_0}$ and $\text{sgn}(v) = \text{sgn}(z)$, the elastoplastic model is equivalent to the LuGre model.

Leuven Friction with Maxwell-slip Elements In the LuGre model, the switching from the presliding to sliding regime occurs when the input force reaches the level of the static friction coefficient. For more precise modeling of the presliding–sliding transition, the Leuven model modifies the original LuGre model by introducing in

Fig. 9.10 Maxwell-slip elements

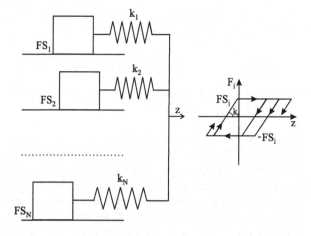

it a hysteresis function (F_h) with nonlocal behavior [72, 140]. A simplified form of this model is given by

$$\frac{dz}{dt} = v\left(1 - \frac{F_h(z)}{g(v)\,\mathrm{sgn}(v)}\right),$$

$$F_f = F_h(z) + \sigma_1\frac{dz}{dt} + F_V v, \tag{9.59}$$

where $g(v)$ is the nonlinear function from the LuGre model used to describe the Stribeck phenomenon. Note that to avoid the singularity of the model the sgn function has to take the value 1 for $v >= 0$ while for $v < 0$ yields $\mathrm{sgn}(v) = -1$.

In steady-state ($\frac{dz}{dt} = 0$) we obtain $F_h(z) = g(v)\,\mathrm{sgn}(v)$, hence the Tustin model (9.37a) is re-obtained.

The hysteresis force can be defined as the sum of forces generated by the parallel connection of N Maxwell-slip elements (see Fig. 9.10)

$$F_h(z) = \sum_{i=1}^{N} F_i(z). \tag{9.60}$$

The behavior of each Maxwell-slip element is described by two state equations. If the elementary model sticks, it behaves like a spring, thus the elementary friction force can be modeled as

$$\text{Stick phase:} \quad \frac{dF_i(z)}{dt} = k_i\frac{dz}{dt},$$

$$\frac{dz}{dt} = v, \tag{9.61}$$

where k_i is the stiffness of the elementary model. The elementary model will slip if the friction force of each element reaches the maximum value of the force FS_i, that it can sustain. Beyond this point, the elementary friction force equals the maximum force FS_i and its value remains constant

$$\text{Slip phase:} \quad \frac{dF_i(z)}{dt} = 0. \tag{9.62}$$

When the motion is reversed, the elementary model will stick again until the force reaches the value $-FS_i$.

Generalized Maxwell Slip Model (GMS) This model also uses N Maxwell elements in parallel connection as in the case of the Leuven model. The Leuven model tries to fit the hysteresis property into the steady-state sliding property and the friction between the elements and the base surfaces is described with Coulomb-like friction model with hysteresis. The GMS model replaces the Coulomb law at slip phase by a more complex friction model, namely with a single state friction model [3]. Let v be the velocity input of the model and z_i $(1 \le i \le N)$ be the ith element of the state vector z. The motion of the elements are described with same dynamic equations but with different parameters.

In stick phase, the dynamics is given by

$$\frac{dz_i}{dt} = v. \tag{9.63}$$

The element remains sticked until the value of z_i reaches $g_i(v)$. Similarly to the LuGre model, $g_i(v)$ is a nonlinear function that describes the Stribeck phenomenon, see, for example, (9.50b).

In slip phase, the dynamics of the model reads as

$$\frac{dz_i}{dt} = \text{sgn}(v)G_i\left(1 - \frac{z_i}{g_i(v)}\right). \tag{9.64}$$

The element remains in slip phase until the velocity goes through zero. The gain parameter G_i determines how fast z_i converges to $g_i(v)$. It can be shown that if $g_i(v)$ is replaced by FS_i the GMS model will reduce to Leuven friction with Maxwell-slip elements.

The friction force is given by the sum of the N elementary state models plus a viscous term

$$F_f = \sum_{i=1}^{N} \left(\sigma_{0i} z_i + \sigma_{1i} \frac{dz_i}{dt}\right) + F_V v. \tag{9.65}$$

The parameters σ_{0i} and σ_{1i} have the same meaning as in the case of the LuGre model.

9.4 Piecewise Linearly Parameterized Friction Model

In order to apply the well-known identification and adaptive control methods for friction compensation and parameter identification it is desirable that the friction

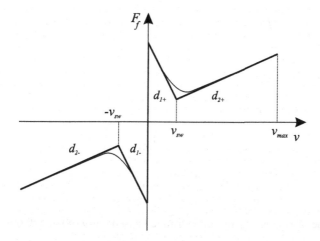

Fig. 9.11 Linearization of Stribeck friction

model could be written in a linearly parameterized form, namely as a scalar product of an unknown parameter vector (θ_f) and a known regressor vector (ξ_f) that depends on velocity $(F_f = \theta_f^T \xi_f)$.

The piecewise linearly parameterized model was developed based on the Tustin model (9.37a), see also [91]. For simplicity, only the positive velocity domain is considered, but similar study can be made for negative velocities. Assume that the mechanical system moves in the $(0, v_{max}]$ velocity domain.

Consider a linear approximation for the exponential curve represented by two lines: d_{1+} which crosses through the $(0, F_f(0))$ point and is tangent to the curve and d_{2+} which passes through the $(v_{max}, F_f(v_{max}))$ point and is tangential to the curve (see Fig. 9.11). These two lines intersect at v_{sw} velocity. In domain $(0, v_{sw}]$, the straight line section d_{1+} can be used to approximate the velocity–friction curve and d_{2+} is used in domain $(v_{sw}, v_{max}]$. The maximum approximation error occurs at the velocity v_{sw} for both linearizations.

Considering only the positive part of the friction model (9.37a), that is, $v > 0$, the obtained equations for d_{1+} and d_{2+}, using Taylor expansions, are

$$d_{1+}: \quad F_{fL1+}(v) = F_S + \left.\frac{\partial F_f(v)}{\partial v}\right|_{v=0} v = F_S + \left(F_V - \frac{F_S - F_C}{v_S}\right)v, \quad (9.66)$$

$$d_{2+}: \quad F_{fL2+}(v) = F_f(v_{max}) + \frac{\partial F_f(v_{max})}{\partial v}(v - v_{max})$$

$$= F_f(v_{max}) + \left(F_V - \frac{F_S - F_C}{v_S}e^{-v_{max}/v_S}\right)(v - v_{max}). \quad (9.67)$$

Thus, the linearization of the exponential friction model with bounded error can be described by two lines in the $(0, v_{max}]$ velocity domain

$$d_{1+}: \quad F_{fL1+}(v) = a_{1+} + b_{1+}v, \quad \text{if } 0 < v \le v_{sw}, \quad (9.68)$$

$$d_{2+}: \quad F_{fL2+}(v) = a_{2+} + b_{2+}v, \quad \text{if } v_{sw} < v \le v_{max}, \qquad (9.69)$$

where

$$a_{1+} = F_S, \qquad (9.70)$$

$$b_{1+} = F_V - \frac{F_S - F_C}{v_S}, \qquad (9.71)$$

$$a_{2+} = F_f(v_{max}) - \left(F_V - \frac{F_S - F_C}{v_S}e^{-v_{max}/v_S}\right)v_{max}, \qquad (9.72)$$

$$b_{2+} = F_V - \frac{F_S - F_C}{v_S}e^{-v_{max}/v_S}. \qquad (9.73)$$

The value of v_{sw} can easily be determined based on (9.68) and (9.69)

$$v_{sw} = \frac{a_{1+} - a_{2+}}{b_{2+} - b_{1+}}. \qquad (9.74)$$

A similar study can be made for negative velocities. Based on linearization, the friction can be modeled as a set of four segments

$$F_{fL}(v) = \begin{cases} a_{1+} + b_{1+}v, & \text{if } v \in (0, v_{sw}], \\ a_{2+} + b_{2+}v, & \text{if } v \in (v_{sw}, v_{max}], \\ a_{1-} + b_{1-}v, & \text{if } v \in [-v_{sw-}, 0), \\ a_{2-} + b_{2-}v, & \text{if } v \in [-v_{max}, -v_{sw-}). \end{cases} \qquad (9.75)$$

9.4.1 Parameter Equivalence with the Tustin Model

There are some relations between the parameters of the original Tustin friction model (9.37a) and the model (9.75) presented above. The parameter a_{1+} represents the static friction term, $a_{1+} = F_S$. The parameter b_{1+}, given by (9.71), is strongly connected to the Stribeck effect, it gives the slope of the Stribeck friction at low velocities.

For high velocities, consider that $v_{max} \to \infty$. It yields

$$a_{2+} = \lim_{v_{max}\to\infty}\left(F_f(v_{max}) - \left(F_V - \frac{F_S - F_C}{v_S}e^{-v_{max}/v_S}\right)v_{max}\right)$$

$$= \lim_{v_{max}\to\infty}\left(F_C + (F_S - F_C)e^{-v_{max}/v_S} + F_V v_{max} - F_V v_{max}\right.$$

$$\left. + (F_S - F_C)\frac{v_{max}/v_S}{e^{v_{max}/v_S}}\right) = F_C, \qquad (9.76)$$

$$b_{2+} = \lim_{v_{max}\to\infty}\left(F_V - \frac{F_S - F_C}{v_S}e^{-v_{max}/v_S}\right) = F_V. \qquad (9.77)$$

Hence, the parameter a_{2+} represents the Coulomb friction parameter, $a_{2+} = F_C$ and b_{2+} represents the viscous friction parameter, $b_{2+} = F_V$.

By (9.74), if $v_{max} \to \infty$ then the value of switching velocity is equal to the Stribeck velocity

$$v_{sw} = \frac{F_S - F_C}{F_V - F_V - (F_S - F_C)/v_S} = v_S. \tag{9.78}$$

9.4.2 Modeling Errors

The Tustin friction model (9.37a) is generally accepted as a reliable static friction model. The maximum deviation of the introduced piecewise linear model (9.75) from (9.37a) occurs at Stribeck velocities, see Fig. 9.11. In the positive velocity regime, the maximum modeling error can be calculated as

$$F_{\Delta max} = F_f(v_S) - F_{fL}(v_S)$$
$$= \left(F_C + (F_S - F_C)e^{-v_S/v_S} + F_V v_S\right) - (a_{2+} + b_{2+} v_S). \tag{9.79}$$

The maximum modeling error is given by

$$F_{\Delta max} = \frac{F_S - F_C}{e}. \tag{9.80}$$

The additive modeling error $(F_\Delta(v))$ is positive and bounded in the positive velocity regime. Similarly, for $v < 0$, $F_\Delta(v)$ is negative and bounded. The upper bound of the modeling error is given by (9.80), $|F_\Delta(v)| \le F_{\Delta max}$.

9.4.3 Incorporating the Dynamic Effects

To take into consideration the dynamic effect during piecewise linear modeling, consider the dynamic LuGre friction model given by (9.50a) and (9.51). Denote z_{ss} the steady-state value of the internal state (z) in the LuGre model. It can be expressed as: $z_{ss} = g(v) \, sgn(v)/\sigma_0$. From (9.50a) and (9.51), yields [110]

$$F_f = \sigma_0 z_{ss} + \sigma_0(z - z_{ss}) + \sigma_1 \frac{dz}{dt} + F_V v$$
$$= g(v) \, sgn(v) + F_V v + F_{fD}(v, z), \tag{9.81}$$

$$F_{fD}(v, z) = \sigma_0(z - z_{ss}) + \left(\sigma_1 \, sgn(v) - \frac{\sigma_0 \sigma_1 z}{g(v)}\right)|v|. \tag{9.82}$$

The term $g(v) \, sgn(v) + F_V v$ represents the static part of the friction model and the rest of the expression describes the dynamic behavior of the friction.

Fig. 9.12 Backlash model

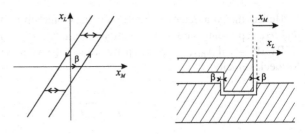

The dynamic part of the model is bounded by velocity dependent upper-bound. Since z, z_{ss} and $g(v) > 0$ are bounded, from the expression of F_{fD} defined in (9.82) yields that there exist two positive finite constants a_D, b_D such that

$$\left| F_{fD}(v, z) \right| < a_D + b_D |v|. \tag{9.83}$$

The friction can be modeled as a sum of a static friction model and a dynamic term. The static term can be written in a linearly parameterized form. The absolute value of the dynamic term is bounded. Its bound can be written in linearly parameterized form.

9.5 Backlash in Mechanical Systems

The backlash appears mainly in gear transmissions and similar mechanical couplings in which the moving parts may temporarily loose the direct contact [107, 142]. It can be described as a nonlinear element with memory which characteristic is shown in Fig. 9.12.

Its model can be explained by considering the mechanical system in Fig. 9.12 in which the L shaped massless object drives the U shaped massless object. The backlash gap size is given by the contact gap between the objects (β). Denote x_M the position of the driving object and x_L the position of the driven object. Let the initial positions of the objects be $x_M = 0$ and $x_L = \beta$, respectively. If the driving object moves to the right, then the driven object will not move until the driving object reaches it. After that, the objects move together. If the driving object stops and changes its direction, then the driven object remains motionless until the contact will not be established again. Hence, in the upward part of the backlash characteristic both x_M and x_L increase and $x_L = x_M - \beta$. At the downward side of the characteristic both x_L and x_M decrease. In the inner segment of the characteristic, x_L remains constant.

Hence, during the motion of mechanical systems with backlash two modes can be distinguished. In the *contact mode* (CM) the driving and driven parts move together. During the *backlash mode* (BM), only the driving part moves, there is no direct contact between the two parts. The backlash mode condition can be formulated by

Fig. 9.13 Backlash response
to sawtooth input

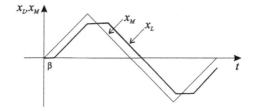

taking into consideration the velocities of both sides

$$\text{(BM):} \quad \text{if } (\dot{x}_M > 0 \text{ and } x_L = x_M - \beta) \text{ or } (\dot{x}_M < 0 \text{ and } x_L = x_M - \beta),$$
$$\text{(CM):} \quad \text{otherwise.} \tag{9.84}$$

Using the conditions above, a compact description of backlash is

$$\dot{x}_L = \begin{cases} \dot{x}_M, & \text{if (CM),} \\ 0, & \text{if (BM).} \end{cases} \tag{9.85}$$

As an example for the influence of backlash on the motion, consider that x_M is a sawtooth signal, as it is shown in Fig. 9.13. It can be seen that there is a lag of the driven side position (x_L) with respect to the driving side and during direction change the peak of the position signal is distorted.

The model above is the so called *friction driven* backlash model, since the driven part does not move during backlash mode. When the backlash gap is not closed, the driven part retains its position as if it kept in place by friction. However, in the model, no knowledge of friction is required.

Another approach for backlash description is the *inertia driven* backlash model where during backlash mode the driven part could move with constant velocity, the friction in the backlash gap is assumed to be zero

$$\begin{cases} \dot{x}_L = \dot{x}_M, & \text{if (CM),} \\ \ddot{x}_L = 0, & \text{if (BM).} \end{cases} \tag{9.86}$$

These models predict a discontinuous velocity transition from backlash mode to contact mode. For example, in the friction driven model, there will be an instantaneous jump in the driven part velocity from zero to the drive side velocity. It is why for simulation purposes the models above cannot be applied directly.

To develop an approximate backlash model for simulation, consider that the load side is driven by a control force τ. Denote with J_M the mass of the driving part and with J_L the mass of the driven part. With the notations, the dynamics of the system with friction driven backlash model is given by

$$\begin{cases} \ddot{x}_M = \frac{\tau}{J_L}, \dot{x}_L = 0, & \text{if (BM),} \\ \ddot{x}_M = \ddot{x}_L = \frac{\tau}{J_L+J_M}, & \text{if (CM).} \end{cases} \tag{9.87}$$

Introduce the following state variable: $x_B = x_L - x_M$.

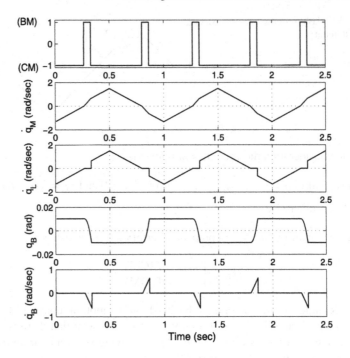

Fig. 9.14 Backlash simulation

For simulation, (BM) and (CM) conditions are reformulated as follows

$$
\begin{aligned}
&\text{(BM)}\quad \text{if } (\dot{x}_M > 0 \text{ and } x_B \le -\beta) \text{ or } (\dot{x}_M < 0 \text{ and } x_B \ge \beta), \\
&\text{(CM)}\quad \text{otherwise.}
\end{aligned}
\tag{9.88}
$$

The equalities in the condition (9.84) have to be changed to inequalities since due to numerical approximations x_B may exceed the value of β, thus a small overshoot can appear.

A mechanical impedance is introduced into the model for contact mode which aim is to keep the value of $x_B = \pm\beta$, depending on the sign of $\dot{x}_M$ while $\dot{x}_B = 0$. The modified model that can be applied for simulation of the mechanical systems with backlash reads as

$$
\text{(BM)}\quad
\begin{cases}
\ddot{x}_M = \frac{\tau}{J_M}, \\
\ddot{x}_B = -\frac{\tau}{J_M},
\end{cases}
\qquad
\text{(CM)}\quad
\begin{cases}
\ddot{x}_M = \frac{\tau}{J_L + J_M}, \\
\ddot{x}_B = -K_f \dot{x}_B - K_s(x_B - \mathrm{sgn}(\dot{x}_M)\beta),
\end{cases}
\tag{9.89}
$$

where K_f and K_s are the viscous friction constant and the spring constant in the mechanical impedance, respectively. In order to have a damped transition from contact mode to backlash mode, the value of K_s has to be smaller that the value of K_f.

Simulation measurements were performed using the model (9.89) with $J_M = J_L = 1$ kg and $\beta = 0.01$ m. The control input τ was chosen as a square signal

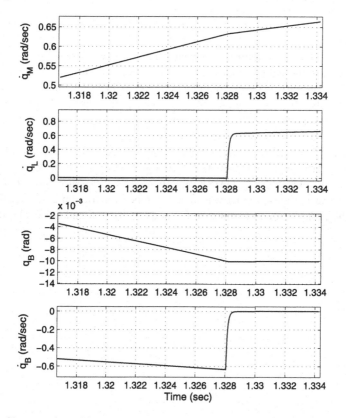

Fig. 9.15 Close-up of transition from (BM) to (CM)

with 1 s period and ±10 N maximum values. The parameters of the mechanical impedance were chosen as $K_f = 10000$ N s/m and $K_s = 100$ N/m.

The simulation results are presented in Figs. 9.14 and 9.15. The simulations show that during backlash mode there is no motion on the driven side and the velocity change in the driving side is greater than in contact mode. During contact mode, the state variable q_M is equal with $\pm\beta$, depending on the sign of velocity and the velocity is zero with high precision. Observe that the during backlash mode–contact mode transition there is a very fast but continuous load side velocity change.

9.6 Closing Remarks

Describing the behavior of systems with nonsmooth nonlinearities is an intensively studied field of the engineering research. Both theoretical and practical results appeared in the last years. From theoretical point of view, it is essential to define the solution of systems with nonsmooth nonlinearities and to elaborate stability theorems for these systems. From practical point of view, it is important to develop models that can be introduced in control algorithms for the direct compensation of non-smooth nonlinearities.

In this chapter, the stability of nonsmooth systems was treated in two different approaches. Firstly, stability theorems were given for switched and piecewise affine systems. The stability theorems for these systems are the extensions of the Lyapunov stability results developed for smooth systems. Another approach for defining the solutions and analyzing the stability of nonsmooth systems is represented by the modeling techniques based on differential inclusions.

Afterward, the practical modeling of friction and backlash in mechanical systems was treated. The most important static (Coulomb, viscous, Stribeck) and dynamic (Dahl, LuGre, elastoplastic, GMS) friction models were summarized. A novel, piecewise linear friction model was also developed for friction compensation purposes. The properties and simulation methods of friction and backlash were also treated.

Chapter 10
Mechanical Control Systems with Nonsmooth Nonlinearities

Overview This chapter introduces a hybrid model for mechanical control systems with Stribeck friction and backlash, using which the stability of these systems can be analyzed. The model is applied for state feedback controller design. Two types of limit cycles, that can appear in these control systems due to wrong controller parameterization, are analyzed in the second part of the chapter, limit cycles around zero velocity and around Stribeck velocities.

10.1 Switched System Model of Mechanical Systems with Stribeck Friction and Backlash

The dynamics of the controlled mechanical system, in which the friction and backlash cannot be neglected, can usually be described by

$$H(q_L)\ddot{q}_L + C(q_L, \dot{q}_L)\dot{q}_L + D(q_L) = B(\tau - \tau_f, q_M, \dot{q}_M, q_L), \tag{10.1}$$

where q_L denotes the n dimensional position vector of the load, n is the degree of freedom (DOF) of the mechanical system, H is the inertia matrix of the system, the vectors $C\,\dot{q}_L$ and D incorporate the effects of centrifugal, Coriolis and gravitational forces, respectively and τ is the vector of control inputs. The vector τ_f contains the friction induced torques/forces that appear in the gears and inside the electrical drive (motor). The effect of the control inputs reaches the load through a backlash-type nonlinearity (B) that also depends on the motor position (q_M), sign of motor velocity and load position. The dynamics of the ith degree of freedom can be determined from Fig. 10.1. The so called inertia driven model [107] is applied to describe the backlash phenomenon. In this model, two regimes are separated: the *contact mode* (CM), when the load is in contact with the motor shaft and the torque developed by the motor acts on the load, and the *backlash mode* (BM), when the direction of motion changes and there is no contact between the motor shaft and the load.

In order to determine the condition for contact mode, denote K_G the gear ratio ($\dot{q}_L = \dot{q}_M/K_G$), and β the backlash gap size. For contact mode, the difference

B. Lantos, L. Márton, *Nonlinear Control of Vehicles and Robots*,
Advances in Industrial Control,
DOI 10.1007/978-1-84996-122-6_10, © Springer-Verlag London Limited 2011

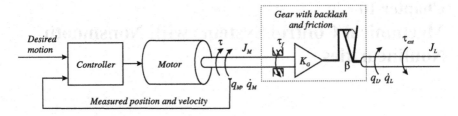

Fig. 10.1 Mechanical control system with friction and backlash

between the motor shaft position and load position, modified with the gear ratio, should be equal to the backlash gap size if the machine moves in the negative direction. Otherwise, the position difference should be equal to the negative gap.

Contact mode (CM):

$$\text{if } \left((\dot{q}_M > 0) \text{ and } (q_L K_G - q_M = -\beta)\right)$$

$$\text{or } \left((\dot{q}_M < 0) \text{ and } (q_L K_G - q_M = \beta)\right). \tag{10.2}$$

Backlash mode (BM):

otherwise.

In backlash mode, the dynamics is given by

$$\text{(BM)} \quad \begin{cases} J_M \ddot{q}_M = \tau - \tau_f, \\ J_L \ddot{q}_L = -\tau_{\text{ext}}. \end{cases} \tag{10.3}$$

Denote J_L the inertia on the load side, J_M the inertia of the motor shaft, τ_f the friction force and τ_{ext} the external torque, respectively. The latter contains the influence of the other degrees of freedom or the environment of the mechanical system on the motion.

In contact mode, the load velocity will be equal to the motor velocity, modified by the gear ratio, and the motion on the motor side is directly influenced by the load

$$\text{(CM)} \quad \begin{cases} (J_M + J_L/K_G^2)\ddot{q}_M = \tau - \tau_f - \tau_{\text{ext}}/K_G, \\ \dot{q}_L = \dot{q}_M/K_G. \end{cases} \tag{10.4}$$

On the motor side, due to the backlash nonlinearity, the inertia of the mechanical system will be different in the backlash mode and in the contact mode. The external torque will have an influence on motor side dynamics only in the contact mode.

To describe the friction in the controlled mechanical system, the model (9.75) introduced in Sect. 9.4 is applied. Assuming that the parameters in the positive velocity domain are equal to the parameters in the negative velocity domain and applying (9.78), the friction model reads as

$$F_{Lf}(\dot{q}_M) = \begin{cases} a_1 + b_1\dot{q}_M, & \text{if } \dot{q}_M \in (0, \dot{q}_S], \\ a_2 + b_2\dot{q}_M, & \text{if } \dot{q}_M \in (\dot{q}_S, \infty), \\ -a_1 + b_1\dot{q}_M, & \text{if } \dot{q}_M \in [-\dot{q}_S, 0), \\ -a_2 + b_2\dot{q}_M, & \text{if } \dot{q}_M \in (-\infty, -\dot{q}_S). \end{cases} \qquad (10.5)$$

Note that $b_1 < 0$ and $b_2 > 0$. The friction will have different behaviors in the low velocity and in the high velocity regimes. If the absolute value of the velocity is smaller than $\dot{q}_S$ the friction may induce unstable behavior and limit cycle.

Using the friction model above and (10.2) and (10.3) in different partitions of the state space, different models describe the controlled system's dynamics (a piecewise affine system with disturbance)

If (BM) and $(|\dot{q}_M| \leq \dot{q}_S)$:

$$\begin{cases} J_M\ddot{q}_M + b_1\dot{q}_M = \tau - a_1\,\mathrm{sgn}(\dot{q}_M), \\ J_L\ddot{q}_L = -\tau_{\text{ext}}. \end{cases} \qquad (10.6)$$

If (BM) and $(|\dot{q}_M| > \dot{q}_S)$:

$$\begin{cases} J_M\ddot{q}_M + b_2\dot{q}_M = \tau - a_2\,\mathrm{sgn}(\dot{q}_M), \\ J_L\ddot{q}_L = -\tau_{\text{ext}}. \end{cases} \qquad (10.7)$$

If (CM) and $(|\dot{q}_M| \leq \dot{q}_S)$:

$$\begin{cases} (J_M + J_L/K_G^2)\ddot{q}_M + b_1\dot{q}_M = \tau - \tau_{\text{ext}}/K_G - a_1\,\mathrm{sgn}(\dot{q}_M), \\ \dot{q}_L = \dot{q}_M/K_G. \end{cases} \qquad (10.8)$$

If (CM) and $(|\dot{q}_M| > \dot{q}_S)$:

$$\begin{cases} (J_M + J_L/K_G^2)\ddot{q}_M + b_2\dot{q}_M = \tau - \tau_{\text{ext}}/K_G - a_2\,\mathrm{sgn}(\dot{q}_M), \\ \dot{q}_L = \dot{q}_M/K_G. \end{cases} \qquad (10.9)$$

If the external torque τ_{ext} is considered as disturbance, the obtained submodels are linear. Hence, the theory of the hybrid, piecewise linear systems can be applied to solve the control of the mechanical system in the presence of friction and backlash [95].

10.2 Motion Control

In high precision position control systems, the friction and backlash cannot be omitted during controller design. The overcompensation of friction leads to limit cycle, its undercompensation may lead to steady-state error [117]. The backlash can also generate limit cycle in the case of wrong controller design [13]. In this section, a fixed structure state feedback controller approach is presented for the control of mechanical systems in the presence of nonlinear Stribeck friction and backlash.

10.2.1 Stabilizing Control

To deal with simultaneous presence of the friction and backlash in the mechanical control systems, soft computing approaches are popular, see, for example, [59, 139]. In these works, neural and fuzzy systems are applied to compensate the effect of these nonsmooth nonlinearities. With known upper bound of the backlash gap size, an adaptive variable structure controller was proposed in [2], that estimates the upper bound of the friction uncertainty. In the paper [100], a two controller based switching control system was proposed to deal with low velocity friction and gear backlash. The effect of backlash and stick-slip friction in heavy-duty hydraulic machines was studied in [105]. A hybrid model based Model Predictive Control (MPC) scheme for backlash compensation was introduced in [124]. In this work, only the viscous friction term was taken into consideration during the modeling.

To solve the stabilization of the system with fixed structure controller, the following strategy is applied: the parameters of the state feedback controllers are designed for one partition of the switched system (using the LQ design approach), then a relation is developed based on which the Lyapunov stability of the control system can be checked in the other partitions.

The LQ state feedback approach of dealing with friction and backlash induced nonlinearities is an alternative for the control approach when a linear controller is combined with nonlinear friction and backlash compensators. However, in the case of nonlinear compensators, the size of the backlash gap and all the friction parameters should be known exactly or adaptive estimation techniques have to be applied, therefore the resulting adaptive controller will have a high computational cost as compared with the LQ controller.

If the load side position and velocity (q_L, $\dot{q}_L$) cannot be measured, the applied controller for the stabilization of the system should rely only on measurements made on the motor side (q_M, $\dot{q}_M$). When the size of backlash gap is unknown, it is also difficult to determine whether the system is in the contact or backlash mode. Hence, for the stabilization, a fixed structure linear state feedback controller is used extended with an integral term to deal with additive disturbances

$$\tau = -K_P q_M - K_V \dot{q}_M - K_I \int_0^t q_M(\xi)\, d\xi. \qquad (10.10)$$

The controller parameters should be designed so that the controller is able to guarantee the stability of the control system in each partition of the state space. To achieve this, the following stability theorem is applied: Let the dynamics of the switched system be given by $\dot{x} = f_i(x)$ with a rule for switching among the submodels. If there exists a common Lyapunov function candidate V for all submodels, which is strictly decreasing in all of the state space partitions, then the switched system is asymptotically stable [15].

According to (10.6)–(10.9) and appropriate choice of the parameters, the state space model of the subsystem, for which the LQ controller is developed, can be described in the form given by

$$
\begin{pmatrix} \dot{q}_M \\ \ddot{q}_M \\ \dot{q}_I \end{pmatrix} = A \begin{pmatrix} q_M \\ \dot{q}_M \\ q_I \end{pmatrix} + B\tau + \begin{pmatrix} 0 \\ -\mathrm{sgn}(\dot{q}_M)a/J - \tau_{\mathrm{ext}}/(JK_G) \\ 0 \end{pmatrix},
$$

$$
A = \begin{bmatrix} 0 & 1 & 0 \\ 0 & -b/J & 0 \\ 1 & 0 & 0 \end{bmatrix}, \qquad B = \begin{bmatrix} 0 \\ 1/J \\ 0 \end{bmatrix},
$$

(10.11)

where q_I denotes the integral of the position input on the motor side: $q_I = \int_0^t q_M(\xi)\,d\xi$. If the model is for backlash mode, the term $\tau_{\mathrm{ext}}/(JK_G)$ will not appear in the disturbance vector.

The standard LQ design procedure is applied to determine the controller parameters. Let Q be a positive definite symmetric matrix and r a strictly positive scalar value. The gain vector of the controller $K = (K_P\,K_V\,K_I)^T$ is calculated as

$$
K = \frac{1}{2}r^{-1}B^T P, \tag{10.12}
$$

where the positive definite symmetric matrix P is the solution of algebraic Riccati equation

$$
PA + A^T P - P B r^{-1} B^T P = -Q. \tag{10.13}
$$

The LQ controller guarantees the stability of the system in the sense of Lyapunov: it can easily be shown, that the time derivative of Lyapunov function candidate

$$
V(x) = x^T P x \tag{10.14}
$$

is negative for $x \neq 0$

$$
\dot{V}(x) < -x^T Q x, \tag{10.15}
$$

where $x = (q_M\,\dot{q}_M\,q_I)^T$ denotes the state vector.

Assume that the controller is determined for a given partition of the state space. It is considered that in the other partitions there is an additive uncertainty for the friction parameter: $b := b + \Delta b$ with regard to the parameter value for which the controller was designed and there is an amplification type uncertainty for the inverse of inertia: $\frac{1}{J} := \frac{\Delta(1/J)}{J}$.

If the controller is designed for the low velocity regime and the absolute value of the velocity of the machine is higher than the Stribeck velocity, the additive friction modeling error is $\Delta b = b_2 - b_1 > 0$, otherwise $\Delta b = b_1 - b_2 < 0$.

If the controller is designed for the contact mode and the backlash mode is active, the inertia modeling uncertainty is $\Delta(1/J) = \frac{J_M + J_L/K_G^2}{J_M} > 1$, otherwise $\Delta(1/J) = \frac{J_M}{J_M + J_L/K_G^2} < 1$.

Hence, in the other partitions the state matrix A and the vector B can be written as the original matrices extended with modeling uncertainties

$$A := A + \Delta A = A + \frac{(1 - \Delta(1/J))b - \Delta(1/J)\Delta b}{J} I_2, \qquad (10.16)$$

$$B := \Delta B B = \Delta(1/J)B, \qquad (10.17)$$

$$\text{where } I_2 = \begin{bmatrix} 0 & 0 & 0 \\ 0 & 1 & 0 \\ 0 & 0 & 0 \end{bmatrix}. \qquad (10.18)$$

Consider that for a given positive Q and r the controller was determined using (10.13) and (10.12). The Lyapunov stability of the control system in the other partition is analyzed based on the Lyapunov function candidate (10.14). The time derivative of the Lyapunov function in the other partitions, using (10.13), can be written as

$$\begin{aligned} \dot{V} &= x^T P \dot{x} + \dot{x}^T P x \\ &= x^T \big((A^T + \Delta A)P + P(A + \Delta A) - \Delta B^2 r^{-1} P B B^T P \big) x \\ &= x^T \big(-Q + \Delta A P + P \Delta A - (\Delta B^2 - 1) r^{-1} P B B^T P \big) x. \qquad (10.19) \end{aligned}$$

Hence, the fixed structure controller can stabilize the switched system in Lyapunov sense if the following relation holds for all partitions of the state space

$$-\Delta A P - P \Delta A + (\Delta B^2 - 1) r^{-1} P B B^T P + Q > 0. \qquad (10.20)$$

With the notations introduced for parameter uncertainties, it yields

$$\begin{aligned} &-\frac{(1 - \Delta(1/J))b - \Delta(1/J)\Delta b}{J}(I_2 P + P I_2) \\ &+ (\Delta(1/J)^2 - 1) r^{-1} P B B^T P + Q > 0. \qquad (10.21) \end{aligned}$$

For the case when no common Lyapunov function can be found, the controller can be designed based on multiple Lyapunov functions [119]. In this case for each partition of the state space, a controller can be designed and additional conditions have to be satisfied to guarantee stability. To implement this controller, the switching condition (the size of the backlash gap) must be exactly known.

10.2.1.1 Case Study

Henceforth, it will be analyzed in which partition is it the best to design the controller in order to have a good chance of guaranteeing the stability of the hybrid control system? To answer this question, the following properties are applied:

- The sum of two symmetric positive definite matrices is also a positive definite symmetric matrix.

- A symmetric matrix is positive definite iff all of its eigenvalues are strictly positive.

In order to determine a simpler relation between the model and controller parameters that can be used to analyze the stability of the control system, consider that Q is a diagonal matrix with same elements in the diagonal: $Q = \text{diag}([q\ q\ q])$, $q > 0$. Divide the matrix Q as follows: $Q := Q + Q_\epsilon = \text{diag}([q\ q\ q]) + \text{diag}([q_\epsilon\ q_\epsilon\ q_\epsilon])$ with $q_\epsilon \ll q$. Note that since P, Q and ΔA are symmetric matrices, the left hand side of the Inequality (10.21) is a symmetric matrix as well.

The Inequality (10.21) holds if the following matrices are positive definite

$$\Delta_1 = \left(\Delta(1/J)^2 - 1\right)r^{-1}PBB^TP + Q_\epsilon$$

$$= \begin{bmatrix} q_\epsilon + \frac{\Delta(\frac{1}{J})^2 - 1}{rJ^2}p_{12}^2 & \frac{\Delta(\frac{1}{J})^2 - 1}{rJ^2}p_{12}p_{22} & \frac{\Delta(\frac{1}{J})^2 - 1}{rJ^2}p_{12}p_{23} \\ \frac{\Delta(\frac{1}{J})^2 - 1}{rJ^2}p_{12}p_{22} & q_\epsilon + \frac{\Delta(\frac{1}{J})^2 - 1}{rJ^2}p_{22}^2 & \frac{\Delta(\frac{1}{J})^2 - 1}{rJ^2}p_{23}p_{22} \\ \frac{\Delta(\frac{1}{J})^2 - 1}{rJ^2}p_{12}p_{23} & \frac{\Delta(\frac{1}{J})^2 - 1}{rJ^2}p_{23}p_{22} & q_\epsilon + \frac{\Delta(\frac{1}{J})^2 - 1}{rJ^2}p_{23}^2 \end{bmatrix}, \tag{10.22}$$

$$\Delta_2 = -\frac{(1 - \Delta(1/J))b - \Delta(1/J)\Delta b}{J}(I_2P + PI_2)$$

$$= \begin{bmatrix} q & -\frac{(1-\Delta(\frac{1}{J}))b-\Delta(\frac{1}{J})\Delta b}{J}p_{12} & 0 \\ -\frac{(1-\Delta(\frac{1}{J}))b-\Delta(\frac{1}{J})\Delta b}{J}p_{12} & q - 2\frac{(1-\Delta(\frac{1}{J}))b-\Delta(\frac{1}{J})\Delta b}{J}p_{22} & -\frac{(1-\Delta(\frac{1}{J}))b-\Delta(\frac{1}{J})\Delta b}{J}p_{23} \\ 0 & -\frac{(1-\Delta(\frac{1}{J}))b-\Delta(\frac{1}{J})\Delta b}{J}p_{23} & q \end{bmatrix}. \tag{10.23}$$

Here, $p_{ij} = p_{ji}$ denotes the element of P from the ith row and jth column. The eigenvalues of Δ_1 and Δ_2 were determined using the `eig` function of MATLAB/*Symbolic Math Toolbox*.

The eigenvalues of Δ_1 are

$$\lambda(\Delta_1) = \begin{pmatrix} q_\epsilon + \frac{\Delta(1/J)^2 - 1}{rJ^2}(p_{12}^2 + p_{22}^2 + p_{23}^2) \\ q_\epsilon \\ q_\epsilon \end{pmatrix}. \tag{10.24}$$

These eigenvalues are positive, if $\Delta(1/J) \geq 1$. If the controller is designed for contact mode, this condition holds.

The eigenvalues of Δ_2 are

$$\lambda(\Delta_2) = \begin{pmatrix} q + \frac{(1-\Delta(1/J))b-\Delta(1/J)\Delta b}{J}(-p_{22} + \sqrt{p_{12}^2 + p_{22}^2 + p_{23}^2}) \\ q + \frac{(1-\Delta(1/J))b-\Delta(1/J)\Delta b}{J}(-p_{22} - \sqrt{p_{12}^2 + p_{22}^2 + p_{23}^2}) \\ q \end{pmatrix}. \tag{10.25}$$

From (10.25), one can conclude that with the proper choice of Δb the sign of the eigenvalues cannot be determined. The positiveness of the eigenvalues can only be guaranteed for high q values. However, since $|-p_{22}+\sqrt{p_{12}^2+p_{22}^2+p_{23}^2}| \leq |-p_{22}-\sqrt{p_{12}^2+p_{22}^2+p_{23}^2}|$ the positiveness of the eigenvalues can be more easily guaranteed if $\frac{(1-\Delta(1/J))b-\Delta(1/J)\Delta b}{J} \leq 0$. Since $\Delta(1/J)$ was chosen greater than 1, the inequality will be guaranteed if the controller is designed for the low velocity regime, where $b < 0$ and $\Delta b > 0$.

The case study above suggests that the controller should be designed for the model (10.8) that describes the behavior of the system in contact mode and for low velocities. However, if the design is performed in this regime it will not guarantee that the hybrid control system will be stable. The condition (10.20) must be checked after the controller design for each partition of the state space.

Accordingly, in order to assure the asymptotic stability of the mechanical control system in the presence of Stribeck friction and backlash the following steps should be performed

- For given positive definite Q and r design P and K, using (10.12) and (10.13), based on the model (10.8).
- Check the relation (10.21) for the following three cases

  1. $\Delta(1/J) = 1$, $\Delta b = b_2 - b_1$.
  2. $\Delta(1/J) = \frac{J_M + J_L/K_G^2}{J_M}$, $\Delta b = 0$.
  3. $\Delta(1/J) = \frac{J_M + J_L/K_G^2}{J_M}$, $\Delta b = b_2 - b_1$.

10.2.2 Extension of the Control Law for Tracking

The controllers developed for mechanical system should also guarantee zero or sufficiently small steady-state error even in the case when the reference position is time varying. In this subsection, the stabilizing controller is extended to the tracking problem. Assume that the reference trajectory ($q_{Mref} = q_{Mref}(t)$) is a twice differentiable function of time. Formulate the control law

$$\tau = J\ddot{q}_{Mref} + b\dot{q}_{Mref}$$
$$+ K_P(q_{Mref} - q_M) + K_V(\dot{q}_{Mref} - \dot{q}_M)$$
$$+ K_I \int_0^t \left(q_{Mref}(\xi) - q_M(\xi)\right) d\xi. \tag{10.26}$$

The same controller parameters (K_P, K_V, K_I) are applied as for stabilization.

With the notation $e = q_{Mref} - q_M$ for tracking error, the dynamics of the controlled system is given by

$$\Delta(1/J)J\ddot{e} + (b + \Delta b + K_V)\dot{e} + K_P e + K_I \int_0^t e(\xi)\, d\xi$$

$$= -(a + \Delta a)\,\text{sgn}(\dot{q}_M) + \tau_{\text{ext}}/K_G + \Delta b \dot{q}_{Mref}$$
$$+ J\big(\Delta(1/J) - 1\big)\ddot{q}_{Mref}. \tag{10.27}$$

For $\dot{q}_M \neq 0$ and τ_{ext}, differentiable yields

$$\Delta(1/J)J\dddot{e} + (b + \Delta b + K_V)\ddot{e} + K_P \dot{e} + K_I e$$
$$= \dot{\tau}_{\text{ext}}/K_G + \Delta b \ddot{q}_{Mref} + J\big(\Delta(1/J) - 1\big)\dddot{q}_{Mref}. \tag{10.28}$$

Assuming that the system is stable, the steady-state value of the tracking error can be calculated as

$$\lim_{t \to \infty} e(t) = \lim_{s \to 0} \left(\frac{1}{K_I K_G} s^2 \tau_{\text{ext}}(s) + \frac{\Delta b}{K_I} s^3 q_{Mref}(s) + \frac{J(\Delta(1/J) - 1)}{K_I} s^4 q_{Mref}(s) \right). \tag{10.29}$$

The zero steady-state tracking error can only be guaranteed for constant external disturbance torque. In this case, the controller developed for the contact mode and the low velocity regime can guarantee zero steady-state error in the high velocity regime if the trajectory is first order polynomial. In the low velocity regime, the tracking of second order trajectories with zero steady-state error can also be guaranteed both in the contact and the backlash mode.

10.2.3 Simulation Results

Simulations were performed in the MATLAB/Simulink environment to demonstrate the applicability of the theoretical results. The dynamics of the controlled system was simulated by implementing the models (10.3) and (10.4). For the implementation of the friction torque, the original Stribeck model was used, given by (9.38). The parameters of the controlled mechanical system were taken as

- Inertia of the load: $J_L = 0.5 \text{ kg m}^2$.
- Inertia on the motor side: $J_M = 0.01 \text{ kg m}^2$.
- Backlash gap: $\delta = 0.001 \text{ rad}$.
- Gear ratio: $K_G = 5$.
- Viscous friction coefficient: $F_V = 0.1 \text{ N m/rad}$.
- Coulomb friction coefficient: $F_C = 0.5 \text{ N m}$.
- Static Friction coefficient: $F_S = 1.5 \text{ N m}$.
- Stribeck velocity: $\dot{q}_S = 0.1 \text{ rad/s}$.

The parameters of the linearized friction model (10.5) were determined based on relations (9.70)–(9.71) and (9.76)–(9.78) respectively.

Two control experiments were performed. In both cases, the initial motor and load positions were taken as 0 and the reference position was taken $q_{Mref} = 1$ rad.

In the first experiment, the LQ controller was determined for backlash mode (BM) and high velocity regimes ($|\dot{q}_M| > \dot{q}_S$). For controller design, the following

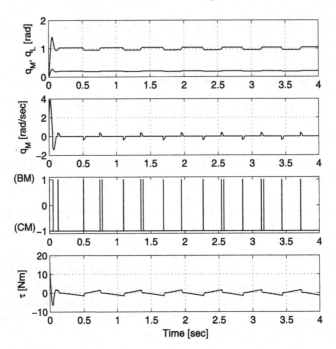

Fig. 10.2 Simulation results (LQ control designed for (BM) and $|\dot{q}_M| > \dot{q}_S$)

Fig. 10.3 The limit cycle in phase plane (LQ control designed for (BM) and $|\dot{q}_M| > \dot{q}_S$)

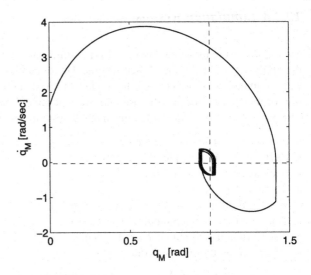

matrices were applied: $Q = \text{diag}([200\ 4\ 200])$, $r = 1$. For these parameters, the obtained control law is

$$\tau = 16.09(q_{Mref} - q_M) - 1.98\dot{q}_M + 14.14 \int_0^t \left(q_{Mref} - q_M(\xi)\right) d\xi. \qquad (10.30)$$

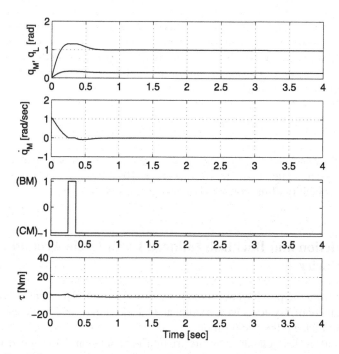

Fig. 10.4 Simulation results (LQ control designed for (CM) and $|\dot{q}_M| < \dot{q}_S$)

Fig. 10.5 Asymptotically stable trajectory in phase plane (LQ control designed for (CM) and $|\dot{q}_M| < \dot{q}_S$)

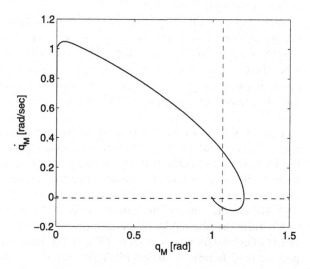

With the designed controller, the relation (10.21) was verified for the other three regions of the state space. It was found that for the low velocity regimes the relation (10.21) does not hold, the symmetric matrix on the left hand side of the inequality has negative eigenvalues. Simulation results are shown in Figs. 10.2 and 10.3. Due to the instability in the low velocity regime, the motor position oscillates around the reference position and a limit cycle appears.

In the second experiment, the LQ controller was designed with the same Q and r matrices as in the first experiment, but for the contact mode (CM) and low velocity regimes. The obtained control law is

$$\tau = 22.08(q_{Mref} - q_M) - 20.06\dot{q}_M + 14.14 \int_0^t \left(q_{Mref} - q_M(\xi)\right) d\xi. \quad (10.31)$$

With this controller the relation (10.21) holds for all regions of the controlled switched system. The simulation results (Figs. 10.4 and 10.5) also show that the asymptotic stability of the control system is guaranteed.

10.3 Friction and Backlash Induced Limit Cycle Around Zero Velocity

It is clear from the previous discussions that wrong controller parameterization in mechanical control systems with friction and backlash may lead to undesired oscillations around the reference position.

The limit cycles generated by the friction phenomenon was studied by many authors, see, for example, [50] and [108]. The chaotic nature of these oscillations were analyzed for example in [159] and [12]. The bifurcations that occur in flexible rotor systems were studied in [157].

In the paper [49], limit cycles induced by static and dynamic friction models were compared. The friction introduced chaotic behavior of system responses was also treated by many authors. In the paper [14], the bifurcations generated by friction, when there is a transition from slip to steady sliding, has been studied for systems having linear spring term. The nonlinear friction induced chaotic behavior in a single link manipulator was presented in [155]. Backlash induced limit cycles were studied in [145].

In most of these works, it was assumed that in the mechanical system there is a spring element and/or it is excited with sinusoidal input signal. In the current work, the limit cycles are treated, that appear without elastic component in system dynamics and with constant reference signal. A method is presented to determine a limit cycle condition between the mechanical plant parameters and controller parameters [94]. The chaotic nature of the limit cycles is also analyzed.

Consider that the model of the mechanical system is described by (10.3) and (10.4) where the friction is given by (9.37a). For limit cycle analysis, the reference position is set to zero. Assume a PID type control law similar to (10.10)

$$\tau = -K_P\left(q_M(t) + T_D\dot{q}_M + \frac{1}{T_I}\int_0^t q_M(\xi)\,d\xi\right), \quad (10.32)$$

where $K_P > 0$ is the proportional gain, $T_D > 0$ is the derivate time constant, $T_I > 0$ is the integral time constant.

Consider the linearization of the nonlinear Stribeck friction model (9.37a) in the vicinity of zero velocities similarly as in (9.66)

$$\tau_{fL0} = \tau_f(0) + \left.\frac{\partial \tau_f}{\partial \dot{q}_M}\right|_{\dot{q}_M \to 0;\ \dot{q}_M \neq 0} \dot{q}_M = F_S \operatorname{sgn}(\dot{q}_M) + \left(F_V - \frac{F_S - F_C}{\dot{q}_S}\right)\dot{q}_M.$$

(10.33)

If the motor side of the mechanical system is considered, the inertia of the system (J) is different in backlash mode and contact mode but other terms of the system model remain the same.

$$J = \begin{cases} J_M, & \text{if (BM)}, \\ J_M + J_L/K_G^2, & \text{if (CM)}. \end{cases}$$

(10.34)

With the linearized model, the dynamics of the closed loop system around zero velocities can be written as

$$J\ddot{q}_M = -K_P q_M - K_P T_D \dot{q}_M - \frac{K_P}{T_I}\int_0^t q_M(\xi)\,d\xi - F_S \operatorname{sgn}(\dot{q}_M)$$

$$- \left(F_V - \frac{F_S - F_C}{\dot{q}_S}\right)\dot{q}_M.$$

(10.35)

By derivating both sides of the above equation for $\dot{q}_M \neq 0$, the characteristic equation of the dynamics can be obtained

$$Js^3 + \left(K_P T_D + F_V - \frac{F_S - F_C}{\dot{q}_S}\right)s^2 + K_P s + \frac{K_P}{T_I} = 0.$$

(10.36)

The location of the roots of the above characteristic equation can be obtained by writing the state space model of the system in controllability representation and locating the system matrix eigenvalues using the Gershgorin theorem [151]. The Gershgorin disks for the control system are presented in Fig. 10.6. Both in backlash and contact mode, the characteristic equation will have two roots in the unit circle of the complex plane. The third pole is located inside a circle with the following center and radius

$$P\left(-\frac{K_P T_D + F_V - \frac{F_S - F_C}{\dot{q}_S}}{J}, 0\right), \qquad R = \frac{K_P}{J}\left(1 + \frac{1}{T_I}\right),$$

(10.37)

where the value of the inertial parameter is different in contact mode and in backlash mode according to (10.34).

In order to determine an approximate relation for stable limit cycles in the mechanical control system, the characteristic equation of the linearized model is considered. For stable limit cycle, it should have two complex conjugate poles on the imaginary axis and one stable real pole. Hence, the characteristic equation should have the following general form

$$(s + \gamma)(s^2 + \omega^2) = 0,$$

(10.38)

Fig. 10.6 Location of the roots of the closed loop system

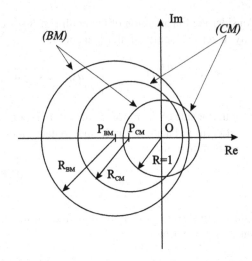

where $\gamma, \omega > 0$.

By comparing the relations (10.36) and (10.38), the parameters γ and ω can be identified

$$\gamma = \frac{K_P T_D + F_V - \frac{F_S - F_C}{\dot{q}_S}}{J}, \tag{10.39}$$

$$\omega^2 = \frac{K_P}{J}. \tag{10.40}$$

For stable limit cycles ($\gamma > 0$), the controller parameters have to satisfy the following relation

$$K_P T_D > \frac{F_S - F_C}{\dot{q}_S} - F_V. \tag{10.41}$$

The relation above clearly shows that the amplification of the derivative term has a great influence on the stability of the control systems with friction. This fact was firstly studied in the early paper [36].

Based on the relations (10.36), (10.38), (10.39) and (10.40), the limit cycle condition can be formulated as

$$K_P T_D + F_V - \frac{F_S - F_C}{\dot{q}_S} = \frac{J}{T_I}. \tag{10.42}$$

Henceforward, the controller parameters that satisfy the relation (10.42) will be denoted as: $K_{PLC}, T_{DLC}, T_{ILC}$.

By introducing the condition (10.42) into (10.36), the characteristic equation of the control system having limit cycle can be written as

$$\left(s + \frac{1}{T_I}\right)(Js^2 + K_P) = 0. \tag{10.43}$$

Note that the limit cycle condition (10.42) differs in the backlash mode and contact mode because the inertial parameter (J) is different in different regimes.

The limit cycle conditions above are approximate relations, since they rely on linearized models. However, these relations between the controller and system parameters can be used to locate the limit cycles in the positioning system.

10.3.1 Chaotic Measures for Nonlinear Analysis

The controlled mechanical system is nonlinear, hence the chaotic nature of the oscillating states should also be treated. It is well known that a third order continuous nonlinear dynamical system may show chaotic behavior. Because of the integral therm that appears in the controller the order of the investigated closed loop system dynamics is three.

The chaotic motion is bounded, irregular and aperiodic. There is no universally accepted definition for chaotic signals but if the signal have some properties, that are specific for chaotic motion, the signal is accepted to be chaotic [35].

The main property of chaotic data, sensitive dependence on initial conditions, means that two trajectories starting very closely together will rapidly diverge from each other, and thereafter have different futures. The practical implication is that long-term prediction becomes impossible in a system like this, where small uncertainties are amplified fast. On the other hand, the rate of the divergence in the evolution of the trajectory represents the degree of chaoticity of the data. Suppose $q(t_0)$ is a point on a certain state trajectory at time t_0, and consider a nearby point, say $q(t_0) + \delta_q(t_0)$, where $|\delta_q(t_0)|$ represents the distance between those two points at the initial time t_0. These trajectories separate exponentially fast, that is, $|\delta_q(t)| = |\delta_q(t_0)|e^{\lambda t}$. The value of λ is called the *Lyapunov exponent*. For chaotic signals, the Lyapunov exponent is greater than zero ($\lambda > 0$).

The *fractal dimension* is a statistical quantity that gives an indication of how completely a fractal appears to fill space, as one zooms down to finer and finer scales. This value can also be determined for a chaotic time series. There are many specific definitions of fractal dimension and none of them should be treated as the universal one. The correlation dimension (ν) is a geometric measure of the attractor that tells how strongly points on the attractor are spatially correlated with one another in phase space. For a time series of signal $x = \{x_0, x_1, \ldots, x_n\}$, the correlation integral $C(r)$ can be calculated as $C(r) = \lim_{n \to \infty} \frac{g}{N^2}$, where g is the total number of pairs of points which have a distance between them that is less than distance r. As the number of points tends to infinity, and the distance between them tends to zero, the correlation integral, for small values of r, is: $C(r) \sim r^{\nu}$. If the number of points is sufficiently large and evenly distributed, the value of ν represents the correlation dimension. For the chaotic data, the fractal dimension is less than the embedded dimension of the state space.

It is not immediately clear if noise (stochastic process) is not the cause for the aperiodic behavior. In order to verify this, we have to ensure that the aperiodic behavior in the signal is attributed to the presence of the chaoticity. This can be achieved by using a *Surrogate Data Test*. This method first specifies some linear process as a null hypothesis, then generates surrogate data sets, which are consistent with this null hypothesis, and finally computes a discriminating statistics for the original and for each of the surrogate data sets. In order to generate the surrogate data sets, the original data are transformed in such a way that all structures except for the assumed properties are destroyed. The generated surrogate data sets are assumed to mimic only the linear properties of the original data.

The surrogate data test is important in the case when the signal to be analyzed comes from experimental measurements, since in that case it is important to verify whether the irregular behavior of the signal is generated by measurement noise or by chaotic nature of the system. The other two values (Lyapunov exponent and fractal dimension) are generally accepted measures for the chaoticity of a signal.

10.3.2 Simulation Measurements

Simulations were performed in MATLAB/Simulink environment to verify the theoretical results and to calculate the chaotic measures of the limit cycles in the control system. Same system and friction parameters were applied as in Sect. 10.2.3.

For the control of the mechanical system a PID type control algorithm was chosen. The parameters of the controller were chosen to satisfy the relation (10.42): $K_{PLC} = 20$, $T_{ILC} = 2$ s, $T_{DLC} = 0.105$ s. Note that with the controller and controlled plant parameters above the relation (10.41) is also satisfied.

The phase plane behavior of the position and velocity is shown in Fig. 10.7. It can clearly be seen that the position and velocity oscillate around the steady-state value. Due to the system nonlinearities the oscillation presents irregular behavior. Note that the phase plane plot shows similar behavior as in the case of wrongly parameterized LQ controller, presented in the previous subsection.

In order to verify the chaoticity of the signals in the control system, the TSTOOL software package was applied, which is a MATLAB toolbox for nonlinear time series analysis. Since during simulations no random disturbance generating sources were added to the system, the surrogate data test of the analyzed signals was omitted. The toolbox was used to calculate the Lyapunov exponent and the correlation dimension of the position and velocity signals, respectively. For the analysis, the simulation had run for 100 seconds. The signals were sampled with 1 millisecond sampling period. To avoid the effect of transients on the analysis, only the last 75% of the samples were used during the calculation of the chaotic measures.

The Lyapunov exponent and the correlation dimension were determined for different values of the derivative time T_D. From the relation (10.41), it can clearly be seen that the derivative time has a great influence on system stability. However, simulation results show that due to the system nonlinearities, the system remains in limit

Fig. 10.7 Position and
velocity in the phase plane

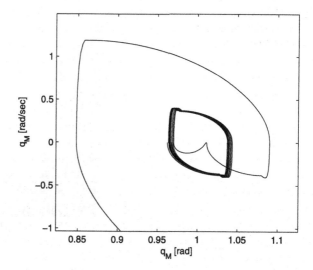

Fig. 10.8 The Lyapunov
exponent for $T_{DLC} = 0.105$ s

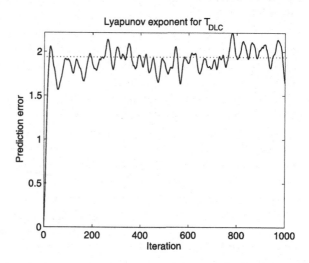

cycle for a wide range of T_D values. During simulation measurements, the derivative time was varied from $0.3T_{DLC}$ to $1.8T_{DLC}$, with $T_{DLC} = 0.105$ s. The values of the other two controller parameters (K_{PLC}, T_{ILC}) were not changed.

The value of the Lyapunov exponent for T_{DLC} and the Lyapunov exponents for different T_D values can be seen in Figs. 10.8 and 10.9, respectively. The maximum Lyapunov exponents are positive and their values are increasing toward the instability region.

The correlation integral and the correlation dimensions for the position and velocity signals can be seen in Figs. 10.10 and 10.11. It can be seen that the value of the correlation dimension is smaller than 2, the embedded dimension of the position-velocity phase space.

Fig. 10.9 Maximum
Lyapunov exponents for
different T_D values

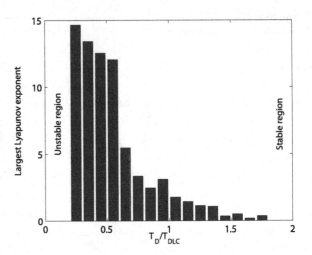

Fig. 10.10 The correlation
integral for $T_{DLC} = 0.105$ s

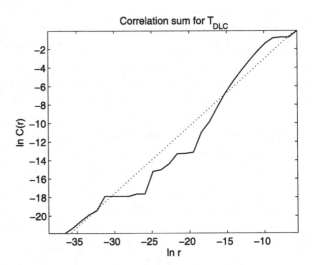

10.4 Friction Generated Limit Cycle Around Stribeck Velocities

For limit cycle analysis, consider the mechanical system's dynamics given by

$$J\ddot{q}_M = \tau - \tau_f(\dot{q}_M), \tag{10.44}$$

where $J > 0$ denotes the inertia of the load, τ represents the control input, $\tau_f(\dot{q}_M)$ is the Stribeck friction.

If the friction is modeled using the piecewise linearly parameterized model (10.5) and the modeling errors upper bounded by (9.80), then the dynamics of the system can be rewritten as

$$J\ddot{q}_M = \tau - b_i\dot{q}_M - (a_i + \tau_\Delta), \tag{10.45}$$

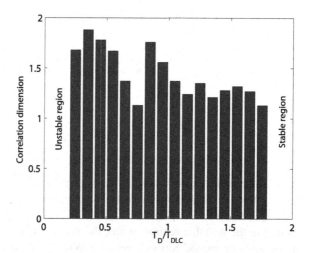

Fig. 10.11 Correlation dimensions for different T_D values

where i depends on the velocity regime, in which the systems moves: $i = 1$ for $|\dot{q}_M| < \dot{q}_S$ and $i = 2$, otherwise. a_i can be incorporated in the additive disturbance bounded term τ_Δ.

Consider that only the position of the plant is measured. The position reference trajectory is considered time varying. A linear controller based on the difference between the measured position (output) and reference position determines the control signal for the plant (see Fig. 10.12).

Let the linear controller given by the transfer function $H_C(s) = \frac{N_{HC}(s)}{D_{HC}(s)}$. In the ith velocity regime, the Laplace transform of the tracking error is (see Fig. 10.12)

$$e(s) = \frac{D_{HC}(s)s(Js + b_i)}{D_{HC}(s)s(Js + b_i) + N_{HC}(s)} q_{Mref}(s)$$
$$- \frac{D_{HC}(s)}{D_{HC}(s)s(Js + b_i) + N_{HC}(s)} \Delta\tau_F(s). \qquad (10.46)$$

To guarantee the stability of the closed loop system, the controller should be chosen such that the real part of the roots of the characteristic equation below are negative in all the four velocity regimes

$$D_{HC}(s)s(Js + b_i) + N_{HC}(s) = 0. \qquad (10.47)$$

In high velocity regimes, $|\dot{q}_M| > \dot{q}_S$, the friction acts as a damping term, which supports the stabilization of the system. If $|\dot{q}_M| < \dot{q}_S$ then $b_i < 0$ and the friction can induce instability. Hence, for certain controller parameterization the control system can be unstable for $|\dot{q}_M| < \dot{q}_S$ and stable otherwise. This fact can lead to limit cycle during tracking control [93].

Consider that the controller parameterization guarantees the stability for $|\dot{q}_M| > \dot{q}_S$ but the closed loop system is unstable for smaller velocities than $\dot{q}_S$. At the same time, the differentiable reference trajectory has the property $0 < \dot{q}_{Mref}$

Fig. 10.12 Block scheme of the control system for limit cycle analysis

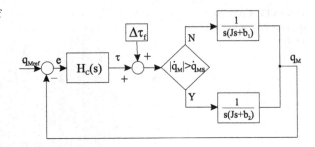

$< \dot{q}_S$. If the initial positive velocity is smaller than the Stribeck velocity, the closed loop system is unstable, the velocity is "thrown out" from the $(0, \dot{q}_S]$ regime. For $\dot{q}_M > \dot{q}_S$, the closed loop system is stable and because $\dot{q}_{Mref} < \dot{q}_S$ the velocity will reenter in the unstable regime, where the velocity rises again. This leads to velocity oscillations around Stribeck velocity. With the same considerations in negative velocity regime, the limit cycle can appear as well.

The limit cycle can be avoided if the controller parameters guarantee the stability of the closed loop system for $|\dot{q}_M| < \dot{q}_S$ velocity domain as well, in which domain the friction has a negative slope given by (9.71).

Case Study—PD Control   Consider that the control signal is calculated by a *PD* type controller, which transfer function is given by

$$H_C(s) = K_P(1 + T_D s),$$ (10.48)

where $K_P > 0$ is the proportional gain, $T_D > 0$ is the derivative time.

With this control algorithm, the characteristic equation of the closed loop system is given by

$$Js^2 + (b_i + K_P T_D)s + K_P = 0.$$ (10.49)

The roots of (10.49) can be calculated as

$$s_{1,2} = \frac{-(b_i + K_P T_D) \pm \sqrt{(b_i + K_P T_D)^2 - 4JK_P}}{2J}.$$ (10.50)

It can be seen from (10.50) that the closed loop system is stable if the following condition holds

$$K_P T_D > -b_i.$$ (10.51)

From the relation above, one can conclude that for $|\dot{q}_M| > \dot{q}_S$ the stability condition holds for any $K_P, T_D > 0$ since $b_i \geq 0$. Otherwise, if $|\dot{q}_M| < \dot{q}_S$ the *PD* control can be unstable if $K_P T_D$ is not chosen high enough, because $b_i < 0$. The limit cycle can be avoided if (10.51) holds in the low velocity regime as well.

In the case of PID control, the characteristic equation is given by (10.36).

10.4.1 Simulation Results

For the simulations, the system model described by the relation (10.44), with $J = 1$ kg is applied. The friction in the system is modeled using the original Tustin friction model (9.38). The friction parameters are chosen as follows: $F_S = 1.5$ N, $F_C = 1$ N, $F_V = 14$ N s/m, $\dot{q}_S = 0.001$ m/s. With these parameters, according to (9.71), the negative slope of the friction curve at low velocities will be $b_1 = -485$ N s/m.

Simulations were performed both with PD and PID type control algorithms. The constant reference velocity was chosen equal to Stribeck velocity. The reference position was obtained by integrating the reference velocity.

The *PD* controller parameters were chosen: $K_P = 10000$, $T_D = 0.02$ s. With these values the relation (10.51) does not hold. As it can be seen in the first group of simulation results in Fig. 10.13, the limit cycle appears, nonlinear oscillations on velocity signal can be observed. For $K_P = 10000$, $T_D = 0.07$ s the stability condition for the low velocities (10.51) will be satisfied, hence the control system will not enter in limit cycle, see the second group of simulation results in Fig. 10.13.

For the *PID* control law, firstly the following parameters were considered: $K_P = 10000$, $T_D = 0.02$ s, $T_I = 5$ s. With these parameters, for $|\dot{q}_M| < \dot{q}_S$, the three roots of the characteristic equation (10.36) are (243.9793 41.2196 −0.1989), hence the limit cycle will appear. By modifying the value of derivative time to $T_D = 0.07$ s the roots are (−147.1213 −67.6779 −0.2009), therefore the theoretical discussion based on the linearized friction model does not predict limit cycles. Simulation results also confirm the theoretical prediction. In Fig. 10.14, it can be seen, that when there is no limit cycle, the PID control guarantees zero steady-state error for the ramp type position reference signal, but when the system enters in limit cycle, it severely influences the position response in a negative sense.

10.4.2 Experimental Measurements

In order to test the limit cycles that can appear in the mechanical control systems, experimental measurements were also performed on a laboratory positioning system. For the experiments, a permanent magnet *DC* servo motor is used. The load of the motor is a metal disk with inertia $J = 0.2$ kg m^2, which is rigidly connected to the shaft of the motor. The position and velocity is measured using a 500 PPT (Pulses Per Turn) rotational encoder. The system is controlled by a personal computer equipped with a data acquisition card which counter input is used for encoder frequency measurement and analogue output for motor control. The mechanical system is connected to the card through an interface circuit, which contains a signal conditioner for the encoder and an H-bridge amplifier for the motor.

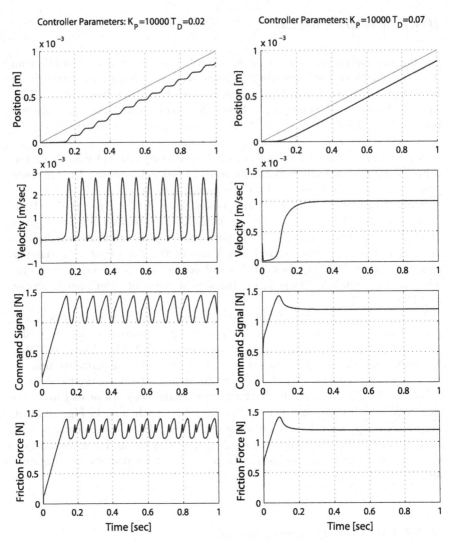

Fig. 10.13 Simulation measurements (PD control)

For position control, a *PD* type algorithm was implemented with 15 ms sampling period, using backward approximation. The reference position trajectory during the experiments was chosen as proportional with time (ramp type). The experiment was performed for two different T_D derivate time with same proportional gain. The measurements clearly show (see Fig. 10.15) that for minor value of T_D the velocity enters in limit cycle, which also has an influence on position response. For the superior value of T_D, the limit cycle disappear.

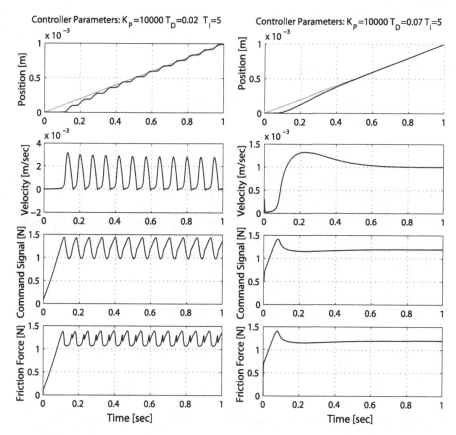

Fig. 10.14 Simulation measurements (PID control)

10.5 Closing Remarks

Nowadays, many types of high resolution position sensors and actuators are available at relatively low cost. With these devices high precision position control tasks became achievable for a wide range of applications. However, in many applications that require precise positioning, it was observed that the control performances are severely influenced by friction and backlash. In this chapter, the effects of these non-smooth nonlinearities on mechanical control systems are analyzed.

First, a piecewise affine model for mechanical systems with Stribeck friction and backlash was introduced. Based on this model a stability condition was developed for mechanical control systems with LQ control law. The stability condition was formulated as a quadratic matrix inequality that depends on the controller design parameters and the parameters of the mechanical system. Simulation results were provided to support the theoretical deductions.

The limit cycles generated by non-smooth nonlinearities were studied in the second part of the chapter. A limit cycle condition was formulated for control systems with both friction and backlash type nonlinearities. A case study was performed to

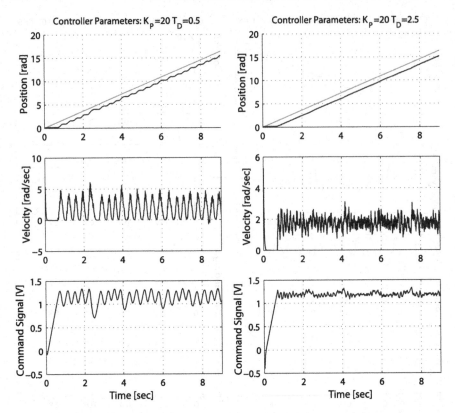

Fig. 10.15 Real-time experiment (PD control) with unstable (*left*) and stable (*right*) behavior

calculate the chaotic measures (Lyapunov exponent, fractal dimension) of the generated limit cycles. These limit cycles appear around zero velocity state.

In the case of position tracking control (time varying position reference signal), oscillations may also appear around Stribeck velocities. These limit cycles can be explained and analyzed based on the piecewise linearized friction model introduced in Chap. 9. Both simulation and real-time measurements were provided to sustain the theoretical results.

Chapter 11
Model Based Identification and Adaptive Compensation of Nonsmooth Nonlinearities

Overview This chapter deals with the identification of nonsmooth mechanical nonlinearities and tracking control of mechatronic systems in the presence of friction. First, a friction and backlash measurement and identification method is presented for robotic manipulators that can be performed in closed loop using velocity controller. It is shown, how the procedure can be extended for hydraulic actuators. Second, a control algorithm is introduced for underactuated systems that applies friction compensation. Afterward, an adaptive friction and payload compensation algorithm is presented for robotic systems. Both the friction identification and friction compensation algorithms are based on the piecewise linear friction model introduced in Sect. 9.4.

11.1 Friction and Backlash Measurement and Identification in Robotic Manipulators

As it was shown in Chap. 10, the friction and backlash can severely influence the performances of mechanical control systems. This is why it is advisable to introduce friction and/or backlash compensation terms into the control algorithm of robotic and mechatronic systems. For effective compensation of these nonsmooth nonlinearities, it is necessary to know the parameters that describe the friction phenomena and the backlash gap size. With known friction model, the robot control algorithm can simply be extended to compensate the effect of friction [17]. With the size of the backlash gap known, the control algorithm can be extended with an inverse backlash element that compensates the backlash-induced effects in the control system [142].

The problem of friction identification was discussed in many studies. In the paper [104], a wavelet network based friction estimation method is proposed. Frequency domain friction identification algorithm is proposed in [29]. The frequency domain identification methods are also popular for backlash identification [102]. Time domain identification method for backlash was proposed in [154].

The frictional parameters may change in time, they depend on external factors such as temperature, humidity, dwell time (the time interval a junction spends in the

B. Lantos, L. Márton, *Nonlinear Control of Vehicles and Robots*,
Advances in Industrial Control,
DOI 10.1007/978-1-84996-122-6_11, © Springer-Verlag London Limited 2011

stuck, when the machine is not moving). The friction parameter variation is a slow process but it could affect the performance of the control system in time. Hence, it is not enough to determine these parameters once, before the robot is putted in operation, the friction should be remeasured periodically and the friction compensation part of the control algorithm retuned. It is why the friction identification has to be a practical algorithm, that uses only sensors necessary for the control of the robotic system. These sensors measure the joint position and velocity (on the motor side) and the motor current, which is necessary to sense the joint overload and for cascade type controllers.

The algorithms proposed in this section, to capture the friction characteristics and backlash gap size in the robot's joints, are based on the measured motor side position, velocity and measured drive motor current [96]. The measurements are performed during controlled robot motion. Both the friction and the backlash are measured during a common measurement procedure. It is considered that the other parameters of the robot arm (lengths, masses, inertias, position of the center of gravity of the segments) are known.

Consider that the robot manipulator is described by (10.1). In this chapter, the load side position of the robot will be denoted by q.

The robot model has two fundamental properties that can be exploited during controller design:

- The inertia matrix $H(q)$ is symmetric and positive definite for every $q \in R^n$.
- The matrix $\dot{H}(q) - 2C(q, \dot{q})$ is skew-symmetric, that is,

$$x^T(\dot{H} - 2C)x = 0, \quad \forall x \in R^n. \tag{11.1}$$

The friction and backlash mainly arise in the gear transmissions that link the robot segments to the drive motors. Consider that the gear ratio in the transmission of the ith joint is N_{2i}/N_{1i}. Consider that the friction (τ_{fi}) in the joints of the robot is described by the nonlinear Stribeck friction model that can be approximated by the piecewise linear model (9.75). A part of the motor generated torque (τ_{Mi}) should compensate the friction force that appears in the gear transmission

$$N_{2i}\tau_{Mi} = N_{1i}(\tau_i + \tau_{fi}(\dot{q}_{Mi})). \tag{11.2}$$

The friction depends on joint velocity, hence it should be measured for different constant velocity values. On the other hand, the frictional parameters differ for positive and negative velocity domains, the measurement should be performed in both velocity domains. The backlash gap can be captured when there is a transition from positive to negative velocity domain or vice versa. Accordingly, to capture one measurement point, the joint motion should be planned in such a way that there are constant velocity regimes both for positive and negative domain and a controlled transition from positive to negative velocities (see Fig. 11.1). Note that these types of velocity profiles (acceleration, constant velocity regime, deceleration) are also extensively used for robot trajectory planning. Hence, the nonsmooth nonlinearities will be identified in such conditions that characterize the robot's motion during normal operation.

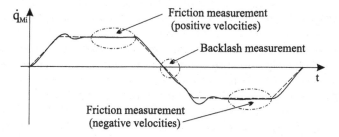

Fig. 11.1 Velocity of the robot's joint during friction and backlash measurement

11.1.1 Friction Measurement and Identification

Assuming that the joint moves in the constant velocity regime ($\ddot{q}_i = 0$) and only the ith joint moves, the motion of the joint is given by

$$C_{ii}\dot{q}_i^2 + D_i(q_i) = \tau_i. \tag{11.3}$$

Since the other joints of the robot do not move, the centrifugal term C_{ii} is constant.

Assuming that the joint is driven by a direct current servo motor, the torque developed is proportional with the motor current. Based on (11.2), the control torques at the input and output of the gear transmission are obtained as

$$\tau_{Mi} = K_{\tau i}i_i, \tag{11.4}$$

$$\tau_i = \frac{N_{2i}}{N_{1i}}K_{\tau i}i_i - \tau_{fi}(\dot{q}_{Mi}), \tag{11.5}$$

where τ_{Mi} is the torque developed by the motor, i_i is the current in the ith joint motor, $K_{\tau i}$ is the torque constant of the drive motor.

Based on (11.3) and (11.5), the relation among the velocity friction force and current is given by

$$\tau_f(\dot{q}_{Mi}) = \frac{N_{2i}}{N_{1i}}K_{\tau i}i_i - C_{ii}\dot{q}_i^2 - D_i(q_i). \tag{11.6}$$

For friction identification, consider that a measurement point is given by the average of the measured velocity and the average of measured current in the constant velocity regime after the transients elapsed: $(\dot{q}_{Mi}, i_i)$.

The velocity dependent friction characteristic can be obtained by repeating the measurement systematically for different velocity values in the operating range.

The parameters of the friction model (9.75) can be obtained using the least squares method. For the ith joint, the cost function is formulated as

$$E = \sum_{i=1}^{NMeas}\left((a_i + b_i\dot{q}_{Mi}) - \left(\frac{N_{2i}}{N_{1i}}K_{\tau i}i_i - D_i(q_i) - C_{ii}\dot{q}_i^2\right)\right)^2. \tag{11.7}$$

Here *NMeas* denote the number of measurements in the corresponding velocity domain.

For $q_{Mi} > 0$, in the low velocity regime, where the friction force decreases with the velocity yields $a_i = a_{i1+}$, $b_i = b_{i1+}$. In the high velocity regime, where the friction force increases with the velocity yields $a_i = a_{i2+}$ and $b_i = b_{i2+}$. The value of the Stribeck velocity can be found at the intersection of the obtained lines. Similar cost function can be defined for negative velocities.

By solving the optimization problem above, the friction parameters for the corresponding velocity domain can be obtained.

11.1.2 Backlash Measurement

The parameter, that characterizes the backlash is the size of the backlash gap (β), the size of the clearance between mating components in the gear transmission that links the driving motor to the robot segment.

The motion of the robot joint transmission in backlash mode is shown in Fig. 11.2. If the motion changes its direction, the joint velocity will be zero both in motor and load side (time moment t_1). In order to achieve zero velocity, both in the motor and load side during direction change, the joint should be in such configuration that the direction of motion is perpendicular to the direction of the gravitational force. The inertia driven motion of the load side (after the motor side stops moving) can be avoided with uniform deceleration, by applying low velocity rate change value. It is beneficial to perform the backlash measurement in the low velocity regime, when the first points of the friction characteristic are measured.

To measure the value of the backlash gap, the measured velocity and current signals during direction change will be used. After the velocity changes, its sign there will be a time period during which only the motor side moves (see Fig. 11.2, time interval t_{12}). It reaches the load side at the time moment t_2, when the drive motor will suffer a shock since the inertia of the mechanics abruptly increases. Accordingly, when the motion changes its direction, the motor current will have a peak and after that its value starts to decrease (backlash mode). When the motor shaft reaches

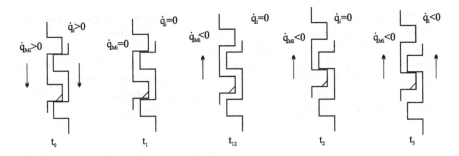

Fig. 11.2 Motion in joint gear transmission during direction change

the load (at the end of the backlash gap), the value of the current will increase again, it will have a local maximum, since the motor starts to drive the load as well.

Two time instants will be captured:

- t_1—the time instant when the machine velocity changes its sign (zero cross detection for the velocity signal).
- t_2—the first moment after t_1 when the current value will have a local minimum (minimum detection for current signal).

The size of the backlash gap can be calculated by integrating the velocity from t_1 to t_2

$$\beta = \int_{t_1}^{t_2} \dot{q}_{Mi}(\sigma) \, d\sigma. \tag{11.8}$$

The integration can be performed using numerical methods. The precision of the numerical integration depends on the chosen sampling period. If the size of the backlash gap is small, smaller sampling period has to be applied.

11.1.3 Velocity Control for Measurements

For friction measurement, the velocity controller should guarantee zero steady state velocity error in the constant velocity regime and it has to deal with modeling uncertainties, since the friction and backlash are considered unknown during the measurements.

To handle the modeling uncertainties, a robust sliding mode controller is developed. To assure zero steady state error for constant reference velocity signal, an integral term is introduced in the controller.

To solve the velocity tracking problem, define the following error metric

$$S_v = \dot{e} + \Lambda \int_0^t \dot{e}(\sigma) \, d\sigma, \quad \text{where } \dot{e} = \dot{q} - \dot{q}_d, \tag{11.9}$$

$\Lambda = \text{diag}([\lambda_1 \lambda_2 \dots \lambda_n]), \lambda_i > 0$.

Consider that in the robot model the effect of friction and backlash appears as an additive bounded uncertainty (d_v)

$$H(q)\ddot{q} + C(q,\dot{q})\dot{q} + D(q) = \tau + d_v. \tag{11.10}$$

It is assumed that the elements of the vector d_v are upper-bounded: $|d_{vi}| \leq D_{Mvi}$.

The control law is formulated as

$$\tau = H(q)(\ddot{q}_d - \Lambda \dot{e}) + C(q,\dot{q})\left(\dot{q}_d - \Lambda \int_0^t \dot{e}(\sigma) \, d\sigma\right) + D(q) - K_{Sv} \text{sat}(S_v/\Phi_v). \tag{11.11}$$

Here Φ denotes the prescribed velocity tracking precision, $\text{sat}(S_v/\Phi_v)$ is a diagonal matrix with the elements: $\text{sat}(S_{vi}/\Phi_{vi})$. $\text{sat}(\cdot)$ is the saturation function and $K_{Sv} = \text{diag}([K_{Sv1}\ K_{Sv2}\ldots K_{Svn}])$, with $K_{Svi} > D_{Mvi}$.

To analyze the proposed control law, consider the following Lyapunov function candidate

$$V = \frac{1}{2}S_v^{\ T}H(q)S_v.$$ (11.12)

The time derivative of V is given by

$$\dot{V} = S_v^{\ T}H(q)\dot{S}_v + \frac{1}{2}S_v^{\ T}\dot{H}(q)S_v.$$ (11.13)

With the control law (11.11), the tracking error dynamics is given by

$$H(q)\dot{S}_v = -C(q,\dot{q})S_v - K_{Sv}\text{sat}(S_v/\Phi_v) + d_v.$$ (11.14)

By substituting (11.14) in (11.13) and applying that $\dot{H}(q) - 2C$ is skew symmetric, the time derivative of Lyapunov function is given by

$$\dot{V} = -S_v^{\ T}K_{Sv}\text{sat}(S_v/\Phi_v) + S_v^{\ T}d_v.$$ (11.15)

Outside the boundary layer Φ_v ($|S_{vi}| > \Phi_v$), we have

$$\dot{V} = -S_v^{\ T}K_{Sv}\text{sgn}(S_v/\Phi_v) + S_v^{\ T}d = -\sum_{i=1}^{n}K_{Svi}|S_{vi}| + S_v^{\ T}d_v.$$ (11.16)

By choosing $K_{Svi} = D_{Mvi} + \eta_i$, where $\eta_i > 0$, we obtain

$$\dot{V} < -\eta_i\sum_{i=1}^{n}|S_{vi}|.$$ (11.17)

The inequality (11.17) implies that with the proposed control law the velocity tracking error metric will converge exponentially fast inside the boundary layer Φ_v.

Since during the measurements only one joint moves, the control law (11.11) will have a simple form that can easily be implemented.

The steps of the identification procedure for a robot joint can be summarized as follows:

- Design the acceleration, constant velocity, deceleration reference velocity profile (see Fig. 11.1) for different constant velocity values.
- Apply the velocity control algorithm (11.11) to track the velocity profiles.
- For low velocity reference values, capture the time instants t_1 and t_2 and determine the size of the backlash gap using (11.8).
- For every reference velocity profile, determine the measurement point $(\dot{q}_i, i_i)$ in the constant velocity regime.
- Compute the joint friction parameters by solving the optimization problem (11.7).

11.1.4 Experimental Measurements

In order to test the applicability of the proposed control and identification methods introduced in this section, a microcontroller based distributed robot control equipment was developed [92]. Here, we present the robot controller in detail. Afterward, friction and backlash identification results are presented that were performed by using the developed control equipment.

11.1.4.1 Experimental Setup

Robot manufacturers usually provide the control system for the sold robots with implemented trajectory planning algorithm, low level control algorithm and robot programming language. These control systems have to satisfy strong requirements related to reliability, control precision, safety. Because of high computational costs of the trajectory planning and robot control algorithm, the used control systems are generally based on professional but expensive industrial computers.

Nowadays microcontrollers and Digital Signal Processors (DSPs) are widely used in many industrial control systems. The main advantage of the microcontrollers is that many interfaces are integrated in the same chip with the processor and accordingly the hardware cost of the control board is low. The power of the DSPs is in their computational capabilities: they can perform mathematical computations, which are necessary in control algorithms, with high speed. Fortunately, there is a tendency in microcontroller industry to join the advantages of these two architectures namely to make controller architectures with high computational capacity and at the same time equipped with useful interfaces. Such controllers are named Digital Signal Controllers (DSC). These types of controllers can effectively be used for the position or velocity control of a single servo motor. They are equipped with interfaces necessary in robot control system:

- Incremental encoder signal processing block (for position and velocity measurements).
- Pulse Width Modulation (PWM) generator module (for motor control).
- Analogue inputs (for resistive position sensors and/or motor current measurement).
- Digital inputs and outputs (for limit switches, pneumatic actuators for gripper control, etc.).

In the case of robots, the control system has to deal with four or more joints, the control of which cannot be done with a single controller. If we would like to use DSC for robot control, each joint should be controlled with one local controller unit. However, in the case of the advanced robot control algorithms, the control law for the i th joint may depend on joint variables of the entire robot. Hence, it is necessary that in every control period the joint information (measured position and velocity) are exchanged between the controllers. Accordingly, fast communication between the local controllers is necessary. It can be done by organizing the controllers in

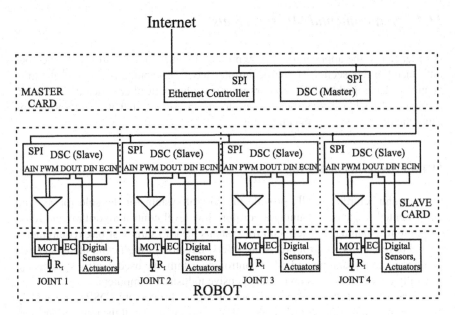

Fig. 11.3 SPI bus based distributed robot control system

a distributed system around a fast communication bus, which can guarantee that the joint information is interchanged between the local units during a single control period (around 1ms). Accordingly, the local controllers must be capable of fast computation and at the same time they must be connected with fast communication bus.

To avoid costly and complex hardware design, the local controllers should be equipped with interfaces necessary for robot control.

The control system should be capable to communicate with a distant PC on which the robot programming language and the trajectory planning algorithm is implemented. The problem can be solved by including an Ethernet controller in the control system through which the control system has Internet access. With direct access to Internet, it is not necessary to put a computer near the robot workspace. The operator of the machine can send predefined trajectory sequences from a distant computer or he/she may command the robot online for example, using haptic device or robot programming language. In this case the motion of the robot can be observed using network cameras.

The schema of the proposed distributed control system can be seen in the Fig. 11.3. The system has a master-slave configuration. All data that should be exchanged between the local controllers (slaves) is sent to a master controller, which redistributes the data to the slaves.

The system is organized around a Serial Peripheral Interface (SPI) bus. This bus is a synchronous serial data link standard named by Motorola that operates in full-duplex mode. A full-duplex system allows communication in both directions, and allows this to happen simultaneously. Devices communicate in master-slave mode where the master device initiates the data frame. Multiple slave devices are allowed with individual slave select lines. The full duplex operation is very useful for robotic application since the local controllers can send their sensory information and at the same time they can receive the reference trajectory.

The chosen Ethernet controller for distant communication can also be connected to the SPI bus; hence, it will also be a slave in the system. The SPI synchronous bus can work with 1–70 MHz clock frequency; hence the communication can be performed with high speed. The architecture of the DSCs also supports the high speed mathematical computation. Accordingly, the computational time is also drastically lower than in the case of classical microcontrollers. Since the applied local controllers are equipped with robot specific interfaces (PWM modules for motor (MOT) control, encoder (EC) interfaces, analogue inputs (AIN) for current measurement), the input/output operations can be performed instantly or parallel with the communication phase.

In each control cycle, the following steps are performed (*Distributed Control Algorithm*):

1. The master controller reads the reference trajectory from the Ethernet controller (slave).
2. a. The master sends the reference trajectories (q_{di}, $\dot{q}_{di}$, $\ddot{q}_{di}$) for the local controllers (slaves).
   b. Simultaneously the local controllers send their joint information (q_i, $\dot{q}_i$) and the drive motor current (i_i) from the current control cycle.
3. The master redistributes the joint information for the local controllers.
4. a. The local controllers calculate the control signals for the joints in parallel.
   b. Simultaneously the master sends the joint information for the human operator through Ethernet controller (for monitoring purpose).

The data structures that are used in the second step of the Distributed Control Algorithm can be seen in the Fig. 11.4. The master sends for the ith slave the reference trajectory data necessary for the computation of the ith control signal.

Simultaneously, the slave sends its joint information (position and velocity) for the master. The auxiliary data (*Aux. data*), that are included at the end of the communication package, contains digital information, for example, control mode, the status of the digital inputs/outputs, digital commands for the pneumatics, grippers.

During the third step of the Distributed Control Algorithm, only the master sends information for the slaves (see Fig. 11.5). In this phase, the joint information will be redistributed for the slaves. Obviously only such joint variables has to be sent for the ith slave that are used in the calculus of the ith component of control signal (τ_i).

In order to handle the communication errors, watchdog-type interrupts should be included in the software of the slave units and the master unit: if no new data arrive in a predefined time, the motion of robot arm must be stopped.

Master Data	q_{d1} (4 bytes)	$\dot{q}_{d1}$ (4 bytes)	$\ddot{q}_{d1}$ (4 bytes)	...	q_{dn} (4 bytes)	$\dot{q}_{dn}$ (4 bytes)	$\ddot{q}_{dn}$ (4 bytes)	Aux. data (4 bytes)
ith Slave Data	q_i (4 bytes)	$\dot{q}_i$ (4 bytes)	i_i (4 bytes)	Aux. data (4 bytes)				

Fig. 11.4 Exchanged data in the second step of the distributed control algorithm

Fig. 11.5 Exchanged data in the third step of the distributed control algorithm

Master Data	q_1 (4 bytes)	$\dot{q}_1$ (4 bytes)	...	q_n (4 bytes)	$\dot{q}_n$ (4 bytes)
ith Slave Data					

11.1.4.2 Experimental Results

A type I SCARA robotic arm was used to test the performances of the developed control system (see Fig. 11.6). The robot has four Degrees Of Freedom, three rotational and one translational joint. There is a strong nonlinear coupling between the first and second joint of the robot. Each joint is equipped with a 12 V DC servo motor and a 1000 Pulses Per Turn (PPT) incremental encoder. Each motor drives the robot segment through a gear head. The main component of the frictional force and the backlash appears in this gear head. The end-effector of the robot is a pneumatically controlled gripper.

Accordingly, to control the robot with the proposed distributed control system, one master and four slave controllers are necessary. The DSCs used as slave processors are dsPIC30F2010 with 80 MHz clock frequency, for Ethernet controller the ENC28J60 circuit was applied. The encoders are connected to the encoder interface of the controllers through a simple signal conditioning circuit. The counting of the encoder pulses are performed inside the controller. The control signal is sent out through the PWM interface of the slave controller which drives an H bridge amplifier. The motors are connected directly to this H bridge drive.

Since the master does not have to deal with sophisticated computation, as master processor a 40 MHz microcontroller was chosen from the PIC18 family. All the controllers (master, slaves and the Ethernet controller) are connected to a common SPI bus with 2.5 MHz clock frequency.

Since the motors are driven through H bridges, the motor current can simply be measured using a high power resistor placed between the lower transistors of the bridge and the ground. With this current sensing method the absolute value of the current can be obtained, which is enough for friction and backlash measurement. The electric signal, proportional with the motor current, is introduced in the slave controllers through their analogue input interface. Beside the friction and backlash identification, the current signal is also used for motor protection: if the motor current exceeds a predefined limit, the motion of the robot is stopped.

With known value of the resistor (R_I in Fig. 11.3) the motor current can simply be calculated ($i_i = U_{INi}/R_I$), where U_{INi} is the measured voltage on the resistor. The control torque developed by the motor can be calculated by multiplying the current value with the torque constant, which is a motor catalogue data, see the relation

Fig. 11.6 SCARA type robotic arm and its control system used for experimental measurements

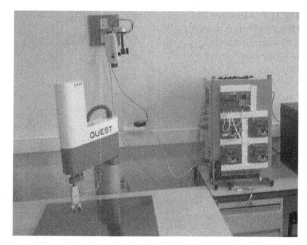

Table 11.1 Robot parameters

Parameter	1st segment and joint	2nd segment and joint
Mass	$m_1 = 5.074$ kg	$m_2 = 8.696$ kg
Inertia (zz)	$I_{zz1} = 0.0825$ kg m^2	$I_{zz2} = 0.1276$ kg m^2
COG position	$l_{c1} = 0.14$ m	$l_{c2} = 0.167$ m
Length	$l_1 = 0.3$ m	$l_2 = 0.3$ m
Gear ratio	$N_2/N_1 = 86/16$	$N_2/N_1 = 40/12$
Motor torque constant	$K_\tau = 0.023$ N m/A	$K_\tau = 0.023$ N m/A

(11.4). Hence, in constant velocity regime the torque developed by the motor can easily be determined.

The dynamic parameters of the first two segments of the robot are summarized in the Table 11.1. The masses and inertias of the third and fourth joint are $m_3 = 1.57$ kg, $I_{zz3} = 658 \times 10^{-6}$ kg m^2 and $m_4 = 0.69$ kg, $I_{zz4} = 56 \times 10^{-6}$ kg m^2, respectively.

During the experiments, the friction and backlash were identified in the first and the second joint. The trajectory was planned such that the joint will move in its entire angle domain (± 160 deg for the first joint, ± 135 deg for the second joint). The friction was identified in the ± 10 deg/s velocity regime. The reference velocity was increased by 0.5 deg/s in each step. A measurement point was determined by simply calculating the mean of the current and the mean of velocity in the constant velocity regime in steady state. The frictional parameters were identified by applying the Least Squares method for the cost function (11.7). The identification results can be observed in the Figs. 11.7 and 11.8. It can be observed that with the proposed identification method the friction model fits well the noisy experimental measurements.

Fig. 11.7 Measured and
identified friction in joint 1
('*'—measurements ,
'-'—fitted model)

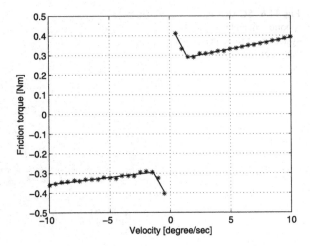

Fig. 11.8 Measured and
identified friction in joint 2
('*'—measurements ,
'-'—fitted model)

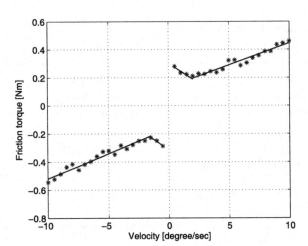

The obtained friction models for the first and the second joints are

$$\tau_{1f}(\dot{q}_1) = \begin{cases} 0.4645 - 0.119\dot{q}_1, & \text{if } \dot{q}_1 \in (0, 1.5], \\ 0.2718 + 0.0106\dot{q}_1, & \text{if } \dot{q}_1 \in (1.5, 10.0], \\ -0.4493 - 0.1085\dot{q}_1, & \text{if } \dot{q}_1 \in [-1.5, 0), \\ -0.2847 + 0.0072\dot{q}_1, & \text{if } \dot{q}_1 \in [-10.0, -1.5). \end{cases} \tag{11.18}$$

$$\tau_{2f}(\dot{q}_2) = \begin{cases} 0.3013 - 0.0562\dot{q}_2, & \text{if } \dot{q}_2 \in (0, 1.8], \\ 0.1334 - 0.0312\dot{q}_2, & \text{if } \dot{q}_2 \in (1.8, 10.0], \\ -0.3141 - 0.0594\dot{q}_2, & \text{if } \dot{q}_2 \in [-1.9, 0), \\ -0.1623 + 0.0359\dot{q}_2, & \text{if } \dot{q}_2 \in [-10.0, -1.9). \end{cases} \tag{11.19}$$

In the above relations, the dimension of the friction generated torque (seen from
the motor side) is N m, the dimension of the velocity is deg/s.

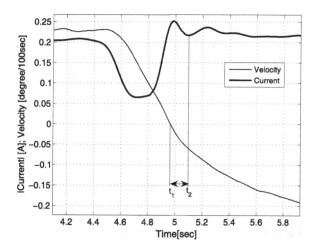

Fig. 11.9 Backlash identification in joint 1

The backlash measurement was performed in the first 10 measurement steps. Firstly, the time moments t_1 (zero velocity) and t_2 (first minimum in current signal after velocity sign change) were captured (see Fig. 11.9) and afterward the velocity was integrated between these time instants using the trapezoidal rule. In the first joint, the mean of the calculated values gives the backlash gap $\beta_{Mean} = 0.47$ deg. The mean deviation of the obtained value from the measured values was also calculated using the formula $\Delta\beta = (\sum_{i=1}^{10} (|\beta_{Mean} - \beta_{iMeasured}|/\beta_{Mean})/10) \cdot 100$. Its value was found as $\Delta\beta = 6.3\%$. In the second joint, the following values were found: $\beta_{Mean} = 0.35$ deg; $\Delta\beta = 11.9\%$.

11.2 Friction Measurement and Identification in Hydraulic Actuators

Hydraulic actuators are widely used in many industrial systems including construction machines (excavators) and heavy-duty manipulators. The walls of chambers and pistons of these actuators are lubricated. The piston does not move until the tangential control force between the piston and the cylinder reaches the level of static friction (phenomenon of sticking). Once the motion starts, the friction force varies as a function of velocity.

To deal with friction in hydraulic actuators, both identification and compensation algorithms were proposed. The paper [162] describes how the Coulomb and viscous friction coefficients, beside other parameters of the dynamic model, can be identified in a hydraulically-driven excavator using Least Squares and generalized Newton methods. An online Coulomb friction observer, combined with velocity observer was proposed in [141]. The paper [18] proposes an identification method for viscous and Coulomb friction parameters in hydraulic cylinders based on pressure measurement. A friction identification method was introduced in [163] for a parallel

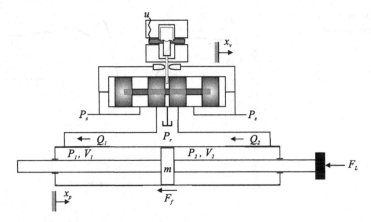

Fig. 11.10 Schematic of hydraulic actuator

hydraulically-driven robot based on a simplified form of the Stribeck model. Evolutionary algorithms were applied in [160] to identify the parameters of a hydraulic servo with flexible load.

In this section, a simply implementable and efficient friction identification algorithm is presented for finding all the parameters of the static, velocity dependent friction model.

11.2.1 Mathematical Model of Hydraulic Actuators

A schematic of a valve-controlled hydraulic actuator is shown in Fig. 11.10. The mathematical model, that relates the control input u to the actuator position x_p depends on the dynamics of the valve spool, the nonlinear flows through the valve control ports, the compressibility of the fluid, as well as the mechanics of the piston motion.

For a two-stage flapper-nozzle servovalve, the relationship between the input current of the torque motor and the position of the valve spool is nonlinear. However, in the design of hydraulic servo-systems, detailed analysis of the nonideal valve characteristics is generally not required. For control system design, the servovalve spool dynamics are often approximated using linear transfer function as model, that captures the main characteristics of the valve dynamics in the low frequency range. Here, a second-order model is used to model the dynamics of the servovalve spool position. Thus, the relationship between the position of the valve spool x_v and the current u applied to the torque motor is

$$x_v = -\frac{1}{\omega_v^2}\ddot{x}_v - \frac{2\xi_v}{\omega_v}\dot{x}_v + k_v u, \tag{11.20}$$

where k_v is the valve spool position gain, and parameters ω_v and ξ_v are the natural frequency and the damping ratio, that best represent the servovalve magnitude and phase characteristics in the low frequency range. The values of this second-order model parameters can easily be estimated from valve frequency response curves given in manufacturer's catalogs.

For a valve with a critically-lapped spool having matched and symmetrical orifices, control flows Q_1 and Q_2 follows the turbulent orifice equation. The nonlinear governing equations can be written in the following compact form [101]

$$Q_1 = K_v \omega x_v \sqrt{\frac{P_s - P_r}{2} + \frac{x_v}{|x_v|}\left(\frac{P_s - P_r}{2} - P_1\right)}, \qquad (11.21)$$

$$Q_2 = -K_v \omega x_v \sqrt{\frac{P_s - P_r}{2} + \frac{x_v}{|x_v|}\left(P_2 - \frac{P_s - P_r}{2}\right)}. \qquad (11.22)$$

In the above equations, $K_v = C_v \sqrt{2/\rho_{\text{oil}}}$ is the valve flow gain, which depends on the orifice coefficient of discharge C_v and the density ρ_{oil} of the hydraulic fluid. Parameter ω is the width of the rectangular port cut into the valve bushing through which the fluid flows. The supply and tank pressures are denoted by P_s and P_r, respectively. P_1 and P_2 refer to the hydraulic pressures in each of the actuator chambers.

The continuity equations for compressible fluid volumes describe the time rate of change of the pressures in actuator chambers. The continuity equation considers both the deformation and the compressibility flows and can be written for each cylinder half as

$$\dot{P}_1 = \frac{\beta_h}{Ax_p - V_1}(Q_1 - Av_p), \qquad (11.23)$$

$$\dot{P}_2 = \frac{\beta_h}{A(L - x_p) + V_2}(Q_2 + Av_p). \qquad (11.24)$$

In the above equations, A is the annulus areas of the piston; parameter β_h denotes the effective bulk modulus of the hydraulic fluid. The volumes of oil contained in the connecting lines between the servovalve and the actuator are given by V_1 and V_2. The length of the actuator stroke is denoted by L.

In order to complete the mathematical model of the hydraulic servo-actuator, the mechanics governing the motion of the actuator are now considered. By carrying out a force balance on the piston, it is found that the motion of the actuator can be described by the following equation

$$m\dot{v}_p = A(P_1 - P_2) - F_f - F_L. \qquad (11.25)$$

Here m is the combined mass of the piston, actuator rods and load. Term $A(P_1 - P_2)$ is the force generated by the actuator, while force F_L refers to the external load disturbance. The force acting between the piston and the cylinder walls due to friction is denoted by F_f.

The nonlinear model of the entire hydraulic actuator in state space can now be written based on equations (11.20)–(11.25)

$$
\begin{cases}
\dot{x}_p = v_p, \\
\dot{v}_p = \frac{A}{m}(P_1 - P_2) - \frac{1}{m}(F_f + F_L), \\
\dot{P}_1 = \frac{\beta_h}{Ax_p - V_1}\left(K_v \omega x_v \sqrt{\frac{P_s - P_r}{2} + \frac{x_v}{|x_v|}(\frac{P_s - P_r}{2} - P_1)} - Av_p\right), \\
\dot{P}_2 = \frac{\beta_h}{A(L - x_p) + V_2}\left(-K_v \omega x_v \sqrt{\frac{P_s - P_r}{2} + \frac{x_v}{|x_v|}(P_2 - \frac{P_s - P_r}{2})} + Av_p\right), \\
\dot{x}_v = v_v, \\
\dot{v}_v = -\omega_v^2 x_v - 2\xi_v \omega_v v_v + k_v \omega_v^2 u.
\end{cases}
\qquad (11.26)
$$

11.2.2 Friction Measurement and Identification

The proposed friction measurement method [99] is based on measurements of pressure chambers (P_1 and P_2) and the velocity of the actuator's rod (v_p). The value of friction is measured at different steady-state velocity values. Additionally, since the frictional parameters may be different for positive and negative velocities, the measurements should be performed in positive and negative velocity domains, separately.

With reference to (11.25) when the rod moves at a constant velocity and without external load, the friction force easily relates to the pressure difference in the actuator

$$
F_f = A(P_1 - P_2). \qquad (11.27)
$$

Assume that during its operation the actuator's velocity domain is $[-v_{pmax}, v_{pmax}]$. From the operation domain, N uniformly distributed velocity values are taken both from positive and negative velocity domains. The measurement algorithm is summarized below.

For all v_{pi} velocity values repeat:

- Stabilize the velocity to v_{pi}, that is, allow the velocity to reach the steady state.
- Calculate the average of the measured velocity and the pressure difference in the chambers over a given time to remove the effect of measurement noise.
- Calculate the friction force (F_{fi}) corresponding to the velocity v_{pi} using (11.27).
- Save the measurement data points (v_{pi}, F_{fi}).

Here i denotes the measurement index.

To identify friction parameters, the friction curve obtained from measurements was approximated using the model (9.75). Considering only positive velocity regions, the first M measurement points, out of total N measurements, are approximated with the first line and the last $N - (M + 1)$ measurement points are approximated with the second line. For positive velocities the approximation error (i.e., the average distance between the measurement points and friction model), is given by

$$E = \frac{1}{N} \left(\sum_{i=1}^{M} \left| F_f(v_{pi}) - \left[(a_{1+}) + (b_{1+}) v_{pi} \right] \right| \right.$$
$$\left. + \sum_{j=M+1}^{N} \left| F_f(v_{pj}) - \left[(a_{2+}) + (b_{2+}) v_{pj} \right] \right| \right). \tag{11.28}$$

The parameters of the piecewise linearized friction model are obtained using the following steps:

- *First iteration* ($M = 2$): Fit the first line to the first two points and the second line to the remaining $N - 2$ points by applying the Least Squares algorithm (e.g., using the `polyfit` MATLAB function). Determine the approximation error E using (11.28).
- *Second iteration* ($M = 3$): Fit the first line on the first three points and the second line to the remaining $N - 3$ points by applying the Least Squares algorithm. Determine the approximation error.
- ...
- $N - 2$ (*last*) *iteration* ($M = N - 2$): Fit the first line to the first $N - 2$ points and the second line to the last two points by applying the Least Squares algorithm. Determine the approximation error.

The optimal pair of segments are those which produce the minimum approximation error as per (11.28).

The iterative algorithm presented here solves an optimization with constraints. In the optimization problem the cost function is given by (11.28) and the constraints are: the first line has to cross the first measurement, the second line has to cross the last measurement and the two lines should have to intersect each other in a measurement point.

11.2.3 Experimental Measurements

The schematic of the experimental setup is shown in Fig. 11.11. The entire system is powered by a hydraulic pump, which produces continuous and stable high-pressure hydraulic fluid (up to 18.27 MPa, i.e., 2650 psi) to the actuator. The two chambers of the actuator are noted as chamber 1 and chamber 2. The actuator is connected to and controlled by a Moog D765 servovalve. This servovalve receives control signals from a PC equipped with a DAS-16 data acquisition board and a Metrabyte M5312 encoder card for displacement and velocity measurement. When operated at 20.7 MPa, Moog D765 valve can supply the actuator with hydraulic fluid at a rate of 0.566 l/s. During experiments, the operating pressure was set to approximately 13.8 MPa (2000 psi).

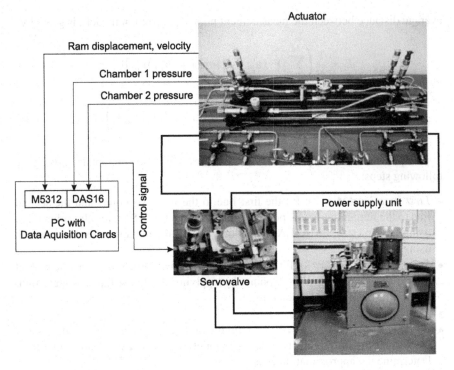

Fig. 11.11 Experimental setup

The measurement algorithms are implemented on a PC with an Intel Pentium III processor. The displacement of the actuator can be measured using Metrabyte M5312 quadrature incremental encoder card. With its rotary optical encoder, M5312 reaches a position measurement resolution of 0.03 mm per increment. Other necessary system states are measured by transducers mounted on the hydraulic circuit and transmitted to a DAS-16 board. The DAS-16 board also transmits control signals from the PC to the Moog D765 valve.

The parameters of the actuator (excluding the friction parameters) used in the identification procedure are given in Table 11.2.

To implement the friction identification algorithm presented in the previous subsection, a square control signal u, having 3 s period, was applied to the system for 120 s. The amplitude of the control signal was increased with a 0.05 V step in every period. The actuator velocity and chamber pressures were measured using a sampling period of 2 ms.

Typical measurement results are shown in the Fig. 11.12. Figure 11.13 shows the close up of the results. It is seen that the measurements are influenced by noise and the pressure difference signal has large transients. During identification in each measurement period, only the second half of the collected pressure—velocity measurement data points (circled in Fig. 11.13) were taken into consideration. This is

Table 11.2 List of nominal system parameters under normal operation

Parameter	Notation	Value
Supply pressure	P_s	≤ 17 MPa
Tank pressure	P_r	<0.25 MPa
Mass of piston, rods and load	m	12.3 kg
Piston annulus area	A	6.63 cm^2
Actuator stroke	L	6.96 cm
Line volumes	V_1, V_2	88.7 cm^3
Hydraulic density	ρ_{oil}	847 kg/m^3
Fluid bulk modulus	β_h	689 MPa
Servovalve discharge coefficient	C_v	0.6
Servovalve flow gain	K_v	2.92×10^{-3} m$^{\frac{3}{2}}/\sqrt{\text{kg}}$
Maximum valve spool displacement	x_{vm}	± 0.279 mm
Servovalve orifice area gradient	ω	20.75 mm^2/mm
Servovalve spool position gain	k_v	0.0279 mm/V
Servovalve natural frequency	ω_v	175 Hz
Servovalve damping ratio	ξ_v	0.65 cm^3

to ensure that the transient responses are excluded. Additionally, the measurements were averaged to mask the effect of measurement noise.

The application of the proposed friction parameter identification algorithm is demonstrated in Fig. 11.14. The measured friction characteristic is obtained from measurements presented in Fig. 11.12. Figure 11.14 shows the second iteration, the one before the last iteration, the evolution of the approximation error versus each iteration, and the iteration corresponding to the case for which the approximation error is minimal. In this case, the eleventh iteration was found to be the optimal one for which $E = 17.175$ N.

The algorithm can be applied in the same manner for the negative velocity regime. The obtained friction model was found to be as follows

$$
F_f(v) = \begin{cases}
1015.4 - 5996.6v_p, & \text{if } 0 < v_p < 0.0696, \\
513.18 + 1219.1v_p, & \text{if } v_p \geq 0.0696, \\
-739.69 - 2545.4v_p, & \text{if } -0.0805 < v_p < 0, \\
-426.96 + 1339.4v_p, & \text{if } v_p \leq -0.0805.
\end{cases}
\tag{11.29}
$$

To validate the proposed identification method, the experimental measurements were compared with simulation measurements. In the simulation model, built in MATLAB/Simulink environment, the actuator's dynamics was described by the relation (11.26) with the parameters given in Table 11.2. The friction in actuator was modeled using (11.29). The comparison was done with square signal inputs u applied to both the experimental test rig and the simulation model with the identified friction model.

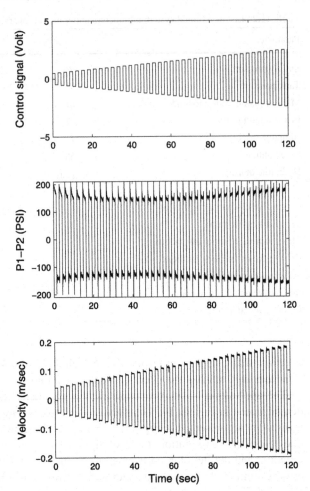

Fig. 11.12 Measurement during identification

The comparison of the simulated velocity output and the measured velocity is seen in the Fig. 11.15. The plot shows the velocity response with the identified friction, for the case where the friction parameters were −15% off the identified values and in the absence of the friction. The comparisons of the measurements and simulation results show firstly the importance of the identification of friction in hydraulic actuators, and secondly the validation of the proposed technique in indentifying proper values for the parameters of the friction model.

For numerical evaluation of the modeling precision, the following formula for the approximation error was assumed

$$E_M = \frac{\sum_{i=1}^{N} |v_{piSimulated} - v_{piMeasured}|}{\sum_{i=1}^{N} |v_{piMeasured}|} 100 \ (\%). \tag{11.30}$$

Fig. 11.13 Closed-up results (42–48 second)

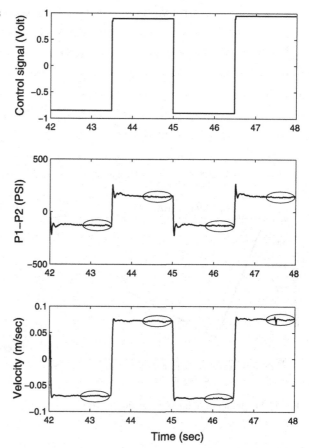

The value of the modeling error, when using the friction model identified in this work, was found as $E_M = 3.66\%$, which is acceptable in many industrial applications.

11.3 Nonlinear Control of a Ball and Beam System Using Coulomb Friction Compensation

Recent research and development in the field of humanoid robotics has motivated great interest in the control of underactuated systems. Well-known examples of underactuated robots include the double inverted pendulum, underactuated two-link robots [137] and the ball and beam systems. Underactuated systems can be inherently unstable, and may also possess complicated internal dynamics. Furthermore, underactuated robotic systems may not be feedback linearizable. In this section, a nonlinear control method is presented for two degree of freedom underactuated sys-

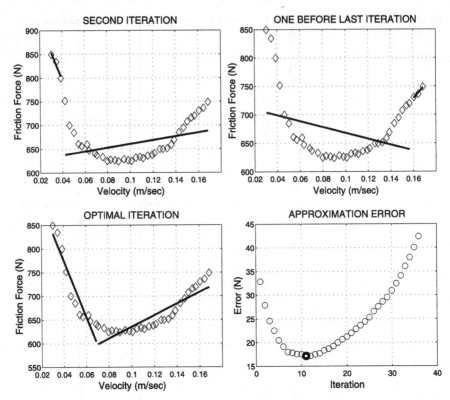

Fig. 11.14 Iterations of the identification algorithm

tems [98]. In order to achieve zero steady state error, the proposed model based control law will be extended with a Coulomb friction compensator term.

The ball and beam system is a classical underactuated mechanical system with two DOF, see Fig. 11.16. The beam rotates in the vertical plane driven by the torque at the center of rotation (usually generated by an electrical drive). The ball rolls freely along the beam and in contact with it. Despite its mechanical simplicity, the ball and beam system presents significant challenges under large motion dynamics; the system is nonlinear, unstable, and has complex structural properties, including a poorly defined relative degree.

Assume the friction appearing in the transmission between the drive and the beam is Coulomb, the equations of motion of the plant are given by

$$\dot{x}_1 = x_2,$$
$$\dot{x}_2 = B(x),$$
$$\dot{x}_3 = x_4,$$
$$\dot{x}_4 = U,$$

(11.31)

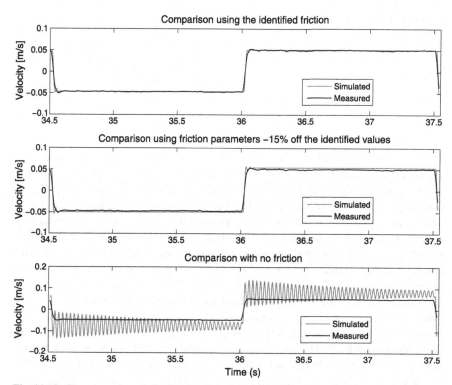

Fig. 11.15 Closed-up view of simulated and measured velocities

where

$$B(x) = \frac{M}{MR^2 + J_b}\left(x_1 x_4^2 - g \sin(x_3)\right). \tag{11.32}$$

and the input torque τ is related to U by

$$\tau = 2M x_1 x_2 x_4 + M g x_1 \cos(x_3) + \left(M x_1^2 + J + J_b\right)U + F_C \operatorname{sgn}(x_4). \tag{11.33}$$

The following notations are used: x_1 for the position of the ball position along the beam; x_2 for ball velocity; x_3 for beam angle, x_4 for beam angular velocity; τ for input torque. Parameters of the model are: J for beam moment of inertia, J_b for ball moment of inertia, M for ball mass, R for ball radius, g for gravitational acceleration, F_C for Coulomb friction coefficient. The state variable is given by the vector $x = (x_1 \ x_2 \ x_3 \ x_4)^T$.

Because the ball and beam dynamics are nonlinear, natural choices to formulate the control law may be either the input-output linearization method or the feedback linearization method. Unfortunately, the relative degree of the system is not well defined because of the centrifugal term $x_1 x_4^2$ in the expression of $B(x)$ [126]. Thus, the conventional input-output linearization approach is not applicable for this system. The standard Jacobian linearization method can approximate well the nonlinear model only when the states are near the linearization point.

Fig. 11.16 Ball and beam schematics

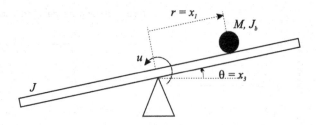

The control of ball and beam dynamics has been recently examined by many researchers. In [48], the control problem was locally solved using approximate input-output feedback linearization. Using nonlinear transformations and neglecting the centrifugal term $x_1 x_4^2$ in the model (11.32), the system has well-defined relative degree. Independently, a fuzzy systems approach has been proposed to blend two control approaches: exact feedback linearization when the system is far from the singularities ($x_1 x_4^2 = 0$), and approximate linearization near the singularities. The control law is a weighted average of the two control laws that were defined over different subspaces [44]. In [148], the idea was further developed using a switching algorithm to choose between approximate and exact feedback linearization algorithms. Instead of switching, a variable structure switching algorithm is also proposed to avoid the singular states [51]. In [6], a model based control algorithm was proposed for the ball and beam system which exploits that the plant model satisfies the so called matching condition. This condition guarantees that the control law will induce some specified structure on the closed loop system. In the paper [109], passivity-based controller was developed for underactuated mechanical systems which also was tested on the ball and beam system.

11.3.1 Adaptive Friction Identification

It is well known that if precise positioning is desired in mechanical control systems, then friction effects must also be taken into consideration in the control algorithm. If friction compensation is omitted, then steady state error can appear. In the case of the ball and beam system, the ball will not reach precisely the desired equilibrium point (center of the beam). The dominant friction appears in the mechanical transmission between the driving motor and the beam.

In this section, an adaptive control approach is used to identify the friction characteristic. Specifically, the unknown friction parameter is estimated using an adaptive control law designed for a trajectory tracking task. The steady state value of the estimated parameter is then used for fixed friction compensation in the model based nonlinear control algorithms.

Assuming that the friction can be described with a Coulomb friction model, the dynamics of the beam without the ball is given by

$$J \dot{x}_4 = \tau - F_C \, \text{sgn}(x_4). \tag{11.34}$$

In the model above, the beam moment of inertia J is known, but the Coulomb friction coefficient F_C is unknown. Denote with $\hat{F}_C$ the unknown (estimated) parameter and let the estimation error be $\tilde{F}_C = F_C - \hat{F}_C$.

Define the tracking error metric for the beam

$$S_B = (x_4 - \dot{x}_{3ref}) + \lambda_B(x_3 - x_{3ref}), \qquad (11.35)$$

where x_{3ref} is the prescribed beam angle, a twice differentiable function of time.

Choose the control law

$$\tau = \hat{F}_C \, \text{sgn}(x_4) + J\left(\ddot{x}_{3ref} - \lambda_B(x_4 - \dot{x}_{3ref})\right) - K_S S_B, \quad K_S > 0. \qquad (11.36)$$

To obtain the unknown friction parameter, choose the adaptation law

$$\dot{\hat{F}} = -\gamma_F S_B \, \text{sgn}(x_4). \qquad (11.37)$$

For stability analysis of the adaptive control law, define the following Lyapunov function candidate

$$V_B = \frac{1}{2}S_B^2 + \frac{1}{2\gamma_F}\tilde{F}_C^2, \quad \gamma_F > 0. \qquad (11.38)$$

It can be shown that with control law (11.36) and adaptation law (11.37), the time derivative of the Lyapunov function is negative definite

$$\dot{V}_B = -K_S S_B^2 < 0, \quad \forall S_B \neq 0. \qquad (11.39)$$

According to (11.39), the tracking error metric converges to zero: $S_B \to 0$ as $t \to \infty$. From the Lyapunov analysis, one can not directly conclude that the friction parameter estimation error converges to zero. However, from the closed loop system dynamics

$$\dot{S}_B + K_S S_B = -F_C \, \text{sgn}(x_4) + \hat{F}_C \, \text{sgn}(x_4) \qquad (11.40)$$

it can be shown that if S_B and $\dot{S}_B$ converge to zero, then the value of the estimated friction parameter will converge to the real friction parameter value if $x_4 \neq 0$.

A nominal, fixed value of the friction coefficient is required for the robust control design methods. A value for the nominal friction coefficient is computed as the mean value of the estimated parameter in steady state. This mean value can be used in the friction compensator terms of various control laws developed for the underactuated system.

11.3.2 Nonlinear Control Algorithm for the Ball and Beam System

To develop a control law for the ball and beam system, define the following combined error metric

$$S = \alpha(x_2 + \lambda_1 x_1) + (x_4 + \lambda_2 x_3), \qquad (11.41)$$

where $\alpha, \lambda_1, \lambda_2 > 0$.

The first term component of the error is meant to stabilize the ball in the center of the beam. The second term is introduced to avoid the large motions of the beam that could introduce instability during the control.

The control task can be formulated as follows: find a control law U such that $S(t) \rightarrow 0$ as $t \rightarrow \infty$.

It is well known that the function $S(t)$ will converge to zero if it satisfies

$$\dot{S} + K_S S = 0, \quad K_S > 0. \tag{11.42}$$

Calculate the time derivative of S

$$\dot{S} = \alpha \big(B(x) + \lambda_1 x_2 \big) + (U + \lambda_2 x_4). \tag{11.43}$$

Formulate the control law as follows

$$U = -\alpha \big(B(x) + \lambda_1 x_2 \big) - \lambda_2 x_4 - K_S S. \tag{11.44}$$

Substituting (11.44) into (11.43) yields (11.42). Hence, with U given by (11.44) the proposed control problem is solved.

The control law (11.44) guarantees that the weighted sum of the states converges to zero. Moreover, the parameters $\lambda_1 > 0$, $\lambda_2 > 0$, $\alpha > 0$ do not depend on the controlled system dynamics. Therefore, the convergence of the combined error metric (11.41) holds for arbitrarily chosen controller parameters and accordingly all the states of the system must converge to zero.

The main advantage of this approach is that the plant nonlinearity appears as an additive term in the dynamics (11.43) of the error metric. Hence, it can easily be compensated with the term "$-\alpha B(x)$" of the control law (11.44) and the stabilization of the function S can be guaranteed. The remaining terms of the control law (11.44), "$-\alpha \lambda_1 x_2 - \lambda_2 x_4 - K_S S$", are equivalent to a linear state feedback.

The stabilizing control law can be extended with the $\hat{F}_C \operatorname{sgn}(x_4)$ term to compensate for friction. During the stabilizing control, the mean value of the estimated parameter $\hat{F}_C$ in steady state is applied, which can be obtained from an *a priori* identification process as described in the previous subsection.

11.3.3 Experimental Evaluations

11.3.3.1 Experimental Setup

The proposed control algorithm was tested on the AMIRA BW500 ball and beam system controlled by a computer equipped with a dSpace DS1102 control board (see Fig. 11.17). The angular position of the beam is measured by an incremental encoder. The position of the ball is sensed by a CCD camera placed over the beam. The exact position of the ball along the beam is computed by fusing the CCD camera signal and the angular position signal. With these measurements, the achievable

Fig. 11.17 The experimental ball and beam system

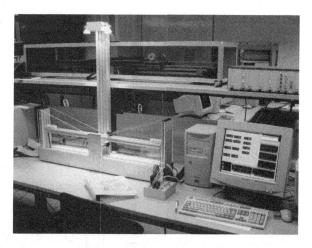

position resolution of the ball on the beam is 1 mm. The ball velocity and the beam angular velocity are obtained by differentiating the corresponding position signals; high frequency noise is limited by using second order low-pass digital filters.

The parameters of the controlled mechanical system are as follows: beam moment of inertia $J = 13.875 \times 10^{-2}$ kg m^2, ball mass $M = 0.025$ kg, ball radius $R = 0.02$ m, gravitational acceleration $g = 9.81$ m/s^2. The ball moment of inertia is computed using the analytical relation $J_b = (2/5)MR^2$.

The control algorithms were implemented on the dSpace DS1102 board which is equipped with a TMS-320 type digital signal processor (DSP). To implement the control algorithm, a MATLAB/Simulink model was developed. A part of the control algorithm (input calibration and filtering) was programmed using Simulink blocks while the control laws are programmed in C language and they are introduced in the Simulink model as S-functions. The built model is compiled and downloaded to the DSP. The sample time was chosen 10ms and the fixed step Euler method was used for numerical integration.

11.3.3.2 Experimental Measurements: Friction Estimation

The adaptive controller described in Sect. 11.3.1 was used to identify the friction model. A sinusoidal reference trajectory was chosen: $x_{3ref} = 0.1\sin(2\pi t)$, and the following controller parameters were used: $\lambda_B = 10$, $K_S = 2$, $\gamma_F = 0.05$. The results are shown in Fig. 11.18 containing the identified friction coefficient and beam tracking error histories. The normalized mean value of the friction coefficient in sinusoidal quasi steady state was computed to be $F_C = 0.19$. For normalization, it was assumed that the maximum value of the control signal is 1. This value is used in the remaining robust controller experiments.

Fig. 11.18 Adaptive friction compensation in the beam

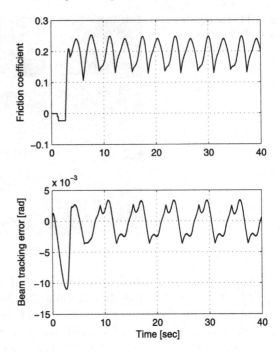

11.3.3.3 Experimental Measurements: LQ Control

In the first experiment, the subspace stabilization controller was compared with the standard LQ control. First, an LQ controller was implemented for the original system (11.31), linearized using the following relation: $B(x) \approx \frac{Mg}{MR^2+J_b}x_3$. The LQ controller parameters were obtained using the system parameters enumerated in Sect. 11.3.3.1. The Q and R matrices were taken as follows: $Q = \text{diag}([1000\ 0.1\ 0.1\ 0.1])$, $R = 1$. The obtained controller parameters, using the lqr MATLAB function, are

$$K_{LQR} = (K_1\ K_2\ K_3\ K_4) = (31.62\ 21.53\ 50.35\ 10.04). \tag{11.45}$$

The experimental results are presented in the Fig. 11.19.

11.3.3.4 Experimental Measurements: Nonlinear Combined Error Metric Control

In the second experiment, the nonlinear combined error metric control law was tested. The applied the control law was (11.44).

The gains in linear part of the control law were chosen in such a way to be equivalent with the LQ controller gains given in (11.45). Based on (11.41) and (11.44), the equivalent gains can be computed using the following relations $K_1 = K_S \alpha \lambda_1$, $K_2 = K_S \alpha$, $K_3 = K_S \lambda_2$, $K_4 = K_S$.

The experimental results are presented in the Fig. 11.20. The experiment shows that the nonlinear combined error metric control approach guarantees better transient response than the standard linear LQ control. Settling time is shorter, and overshoot is also much more reduced.

The choice of the sampling period is critical during controller implementation. If the sampling period is too large, the closed loop system may became uncontrollable. The sampling period can be chosen by taking into account the bandwidth and time constants of the closed loop system. Reasonable sampling rates are 10 to 30 times the closed loop bandwidth, assuming the dynamics can be approximated by a first order linear system [11]. The sampling frequency in the experiments was 100 Hz (approximately ten times the closed loop bandwidth).

11.4 Adaptive Payload and Friction Compensation in Robotic Manipulators

The current section presents a tracking control algorithm for robotic manipulators with partially known parameters. A reformulated robot model with friction is introduced based on which the robot control algorithms can be extended for adaptive compensation of unknown friction and payload parameters [97].

Robot control algorithms should guarantee high tracking precision both in the low velocity regime, as the start configuration and the desired configuration are close to each other, and in high velocity regime too. To assure high tracking precision in the acceleration regimes as well, the dynamic parameters of the robot have to be taken into the consideration in the control algorithm, their values should be known exactly. The parameters of the dynamic model (masses, inertias, length of the links, position of the center of gravity (COG) of the links) are often catalog data which are given by the manufacturer of the robotic system. However, in many robotic applications the payload parameters (mass, inertia) are unknown and varying. If the value of these parameters are comparable with the masses/inertias of the robot segments, the payload has to be taken into the consideration in the control algorithm. The friction also has a great influence on the motion of robotic manipulators. The friction phenomena can be described with models whose parameters are slowly time varying, depending on external factors such as the applied lubricant, the temperature and humidity of the environment in which the robot is used. Accordingly, the a priori identification of friction will not describe precisely the frictional phenomena for a long time, for precise tracking and positioning the friction compensator term of the robot control algorithm has to be retuned.

The tracking algorithms for robotic manipulators in the presence of model uncertainties were permanently in the focus of the researches in the last two decades. The proposed solutions try to obtain more accurate system parameters online to improve the quality of the used model in control. Early results can be found in the paper [132] in which it was explored that the robot model can be written in linearly parameterized form. The behavior of commercial PD and PID compensation techniques for

Fig. 11.19 LQ control (with friction compensation)

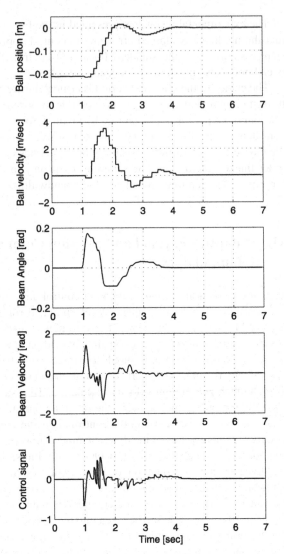

robot control systems in the presence of stick-slip friction was studied in [143]. PID type robot control algorithms can be extended with switching terms to improve the robustness of the control. Such variable structure PID controllers were reported in the papers [58] and [111]. To compensate the effect of friction, disturbance observer was proposed in [28]. Adaptive control algorithms which incorporates online estimation laws for static friction parameters were proposed in [5] and [52]. Online estimation of the unknown state and some of the parameters in the dynamic Lu-Gre friction model in robotic systems were reported in [147] and [110]. The paper [89] presents a model-based compensator for canceling friction in the tendon-driven

Fig. 11.20 Nonlinear
combined error metric control
(with friction compensation)

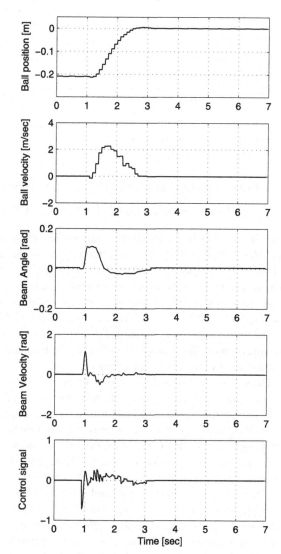

joints of a haptic-feedback teleoperator. A friction compensator method for robots, for the case when joint torque measurements are available, was reported in [144].

The dynamics of an n DOF rigid manipulator with friction is described by the following relation

$$H(q)\ddot{q} + C(q,\dot{q})\dot{q} + D(q) = \tau - \tau_f. \tag{11.46}$$

In the equation above, τ_f is the vector of the frictional forces/torques that is considered to be described by the LuGre model in the form (9.81). The static part of the model is approximated with the relation (9.75): $\tau_{fLi}(\dot{q}_i) = \theta_{fLi}^T \xi_{fLi}$, where

$\xi_{fLi} = (1 \ \dot{q}_i)^T$. The parameter vector θ_{fLi} takes different values in different velocity regimes in different velocity domain, according to (9.75).

The dynamic part is bounded as in (9.83): $|\tau_{fDi}| < \theta_{fDi}^T \xi_{fDi}$, where $\theta_{fDi} = (a_{Di} \ b_{Di})^T$; $\xi_{fDi} = (1 \ |\dot{q}_i|)^T$.

Assume that the dimension of the payload can be neglected related to the dimension of the robot link. However, it does not mean that the mass (m_{PL}) and inertia of the payload do not influence the motion of the robot. Assume that the payload is rigid body and it has a sphere-like shape, hence $I_{PLxx} = I_{PLyy} = I_{PLzz} = I_{PL}$. Rewrite the model in the following form

$$\left(H_R(q) + m_{PL} H_{mPL}(q) + I_{PL} H_{IPL}(q)\right)\ddot{q} + \left(C_R(q,\dot{q}) + m_{PL} C_{mPL}(q,\dot{q})\right)\dot{q}$$
$$+ D_R(q) + m_{PL} D_{mPL}(q) + d = \tau - (\tau_{fL} + \tau_{fD}). \tag{11.47}$$

The H_R, C_R, D_R matrices are the terms in the robotic arm model without payload. The H_{mPL}, C_{mPL}, D_{mPL} and H_{IPL} terms contain the parts from the robot model (11.46) that multiply the payload mass and inertia, respectively. It was exploited that the product $m_{PL} \cdot I_{PL}$ can never appear in the robot model. With the known parameters of the robotic arm and the joint variables q and $\dot{q}$ measured, the H_R, C_R, D_R, H_{mPL}, C_{mPL}, D_{mPL}, H_{IPL} terms are considered known.

The term d represents a bounded disturbance vector that incorporates the modeling uncertainties, for example the effect of the backlash that may appear in the gear transmissions of the joints, satisfying $|d| < D_M$.

The vector τ contains the torques exerted by the motors, the vectors τ_{fL} and τ_{fD} denote static and dynamic part of friction forces/torques.

Generally the model based and robust robot control algorithms applied to solve the tracking control of robotic systems are based on the model (11.46) and they directly apply the terms H, C, D for the compensation of nonlinearities [76]. If the model parameters are unknown, the left-hand side of the model (11.46) is rewritten in linearly parameterized form. The control method presented here proposes an extension for the model based and robust tracking control schemes in the case when the payload parameters and joint friction are unknown but, in order to achieve the desired control precision, their effects cannot be neglected. In this case, without reformulating the model and rewriting the control law, the model based control algorithm can simply be extended with the payload and friction compensation part.

The *control task*: design a control input $\tau = (\tau_1 \ \tau_2 \dots \tau_n)^T$ such that the joint position $q = (q_1 \ q_2 \dots q_n)^T$ track the desired trajectory $q_d = (q_{1d} \ q_{2d} \dots q_{nd})^T$ with given precision. The desired joint trajectories q_{di} are known bounded functions of time with bounded, known first and second order derivatives.

Define the error metric S_i that describes the desired dynamics of the error system for the ith joint: $S_i(t) = (\frac{d}{dt} + \lambda_i)e_i$, $e_i = q_i - q_{di}$, where λ_i is strictly positive constant. For the entire robot, the error metric can be defined as

$$S = (\dot{q} - \dot{q}_d) + \Lambda(q - q_d), \tag{11.48}$$

where $\Lambda > 0$ is a diagonal matrix with positive elements.

Based on the error metric, the control problem can be reformulated: design a control input τ such that $|S_i(t)| < \Phi_i$ if $t \to \infty$, $\forall i = 1 \ldots n$. Define the vector $\Phi = (\Phi_1 \ \Phi_2 \ldots \Phi_n)^T$, $\Phi_i > 0$ that contains the prescribed precisions for the joints. Using the reformulated model (11.47), the dynamics of the error metric reads

$$
\begin{aligned}
H\dot{S} &= H\big(\ddot{q} - \ddot{q}_d + \Lambda(\dot{q} - \dot{q}_d)\big) \\
&= H\big(-\ddot{q}_d + \Lambda(\dot{q} - \dot{q}_d)\big) - C(q,\dot{q})\dot{q} - D(q) + \tau - \tau_f(\dot{q}) \\
&= -CS + H_R\big(-\ddot{q}_d + \Lambda(\dot{q} - \dot{q}_d)\big) + C_R\big(-\dot{q}_d + \Lambda(q - q_d)\big) - D_R \\
&\quad + m_{PL}\xi_{mPL} + I_{PL}\xi_{IPL} + \tau - \tau_{fL} - \tau_{fD} - d,
\end{aligned}
$$

where $\xi_{IPL} = H_{IPL}\big(-\ddot{q}_d + \Lambda(\dot{q} - \dot{q}_d)\big)$,

$$
\xi_{mPL} = H_{mPL}\big(-\ddot{q}_d + \Lambda(\dot{q} - \dot{q}_d)\big) + C_{mPL}\big(-\dot{q}_d + \Lambda(q - q_d)\big) - D_{mPL}.
$$

$$(11.49)$$

Control Law The payload and friction parameters are unknown, only their estimated values can be used in the control algorithm. Denote $\hat{m}_{PL}$ and $\hat{I}_{PL}$ the estimated payload parameters, $\hat{\theta}_{fLi}$ the estimated static friction parameters and $\hat{\theta}_{fDi}$ the estimated bound of dynamic friction parameters in the ith joint.

In function of known robot arm parameters and estimated friction and payload parameters, the control signal τ is defined as follows

$$
\begin{aligned}
\tau &= -H_R\big(-\ddot{q}_d + \Lambda(\dot{q} - \dot{q}_d)\big) - C_R\big(-\dot{q}_d + \Lambda(q - q_d)\big) + D_R - K_S S \\
&\quad - \hat{m}_{PL}\xi_{mPL} - \hat{I}_{PL}\xi_{IPL} + \hat{\tau}_{fL} - \hat{\tau}_{fD} . * \operatorname{sat}(S/\Phi) - D_M . * \operatorname{sat}(S/\Phi),
\end{aligned}
$$

where $\hat{\tau}_{fLi} = \hat{\theta}_{fLi}^T \xi_{fLi}$, $\hat{\tau}_{fDi} = \hat{\theta}_{fDi}^T \xi_{fDi}$,

$$
\operatorname{sat}(S/\Phi) = \big(\operatorname{sat}(S_1/\Phi_1) \ \operatorname{sat}(S_2/\Phi_2) \ldots \operatorname{sat}(S_n/\Phi_n)\big)^T.
$$

$$(11.50)$$

Here $.*$ denotes elementwise multiplication according to MATLAB convention.

The last two terms in the control law are introduced to compensate the effect of dynamic friction behavior and uncertainties of modeling, respectively. For the estimation of unknown parameters, the following adaptation laws are applied

$$
\begin{aligned}
\dot{\hat{m}}_{PL} &= \gamma_m S^T \xi_{mPL}, \qquad \dot{\hat{I}}_{PL} = \gamma_I S^T \xi_{IPL}, \\
\dot{\hat{\theta}}_{fLi} &= \Gamma_{fLi} \xi_{fLi} S_i, \qquad \dot{\hat{\theta}}_{fDi} = \Gamma_{fDi} \xi_{fDi} |S_i|,
\end{aligned}
$$

$$(11.51)$$

where $\gamma_m, \gamma_I > 0$ and $\Gamma_{fLi}, \Gamma_{fDi}$ are diagonal matrices with positive elements in the diagonal.

D_M represents the upper bound of modeling uncertainties and its value can be tuned by knowing the upper bound of the precision for the robot parameters and the resolution of the position and velocity measurements. Based on these values the upper bound of the modeling can be estimated.

Lyapunov Analysis of the Control   In order to analyze the convergence of the tracking error metric, define the following Lyapunov function candidate

$$V(t) = \frac{1}{2} S^T H(q) S + \frac{\gamma_m^{-1}}{2} \tilde{m}_{PL}^2 + \frac{\gamma_I^{-1}}{2} \tilde{I}_{PL}^2 + \frac{1}{2} \sum_{i=1}^{n} \tilde{\theta}_{fLi}^T \Gamma_{fLi}^{-1} \tilde{\theta}_{fLi}$$

$$+ \frac{1}{2} \sum_{i=1}^{n} \tilde{\theta}_{fDi}^T \Gamma_{fDi}^{-1} \tilde{\theta}_{fDi}. \tag{11.52}$$

Here the estimation errors are defined as $\tilde{m}_{PL} = m_{PL} - \hat{m}_{PL}$, $\tilde{I}_{PL} = I_{PL} - \hat{I}_{PL}$, $\tilde{\theta}_{fLi} = \theta_{fLi} - \hat{\theta}_{fLi}$, $\tilde{\theta}_{fDi} = \theta_{fDi} - \hat{\theta}_{fDi}$.

By assuming that the unknown real parameters are constant, the time derivative of (11.52) can be calculated as

$$\dot{V}(t) = S^T H(q) \dot{S} + \frac{1}{2} S^T \dot{H}(q) S - \tilde{m}_{PL} \gamma_m^{-1} \dot{\hat{m}}_{PL} - \tilde{I}_{PL} \gamma_I^{-1} \dot{\hat{I}}_{PL}$$

$$- \sum_{i=1}^{n} \tilde{\theta}_{fLi}^T \Gamma_{fLi}^{-1} \dot{\hat{\theta}}_{fLi} - \sum_{i=1}^{n} \tilde{\theta}_{fDi}^T \Gamma_{fDi}^{-1} \dot{\hat{\theta}}_{fDi}. \tag{11.53}$$

Substitute the expression of the control law (11.50) in the equation of error dynamics (11.49)

$$H\dot{S} = -CS + \tilde{m}_{PL} \xi_{mPL} + \tilde{I}_{PL} \xi_{IPL} - \tilde{\tau}_{fL} + \left(\tau_{fD} - \hat{\tau}_{fD}. * \mathrm{sat}(S/\Phi) \right) - K_S S$$

$$- d - D_M. * \mathrm{sat}(S/\Phi), \quad \text{where } \tilde{\tau}_{fLi} = \tilde{\theta}_{fLi}^T \xi_{fLi}. \tag{11.54}$$

Outside the boundary layer $(|S_i| > \Phi \ \forall i = 1, \ldots, n)$ we have $\mathrm{sat}(S/\Phi) = \mathrm{sgn}(S)$.

Introduce the error dynamics (11.54) and the adaptation laws (11.51) into (11.53) and apply the property (11.1) of the robot model, then the time derivative of the Lyapunov function reads as

$$\dot{V}(t) = -S^T K_S S - \sum_{i=1}^{n} \left(\tau_{fDi} S_i + \hat{\theta}_{fDi}^T \xi_{fDi} |S_i| \right) - \sum_{i=1}^{n} \left(\tilde{\theta}_{fDi}^T \xi_{fDi} |S_i| \right)$$

$$- S^T d - |S|^T D_M. \tag{11.55}$$

Applying that $-d_i S_i < D_M |S_i|$, if $|d_i| < D_M$, yields

$$\dot{V}(t) < -S^T K_S S - \sum_{i=1}^{n} (\tau_{fDi} S_i - \theta_{fDi}^T \xi_{fDi} |S_i|). \tag{11.56}$$

According to (9.83), the second term in (11.56) is negative. It yields

$$\dot{V}(t) < -S^T K_S S. \tag{11.57}$$

From (11.57), one can conclude that the function V is nonincreasing, hence the parameter estimation errors and the tracking error metric S are bounded ($S \in L_\infty^n$), moreover $V(\infty)$ is finite. According to (11.54), since the inertia matrix H is strictly positive with bounded norm and the tracking error, parameter estimation errors, regressor vectors and the disturbance are bounded, hence the derivate of the tracking error is also bounded ($\dot{S} \in L_\infty^n$).

From (11.57), follows $\dot{V}(t) < -K_{Smin}S^T S$, where K_{Smin} is the smallest element of the strictly positive diagonal matrix K_S. By integrating both sides and taking into consideration that $V(0)$ is finite, yields that $S \in L_2^n$

$$\lim_{t \to \infty} \int_0^t S^T S \le \frac{V(\infty) - V(0)}{K_{Smin}}. \tag{11.58}$$

Outside the boundary layer (Φ) it is evident that $S \in L_2^n$, $S \in L_\infty^n$, $\dot{S} \in L_\infty^n$. Hence, S goes to zero as t goes to ∞. Thus, S reaches to boundary layer within finite time.

Note that the adaptation laws can be robustified, for example, using the switching-σ method [27, 54]. With this modification, the drift of the estimated parameters can be avoided during control in the presence of large parameter uncertainties. The σ modification is necessary in the case of the estimates for the upper bounds of the dynamic friction ($\hat{\theta}_{FDi}$). Due to the integrative nature of the classical adaptation laws, the estimated parameters $\hat{a}_{Di}$ would always increase because their adaptation is driven by $|S_i|$ which is always positive. The σ modification transforms the integrative character of the classical adaptation laws into a stable first order dynamics; hence the parameters will remain bounded.

11.4.1 Simulation Results—Adaptive Friction Compensation in the Presence of Backlash

Firstly, the control algorithm is tested on a 1 DOF positioning system. Moreover, it is considered that there is a backlash type element between the drive and the load. The dynamics of the controlled system is described by

$$J\ddot{q} = \tau - \tau_f(\dot{q}) + d, \tag{11.59}$$

where $J > 0$ denotes the inertia (or mass) of the load, τ represents the control input and d is unmeasurable bounded additive disturbance, $|d| < D_F$. The dynamic friction effects are not taken into consideration during these simulation experiments, it is considered that the friction is described by the relation (9.37a) that can be approximated with the piecewise linearly parameterized model (9.75). The parameters of the system (mass/inertia of the load, and the frictional parameters) are considered unknown.

Formulate the following control law in function of estimated parameters

$$\tau = \hat{J}\big(\ddot{q}_d + \lambda\dot{e}(t)\big) + \hat{\theta}_{fL}^T \xi_{fL}(\dot{q}) + k_S S, \tag{11.60}$$

Fig. 11.21 Mean position tracking error in function of backlash gap size

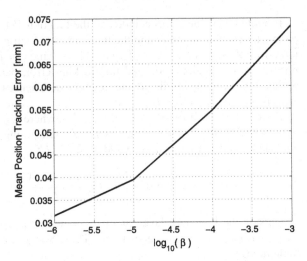

where $k_S > D_F/\Phi$, $e(t) = q_d(t) - q(t)$, $S(t) = (\frac{d}{dt} + \lambda)e(t)$. The unknown friction vector parameter and regressor vector are: $\hat{\theta}_f = (\hat{a}_i \ \hat{b}_i)^T$, $\xi_f(\dot{q}) = (1 \ \dot{q})^T$. The index i depends on velocity regime, where the machine moves, according to (9.75).

The adaptation laws for the unknown load and frictional parameters are formulated as

$$\dot{\hat{J}} = \gamma_J S(t)(\ddot{q}_d + \lambda\dot{e}(t)), \qquad \dot{\hat{\theta}}_{fL} = \Gamma_f S(t)\xi_f(\dot{q}). \tag{11.61}$$

The parameters of the controlled system were chosen as: the mass on the motor side is $J = 0.1$ kg, the mass on the load side is $J_L = 10$ kg, the value of the gear amplification is considered $K_G = 1$. The friction model parameters were chosen to be: $F_S = 1.5$ N, $F_C = 1$ N, $F_V = 14$ N s/m, $v_S = 0.001$ m/s.

In the controller, the model (9.75) has been applied. Based on the frictional parameters enumerated above, the parameters of the linearly parameterized model can be determined as: $a_{1+} = -a_{1-} = 1.5$ N, $b_{1+} = b_{1-} = -480$ N s/m, $a_{2+} = -a_{2-} = 1$ N s/m, $b_{2+} = b_{2-} = 14$ N s/m.

The prescribed trajectory is a sinusoidal one with small amplitude which assures that the mechanical system moves both in positive and negative velocity regime near zero velocities.

$$\dot{q}_d(t) = 0.02\sin\left(\frac{2\pi}{5}t\right), \qquad q_d(t) = \int_0^t \omega_d(\tau)\,d\tau, \qquad \ddot{q}_d(t) = \frac{d\omega_d(t)}{dt}. \tag{11.62}$$

The control objective is to track the prescribed position given in the Equation above, such that $S(t) \leq 10^{-3}$. The parameters of the controller have been chosen as follows: $\lambda = 10$, $k_S = 0.1$, $D_M = 1$, $\Phi = 10^{-3}$.

Simulation were performed with and without backlash in the controlled system (Figs. 11.21, 11.22 and 11.23). In the Fig. 11.23, B indicates whether the system is in backlash mode ($B = 2$) or in contact mode ($B = -2$).

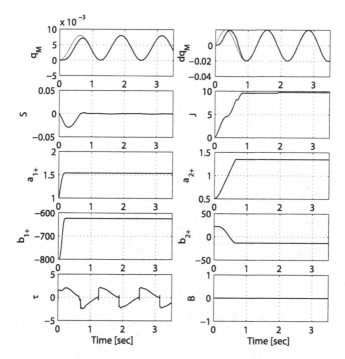

Fig. 11.22 Adaptive friction compensation—no backlash

At the beginning of the simulations, in the control law all friction parameters have been departed with 50% from their real values. The estimated load mass at the beginning of the adaptation was taken to be zero ($\hat{J}(0) = 0$).

Simulation results show that in the presence of backlash the settling time increases. To evaluate the effect of backlash on the control precision, the mean value of the absolute position error ($|e(t)|$) was computed for different backlash gap sizes (β) during the first 3.5 seconds of the control. The results can be seen in the Fig. 11.21. With increasing backlash gap size, the mean error also increases logarithmically.

11.4.2 Experimental Measurements

The applicability of the adaptive control algorithm introduced in this section was tested on the SCARA robot control system presented in the Sect. 11.1.4.2. The dynamic model of the SCARA type manipulators is well known from the robot control literature, see, for example, [81]. The parameters of the dynamic model of the applied robot are also given in the Sect. 11.1.4.2.

During the experiments, it was considered that the third and the fourth links of the robot represent the payload. Hence, the adaptive friction and payload compensation algorithm was implemented for the first and second joint of the robot which is

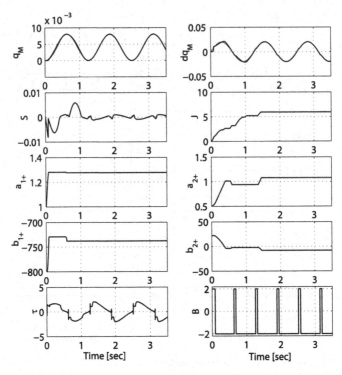

Fig. 11.23 Adaptive friction compensation—with backlash ($\beta = 10^{-4}$ m)

modeled here as a 2 DOF horizontal robot with revolute joints. In this approach the terms of the robot model, based on which the control algorithm was implemented, are

$$H_R = \begin{bmatrix} l_{c1}^2 m_1 + (l_1^2 + 2l_1 l_{c2}\cos(q_2) + l_{c2}^2)m_2 + I_1 + I_2 & (l_{c2}^2 + l_1 l_{c2}\cos(q_2))m_2 + I_2 \\ (l_{c2}^2 + l_1 l_{c2}\cos(q_2))m_2 + I_2 & l_{c2}^2 m_2 + I_2 \end{bmatrix},$$

$$H_{mPL} = \begin{bmatrix} l_1^2 + 2l_1 l_2\cos(q_2) + l_2^2 & l_2^2 + l_1 l_2\cos(q_2) \\ l_2^2 + l_1 l_2\cos(q_2) & l_2^2 \end{bmatrix},$$

$$H_{IPL} = \begin{bmatrix} 1 & 1 \\ 1 & 1 \end{bmatrix},$$

$$C_R = \begin{bmatrix} -2l_1 l_{c2}\sin(q_2)\dot{q}_2 m_2 & -l_1 l_{c2}\sin(q_2)\dot{q}_2 m_2 \\ l_1 l_{c2}\sin(q_2)\dot{q}_1 m_2 & 0 \end{bmatrix},$$

$$C_{mPL} = \begin{bmatrix} -2l_1 l_{c2}\sin(q_2)\dot{q}_2 & -l_1 l_2\sin(q_2)\dot{q}_2 \\ l_1 l_2\sin(q_2)\dot{q}_1 & 0 \end{bmatrix}.$$

During the control experiment, the first and second joints were moved simultaneously. The reference velocity profiles were designed in such way that the joints moved both in positive and negative velocity regimes near zero velocities where the

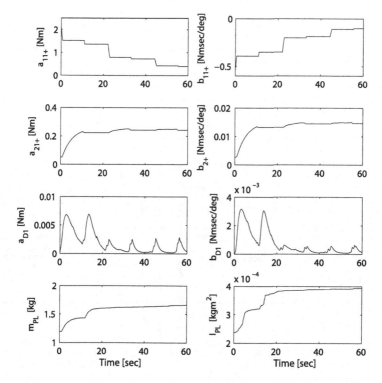

Fig. 11.24 Experimental measurements—parameter adaptation

effect of the dynamic friction is more accentuated. The maximum reference velocity was chosen 5 deg/s. During direction change, the value of acceleration was chosen 5 deg/s^2; hence, the velocity goes in 2 s from -5 deg/s to $+5$ deg/s and the joint position reference trajectory is a continuous function (second order polynomial).

The control algorithm was realized with 10 ms sampling period. The control law was implemented with $K_S = \text{diag}([150\ 150])$ and $\Lambda = \text{diag}([10\ 10])$. The bound for tracking precisions were chosen as $\Phi = (5\ 5)^T$. The adaptation laws were implemented using Euler approximation. To determine the adaptation gains, the following strategy was applied: the gains were increased slowly until the estimated parameter signals showed the fastest response without overshot.

The adaptation of the static friction parameters for the first joint in the positive velocity domain, the upper bound of the dynamic friction and the payload parameters are presented in Fig. 11.24. It can be observed that the parameters are tuned only when the joint moves in the corresponding velocity regime. During its motion, the robot spends more time in the high velocity regime (over Stribeck velocities) hence the parameters a_{2+} and b_{2+} show faster convergence. The upper bound of the dynamic frictional terms takes higher values and accordingly modifies control signal when there is a direction change and the influence of the dynamic frictional effects are accentuated. The estimates for the upper bounds of the dynamic parameters (a_{D1}, b_{D1}) converge to zero when the tracking error converges to zero, because

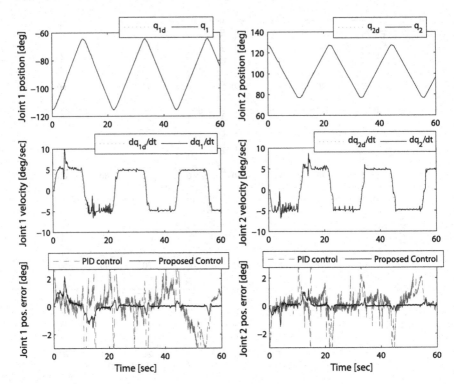

Fig. 11.25 Experimental measurements—desired and real motion

in their case the σ modification was applied in the adaptation law. Figure 11.25 presents the robot motion during the experiment. After parameter convergence, both the position and velocity reaches its reference value with high precision.

The same tracking control task was also solved using a high gain PID controller. The comparison of position tracking error (proposed control, PID control) is shown in Fig. 11.25. It can be seen that the control algorithm proposed in this paper clearly outperforms the high gain PID control algorithm.

11.5 Closing Remarks

To eliminate the undesirable effects of nonsmooth nonlinearities, the motion control algorithms have to be extended with compensator terms. If reliable models with precisely identified parameters are available, fixed structure compensator terms can be applied. If the parameters of the nonsmooth nonlinearities are unknown or slowly time varying adaptive compensation methods can be applied.

In the first part of the chapter, a friction and backlash identification method was introduced for robotic systems. The method is based on drive motor current and velocity measurements respectively. By performing current measurement at different constant velocities, the friction force-joint velocity profile can be measured. An identification method was also proposed to obtain the parameters of the friction model based on these measurements. By capturing the moment of velocity sign change and the current behavior after the direction change, the size of the backlash gap can also be measured. The friction identification method was extended for hydraulic actuators. In that case, instead of the current, the differential pressure in the actuator chambers have to be measured at different constant velocity values. Real-time experimental results are provided to demonstrate the applicability of the proposed identification method.

The model based friction compensation was demonstrated on the ball and beam underactuated system. To solve the control problem, a novel error metric was introduced. Based on this error metric, a nonlinear control law was designed which guarantees closed loop stability. For precise positioning, the nonlinear control algorithm was extended with a friction compensation term. To show the performances of the control algorithm, it was compared with an LQ state feedback controller in the frame of real-time experiments. The proposed nonlinear control algorithm clearly outperformed the LQ controller.

An adaptive tracking control algorithm was proposed for robotic systems with unknown friction in the joints and unknown payload parameters. It was shown, how the robot model have to be modified in order to separate the effect of the payload in the robot system dynamics. Using Lyapunov analysis, it was shown that the proposed adaptive tracking algorithm guarantees the convergence of the joint tracking errors inside a predefined boundary layer and the signals in the control system are bounded. The experimental measurements show that using the proposed control algorithm high position tracking accuracy can be achieved.

Chapter 12
Conclusions and Future Research Directions

Overview This final chapter summarizes the introduced modeling and control techniques which were presented in this book. The need of development of such control algorithms which take into consideration the smooth and nonsmooth nonlinearities that appear in the model of robotic systems and vehicles was a motivating factor of this work. Such algorithms were presented that are theoretically founded and at the same time are implementable real-time. The second part of the chapter enumerates some possible future research directions.

12.1 Summary

Modeling, path design, control, and state estimation of robots and vehicles are central problems of control engineering.

First, the most often used nonlinear control methods (input–output linearization, flatness control, backstepping, sliding control, receding horizon control) and stability results (Lyapunov stability, LaSalle theorem, Barbalat lemma, passivity theory) were surveyed.

A unified method was presented for finding the models of robots, cars, airplanes, and ships. The origin was the rigid body to which the special effects of the different system types were merged. These models were the basis of control design.

For the control of industrial robots, first the decentralized three loop cascaded control was considered and suggestions were given how to design controllers. From the model based control methods, the nonlinear decoupling, both in joint and Cartesian space, was shown and the latter was generalized for hybrid force and position control in compliance problems. For parameter estimation of the dynamic model, selftuning adaptive control was suggested. For realization of paths with sharp corners, backstepping control was used.

For ground cars, the typical control problems were investigated in the frame of a Collision Avoidance System. For path design, the elastic band method was chosen. Input–output linearization and Receding Horizon Control (RHC) was used for control law. State estimation based on two antenna GPS and IMU was applied and two approaches were shown for software implementations.

B. Lantos, L. Márton, *Nonlinear Control of Vehicles and Robots*,
Advances in Industrial Control,
DOI 10.1007/978-1-84996-122-6_12, © Springer-Verlag London Limited 2011

For the control of airplanes, a simple control approach was chosen which applies internal disturbance observer for robust stabilization and external RHC for optimizing the command signals. The internal controller can be changed to any other (H_2/H_∞) controller and constraints can be prescribed.

For indoor quadrotor helicopter control, all usual parts of the design were considered. For control proposes, two dynamic models were derived. The main concept of sensory system consisting of IMU and vision system was elaborated, including calibration of the sensors. The state estimator integrated both sensor types and applied two level extended Kalman filters. The chosen control for the underactuated system was backstepping control. The details of the embedded control system were discussed and it was shown how to test the complex system in real-time before the first flight.

For ship control path design methods, state estimation based on GPS and IMU, and popular control methods were considered. Control methods like acceleration control with PID, nonlinear decoupling and backstepping control in 6 DOF were discussed with some comments to control allocation.

For formation control of vehicles, two approaches were shown. For stabilization of the formation of ground vehicles, the potential field method was presented. For formation control of surface ships, the method based on passivity theory and filtered gradient technique was discussed.

Afterward, the problem of modeling, identification and compensation of nonsmooth nonlinearities in mechanical control systems was addressed. The most important nonsmooth nonlinearities that can influence the control performances in mechanical control systems are the friction and the backlash. Firstly, general models and the stability of nonsmooth systems were discussed. Afterward, the most important static and dynamic friction models were enumerated. A novel piecewise linearly parameterized friction model was introduced that simplifies the identification and compensation of friction in electrically and hydraulically driven robotic systems. The modeling of backlash in dynamic mechanical systems was also treated. A simulation technique was proposed for mechanical control systems with backlash.

The friction and backlash may induce limit cycles in mechanical systems. Based on the theory of hybrid systems, the stability of mechanical control systems with Stribeck friction and backlash was analyzed. Conditions were formulated for guaranteed stability for the case when the control algorithm was designed with LQ techniques. Afterward, the friction and backlash induced limit cycles were analyzed. It was shown that if the control algorithm contains integral term the limit cycle may show chaotic behavior. Specifically, for nonconstant position reference, the friction may also induce limit cycles around Stribeck velocities.

The effect of friction and backlash on control performances can be attenuated or even eliminated if the control algorithm contains compensation terms for these nonlinearities. In order to obtain the parameters of these nonsmooth nonlinearities, identification algorithms were introduced. The proposed identification algorithms are based only on measurements from sensors that anyway have to be applied for the control of mechanical systems: velocity, position measurements, current measurements (electrical actuators), pressure measurements (hydraulic actuators). Model based control algorithm was proposed for the underactuated ball and beam system that applies friction compensation term. Since the frictional parameters are slowly time varying, friction models with off-line identified parameters may be insufficient for precise friction compensation. For such cases, an adaptive friction and payload compensation algorithm was introduced for robotic systems. Experimental results were also presented to show the applicability of the proposed identification and compensation methods.

12.2 Future Research Directions

The discussed problems (path design, modeling, control, state estimation) will remain central problems of researches in many institutions all over the world. Only some part of them will be picked out in the sequel.

An important problem for future research may be the identification of small size autonomous flying robots (UAVs). An approximation for modeling the special effects a set of nonlinear signal relations (products, higher powers) can be chosen with unknown weighting parameters. Many signals in the model are usually not measurable but should be estimated from measurable ones obtained from GPS, IMU etc.

An important problem is the real-time smooth path design with limited acceleration if moving obstacles are present. The result can be applied in car control using GPS, IMU and vision system.

An other direction may be the extension of indoor helicopter control method for outdoor ones using the results of the elaborated path design and the above sensory system.

Perhaps formation control will be the most important research field in autonomous robot control. Here, all concepts of the above directions appear in a complex form. Communication strategies are needed if sensors can fall out for short periods. Formation strategies should be dynamically changed if suddenly new obstacles appear. The formation stabilization has to be robust and suitable for embedded realization.

It should not be forgotten that ground, aerial, and marine robots are underactuated systems hence soft computing methods can help in both modeling and control.

An important field is the design of low cost but precise servo systems for the large number of actuators taking part in the control realization. These actuators are severely influenced by non-smooth nonlinearities such as friction and backlash.

It is well known that the friction depends on many external factors such that temperature, humidity of the environment. Moreover, the temperature of the lubricant also increases if the machine moves as a consequence of the heat dissipative property of friction phenomenon. Due to these factors, the properties of the lubricant change, and the parameters of the friction model alter. In this book this modification was handled using adaptive control techniques. However, if reliable models could be found that describe the dependency of friction on this factors, the adaptive compensation may be omitted and model based or state estimation based friction compensation techniques can be applied. With this approach, the friction compensation part of the control algorithm would be more precise and at the same time its complexity would be reduced.

The limit cycles generated by friction or backlash are well studied topics in the control literature. In this book the combined action of these nonlinearities was investigated including the chaotic properties of the limit cycles that were generated by the simultaneous presence of friction and backlash. Further investigation needed for the limit cycles in the presence of friction and other nonlinearities, that may appear in the model of a robotic system.

The list of future research directions is large and we hope the book can support the reader to find his/her research perspectives.

Appendix A
Kinematic and Dynamic Foundations of Physical Systems

Overview In this Appendix, we summarize the different characterizations of the orientation in kinematic models of physical systems (robots, airplanes, helicopters etc.), the description of position and orientation by homogeneous transformation, the idea of inertia matrix and the Lagrange and Euler Equations playing important role in setting up dynamic models.

A.1 Orientation Description Using Rotations and Quaternion

In the navigation of robots and vehicles, the common practice is to use homogeneous transformations for describing the pose (position and orientation) of moving objects and rotations or quaternions for characterizing their orientation.

A.1.1 Homogeneous Transformations

Let us consider first two orthonormed coordinate systems (frames) denoted by K and K' and let their basis vectors be i, j, k and i', j', k', respectively (see Fig. A.1).

Since $r = r' + p$ and $r, p \in K$ and $r' \in K'$ therefore

$$r = \rho_x i + \rho_y j + \rho_z k, \qquad p = p_x i + p_y j + p_z k,$$
$$r' = \rho'_x i' + \rho'_y j' + \rho'_z k', \qquad i' = a_{11} i + a_{21} j + a_{31} k, \tag{A.1a}$$
$$j' = a_{12} i + a_{22} j + a_{32} k, \qquad k' = a_{13} i + a_{23} j + a_{33} k,$$

$$r = (a_{11}\rho'_x + a_{12}\rho'_y + a_{13}\rho'_z + p_x)i$$
$$+ (a_{21}\rho'_x + a_{22}\rho'_y + a_{23}\rho'_z + p_y)j$$
$$+ (a_{31}\rho'_x + a_{32}\rho'_y + a_{33}\rho'_z + p_z)k. \tag{A.1b}$$

B. Lantos, L. Márton, *Nonlinear Control of Vehicles and Robots*,
Advances in Industrial Control,
DOI 10.1007/978-1-84996-122-6, © Springer-Verlag London Limited 2011

Fig. A.1 The frame concept

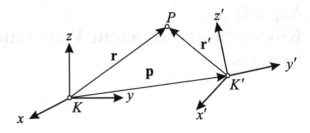

These relations can also be described in matrix form:

$$\begin{pmatrix} \rho_x \\ \rho_y \\ \rho_z \end{pmatrix} = \begin{pmatrix} a_{11} & a_{12} & a_{13} \\ a_{21} & a_{22} & a_{23} \\ a_{31} & a_{32} & a_{33} \end{pmatrix} \begin{pmatrix} \rho'_x \\ \rho'_y \\ \rho'_z \end{pmatrix} + \begin{pmatrix} p_x \\ p_y \\ p_z \end{pmatrix}, \tag{A.2}$$

$$\underbrace{\uparrow i' \ \uparrow j' \ \uparrow k'}_{\substack{\text{described in} \\ i,j,k \ \text{basis}}}$$

$$r = Ar' + p. \tag{A.3}$$

Adding to them the identity $1 = 1$ then the change of the orientation A between K and K' and the position vector p between the origins can be reduced to a homogeneous transformation T:

$$\begin{pmatrix} \rho_x \\ \rho_y \\ \rho_z \\ 1 \end{pmatrix} = \begin{pmatrix} a_{11} & a_{12} & a_{13} & p_x \\ a_{21} & a_{22} & a_{23} & p_y \\ a_{31} & a_{32} & a_{33} & p_z \\ 0 & 0 & 0 & 1 \end{pmatrix} \begin{pmatrix} \rho'_x \\ \rho'_y \\ \rho'_z \\ 1 \end{pmatrix}, \tag{A.4}$$

$$\begin{pmatrix} r \\ 1 \end{pmatrix} = \left(\begin{array}{c|c} A & p \\ \hline 0^T & 1 \end{array} \right) \begin{pmatrix} r' \\ 1 \end{pmatrix} =: \begin{pmatrix} l & m & n & p \\ 0 & 0 & 0 & 1 \end{pmatrix} \begin{pmatrix} r' \\ 1 \end{pmatrix} =: T \begin{pmatrix} r' \\ 1 \end{pmatrix}, \tag{A.5}$$

$$\begin{pmatrix} r \\ 1 \end{pmatrix} = T \begin{pmatrix} r' \\ 1 \end{pmatrix}. \tag{A.6}$$

Here $(r^T \ 1)^T \in R^4$ and $(r'^T \ 1)^T \in R^4$ denote the homogeneous coordinates of $r \in R^3$ and $r' \in R^3$.

The homogeneous transformation T has two important properties:

(i) T gives the orientation and position of K' relative to K.
(ii) T makes it possible to determine the coordinates in K of the point P if we know its coordinates in K'.

It is useful to provide the homogeneous transformation T with indices in order to emphasize that the components of l, m, n, p belonging K' should be given in the basis of K: $T_{K,K'}$. It is apparent that the order of the indices is important in the interpretation.

The sequence of homogeneous transformations can be described by the product of homogeneous transformations:

$$T_1 T_2 = \begin{bmatrix} A_1 & p_1 \\ 0^T & 1 \end{bmatrix} \begin{bmatrix} A_2 & p_2 \\ 0^T & 1 \end{bmatrix} = \begin{bmatrix} A_1 A_2 & A_1 p_2 + p_1 \\ 0^T & 1 \end{bmatrix}. \tag{A.7}$$

A.1.2 Orientation Description Using Rotations

Let us consider first the change of *orientation*. Let us assume that K and K' originally coincided but K' has been rotated around some axis of K by the angle φ and consider the state after the rotation. In order to simplify the relations, the following convention will be introduced hereafter:

$$C_\varphi := \cos\varphi, \quad S_\varphi := \sin\varphi, \quad C_{\alpha\beta} := \cos(\alpha + \beta), \quad C_{12} := \cos(q_1 + q_2), \text{ etc.} \tag{A.8}$$

Elementary rotations
Using (A.2), the elementary rotations are as follows:

$$Rot(z, \varphi) = \underbrace{(i'_{rot}\ j'_{rot}\ k'_{rot})}_{\text{expressed in } i,j,k} = \begin{bmatrix} \cos\varphi & -\sin\varphi & 0 \\ \sin\varphi & \cos\varphi & 0 \\ 0 & 0 & 1 \end{bmatrix} = \begin{bmatrix} C_\varphi & -S_\varphi & 0 \\ S_\varphi & C_\varphi & 0 \\ 0 & 0 & 1 \end{bmatrix}, \tag{A.9}$$

$$Rot(y, \varphi) = (i'_{rot}\ j'_{rot}\ k'_{rot}) = \begin{bmatrix} C_\varphi & 0 & S_\varphi \\ 0 & 1 & 0 \\ -S_\varphi & 0 & C_\varphi \end{bmatrix}, \tag{A.10}$$

$$Rot(x, \varphi) = (i'_{rot}\ j'_{rot}\ k'_{rot}) = \begin{bmatrix} 1 & 0 & 0 \\ 0 & C_\varphi & -S_\varphi \\ 0 & S_\varphi & C_\varphi \end{bmatrix}. \tag{A.11}$$

Rodriguez formula
Let t be a general direction vector, $\|t\| = 1$. Let us assume that K and K' originally coincided, but K' has been rotated around t by the angle φ.

Let x be the point to be rotated having projections $x_t = t\langle t, x \rangle$ and $x_{t\perp} = x - x_t = x - t\langle t, x \rangle$. Here $\langle t, x \rangle = t^T x = x^T t$ denotes the scalar product. Let the rotated image of $x_{t\perp}$ in the plane orthogonal to t be denoted by $x_{t\perp}^{rot}$. Since x_t does not change during rotation, hence $x^{rot} = x_t + x_{t\perp}^{rot}$. On the other hand, $x_{t\perp}$ and $t \times x_{t\perp}$ are two orthogonal directions in the plane orthogonal to t, furthermore $\|t \times x_{t\perp}\| = \|t\| \cdot \|x_{t\perp}\| \cdot \sin(90°) = \|x_{t\perp}\|$, hence $x_{t\perp}^{rot} = x_{t\perp}\cos\varphi + t \times x_{t\perp}\sin\varphi$, and as a consequence $x^{rot} = t\langle t, x \rangle + (x - t\langle t, x \rangle)\cos\varphi + t \times x \sin\varphi$.

The rotation is a linear transformation $\mathcal{R}ot(t, \varphi)$ whose matrix is $Rot(t, \varphi)$:

$$x^{rot} = \mathcal{R}ot(t, \varphi)x = x\cos\varphi + t\langle t, x \rangle(1 - \cos\varphi) + t \times x \sin\varphi, \tag{A.12}$$

$$Rot(t, \varphi) = C_\varphi I + (1 - C_\varphi)[t \circ t] + S_\varphi[t \times], \tag{A.13}$$

which is the so-called *Rodriguez formula*. Using the matrices of the dyadic product $[t \circ t] = t \cdot t^T$ and the vector product $[t \times]$ (see later in (A.27) and (A.28)), it yields

$$Rot(t, \varphi) = \begin{bmatrix} l_x & m_x & n_x \\ l_y & m_y & n_y \\ l_z & m_z & n_z \end{bmatrix} = C_\varphi I + (1 - C_\varphi)[t \circ t] + S_\varphi[t \times]$$

$$= \begin{bmatrix} C_\varphi + (1 - C_\varphi)t_x t_x & (1 - C_\varphi)t_x t_y - S_\varphi t_z & (1 - C_\varphi)t_x t_z + S_\varphi t_y \\ (1 - C_\varphi)t_x t_y + S_\varphi t_z & C_\varphi + (1 - C_\varphi)t_y t_y & (1 - C_\varphi)t_y t_z - S_\varphi t_x \\ (1 - C_\varphi)t_x t_z - S_\varphi t_y & (1 - C_\varphi)t_y t_z + S_\varphi t_x & C_\varphi + (1 - C_\varphi)t_z t_z \end{bmatrix}.$$
$$\tag{A.14}$$

The inverse of the rotation is the rotation around t by the angle $-\varphi$. Since in (A.13) $[t \circ t]$ is symmetric, $[t \times]$ is skew-symmetric, $\cos(-\varphi) = \cos(\varphi)$ and $\sin(-\varphi) = -\sin(\varphi)$ hence

$$\left[Rot(t, \varphi)\right]^{-1} = \left[Rot(t, \varphi)\right]^T. \tag{A.15}$$

Therefore in case of orthonormed coordinate systems, the inverse of the orientation matrix A is the transpose of the matrix A from which it follows that the inverse of the homogeneous matrix T can be easily computed:

$$A^{-1} = A^T, \tag{A.16}$$

$$T = \left[\begin{array}{c|c} A & p \\ \hline 0^T & 1 \end{array}\right]^{-1} = \left[\begin{array}{c|c} A^T & -A^T p \\ \hline 0^T & 1 \end{array}\right]. \tag{A.17}$$

The orientation part A in the homogeneous transformation T contains $3 \times 3 = 9$ elements. However, they are not independent because i', j', k' is an orthonormed basis. Therefore, $\|i'\| = \|j'\| = \|k'\| = 1$ and $\langle i', j' \rangle = \langle i', k' \rangle = \langle j', k' \rangle = 0$, which are 6 conditions for the 9 elements. Thus, the orientation can be described by $9 - 6 = 3$ free parameters. In the case of Rodriguez formula, these are t and φ where t is a unit vector having only 2 free parameters.

Euler and RPY angles
Beside Rodriguez formula there are often used other orientation descriptions, too. If K and K' are given and have common origins, then we can choose another exemplar of K and starting from it we can reach K' by using a sequence of elementary rotations. The rotated axis of the second exemplar can be denoted by prime and double-prime and the rotation angles by φ, ϑ, ψ, respectively. Using the usual terminology in *robotics*, we can speak about z, y', z'' rotations called *Euler angles*, and z, y', x'' rotations called *roll, pitch, yaw* (RPY) angles. They are often called also as Euler z–y–z angles and Euler z–y–x angles, respectively. By these conventions

$$A_{K,K'} = Euler(\varphi, \vartheta, \psi) = Rot(z, \varphi)Rot(y', \vartheta)Rot(z'', \psi)$$

$$
= \begin{bmatrix} C_\varphi & -S_\varphi & 0 \\ S_\varphi & C_\varphi & 0 \\ 0 & 0 & 1 \end{bmatrix} \begin{bmatrix} C_\vartheta & 0 & S_\vartheta \\ 0 & 1 & 0 \\ -S_\vartheta & 0 & C_\vartheta \end{bmatrix} \begin{bmatrix} C_\psi & -S_\psi & 0 \\ S_\psi & C_\psi & 0 \\ 0 & 0 & 1 \end{bmatrix}
$$

$$
= \begin{bmatrix} C_\varphi C_\vartheta C_\psi - S_\varphi S_\psi & -C_\varphi C_\vartheta S_\psi - S_\varphi C_\psi & C_\varphi S_\vartheta \\ S_\varphi C_\vartheta C_\psi + C_\varphi S_\psi & -S_\varphi C_\vartheta S_\psi + C_\varphi C_\psi & S_\varphi S_\vartheta \\ -S_\vartheta C_\psi & S_\vartheta S_\psi & C_\vartheta \end{bmatrix}, \tag{A.18}
$$

$$
A_{K,K'} = RPY(\varphi, \vartheta, \psi) = Rot(z, \varphi)Rot(y', \vartheta)Rot(x'', \psi)
$$

$$
= \begin{bmatrix} C_\varphi & -S_\varphi & 0 \\ S_\varphi & C_\varphi & 0 \\ 0 & 0 & 1 \end{bmatrix} \begin{bmatrix} C_\vartheta & 0 & S_\vartheta \\ 0 & 1 & 0 \\ -S_\vartheta & 0 & C_\vartheta \end{bmatrix} \begin{bmatrix} 1 & 0 & 0 \\ 0 & C_\psi & -S_\psi \\ 0 & S_\psi & C_\psi \end{bmatrix}
$$

$$
= \begin{bmatrix} C_\varphi C_\vartheta & C_\varphi S_\vartheta S_\psi - S_\varphi C_\psi & C_\varphi S_\vartheta C_\psi + S_\varphi S_\psi \\ S_\varphi C_\vartheta & S_\varphi S_\vartheta S_\psi + C_\varphi C_\psi & S_\varphi S_\vartheta C_\psi - C_\varphi S_\psi \\ -S_\vartheta & C_\vartheta S_\psi & C_\vartheta C_\psi \end{bmatrix}. \tag{A.19}
$$

In the navigation practice of vehicles, the names of the angles for z and x rotations are exchanged in the transformation RPY and simply called Euler angles: $A_{K,K'} = RPY(\psi, \vartheta, \varphi)$ which can easily be determined from (A.19). It is also a common practice to use $A_{K,K'}^T = RPY^T(\psi, \vartheta, \varphi) =: S(\varphi, \vartheta, \psi)$, but in this case S transforms vectors from K into K'. Notice that in the navigation practice φ is the roll angle and ψ is the yaw angle, respectively.

A.1.3 Orientation Description Using Quaternion

Orientations can also be characterized with *quaternions*. The quaternions (Q) provide extension of the three dimensional vector space (R^3) in which multiplication ($*$), multiplication by scalar, conjugate and norm are defined. Let $q \in Q$, $w \in R^3$, $s, \alpha \in R^1$, then by definition

$$
q = (s, w),
$$

$$
\alpha q = (\alpha s, \alpha w),
$$

$$
q_1 + q_2 = (s_1, w_1) + (s_2, w_2) = (s_1 + s_2, w_1 + w_2),
$$

$$
q_1 * q_2 = \big(s_1 s_2 - \langle w_1, w_2 \rangle, w_1 \times w_2 + s_1 w_2 + s_2 w_1 \big), \tag{A.20}
$$

$$
\tilde{q} = (s, -w),
$$

$$
\|q\|^2 = q * \tilde{q} = \big(\|w\|^2 + s^2, 0 \big) = \tilde{q} * q.
$$

Regarding applicability the most important property is perhaps that

$$
\|q_1 * q_2\| = \|q_1\| \cdot \|q_2\|. \tag{A.21}
$$

Since $s \in R^1$ and $(s, 0) \in Q$, furthermore $w \in R^3$ and $(0, w) \in Q$ may be identified hence R^1 and R^3 are embedded into Q. According to the multiplicative property of the quaternion norm, it yields

$$\min_{\|q\|=1} \|q * r * \tilde{q}\| = \min_{\|q\|=1} \left(\|q * r\| \cdot \|q\| \right) = \min_{\|q\|=1} \|q * r\|. \tag{A.22}$$

The quaternion q and the orientation matrix A are in tight connection. By Rodriguez formula, every rotation A can be described by a unit rotation direction vector t and a rotation angle φ:

$$Ar = \cos \varphi r + (1 - \cos \varphi) \langle t, r \rangle r + \sin \varphi t \times r. \tag{A.23}$$

On the other hand, we can introduce the quaternion

$$q := \left(\cos \frac{\varphi}{2}, \sin \frac{\varphi}{2} t \right), \tag{A.24}$$

where $\|q\| = \|t\| = 1$, from which by using the Rodriguez formula it follows

$$q * (0, r) * \tilde{q} = (0, Ar), \tag{A.25}$$

and thus $q * r * \tilde{q}$ and Ar can be identified. It is clear that if the orientation is characterized by the quaternion $q = (s, w)$ then A can be determined from it since

$$\varphi = 2 \arccos(s), \qquad t = w / \sin \left(\frac{\varphi}{2} \right) \quad \Rightarrow \quad A = C_\varphi I + (1 - C_\varphi)[t \circ t] + S_\varphi[t \times], \tag{A.26}$$

where

$$[t \circ t] = \begin{bmatrix} t_x t_x & t_x t_y & t_x t_z \\ t_y t_x & t_y t_y & t_y t_z \\ t_z t_x & t_z t_y & t_z t_z \end{bmatrix}, \tag{A.27}$$

$$[t \times] = \begin{bmatrix} 0 & -t_z & t_y \\ t_z & 0 & -t_x \\ -t_y & t_x & 0 \end{bmatrix}. \tag{A.28}$$

A.1.4 Solution of the Inverse Orientation Problem

If the orientation A is known

$$A = \begin{bmatrix} l_x & m_x & n_x \\ l_y & m_y & n_y \\ l_z & m_z & n_z \end{bmatrix}, \tag{A.29}$$

and it is necessary to characterize it by rotations (inverse orientation problem) then the following methods can be used for the three different characterizations.

Solution of the inverse Rodriguez problem

$$C_\varphi = \frac{l_x + m_y + n_z - 1}{2},$$

$$S_\varphi = \frac{+\sqrt{(m_z - n_y)^2 + (n_x - l_z)^2 + (l_y - m_x)^2}}{2}, \tag{A.30a}$$

$$\varphi = \text{atan} \, 2(S_\varphi, C_\varphi) \in (-\pi, \pi],$$

$$t_x = \sqrt{\frac{l_x - C_\varphi}{1 - C_\varphi}} \, \text{sign}(m_z - n_y),$$

$$t_y = \sqrt{\frac{m_y - C_\varphi}{1 - C_\varphi}} \, \text{sign}(n_x - l_z), \tag{A.30b}$$

$$t_z = \sqrt{\frac{n_z - C_\varphi}{1 - C_\varphi}} \, \text{sign}(l_y - m_x).$$

The problem has infinite many solutions if $\varphi = 0$ ($C_\varphi = 1$). In reality, if the rotation angle is null then any axis t is suitable.

Solution of the inverse Euler problem

(i) $n_x^2 + n_y^2 \neq 0 \Rightarrow S_\vartheta \neq 0$

$$\varphi = \arctan\left(\frac{n_y}{n_x}\right) + \left\{\begin{array}{c} 0 \\ \pi \\ -\pi \end{array}\right\} \in (-\pi, \pi], \tag{A.31a}$$

$$\left.\begin{array}{l} S_\vartheta = C_\varphi n_x + S_\varphi n_y \\ C_\vartheta = n_z \end{array}\right\} \quad \vartheta = \text{atan} \, 2(S_\vartheta, C_\vartheta), \tag{A.31b}$$

$$\left.\begin{array}{l} S_\psi = -S_\varphi l_x + C_\varphi l_y \\ C_\psi = -S_\varphi m_x + C_\varphi m_y \end{array}\right\} \quad \psi = \text{atan} \, 2(S_\psi, C_\psi). \tag{A.31c}$$

(ii) *Singular configuration* $n_x = n_y = 0 \Rightarrow S_\vartheta = 0$
  (a) $n_z = C_\vartheta = 1 \Rightarrow \vartheta = 0 \, (z = z')$

$$\left.\begin{array}{l} C_{\varphi\psi} = l_x \\ S_{\varphi\psi} = l_y \end{array}\right\} \quad \varphi + \psi = \text{atan} \, 2(S_{\varphi\psi}, C_{\varphi\psi}) \quad \Rightarrow \quad \varphi + \psi. \tag{A.32}$$

(b) $n_z = C_\vartheta = -1 \Rightarrow \vartheta = \pi \, (z = -z')$

$$\left.\begin{array}{l} S_{\psi-\varphi} = m_x \\ C_{\psi-\varphi} = m_y \end{array}\right\} \quad \psi - \varphi = \text{atan} \, 2(S_{\psi-\varphi}, C_{\psi-\varphi}) \quad \Rightarrow \quad \psi - \varphi. \tag{A.33}$$

Solution of the inverse RPY problem

(i) $l_x^2 + l_y^2 \neq 0 \Rightarrow C_\vartheta \neq 0$

$$\varphi = \arctan\left(\frac{l_y}{l_x}\right) + \left\{\begin{matrix} 0 \\ \pi \\ -\pi \end{matrix}\right\} \in (-\pi, \pi], \qquad (\text{A.34a})$$

$$\left.\begin{matrix} S_\vartheta = -l_z \\ C_\vartheta = S_\varphi l_y + C_\varphi l_x \end{matrix}\right\} \quad \vartheta = \text{atan}\, 2(S_\vartheta, C_\vartheta), \qquad (\text{A.34b})$$

$$\left.\begin{matrix} S_\psi = S_\varphi n_x - C_\varphi n_y \\ C_\psi = -S_\varphi m_x + C_\varphi m_y \end{matrix}\right\} \quad \psi = \text{atan}\, 2(S_\psi, C_\psi). \qquad (\text{A.34c})$$

(ii) *Singular configuration* $l_x = l_y = 0 \Rightarrow C_\vartheta = 0$
   (a) $l_z = -S_\vartheta = 1 \Rightarrow \vartheta = -\pi/2$

$$\left.\begin{matrix} S_{\varphi\psi} = -n_y \\ C_{\varphi\psi} = -n_x \end{matrix}\right\} \quad \varphi + \psi = \text{atan}\, 2(S_{\varphi\psi}, C_{\varphi\psi}) \quad \Rightarrow \quad \varphi + \psi. \qquad (\text{A.35})$$

(b) $l_z = -S_\vartheta = -1 \Rightarrow \vartheta = \pi/2$

$$\left.\begin{matrix} S_{\psi-\varphi} = m_x \\ C_{\psi-\varphi} = m_y \end{matrix}\right\} \quad \psi - \varphi = \text{atan}\, 2(S_{\psi-\varphi}, C_{\psi-\varphi}) \quad \Rightarrow \quad \psi - \varphi. \qquad (\text{A.36})$$

A.2 Differentiation Rule in Moving Coordinate System

Let K_0 be a fixed (non moving) coordinate system then the time derivatives of its i_0, j_0, k_0 unit vectors are zero. Let furthermore K a moving coordinate system. The unit vectors of K can be expressed in the basis of K_0 according to (A.1a). Taking the time derivatives (denoted here by primes) and using the fact that the time derivatives of the unit vectors of K_0 are zero it yields

$$\begin{aligned} i' &= a'_{11} i_0 + a'_{21} j_0 + a'_{31} k_0, \\ j' &= a'_{12} i_0 + a'_{22} j_0 + a'_{32} k_0, \\ k' &= a'_{13} i_0 + a'_{23} j_0 + a'_{33} k_0. \end{aligned} \qquad (\text{A.37})$$

Since $A := A_{K_0 K}$ and $A^{-1} = A_{K, K_0} = A^T$, hence

$$\begin{aligned} i_0 &= a_{11} i + a_{12} j + a_{13} k, \\ j_0 &= a_{21} i + a_{22} j + a_{23} k, \\ k_0 &= a_{31} i + a_{32} j + a_{33} k. \end{aligned} \qquad (\text{A.38})$$

$$i' = a'_{11}(a_{11}i + a_{12}j + a_{13}k) + a'_{21}(a_{21}i + a_{22}j + a_{23}k)$$
$$+ a'_{31}(a_{31}i + a_{32}j + a_{33}k)$$
$$= (a_{11}a'_{11} + a_{21}a'_{21} + a_{31}a'_{31})i + (a_{12}a'_{11} + a_{22}a'_{21} + a_{32}a'_{31})j$$
$$+ (a_{13}a'_{11} + a_{23}a'_{21} + a_{33}a'_{31})k. \tag{A.39}$$

Notice that the factors before i, j, k are the components of the column vector of the following product:

$$\begin{bmatrix} a_{11} & a_{21} & a_{31} \\ a_{12} & a_{22} & a_{32} \\ a_{13} & a_{23} & a_{33} \end{bmatrix} \begin{pmatrix} a'_{11} \\ a'_{21} \\ a'_{31} \end{pmatrix} = A^T \begin{pmatrix} a'_{11} \\ a'_{21} \\ a'_{31} \end{pmatrix} = A^{-1} \begin{pmatrix} a'_{11} \\ a'_{21} \\ a'_{31} \end{pmatrix}. \tag{A.40}$$

On the other hand the Rodriguez formula can be applied for $i + di, j + dj, k + dk$ by which for infinitesimal small rotation angle $d\varphi$ it can be written for example for i

$$i + di = i + d\varphi t \times i. \tag{A.41}$$

Let us define the angular velocity ω in the following way:

$$\omega := \frac{td\varphi}{dt} = \omega_x i + \omega_y j + \omega_z k = {}^0\omega_x i_0 + {}^0\omega_y j_0 + {}^0\omega_z k_0, \tag{A.42}$$

then

$$i' = \omega \times i,$$
$$j' = \omega \times j, \tag{A.43}$$
$$k' = \omega \times k,$$
$$[\omega \times] = A^{-1}A' \quad \text{and} \quad [{}^0\omega \times] = A'A^{-1}. \tag{A.44}$$

Denote respectively r, r' and r'' the position, velocity and acceleration of the point P where $r_x i_0 + r_y j_0 + r_z k_0 = \rho_x i + \rho_y j + \rho_z k + p_x i_0 + p_y j_0 + p_z k_0$. Then

$$r = A\rho + p,$$
$$r' = A\rho' + A'\rho + p', \tag{A.45}$$
$$r'' = A\rho'' + 2A'\rho' + A''\rho + p'',$$

where the coordinates of r, p are given in K_0 while those of ρ are given in K.

Since $AA^{-1} = I \Rightarrow A'A^{-1} + A(A^{-1})^{-1} = 0 \Rightarrow (A^{-1})' = -A^{-1}A'A^{-1}$ hence

$$[\omega \times]' = (A^{-1}A')' = -A^{-1}A'A^{-1}A' + A^{-1}A''$$
$$= -[\omega \times][\omega \times] + A^{-1}A'', \tag{A.46}$$
$$\varepsilon := \omega'_x i + \omega'_y j + \omega'_z k + \omega_x i' + \omega_y j' + \omega_z k'$$

$$= \omega_x' i + \omega_y' j + \omega_z' k + \omega \times \omega$$

$$= \omega_x' i + \omega_y' j + \omega_z' k, \tag{A.47}$$

where $\varepsilon := \omega'$ is the angular acceleration. The r' velocity and r'' acceleration of the point P can also be expressed in the moving coordinate system K:

$$A^{-1} r' = \rho' + A^{-1} A' \rho + A^{-1} p', \tag{A.48}$$

$$A^{-1} r'' = \rho'' + 2A^{-1} A' \rho + A^{-1} A'' \rho + A^{-1} p''$$

$$= \rho'' + 2\omega \times \rho' + \varepsilon \times \rho + \omega \times (\omega \times \rho) + A^{-1} p''. \tag{A.49}$$

Denote respectively $v_K := A^{-1} p'$, $a_K := A^{-1} p''$, $\omega_K := \omega$ and $\varepsilon_K := \varepsilon$ the velocity, acceleration, angular velocity and angular acceleration of the origin of the moving frame K then the velocity v_ρ and acceleration a_ρ of the point P satisfy the *differentiation rule*

$$v_\rho = v_K + \omega_K \times \rho + \rho', \tag{A.50}$$

$$a_\rho = a_K + \varepsilon_K \times \rho + \omega_K \times (\omega_K \times \rho) + \rho'' + 2\omega_K \times \rho'. \tag{A.51}$$

Especially, if P is any point of a rigid body, then $\rho' = \rho'' = 0$, $\omega_\rho = \omega_K$, $\varepsilon_\rho = \varepsilon_K$ and the relations assume simple forms.

A.3 Inertia Parameters of Rigid Objects

Let us consider the expression $[\rho \times]^T [\rho \times]$ that will play an important role in the kinetic energy:

$$[\rho \times]^T [\rho \times] = \begin{bmatrix} 0 & \rho_z & -\rho_y \\ -\rho_z & 0 & \rho_x \\ \rho_y & -\rho_x & 0 \end{bmatrix} \begin{bmatrix} 0 & -\rho_z & \rho_y \\ \rho_z & 0 & -\rho_x \\ -\rho_y & \rho_x & 0 \end{bmatrix}$$

$$= \begin{bmatrix} \rho_y^2 + \rho_z^2 & -\rho_x \rho_y & -\rho_x \rho_z \\ -\rho_x \rho_y & \rho_x^2 + \rho_z^2 & -\rho_y \rho_z \\ -\rho_x \rho_z & -\rho_y \rho_z & \rho_x^2 + \rho_y^2 \end{bmatrix}. \tag{A.52}$$

If ρ is any point of the segment (rigid body) $[s]$ and dm is the infinitesimal mass belonging to the point, then the inertia tensor is the linear transformation $\mathcal{I} x := - \int_{[s]} \rho \times (\rho \times x)\, dm$, and its positive definite matrix is the inertia matrix $\mathbf{I}$:

$$\mathbf{I} := \int_{[s]} [\rho \times]^T [\rho \times]\, dm = \begin{bmatrix} I_x & -I_{xy} & -I_{xz} \\ -I_{xy} & I_y & -I_{yz} \\ -I_{xz} & -I_{yz} & I_z \end{bmatrix}, \tag{A.53}$$

Fig. A.2 Transformation of the inertia matrix

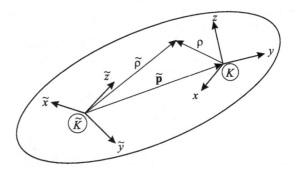

where for example

$$I_x = \int_{[s]} \left(\rho_y^2 + \rho_z^2 \right) dm, \tag{A.54}$$

$$I_{xy} = \int_{[s]} \rho_x \rho_y \, dm, \tag{A.55}$$

are the inertia moments belonging to the x axis and to the $[x, y]$ plane, respectively. The mass and the center of mass of the rigid body are defined as

$$m = \int_{[s]} dm \quad \text{and} \quad m\rho_c = \int_{[s]} \rho \, dm. \tag{A.56}$$

The product $m\rho_c$ is called the first moment of the rigid body.

Let K and $\tilde{K}$ be two orthonormed frames and investigate the dependence of the inertia matrix on the choice of the reference point (origin of the coordinate system), see Fig. A.2.

Since $\tilde{\rho} = \tilde{p} + \rho$ in coordinate independent form and

$$[\tilde{\rho} \times][\tilde{\rho} \times] = [\tilde{p} \times][\tilde{p} \times] + [\tilde{p} \times][\rho \times] + [\rho \times][\tilde{p} \times] + [\rho \times][\rho \times],$$

hence we receive after integration the *Huygens–Steiner formula* in tensor form

$$\mathcal{I}_{\tilde{\rho}} = \mathcal{I}_\rho - m[\tilde{p} \times][\tilde{p} \times] - [\tilde{p} \times][m\rho_c \times] - [m\rho_c \times][\tilde{p} \times], \tag{A.57}$$

and in matrix form

$$\mathbf{I}_{\tilde{\rho}} = A_{\tilde{K}K} \mathbf{I}_\rho A_{\tilde{K}K}^T - m[\tilde{p} \times][\tilde{p} \times]$$
$$- [\tilde{p} \times][(mA_{\tilde{K}K}\rho_c) \times] - [(mA_{\tilde{K}K}\rho_c) \times][\tilde{p} \times]. \tag{A.58}$$

Especially, if $K := K_c$ is located in the center of mass and $\tilde{K} := K$ is optional, then $m\rho_c = \int \rho \, dm := 0$, $\tilde{p} := \rho_c$,

$$\mathbf{I} = A_{K,K_c} \mathbf{I}_c A_{K,K_c}^T - m[\rho_c \times][\rho_c \times]$$
$$= A_{K,K_c} \mathbf{I}_c A_{K,K_c}^T + m\|\rho_c\|^2 I - m[\rho_c \circ \rho_c], \tag{A.59}$$

Fig. A.3 Resulting inertia
matrix of composite system

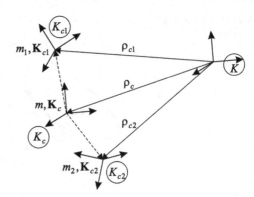

which makes it possible to compute the inertia matrix for the origin of K if the center
of mass ρ_c and the inertia matrix K_c for the center of mass are known. Although the
origin of the frame K_c is in the center of mass, its axes x_c, y_c, z_c are not necessarily
the main axes.

Apply now this result for the composite system consisting of two rigid bodies, see Fig. A.3. Let the inertia parameters of the two bodies be m_1, ρ_{c1}, K_{c1} and
m_2, ρ_{c2}, K_{c2}. The inertia matrices are given for the centers of mass. (The frame K
in case of robot may be that in the Denavit–Hartenberg form. The bodies may be the
robot link and the motor mounted on the link which moves the next following link.)
The centers of mass are given in the frame K and assume that the frames K and K_c
are parallel. Apply (A.59) by the choice $\mathbf{I}_{c1}$, K_c, $\rho_{c1} - \rho_c$, $A_{K_c,K_{c1}} = A_{K,K_{c1}}$ and
$\mathbf{I}_{c2}$, K_c, $\rho_{c2} - \rho_c$, $A_{K_c,K_{c2}} = A_{K,K_{c2}}$, respectively.
Then the resulting system has the inertia parameters $m, \rho_c, \mathbf{I}_c$ as follows:

$$m = m_1 + m_2 \quad \text{and} \quad \rho_c = \frac{m_1 \rho_{c1} + m_2 \rho_{c2}}{m_1 + m_2}, \tag{A.60}$$

$$\mathbf{I}_c = A_{K,K_{c1}} \mathbf{I}_{c1} A_{K,K_{c1}}^T - m_1 \big[(\rho_{c1} - \rho_c) \circ (\rho_{c1} - \rho_c) \big] + m_1 \|\rho_{c1} - \rho_c\|^2 I$$
$$+ A_{K,K_{c2}} \mathbf{I}_{c2} A_{K,K_{c2}}^T - m_2 \big[(\rho_{c2} - \rho_c) \circ (\rho_{c2} - \rho_c) \big] + m_2 \|\rho_{c2} - \rho_c\|^2 I. \tag{A.61}$$

A.4 Lagrange, Appell and Newton–Euler Equations

Physical principles for finding the dynamic model of moving systems can be derived
from simple results for system of particles and their generalization for rigid bodies.

Consider a mass point system consisting of N particles. Denote the mass and
position vector of the l-th particle m_l and r_l, respectively. Let the coordinates of
the position vectors be $x_1, x_2, x_3, \ldots, x_{3N-2}, x_{3N-1}, x_{3N}$ in a fixed orthonormed
coordinate system. Assume that the coordinates satisfy k holonomy and stationary
constraints given by the scalar equations

$$f_\nu(r_1, \ldots, r_N) = 0, \quad \nu = 1, \ldots, k. \tag{A.62}$$

By d'Alambert's law, the dynamic of the system is described by

$$F_l - m_l \ddot{r}_l + R_l = 0, \quad l = 1, \ldots, N, \tag{A.63}$$

where F_l is the active force and R_l is the resulting reaction force acting on m_l. The degree of freedom (DOF) of the system is $n = 3N - k$. Therefore, there are $q_1, \ldots, q_n$ independent scalar parameters, the so called generalized coordinates, so that r_l can be written in the form

$$r_l = r_l(q_1, \ldots, q_n), \quad l = 1, \ldots, N. \tag{A.64}$$

For deriving the motion equations, the principle of virtual work can be applied. Let x be the position of the system in time instant t. If x' is another possible position of the system at the same time instant t, then the system can be moved from x into x' affected to the movement $\delta x = x' - x$ at the same time instant t. The deviation δx is called virtual movement opposite to the dx movement within dt time where forces and constraints can be changed.

Give the system the virtual movement δx and multiply the equations by δr_l belonging to it then $\langle F_l - m_l \ddot{r}_l + R_l, \delta r_l \rangle = 0$, $l = 1, \ldots, N$, and by adding them it yields $\sum_{l=1}^{N} \langle F_l - m_l \ddot{r}_l + R_l, \delta r_l \rangle = 0$. Assuming ideal constraints $\sum_{l=1}^{N} \langle R_l, \delta r_l \rangle = 0$,

$$\sum_{l=1}^{N} \langle F_l - m_l \ddot{r}_l, \delta r_l \rangle = 0 \quad \Leftrightarrow \quad \sum_{l=1}^{N} \langle F_l, \delta r_l \rangle = \sum_{l=1}^{N} \langle m_l \ddot{r}_l, \delta r_l \rangle. \tag{A.65}$$

It follows from (A.64) that

$$\delta r_l = \sum_{l=1}^{N} \frac{\partial r_l}{\partial q_i} \delta q_i, \tag{A.66}$$

$$\sum_{i=1}^{n} \left(\sum_{l=1}^{N} \left\langle F_l, \frac{\partial r_l}{\partial q_i} \right\rangle \right) \delta q_i = \sum_{i=1}^{n} \left(\sum_{l=1}^{N} \left\langle m_l \ddot{r}_l, \frac{\partial r_l}{\partial q_i} \right\rangle \right) \delta q_i. \tag{A.67}$$

Let us define the generalized force Q_i belonging to the generalized coordinate q_i by

$$Q_i := \sum_{l=1}^{N} \left\langle F_l, \frac{\partial r_l}{\partial q_i} \right\rangle, \tag{A.68}$$

then, because the δq_i virtual displacements are independent, it follows

$$Q_i = \sum_{l=1}^{N} \left\langle m_l \ddot{r}_l, \frac{\partial r_l}{\partial q_i} \right\rangle, \tag{A.69}$$

which can be considered an *abstract physical principle*.

A.4.1 Lagrange Equation

The kinetic energy of the mass point system consisting of N particles is

$$K = \frac{1}{2} \sum_{l=1}^{N} m_l \dot{r}_l^2. \tag{A.70}$$

Since by (A.64)

$$\dot{r}_l = \sum_{i=1}^{n} \frac{\partial r_l}{\partial q_i} \dot{q}_i \quad \text{and} \quad \frac{\partial \dot{r}_l}{\partial \dot{q}_i} = \frac{\partial r_l}{\partial q_i}, \tag{A.71}$$

$$\ddot{r}_l = \sum_{i=1}^{n} \left\{ \frac{\partial r_l}{\partial q_i} \ddot{q}_i + \sum_{j=1}^{n} \frac{\partial^2 r_l}{\partial q_i \partial q_j} \dot{q}_i \dot{q}_j \right\}, \tag{A.72}$$

hence

$$\frac{\partial K}{\partial \dot{q}_i} = \sum_{l=1}^{N} \left\langle m_l \dot{r}_l, \frac{\partial \dot{r}_l}{\partial \dot{q}_i} \right\rangle = \sum_{l=1}^{N} \left\langle m_l \dot{r}_l, \frac{\partial r_l}{\partial q_i} \right\rangle, \tag{A.73}$$

$$\frac{d}{dt} \frac{\partial K}{\partial \dot{q}_i} = \sum_{l=1}^{N} \left\langle m_l \ddot{r}_l, \frac{\partial r_l}{\partial q_i} \right\rangle + \sum_{l=1}^{N} \left\langle m_l \dot{r}_l, \sum_{j=1}^{n} \frac{\partial^2 r_l}{\partial q_i \partial q_j} \dot{q}_j \right\rangle, \tag{A.74}$$

$$\frac{\partial K}{\partial q_i} = \sum_{l=1}^{N} \left\langle m_l \dot{r}_l, \frac{\partial \dot{r}_l}{\partial q_i} \right\rangle = \sum_{l=1}^{N} \left\langle m_l \dot{r}_l, \sum_{j=1}^{n} \frac{\partial^2 r_l}{\partial q_j \partial q_i} \dot{q}_j \right\rangle. \tag{A.75}$$

Using the fact that for smooth functions the mixed second order partial derivatives satisfy

$$\frac{\partial^2 x_v}{\partial q_i \partial q_j} = \frac{\partial^2 x_v}{\partial q_j \partial q_i},$$

hence

$$\frac{d}{dt} \frac{\partial K}{\partial \dot{q}_i} - \frac{\partial K}{\partial q_i} = \sum_{l=1}^{N} \left\langle m_l \ddot{r}_l, \frac{\partial r_l}{\partial q_i} \right\rangle = Q_i. \tag{A.76}$$

This relation remains also valid for joined rigid body systems as a consequence of the property of integrals. The generalized force Q_i belonging to the generalized coordinate q_i can be decomposed into the driving torque (force) τ_i and the effect of the potential energy $P(q)$ which latter is $-\partial P/\partial q_i$.

Hence, by the help of the kinetic energy K, the potential energy P, the Lagrange function $L := K - P$ and the driving torque (force) τ_i the results can be summarized in the following two forms of the *Lagrange Equation*:

$$\frac{d}{dt}\frac{\partial K}{\partial \dot{q}_i} - \frac{\partial K}{\partial q_i} + \frac{\partial P}{\partial q_i} = \frac{d}{dt}\frac{\partial L}{\partial \dot{q}_i} - \frac{\partial L}{\partial q_i} = \tau_i. \tag{A.77}$$

Let us consider a rigid body with the coordinate system fixed to it. Denote v and ω the velocity and the angular velocity of the origin of the coordinate system. Then for any point ρ of the rigid body it yields

$$\begin{aligned}
\langle v_\rho, v_\rho \rangle &= \langle v + \omega \times \rho, v + \omega \times \rho \rangle \\
&= \langle v, v \rangle + 2\langle v, \omega \times \rho \rangle + \langle \omega \times \rho, \omega \times \rho \rangle \\
&= \langle v, v \rangle + 2\langle \rho, v \times \omega \rangle + \langle [\rho\times]^T [\rho\times]\omega, \omega \rangle.
\end{aligned}$$

Therefore, the kinetic energy K of the rigid body has the form

$$K = \frac{1}{2}\langle v, v \rangle m + \frac{1}{2}\langle \mathbf{I}\omega, \omega \rangle + \langle m\rho_c, v \times \omega \rangle, \tag{A.78}$$

where ρ_c is the center of mass and $\mathbf{I}$ is the inertia matrix belonging to the origin of the coordinate system. Especially, if the origin of the frame coincides with the center of mass, then $\rho_c = 0$ and hence the kinetic energy of the rigid body simplifies to

$$K = \frac{1}{2}\langle v_c, v_c \rangle m + \frac{1}{2}\langle \mathbf{I}_c\omega, \omega \rangle, \tag{A.79}$$

where $v_c = v + \omega \times \rho_c$ is the velocity of the center of mass and $\mathbf{I}_c$ is the inertia matrix belonging to the center of mass.

A.4.2 Appell Equation

The "acceleration" energy or Gibbs function of a mass point system in case of N particles is

$$G = \frac{1}{2}\sum_{l=1}^{N} m_l \ddot{r}_l^2. \tag{A.80}$$

Since by (A.72)

$$\frac{\partial \ddot{r}_l}{\partial \ddot{q}_i} = \frac{\partial r_l}{\partial q_i}, \tag{A.81}$$

$$\frac{\partial G}{\partial \ddot{q}_i} = \sum_{l=1}^{N}\left\langle m_l \ddot{r}_l, \frac{\partial \ddot{r}_l}{\partial \ddot{q}_i}\right\rangle = \sum_{l=1}^{N}\left\langle m_l \ddot{r}_l, \frac{\partial r_l}{\partial q_i}\right\rangle, \tag{A.82}$$

Fig. A.4 Fixed K_0 and moving K frames

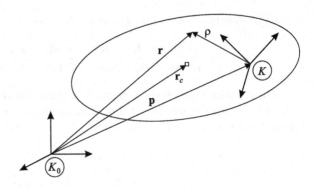

hence using (A.69) we get the *Appell Equation* in very simple form:

$$\frac{\partial G}{\partial \ddot{q}_i} = Q_i \quad \text{and} \quad \frac{\partial G}{\partial \ddot{q}_i} + \frac{\partial P}{\partial q_i} = \tau_i. \tag{A.83}$$

For rigid body, the expression $\langle a_\rho, a_\rho \rangle$ has the form

$$\langle a_\rho, a_\rho \rangle = \langle a + \varepsilon \times \rho + \omega \times (\omega \times \rho), a + \varepsilon \times \rho + \omega \times (\omega \times \rho) \rangle$$
$$= \langle a, a \rangle + \langle [\rho \times]^T [\rho \times] \varepsilon, \varepsilon \rangle + 2 \langle \rho, a \times \varepsilon \rangle$$
$$+ 2 \langle \rho, [\omega \times][\omega \times] a \rangle - 2 \langle \varepsilon, ([\rho \times]^T [\rho \times] \omega) \times \omega \rangle + \cdots ,$$

where the not detailed term does not depend on $\ddot{q}$ and hence the simplified Gibbs function is as follows:

$$G = \frac{1}{2} \langle a, a \rangle m + \frac{1}{2} \langle \mathbf{I}\varepsilon - 2(\mathbf{I}\omega) \times \omega, \varepsilon \rangle$$
$$+ \langle m\rho_c, a \times \varepsilon + \omega \times (\omega \times a) \rangle. \tag{A.84}$$

Especially, if the origin of the frame coincides with the center of mass, then $\rho_c = 0$ and

$$G = \frac{1}{2} \langle a_c, a_c \rangle m + \frac{1}{2} \langle \mathbf{I}_c \varepsilon - 2(\mathbf{I}_c \omega) \times \omega, \varepsilon \rangle, \tag{A.85}$$

where $a_c = a + \varepsilon \times \rho_c + \omega \times (\omega \times \rho_c)$ is the acceleration of the center of mass and $\mathbf{I}_c$ is the inertia matrix belonging to the center of mass.

A.4.3 Newton–Euler Equations

Let K_0 and K be a fixed and a moving frame respectively, and the latter is fixed to the moving rigid body. Let p be the vector pointing from the origin of K_0 to the origin of K, see Fig. A.4.

Instead of rigid body we consider first a system of particles. For the mass point system, (A.63) can be applied and we can take the following sum:

$$\sum_{l=1}^{N} m_l \ddot{r}_l = \sum_{l=1}^{N} F_l + \sum_{l=1}^{N} R_l. \tag{A.86}$$

Using the Newton and Euler principle, we can assume that *the sum of the internal forces and the sum of the moment of internal forces to any point are zero.* Let the sum of the external forces be F_{ext}. Denote m and r_c the resulting mass and the center of mass then it follows from (A.86) the *Newton Equation*:

$$m\ddot{r}_c = F_{\text{ext}}. \tag{A.87}$$

Let the angular moment be defined as

$$\Pi_K = \sum_{l=1}^{N} (r_l - p) \times m_l \dot{r}_l, \tag{A.88}$$

then

$$\frac{d\Pi_K}{dt} = \sum_{l=1}^{N} (\dot{r}_l - \dot{p}) \times m_l \dot{r}_l + \sum_{l=1}^{N} (r_l - p) \times m_l \ddot{r}_l$$

$$= -\dot{p} \times \sum_{l=1}^{N} m_l \dot{r}_l + \sum_{l=1}^{N} (r_l - p) \times m_l \ddot{r}_l$$

$$= -\dot{p} \times m\dot{r}_c + \sum_{l=1}^{N} (r_l - p) \times m_l \ddot{r}_l, \tag{A.89}$$

$$\frac{d\Pi_K}{dt} + \dot{p} \times m\dot{r}_c = \sum_{l=1}^{N} (r_l - p) \times m_l \ddot{r}_l.$$

Denote $N_{K,\text{ext}}$ the moment of external forces:

$$N_{K,\text{ext}} = \sum_{l=1}^{N} (r_l - p) \times F_l, \tag{A.90}$$

then it follows from the assumptions that

$$\frac{d\Pi_K}{dt} + \dot{p} \times m\dot{r}_c = N_{K,\text{ext}}. \tag{A.91}$$

In case of rigid body

$$\Pi_K = \int_{[s]} \rho \times v_\rho \, dm, \tag{A.92}$$

hence by using

$$\rho \times v_\rho = \rho \times (v + \omega \times \rho) = \rho \times v + \rho \times (\omega \times \rho)$$
$$= \rho \times v - \rho \times (\rho \times \omega)$$

it follows

$$\Pi_K = m\rho_c \times v + \mathbf{I}\omega. \tag{A.93}$$

By using the properties of integrals and the differentiation rule in moving frames, it follows

$$\frac{d\Pi_K}{dt} = mv_c \times v + mr_c \times a + \mathbf{I}\varepsilon + \omega \times (\mathbf{I}\omega),$$

$$\frac{d\Pi_K}{dt} + v \times mv_c = mv_c \times v + m\rho_c \times a + \mathbf{I}\varepsilon + \omega \times (\mathbf{I}\omega) + v \times mv_c, \tag{A.94}$$

$$\mathbf{I}\varepsilon + \omega \times (\mathbf{I}\omega) + m\rho_c \times a = N_{K,\text{ext}}.$$

Especially, if the origin of the frame K is in the center of mass of the rigid body, then $\rho_c = 0$ and the *Euler Equation* simplifies to the form

$$\mathbf{I}_c\varepsilon + \omega \times (\mathbf{I}_c\omega) = N_{K_c,\text{ext}}. \tag{A.95}$$

A.5 Robot Kinematics

Many industrial robots wide-spread in practice are so called open chain rigid robots without branching. Their kinematic (geometric) models are often described with the original Denavit-Hartenberg (DH) form. Beside the classical original DH-form, in order to simplify the computation of the dynamic model, a second modified form is also used in the practice, but we consider here only the original form. The name "rigid robot" is an idealization and it means that the flexible deformations in the robot links and joints can be neglected in the domain of the guaranteed precision of the robot.

A.5.1 Denavit-Hartenberg Form

The robot consists of links (segments) which are connected by one degree of freedom (1-DOF) joints. In case of the original DH-form, the joint axes have z-axes direction. The joint variable is d (for prismatic or translational joint T) or ϑ (for revolute or rotational joint R). The DH-parameters are d, a (distances) and ϑ, α (angles).

Let us assume that the frame K_{i-1} has already been chosen. In Fig. A.5, we have drawn the z_{i-1} and x_{i-1} axes where the direction of z_{i-1} is identical with the

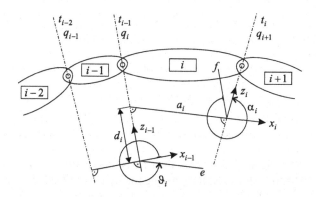

Fig. A.5 The original Denavit-Hartenberg form

direction of the joint axis t_{i-1} connecting the links $\boxed{i-1}$ and $\boxed{i}$. Let t_i be the direction of the joint axis connecting the links $\boxed{i}$ and $\boxed{i+1}$. Let the x_i axis of the frame K_i be orthogonal to both t_{i-1} and t_i. Let the origin of K_i be that point where x_i and t_i intersect and z_i is aligned with t_i. Let e be parallel to x_i through the origin of K_{i-1}, and let f parallel with z_{i-1} through the origin of K_i. Then in case of general (nonintersecting and nonparallel) t_{i-1} and t_i joint axes

   (i) ϑ_i is the rotation angle around z_{i-1}. axis which rotates x_{i-1} axis into e ($e \parallel x_i$).
   (ii) d_i is the displacement along z_{i-1} axis which moves line e into x_i.
   (iii) a_i is the displacement along x_i axis which moves the intersection of z_{i-1} and x_i into the origin of K_i.
   (iv) α_i is the rotation angle around x_i axis which rotates line f into z_i.

The sign of distances is positive if the displacement was performed in $+z_{i-1}$- or $+x_i$-direction. Similarly, the sign of rotation angles are positive if the rotation around z_{i-1} or x_i is positive according to the right screw rule. The homogeneous transformation between K_{i-1} and K_i is given by

$$T_{i-1,i} = Rot(z, \vartheta_i) \cdot Trans(z, d_i) \cdot Trans(x, a_i) \cdot Rot(x, \alpha_i)$$

$$
= \begin{bmatrix} C_{\vartheta_i} & -S_{\vartheta_i} & 0 & 0 \\ S_{\vartheta_i} & C_{\vartheta_i} & 0 & 0 \\ 0 & 0 & 1 & 0 \\ 0 & 0 & 0 & 1 \end{bmatrix} \cdot \begin{bmatrix} 1 & 0 & 0 & a_i \\ 0 & 1 & 0 & 0 \\ 0 & 0 & 1 & d_i \\ 0 & 0 & 0 & 1 \end{bmatrix} \cdot \begin{bmatrix} 1 & 0 & 0 & 0 \\ 0 & C_{\alpha_i} & -S_{\alpha_i} & 0 \\ 0 & S_{\alpha_i} & C_{\alpha_i} & 0 \\ 0 & 0 & 0 & 1 \end{bmatrix}
$$

$$
= \begin{bmatrix} C_{\vartheta_i} & -S_{\vartheta_i} C_{\alpha_i} & S_{\vartheta_i} S_{\alpha_i} & a_i C_{\vartheta_i} \\ S_{\vartheta_i} & C_{\vartheta_i} C_{\alpha_i} & -C_{\vartheta_i} S_{\alpha_i} & a_i S_{\vartheta_i} \\ 0 & S_{\alpha_i} & C_{\alpha_i} & d_i \\ 0 & 0 & 0 & 1 \end{bmatrix} . \tag{A.96}
$$

If t_{i-1} and t_i are parallel, then they have an infinity of common normals. Then in case if rotational joint $d_i = 0$ can be chosen and in case of translational joints the origins of K_i and K_{i+1} may be identical.

If t_{i-1} and t_i intersect, then we can chose $a_i = 0$ and x_i is parallel with $k_{i-1} \times k_i$.

In case of link $\boxed{0}$ the direction of z_0 is equal to t_0 and x_0 may have any direction orthogonal to z_0.

In case of the last link $\boxed{m}$ the direction of z_m is optional (there is no more joint axis) and x_m is orthogonal to the directions of z_{m-1} and z_m.

In some special cases (first x-axis of Cartesian TTT positioning or last axis x of RPY orientation) it may be profitable to align not the z- but the x-axis with the joint axis. The homogeneous transformation $T_{i-1,i}$ still remains valid but the joint variable is changed to a_i or α_i.

Robot control softwares, especially their parts relating to path design, often use the original DH-form. However, for deriving the dynamic model the *modified DH-form* may be more advantageous. The main difference is that in the case of the modified DH-form the origin of the frame K_i is placed on the joint axis (now it may be denoted by t_i) that moves the link $\boxed{i}$. This own joint axis is the z_i-axis and the axis x_i is orthogonal to z_i and z_{i+1}. This simplifies the computation of the dynamic model, especially if the Newton–Euler method is used, because the driving torque (force) τ_i acting to link i is immediately the projection of the reaction force f_i (for translational joint) or reaction torque τ_i (for rotational joint) to the t_i-axis.

A.5.2 Direct Kinematic Problem

An open chain rigid robot without branching consists of links $\boxed{0}$, $\boxed{1}$, ..., $\boxed{m}$ and the end effector (gripper, tool etc.). Neighboring links are connected by joints which are either prismatic or revolute ones. A prismatic joint performs translation along the joint axis while a revolute joint allows a rotation around the joint axis. Let d (or exceptionally a) denote the movement of the prismatic joint (T). Similarly, ϑ (or exceptionally α) denotes the angle of rotation of a revolute joint (R). Their common name is the joint variable q.

The name "open chain robot without branching" means that the links can be numbered in such a way that every link is followed by exactly one link $(\boxed{m}$ is followed by the end effector), and the position and orientation of link $\boxed{i}$ depend only on the position and orientation of the preceding links and the joint variable q_i between links $\boxed{i-1}$ and $\boxed{i}$. The preceding links define the position and orientation of the frame K_{i-1} fixed to link $\boxed{i-1}$, while q_i defines the relative position and orientation of the frame K_i fixed to link $\boxed{i}$ relative to K_{i-1} described by the homogeneous transformation $T_{K_{i-1},K_i} := T_{i-1,i}$. The joint axis between links $\boxed{i-1}$ and $\boxed{i}$ is denoted by t_{i-1} (see Fig. A.6).

In analogy to a human, we can speak about trunk $(\boxed{0})$, arm $(\boxed{1}$, $\boxed{2}$, $\boxed{3})$, wrist $(\boxed{4}$, $\boxed{5}$, $\boxed{6})$ and hand (gripper etc.). The end effector is normally exchangeable and it is suitable to speak about the point where it is attached to $\boxed{m}$ and its *tool center point* (TCP).

Fig. A.6 The principal structure of an open chain rigid robot

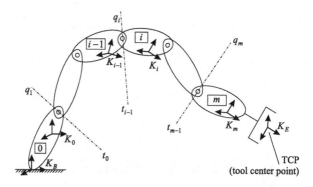

The frame K_{i-1} fixed to link $\boxed{i-1}$ could be anywhere. In practice, we often use the Denavit-Hartenberg convention. The DH-parameters $\vartheta_i, d_i, a_i, \alpha_i$ describe the homogeneous transformation $T_{i-1,i}$ between K_{i-1} and K_i. If we choose a point P_j of link $\boxed{j}$ with coordinates x_j, y_j, z_j in K_j, then the same point will have coordinates $x_{j-1}, y_{j-1}, z_{j-1}$ in K_{j-1} satisfying

$$\begin{pmatrix} x_{j-1} \\ y_{j-1} \\ z_{j-1} \\ 1 \end{pmatrix} = T_{j-1,j} \begin{pmatrix} x_j \\ y_j \\ z_j \\ 1 \end{pmatrix}. \tag{A.97}$$

Using this relation step by step, it follows for the coordinates x_i, y_i, z_i in K_i of the same point

$$\begin{pmatrix} x_i \\ y_i \\ z_i \\ 1 \end{pmatrix} = T_{i,i+1} \cdot T_{i+1,i+2} \cdots T_{j-1,j} \begin{pmatrix} x_j \\ y_j \\ z_j \\ 1 \end{pmatrix} = T_{i,j} \begin{pmatrix} x_j \\ y_j \\ z_j \\ 1 \end{pmatrix}, \tag{A.98}$$

where

$$T_{i,j} = T_{i,i+1} \cdot T_{i+1,i+2} \cdots T_{j-1,j}. \tag{A.99}$$

Here $T_{i,j}$ describes the coordinate transformation and defines the position and orientation of K_j relative to K_i. Especially,

$$T_{0,m} = T_{0,1} \cdot T_{1,2} \cdots T_{m-1,m} = \left(\begin{array}{c|c} A_{0,m} & p_{0,m} \\ \hline 0^T & 1 \end{array} \right), \tag{A.100}$$

$$T_{B,E} = T_{B,0} \cdot T_{0,m} \cdot T_{m,E}. \tag{A.101}$$

Observe that

$$A_{0,m} = A_{0,1} \cdot A_{1,2} \cdots A_{m-1,m}, \tag{A.102}$$

$$A_{0,m}^{-1} = A_{m-1,m}^T \cdots A_{1,2}^T \cdot A_{0,1}^T, \tag{A.103}$$

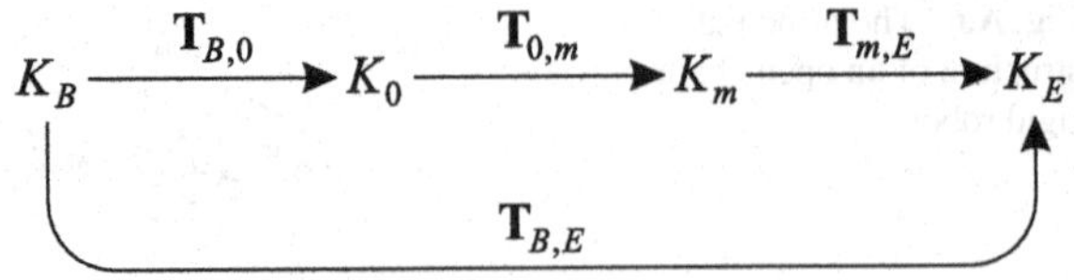

Fig. A.7 The robot graph

$$T_{j,i} := T_{i,j}^{-1} = T_{j-1,j}^{-1} \cdots T_{i+1,i+2}^{-1} \cdot T_{i,i+1}^{-1}, \tag{A.104}$$

$$T_{0,m} = T_{B,0}^{-1} \cdot T_{B,E} \cdot T_{m,E}^{-1}, \tag{A.105}$$

and the last equation suggest characterizing the robot by its *robot graph*, see Fig. A.7.

The robot graph is a directed attributed graph. The nodes of the graph are the frames, the frames are connected by edges, the attributes of the edges are homogeneous transformations and the edges are directed from left to right according to the indexing. We can reach K_m from K_0 in two different ways:

(i) In forward direction in the sequence K_0, K_m.
(ii) In backward direction in the sequence K_0, K_B, K_E, K_m.

If we interpret the motion against an arrow as an inversion of the homogeneous transformation, then we can obtain (A.105), assuming we write down the homogeneous transformation or its inverse along the edges from left to right and multiply them together. The robot graph can be generalized for more complicated cases (robot, conveyor, object, block, camera etc.).

We speak about the *direct or forward kinematic (geometric) problem* if we want to determine, how $T_{B,E}$ or $T_{0,m}$ depends on the joint variable vector $q = (q_1, \ldots, q_m)^T$. Usually, we have to find this relation in symbolic form as a formula using the Denavit-Hartenberg parameters of the robot and (A.100):

$$\text{direct kinematic (geometric) problem:} \quad q \mapsto T_{0,m}. \tag{A.106}$$

The symbolic form is needed in order to find a computational method for the inverse kinematic (geometric) problem described later.

A.5.3 Inverse Kinematic Problem

The goal of the inverse kinematic (geometric) problem is to find the values of the joint variables if the position and the orientation of the end effector are given. The values of the joint variables can be applied as reference signals (time dependent set points) to the servo control of the axes. We shall refer to the problem as

$$\text{inverse kinematic (geometric) problem:} \quad q = \text{solve } T_{0,m}. \tag{A.107}$$

The task needs to solve a set of transcendent equations which usually cannot be solved in symbolic (analytical) form. However, for the practice there is a special case in which analytical solution exists.

Theorem A.1 (Solution of the inverse kinematic problem using decomposition) *If the last three joint axes are revolute ones and intersect in a common point, then the inverse kinematic (geometric) problem can be decomposed in a positioning and an orienting subproblem that can separately be solved.*

Sketch of the proof Let $m = 6$, $a_4 = a_5 = a_6 = 0$ and $\alpha_6 = 0°$ in the Denavit-Hartenberg form of the robot then

$$T_{0,6} = \begin{bmatrix} l_{0,6} & m_{0,6} & n_{0,6} & p_{0,6} \\ 0 & 0 & 0 & 1 \end{bmatrix}$$

$$= T_{0,3} \begin{bmatrix} & 0 \\ A_{3,4} & 0 \\ & d_4 \\ \hline 0^T & 1 \end{bmatrix} \begin{bmatrix} A_{4,5} & 0 \\ \hline 0^T & 1 \end{bmatrix} \begin{bmatrix} * & * & 0 & 0 \\ * & * & 0 & 0 \\ 0 & 0 & 1 & d_6 \\ 0 & 0 & 0 & 1 \end{bmatrix}$$

$$= \begin{bmatrix} l_{0,3} & m_{0,3} & n_{0,3} & p_{0,3} \\ 0 & 0 & 0 & 1 \end{bmatrix} \begin{bmatrix} 1 & 0 & 0 & 0 \\ 0 & 1 & 0 & 0 \\ 0 & 0 & 1 & d_4 \\ 0 & 0 & 0 & 1 \end{bmatrix} \begin{bmatrix} A_{3,6} & 0 \\ \hline 0^T & 1 \end{bmatrix} \begin{bmatrix} 1 & 0 & 0 & 0 \\ 0 & 1 & 0 & 0 \\ 0 & 0 & 1 & d_6 \\ 0 & 0 & 0 & 1 \end{bmatrix}$$

$$= \begin{bmatrix} A_{0,6} & p_{0,3} + d_4 n_{0,3} + d_6 n_{0,6} \\ \hline 0^T & 1 \end{bmatrix}, \tag{A.108}$$

and hence

$$p_{0,3}(q_1, q_2, q_3) + d_4 n_{0,3}(q_1, q_2, q_3) = p_{0,6} - d_6 n_{0,6}, \tag{A.109}$$

$$A_{3,6}(q_4, q_5, q_6) = A_{0,3}^T(q_1, q_2, q_3) \begin{bmatrix} l_{0,6} & m_{0,6} & n_{0,6} \end{bmatrix}. \tag{A.110}$$

So we have divided the original inverse kinematic (geometric) problem into two subproblems. The first one is (A.109) which is a pure positioning subproblem in three dimensions having solution q_1, q_2, q_3. Having solved the first subproblem, we can compute the right side of (A.110) and seek for the solution of the second subproblem in q_4, q_5, q_6 which is a pure orienting subproblem in three dimensions. For many important industrial robots, the second subproblem often results in an inverse Euler or an inverse RPY problem that we have already solved earlier.

If the above conditions are not satisfied, then we can try to separate a sequence of nonlinear equations with only one unknown, however the separability is not always guaranteed. In the general case, iterative numerical methods can be suggested.

A.5.4 Robot Jacobian

The geometric model of the robot is a nonlinear function between the joint variables and the position and orientation of the end effector. By locally linearizing this relation in the actual configuration q, we can obtain the Jacobian matrix $J(q)$ of the robot. The linearization is performed by using differentiation which can be assumed to be the application of the superposition principle.

Let us freeze all the values of the joint variables except q_i in the actual configuration and give q_i a small change dq_i. Then the links $\boxed{i}$, $\boxed{i+1}$, ..., $\boxed{m}$ compose a single rigid body either moving along the joint axis t_{i-1} or rotating around the joint axis t_{i-1} depending on the joint type (T or R). Dividing by the time dt and separating from the effects the joint velocity $\dot{q}_i$ the result is the partial velocity ${}^m d_{i-1}$ and the partial angular velocity ${}^m t_{i-1}$ satisfying

$$
{}^m d_{i-1} = A_{i-1,m}^{-1} \left\{ (1 - \kappa_i) t_{i-1} + \kappa_i t_{i-1} \times p_{i-1,m} \right\},
\tag{A.111}
$$

$$
{}^m t_{i-1} = A_{i-1,m}^{-1} \kappa_i t_{i-1},
\tag{A.112}
$$

where $\kappa_i = 1$ for revolute joints (R) and $\kappa_i = 0$ for prismatic joints (T), respectively.

By the superposition principle the velocity ${}^m v_m$ and the angular velocity ${}^m \omega_m$ of the origin of the frame K_m, expressed in the basis of K_m, are respectively

$$
{}^m v_m = \sum_{i=1}^m {}^m d_{i-1} \dot{q}_i,
\tag{A.113}
$$

$$
{}^m \omega_m = \sum_{i=1}^m {}^m t_{i-1} \dot{q}_i.
\tag{A.114}
$$

If z_{i-1} is the joint axis as in the original DH-form, then $t_{i-1} = k_{i-1}$ and because of $A^{-1} = A^T$ it yields:

In case of rotational joint (R):

$$
\begin{aligned}
{}^m t_{i-1,x} &= l \cdot t_{i-1} = l \cdot k_{i-1} = l_z, \\
{}^m t_{i-1,y} &= m \cdot t_{i-1} = m \cdot k_{i-1} = m_z, \\
{}^m t_{i-1,z} &= n \cdot t_{i-1} = n \cdot k_{i-1} = n_z,
\end{aligned}
\tag{A.115}
$$

$$
\begin{aligned}
{}^m d_{i-1,x} &= l \cdot (t_{i-1} \times p) = l \cdot (k_{i-1} \times p) \\
&= l^T \begin{bmatrix} 0 & -1 & 0 \\ 1 & 0 & 0 \\ 0 & 0 & 0 \end{bmatrix} \begin{pmatrix} p_x \\ p_y \\ p_z \end{pmatrix} = l^T \begin{pmatrix} -p_y \\ p_x \\ 0 \end{pmatrix} = -l_x p_y + l_y p_x,
\end{aligned}
\tag{A.116}
$$

$$
{}^m d_{i-1,y} = m \cdot (t_{i-1} \times p) = m \cdot (k_{i-1} \times p) = -m_x p_y + m_y p_x,
$$

$$
{}^m d_{i-1,z} = n \cdot (t_{i-1} \times p) = n \cdot (k_{i-1} \times p) = -n_x p_y + n_y p_x.
$$

In case of translational joint (T):

$$
{}^m t_{i-1} = 0,
\tag{A.117}
$$

$$
\begin{aligned}
{}^{m}d_{i-1,x} &= l \cdot t_{i-1} = l \cdot k_{i-1} = l_z, \\
{}^{m}d_{i-1,y} &= m \cdot t_{i-1} = m \cdot k_{i-1} = m_z, \\
{}^{m}d_{i-1,z} &= n \cdot t_{i-1} = n \cdot k_{i-1} = n_z.
\end{aligned}
\tag{A.118}
$$

It follows from (A.111)

$$
\begin{aligned}
\frac{\partial T_{i-1,m}}{\partial q_i} &=
\begin{bmatrix}
i & j & k & p \\
0 & 0 & 0 & 1
\end{bmatrix}'
=
\begin{bmatrix}
i' & j' & k' & p' \\
0 & 0 & 0 & 0
\end{bmatrix} \\
&=
\begin{bmatrix}
\kappa_i[t_{i-1}\times]i & \kappa_i[t_{i-1}\times]j & \kappa_i[t_{i-1}\times]k & \kappa_i[t_{i-1}\times]p + (1-\kappa_i)t_{i-1} \\
0 & 0 & 0 & 0
\end{bmatrix} \\
&=
\begin{bmatrix}
\kappa_i[t_{i-1}\times] & (1-\kappa_i)t_{i-1} \\
0^T & 0
\end{bmatrix}
\begin{bmatrix}
i & j & k & p \\
0 & 0 & 0 & 1
\end{bmatrix}.
\end{aligned}
$$

Hence, by introducing the notations

$$
\Delta_{i-1} =
\begin{bmatrix}
[\kappa_i t_{i-1}\times] & (1-\kappa_i)t_{i-1} \\
0^T & 0
\end{bmatrix},
\tag{A.119a}
$$

$$
{}^{m}\Delta_{i-1} =
\begin{bmatrix}
[{}^{m}t_{i-1}\times] & {}^{m}d_{i-1} \\
0^T & 0
\end{bmatrix},
\tag{A.119b}
$$

we can write

$$
\frac{\partial T_{i-1,m}}{\partial q_i} = \Delta_{i-1}T_{i-1,m} = T_{i-1,m}\,{}^{m}\Delta_{i-1},
\tag{A.120a}
$$

$$
{}^{m}\Delta_{i-1} = T_{i-1,m}^{-1}\Delta_{i-1}T_{i-1,m}.
\tag{A.120b}
$$

The results can be formulated in a compact form by introducing the *Jacobian matrix* ${}^{m}J_m$ of the robot:

$$
{}^{m}J_m =
\begin{bmatrix}
{}^{m}d_0 & {}^{m}d_1 \cdots {}^{m}d_{m-1} \\
{}^{m}t_0 & {}^{m}t_1 \cdots {}^{m}t_{m-1}
\end{bmatrix},
\tag{A.121}
$$

$$
\begin{pmatrix}
{}^{m}v_m \\
{}^{m}\omega_m
\end{pmatrix}
= {}^{m}J_m \dot{q}.
\tag{A.122}
$$

Since in the basis of the fixed frame K_0

$$
{}^{0}v_m = A_{0,m}\,{}^{m}v_m \quad \text{and} \quad {}^{0}\omega_m = A_{0,m}\,{}^{m}\omega_m,
\tag{A.123}
$$

hence

$$
{}^{0}J_m =
\begin{bmatrix}
A_{0,m} & 0 \\
0 & A_{0,m}
\end{bmatrix}
{}^{m}J_m,
\tag{A.124}
$$

$$
\begin{pmatrix}
{}^{0}v_m \\
{}^{0}\omega_m
\end{pmatrix}
= {}^{0}J_m \dot{q}.
\tag{A.125}
$$

Here $^m J_m$ and $^0 J_m$ are the robot Jacobian belonging to the origin of the frame K_m, expressed in the basis of K_0 and K_m, respectively.

In cases when the inverse kinematic (geometric) problem cannot analytically be solved in closed form, we can determine $\dot{q}(t)$ from (A.122) or (A.125) in the neighborhood of the actual configuration q, from which we can obtain the value of $q(t)$ by numerical integration (velocity algorithm). In this case, the velocity and the angular velocity along the path should be prescribed and $\dot{q}$ can be determined by using the pseudo-inverse $J_m^{\#}$ of the Jacobian:

$$\dot{q} = {}^0 J_m^{\#} \begin{pmatrix} {}^0 v_m \\ {}^0 \omega_m \end{pmatrix}. \tag{A.126}$$

If $m = 6$ and J_m is nonsingular, then $J_m^{\#} = J_m^{-1}$. Such robot configurations q for which the image space of J_m is not the whole space R^6 (dim $range(J_m) < 6$) will be called singular configurations. In singular configurations, there are directions along or around them the robot cannot be moved or rotated, respectively.

If $m > 6$, then the robot has redundant degree of freedom, and it is advisable to choose from the infinity of solutions that one for which $\|\dot{q}\| \to \min$, because this assures the shortest transient. In this case, the optimum solution is

$$\dot{q} = J_m^T \left(J_m J_m^T \right)^{-1} \begin{pmatrix} v_m \\ \omega_m \end{pmatrix}. \tag{A.127}$$

If $m < 6$, then it is not possible to guarantee the prescribed velocity and angular velocity at the same time, because the robot has too few degrees of freedom. In this case, there are two possibilities:

(i) We can choose coordinates of v_m and ω_m whose number is dim $\dot{q}$. Then we consider only that equations which correspond to the selected coordinates and the other equations will be neglected. For example, if dim $\dot{q} = 2$ and p_x, p_y depend only on q_1 and q_2, then we can choose v_x and v_y and we can determine $\dot{q}_1$ and $\dot{q}_2$ from the first two equations.

(ii) We can choose such a $\dot{q}$ which simultaneously minimizes the equation errors in LS sense in which case this compromise solution is

$$\dot{q} = \left(J_m^T J_m \right)^{-1} J_m^T \begin{pmatrix} v_m \\ \omega_m \end{pmatrix}. \tag{A.128}$$

Because of its importance, we deal in detail with the *inversion of the Jacobian* in the special 6-DOF case provided the last three axes are revolute ones and intersect in a common point (6*), see also (A.108):

$$T_{B,E} = T_{B,0} \cdot T_{0,3} \cdot \begin{bmatrix} 1 & 0 & 0 & 0 \\ 0 & 1 & 0 & 0 \\ 0 & 0 & 1 & d_4 \\ 0 & 0 & 0 & 1 \end{bmatrix} \begin{bmatrix} A_{3,6} & 0 \\ 0^T & 1 \end{bmatrix} \cdot T_{6*,E}, \tag{A.129}$$

$$T_{6*,E} = \begin{bmatrix} 1 & 0 & 0 & 0 \\ 0 & 1 & 0 & 0 \\ 0 & 0 & 1 & d_6 \\ 0 & 0 & 0 & 1 \end{bmatrix} T_{6,E}. \tag{A.130}$$

We assume that the velocity $^B v_E$ and angular velocity $^B \omega_E$ of the end effector are given in the base frame K_B of the robot. Then $\dot{q}$ can be determined in the following steps:

(i) The velocity and the angular velocity can be expressed in K_E:

$$^E v_E = A_{B,E}^{-1}\,^B v_E; \qquad ^E \omega_E = A_{B,E}^{-1}\,^B \omega_E. \tag{A.131}$$

(ii) The velocity and angular velocity of the origin of K_{6*} can be determined as

$$^{6*} v_{6*} = A_{6*,E}\left\{ ^E v_E + {}^E \omega_E \times \left(-A_{6*,E}^{-1} p_{6*,E} \right) \right\}, \tag{A.132}$$

$$^{6*} \omega_{6*} = A_{6*,E}\,^E \omega_E. \tag{A.133}$$

(iii) Since $^{6*} d_{i-1} = 0$, $i = 4, 5, 6*$, hence the Jacobian matrix is a block lower triangular matrix:

$$^{6*} J_{6*} = \left[\begin{array}{c|c} A & 0 \\ \hline B & C \end{array} \right]. \tag{A.134}$$

Using the decomposition $q_A = (q_1, q_2, q_3)^T$, $q_B = (q_4, q_5, q_6)^T$, the inverse velocity problem can be divided into two subproblems in which the inverses of the 3×3 matrices can be computed in closed form without iterations:

$$\dot{q}_A = A^{-1\,6*} v_{6*}, \tag{A.135}$$

$$\dot{q}_B = C^{-1}\left\{ -BA^{-1\,6*} v_{6*} + {}^{6*} \omega_{6*} \right\}. \tag{A.136}$$

Appendix B
Basis of Differential Geometry for Control Problems

Overview   In this Appendix, we summarize the basic ideas of differential geometry (Lie derivatives, submanifold, tangent space, distribution), the Frobenius theorem about the equivalence of fully integrable and involutive distributions, the generalization of reachability and observability for nonlinear systems, the input/output linearization of nonlinear systems and the influence of zero dynamics on stability, all playing important role in the control of nonlinear systems.

B.1 Lie Derivatives, Submanifold, Tangent Space

First, we summarize the fundamentals of differential geometry. Similarly to Vidyasagar [152] and Lantos [73], in the investigations we prefer "coordinate dependent" forms which are easier to understand for engineers. Theoretically more founded approaches and other important methods can be found in the books of Isidori [56, 57], Nijmeier and van der Schaft [106], Sastry [126], and van der Schaft [127]. Let consider first some important definitions.

Continuously differentiable function:
$f : R^n \to R^n$ is a continuously differentiable function ($f \in C^{(1)}$) if $\frac{\partial f_i}{\partial x_j}$ is continuous, $i = 1, \ldots, n$, $j = 1, \ldots, n$.

Smooth function:
The function $f : R^n \to R^n$ is smooth if every f_i component of f has continuous derivatives of all orders with respect to all combinations of its components ($C^{(\infty)}$).

Jacobian matrix:

$$\frac{\partial f}{\partial x} = \begin{bmatrix} \frac{\partial f_1}{\partial x_1} & \cdots & \frac{\partial f_1}{\partial x_n} \\ \vdots & \vdots & \vdots \\ \frac{\partial f_n}{\partial x_1} & \cdots & \frac{\partial f_n}{\partial x_n} \end{bmatrix}. \tag{B.1}$$

Diffeomorphism:

B. Lantos, L. Márton, *Nonlinear Control of Vehicles and Robots*,
Advances in Industrial Control,
DOI 10.1007/978-1-84996-122-6, © Springer-Verlag London Limited 2011

Let $U, V \subset R^n$ be open sets and $f : U \to V$, $f \in C^{(1)}$. Then f is called a diffeomorphism if

(i) $f(U) = V$.
(ii) f is one-to-one ($f(x) = f(y) \Leftrightarrow x = y$).
(iii) The inverse function $f^{-1} : V \to U$ is continuously differentiable ($f^{-1} \in C^{(1)}$).

We say that f is a smooth diffeomorphism if f, f^{-1} are smooth functions.

Inverse function theorem:
Let $f : R^n \to R^n$, $f \in C^{(1)}$, $y_0 = f(x_0)$ and $\frac{\partial f}{\partial x}|_{x=x_0}$ is nonsingular. Then $\exists U(x_0)$, $V(x_0)$ open sets such that f is a diffeomorphism of U onto V. If, in addition, f is smooth, then f^{-1} is also smooth, that is, f is a smooth diffeomorphism.

In what follows, we suppose that the diffeomorphism is always smooth and $X \subset R^n$ is an open set, n is fixed. The following definitions and notations will be regularly used:

(i) Vector field on X: any smooth function $f : X \to R^n$.
(ii) $V(X)$: the set of vector fields over X.
(iii) $S(X)$: the set of smooth real-valued functions $a : X \to R^1$.
(iv) Space of row vectors: $(R^n)^*$.
(v) Form on X: any smooth function $h : X \to (R^n)^*$.
(vi) $F(X)$: the set of forms over X.

Properties:

(i) $a, b \in S(X) \Rightarrow a \pm b, ab \in S(X) \Rightarrow S(X)$ is a ring.
(ii) In case of $a, b \in S(X)$ and $f, g \in V(X)$ yields

$$a(f + g) = af + ag,$$

$$(a + b)f = af + bf,$$

$$(ab)f = a(bf).$$

Hence, $V(X)$ is a module over the ring $S(X)$.

Curve:
c is a curve in X passing through x_0 if $c : (-\alpha, \beta) \to X$ is a smooth function where $-\alpha < 0 < \beta$ and $c(0) = x_0$.

Integral curve:
If $f \in V(X)$, then the initial value problem $\frac{d}{dt}x(t) = f(x(t))$, $x(0) = x_0$ has a unique solution for sufficiently small t (in both direction). The solution $x(t) =: s_{f,t}(x_0)$ is called the integral curve of f passing through the point x_0.

Nonlinear coordinate transformation:
Let $f \in V(X)$, $x_0 \in X$ and T a smooth diffeomorphism over some $U(x_0)$ neighborhood of x_0. Let us introduce the coordinate transformation $y = T(x)$. Let $x(t)$ be the integral curve of f through x_0 and denote J the Jacobian matrix of T. Then the

effect of the coordinate transformation $y = T(x)$ on the curve $x(t)$ is

$$y(t) = T(x(t)),$$

$$\frac{d}{dt}y(t) = \frac{\partial T}{\partial x}\Big|_{x(t)} \frac{dx(t)}{dt} = J|_{x(t)}f(x(t)) = J(T^{-1}(y(t)))(f(T^{-1}(y(t)))),$$

$$\frac{d}{dt}y(t) = f_T(y(t)),$$

(B.2)

$$f_T(y) := J(T^{-1}(y))f(T^{-1}(y)) \quad \Rightarrow \quad f_T = (Jf) \circ T^{-1}.$$

Similarly yields: $s_{f,t} = T^{-1} \circ s_{f_T,t} \circ T \Leftrightarrow T \circ s_{f,t} = s_{f_T,t} \circ T$.

Gradient:

Let $a \in S(X)$ then $\nabla a = (\frac{\partial a}{\partial x_1}, \ldots, \frac{\partial a}{\partial x_n}) =: da$.

Lie derivatives:

(1) Lie derivative of $a \in S(X)$ with respect to the vector field $f \in V(X)$ is $L_f a := \langle da, f \rangle$.

(2) Suppose $f, g \in V(x)$ then the Lie derivative of g with respect to f is the Lie bracket: $L_f g = [f, g] := \frac{\partial g}{\partial x}f - \frac{\partial f}{\partial x}g$.

(3) Suppose $f \in V(X)$ and $h \in F(X)$ then the Lie derivative of h with respect to f is the form defined by $L_f h := f^T(\frac{\partial h^T}{\partial x})^T + h\frac{\partial f}{\partial x}$.

Properties of Lie derivatives:

(1) $(L_f a)(x_0) = \lim_{t\to 0}\frac{1}{t}\{a(s_{f,t}(x_0)) - a(x_0)\}$. Therefore, $L_f a$ can be interpreted as the Lie derivative of a along the integral curves of the vector field f.

(2) $[f, g](x_0) = \lim_{t\to 0}\frac{1}{t}\{[\frac{\partial s_{f,-t}(x)}{\partial x}]_{s_{f,t}(x_0)} \circ g[s_{f,t}(x_0)] - g(x_0)\}$. Therefore, the Lie bracket $[f, g]$ can be interpreted as the directional derivative of the vector field g along the integral curves of f. However, in case of abstract manifolds $g[s_{f,t}(x_0)]$ and $g(x_0)$ cannot simply be subtracted because they belong to different tangent spaces. Hence, an extra factor $[\partial s_{f,-t}(x)/\partial x]_{s_{f,t}(x_0)}$ "pulls back" the vector $g[s_{f,t}(x_0)]$ into the tangent space of X at x_0.

(3) $[f, g](x_0) = \lim_{t\to 0}\frac{1}{t^2}\{[s_{g,-t} \circ s_{f,-t} \circ s_{g,t} \circ s_{f,t}](x_0) - x_0\}$. It means that we start at the point x_0 and follow the integral curve of f for a very short time t, then we follow the integral curve of g for a duration of t, after this we follow the integral curve of f backwards in time for a duration of t, and finally we follow the integral curve of g backwards in time for a duration of t. Since the inverse of $s_{f,t}$ is $s_{f,-t}$ etc., hence if $s_{f,t}$ and $s_{g,t}$ commute for all sufficiently small t then $[f, g]$ is zero, otherwise it is a measure in which extent the solution fails to commute (see Fig. B.1).

(4) $[f, g] = 0 \Leftrightarrow s_{f,t} \circ s_{g,\tau} = s_{g,\tau} \circ s_{f,t}$ for all sufficiently small t, τ.

(5) Lie bracket is invariant under coordinate transformation: $[f, g] = [f_T, g_T]$.

(6) Let $a \in S(X)$ and $f, g \in V(X)$ then $L_{[f,g]}a = L_f(L_g a) - L_g(L_f a)$.

(7) Let $f, g \in V(x)$ and $h \in F(X)$ then $L_f\langle h, g\rangle = \langle L_f h, g\rangle + \langle h, \underbrace{L_f g}_{[f,g]}\rangle$.

Fig. B.1 Illustration of the Lie bracket

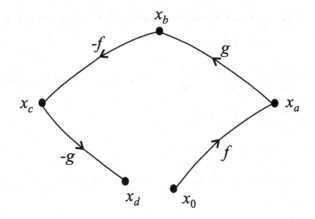

(8) Let $f, g, h \in V(X)$, $a \in S(X)$ and $\alpha, \beta \in R^1$ then the Lie bracket has the following properties:

$$[f, \alpha g + \beta h] = \alpha[f, g] + \beta[f, h],$$

$$[f, g] = -[g, f] \quad \text{(anti-symmetry)},$$

$$[f, [g, h]] + [g, [h, f]] + [h, [f, g]] = 0 \quad \text{(Jacobi identity)},$$

$$[f, ag] = a[f, g] + (L_f a)g \quad \text{and} \quad L_f(ag) = a(L_f g) + (L_f a)g$$

(product rule).

The proof of the properties is simple. For example, the proof of property (6) is based on the symmetry of the Hessian matrix $\partial^2 a / \partial x^2$.

Repeated Lie bracket ("ad" operator):

$$ad_f^0 g := g,$$
$$ad_f^{i+1} g := [f, ad_f^i g],$$

(B.3)

especially $ad_f^1 g = [f, g]$ and $ad_f^2 g = [f, [f, g]]$.

Linearly independent vector fields:

(i) $f_1, f_2, \ldots, f_k \in V(X)$ are linearly independent at the point x_0 if the column vectors $f_1(x_0), f_2(x_0), \ldots, f_k(x_0)$ are linearly independent.

(2) Because of continuity, it follows from property (1) that there exists an open set $U(x_0) \subset R^n$ such that linear independence is valid for every $x \in U(x_0)$.

k-dimensional submanifold:

We say that $M \subset X$ is a k-dimensional submanifold of X ($k < n$) if for every $x_0 \in M$ there exist an open set $U(x_0) \subset X$ and $\phi_{k+1}, \ldots, \phi_n \in S(X)$ such that the following conditions are satisfied:

(1) $\{d\phi_{k+1}(x), \ldots, d\phi_n(x)\}$ are linearly independent, $\forall x \in U(x_0)$.
(2) $U(x_0) \cap M = \{x \in U : \phi_{k+1}(x) = \cdots = \phi_n(x) = 0\}$.

Let notice that the submanifold is defined by $n - k$ nonlinear equations so that the remaining dimension is k. It is easy to show that if the above two conditions are satisfied then the submanifold does not depend on the choice of representatives.

Tangent space:
Let $M \subset X$ be a k-dimensional submanifold of X and choose functions $\phi_{k+1}, \ldots, \phi_n$ $\in S(X)$ satisfying the definition. Then the tangent space of M at the point $x \in M$ is the subspace $TM_x \subset R^n$ which is defined by $TM_x := \{v \in R^n : \langle d\phi_{k+1}(x), v \rangle = \cdots = \langle d\phi_n(x), v \rangle = 0\}$.
We say that $f \in V(X)$ is tangent to M at the point x if $f(x) \in TM_x$, i.e. $\langle d\phi_{k+1}(x), f(x) \rangle = \cdots = \langle d\phi_n(x), f(x) \rangle = 0$.

Properties of the tangent space:

(1) TM_x does not depend on which representative $\phi_{k+1}, \ldots, \phi_n \in S(X)$ was chosen in the definition of M.
(2) $TM_x = (\text{Span}\{d\phi_{k+1}(x), \ldots, d\phi_n(x)\})^{\perp}$ where $\perp$ denotes the orthogonal complement.

Lemma B.1 (Construction algorithm) *Functions having linearly independent differentials can be completed to coordinate transformation.*

Proof Let $\phi_{k+1}, \ldots, \phi_n \in S(X)$ be the functions having linearly independent differentials.

Construction: Pick $v_1, \ldots, v_k \in (R^n)^*$ such that they form a basis together with the linearly independent row vectors $d\phi_{k+1}(x_0), \ldots, d\phi_n(x_0)$. Let $\phi_i(x) := v_i(x - x_0)$, $i = 1, \ldots, k$ then $\{v_1, \ldots, v_k, d\phi_{k+1}(x_0), \ldots, d\phi_n(x_0)\}$ is a basis in $(R^n)^*$.

Coordinate transformation: Define $y_i = (T(x))_i := \phi_i(x), i = 1, \ldots, n$. By the construction the Jacobian matrix, $J = [\frac{\partial T}{\partial x}]_{x_0}$ is nonsingular. By the inverse function theorem $\exists U_0(x_0) \subset U(x_0)$ open set whose coordinates can be chosen $y_1, \ldots, y_n$.

Submanifold using special coordinate choice:
Let $M \subset X$ be a k-dimensional submanifold, $x_0 \in M$, and choose $U(x_0)$ and $\phi_{k+1}, \ldots, \phi_n \in S(X)$ according to the definition. By Lemma B.1, we can choose functions $\phi_1, \ldots, \phi_k \in S(X)$ such that $\phi_1(x_0) = \cdots = \phi_k(x_0) = 0$ and $\{d\phi_1(x_0), \ldots, d\phi_n(x_0)\}$ is a basis in $(R^n)^*$ defining a coordinate transformation. What will the submanifold M looks like in this coordinates? By the definition of the submanifold,

$$U_0 \cap M = \left\{ y \in R^n : y = \begin{pmatrix} y_a \\ 0 \end{pmatrix} : y_a \in N(0) \subset R^k \text{ is open} \right\}. \tag{B.4}$$

Hence, after appropriate choice of smooth coordinate transformation, the k-dimensional submanifold M looks like a k-dimensional "slice" of X.

Tangent space using special coordinate choice:
According to the above coordinate transformation, $\phi_i = y_i$, $d\phi_i = e_i^T$ is the i-th
standard unit vector, $y_0 = T(x_0)$ and

$$TM_{y_0} = \left\{ v \in R^n : v = \begin{pmatrix} v_a \\ 0 \end{pmatrix}, v_a \in R^k \right\} = \left\{ v \in R^n : v_{k+1} = \cdots = v_n = 0 \right\}, \quad \text{(B.5)}$$

$$f \in TM_{x_0} \quad \Rightarrow \quad f_T(y) = \begin{pmatrix} f_a(y) \\ 0_{n-k} \end{pmatrix}. \tag{B.6}$$

B.2 Frobenius Theorem

Notation: $L^k \subset R^n$ is a k-dimensional subspace.

k-dimensional distribution: $\Delta : x \mapsto L^k(x)$
A distribution Δ on X is a map which assigns to each $x \in X$ a k-dimensional subspace of R^n in such a way that the following smoothness conditions are satisfied:
For each $x_0 \in X$, there exist an open set $U(x_0) \subset X$ and $f_1, \ldots, f_k \in V(X)$ such
that

(1) $\{f_1(x), \ldots, f_k(x)\}$ are linearly independent for every $x \in U(x_0)$.
(2) $\Delta(x) = \text{Span}\{f_1(x), \ldots, f_k(x)\}$.

Modifications:

  (i) The k-dimensional distribution is a continuous mapping where the generating
      vector fields may vary with $U(x_0)$. It would be convenient to fix first a gen-
      erating set of vector fields for the entire X and then define the distribution by
      condition (2). However, this can cause problems.
 (ii) Let $f_1, \ldots, f_k \in V(X)$ and define the distribution Δ by (2). It is possible that
      the rank of the matrix $[f_1(x) \cdots f_k(x)]$ is not constant everywhere. Hence, in
      order to define a k-dimensional distribution, it may be convenient to define it as
      the span of k *or more* vector fields, and require that the distribution spanned by
      them has dimension no more then k and each open set $U(x)$ contains at least
      one point y such that $\dim \Delta(y)$ exactly equals k.
(iii) If $\dim \Delta(x) = k$ then x is called a *regular point* of the distribution Δ. An
      alternative formulation may be that x is a regular point of Δ if $\exists U(x)$ open set
      so that the dimension of $\Delta(y)$ is the same for every $y \in U(x)$.
 (iv) We say that $f \in V(X)$ belongs to the distribution Δ if $f(x) \in \Delta(x) \; \forall x \in X$.
      Notation: $f \in \Delta$.

The next lemma expresses the elements of the distribution by the help of the generating basis.

Lemma B.2 *Every element of a distribution can be expressed by the generating
basis where the weights are smooth functions.*

Let Δ be a k-dimensional distribution, $U \subset X$ is an open set, $f_1, \ldots, f_m \in V(X)$, $m \geq k$, span the distribution Δ for every $x \in X$: $\Delta(x) = \mathrm{Span}\{f_1(x), \ldots, f_m(x)\}$. Let $f \in \Delta$ then $\exists \alpha_1, \ldots, \alpha_m \in S(X)$ such that

$$f(x) = \sum_{i=1}^{m} \alpha_i(x) f_i(x). \tag{B.7}$$

Completely integrable distribution:
We say that the distribution $\Delta(x)$ is completely integrable if for every $x \in X$ there exists a submanifold $M_x \subset X$ such that its tangent space TM_x at x equals the distribution: $TM_x = \Delta(x)$. M_x is called the integral manifold of Δ passing through x. When is a distribution completely integrable? The answer to this question will be the Frobenius theorem giving a relation between integrable and involutive distributions.

Involutive distribution:
A distribution Δ is involutive if it is closed for the Lie bracket, i.e. $\forall f, g \in \Delta \Rightarrow [f, g] \in \Delta$.

Involutive property test:

(i) First of all, each distribution contains infinitely many elements: $f \in \Delta$, $a \in S(X) \Rightarrow af \in \Delta$.
(ii) On the other hand, let us assume that $f_1, \ldots, f_m \in V(X)$ spans the k-dimensional distribution ($k \leq m$). Since any $f \in \Delta$ has the form $f(x) = \sum_{i=1}^{m} \alpha_i(x) f_i(x)$, therefore it is enough to test whether there exists $\alpha_{ijl} \in S(X)$, $1 \leq i, j, l \leq m$ such that

$$[f_i, f_j] = \sum_{l=1}^{m} \alpha_{ijl}(x) f_l(x), \quad \forall x \in U(x_0). \tag{B.8}$$

The following lemma makes it possible to transform a vector field in such a way that after an appropriately chosen coordinate transformation its new image will be the first standard unit vector.

Lemma B.3 *Any vector field can be transformed to the first standard unit vector.*
Assume $f \in V(X)$, $x_0 \in X$ and $f(x_0) \neq 0$. Then there exist a neighborhood $U \subset X$ of x_0 and a diffeomorphism $T : U \to X$ such that

$$f_T(y) = (1 \quad 0 \quad \cdots \quad 0)^T, \quad \forall y \in T(U). \tag{B.9}$$

Proof For simplicity, we assume $x_0 = 0$. Let $s_{f,t}$ be the integral curve of the vector field f. Then $s_{f,t}(x_0)$ is the solution of the differential equation $\dot{x}(t) = f(x(t))$, $x(0) = x_0$ evaluated at the time moment t. Choose a nonsingular $(n \times n)$ matrix $M = [f(0)\overline{M}]$, whose first column is $f(0) \neq 0$. Let $x = (x_1, x_2, \ldots x_n)^T \in X$, $\bar{x} = (x_2, \ldots, x_n)^T \in R^{n-1}$ and $q(x) := s_{f,x_1}(M(0 \ \bar{x}^T)^T)$. Determine the integral curve $q(x)$ passing through the point $M(0\ \bar{x}^T)^T$ and consider it at "time moment" x_1. Since X is open, $x_0 = 0 \in X$ and $M(0\bar{x}^T)^T$ is small if $\|x\|$ is sufficiently small therefore $q(x)$ is well defined and $q(x) \in X$, furthermore the Jacobian matrix can easily be determined because

$$\frac{\partial}{\partial t}s_{f,t}(x) = f\big(s_{f,t}(x)\big), \qquad s_{f,0}(x) = x \quad \Rightarrow \quad \left[\frac{\partial}{\partial t}s_{f,t}(x)\right]_{(t,x)=(0,0)} = f(0),$$

$$\left[\frac{\partial}{\partial x}s_{f,t}(x)\right]_{(t,x)=(0,0)} = \left[\frac{\partial}{\partial x}s_{f,0}(x)\right]_{x=0} = I,$$

$$\left[\frac{\partial q}{\partial x_1}\right]_{x=0} = f(M \cdot 0) = f(0),$$

$$\left[\frac{\partial q}{\partial \bar{x}}\right]_{x=0} = I \cdot \overline{M} = \overline{M},$$

$$J = \left[\frac{\partial q}{\partial x}\right]_{x=0} = [f(0) \quad \overline{M}] = M \quad \text{is nonsingular.}$$

Hence, q is a local diffeomorphism at 0. It means that if $y \in X$ is sufficiently near to 0 then there exist a unique $\bar{z} \in R^{n-1}$ and "time moment" τ such that

$$y = s_{f,\tau}\big(M(0\ \bar{z}^T)^T\big). \tag{B.10}$$

Let $T = q^{-1}$ then $T(0) = 0$ and T is also a local diffeomorphism at 0. Let $g \in V(x)$ be the transformed vector field f_T, that is, the vector field f transformed by the coordinate transformation T, then $s_{g,t} = Ts_{f,t}T^{-1}$. Assuming $x \in X$ is sufficiently near to 0, then

$$T^{-1}(x) = q(x) = s_{f,x_1}\big(M(0\ \bar{x}^T)^T\big), \tag{B.11}$$

$$s_{f,t}\big(T^{-1}(x)\big) = s_{f,t}\big(s_{f,x_1}\big(M(0\ \bar{x}^T)^T\big)\big) = s_{f,t+x_1}\big(M(0\ \bar{x}^T)^T\big) =: y. \tag{B.12}$$

Now $T(y) = q^{-1}(y)$ consists of the unique $\tau \in R^1$ and $\bar{z} \in R^{n-1}$ and holds (B.10). Hence, comparing (B.10) and (B.12) follows that

$$T(y) = \begin{pmatrix} x_1 + t \\ \bar{x} \end{pmatrix} \quad \Rightarrow \quad s_{g,t}(x) = \begin{pmatrix} x_1 + t \\ \bar{x} \end{pmatrix}, \tag{B.13}$$

and after differentiation

$$g(x) = \begin{pmatrix} 1 & 0 & \cdots & 0 \end{pmatrix}^T = f_T(y). \tag{B.14}$$

Hence, T is the required diffeomorphism.

Theorem B.1 (Frobenius theorem) *A distribution is completely integrable $\Leftrightarrow$ it is involutive.*

Proof

I. Suppose Δ is a k-dimensional completely integrable distribution.

Then to each $x \in X$ belongs an integral manifold M_x satisfying $TM_x = \Delta(x)$. By the definition of the submanifold, for each $x_0 \in X$ there exist an open set $U(x_0) \subset X$ and smooth functions $\phi_{k+1}, \ldots, \phi_n \in S(X)$ such that the differentials are linearly independent row vectors at the point x_0 and the functions $\phi_{k+1}, \ldots, \phi_n$ are constant at every point of M_{x_0}. Choose the functions $\phi_1, \ldots, \phi_k \in S(X)$ so that the row vectors $\{d\phi_1(x_0), \ldots, d\phi_n(x_0)\}$ build a basis then, according to the former construction, the mapping $T : U(x_0) \to R^n$ is a diffeomorphism over some neighborhood $U_0(x_0) \subset U(x_0)$. Since $f \in \Delta$ is tangent to M_x at each point $x \in U_0(x_0)$ hence, in the new coordinate system, $f_T \in \Delta_T$ has the form

$$f_T(y) = \begin{pmatrix} f_a(y) \\ 0 \end{pmatrix}, \quad \forall y \in T(U_0(x_0)). \tag{B.15}$$

The key to all is that the tangent relation is fulfilled at each point of an *open set*. Since $TM_x = \Delta(x)$, $\forall x \in U_0(x_0)$, hence it follows that Δ_T is precisely the set of vector fields of the form $f_T(y)$. Therefore if $f_T, g_T \in \Delta_T$, then

$$f_T(y) = \begin{pmatrix} f_a(y) \\ 0_{n-k} \end{pmatrix}, \quad g_T(y) = \begin{pmatrix} g_a(y) \\ 0_{n-k} \end{pmatrix}, \tag{B.16a}$$

$$[f_T, g_T] = \frac{\partial g_T}{\partial y} f_T - \frac{\partial f_T}{\partial y} g_T \quad \Rightarrow \quad [f_T, g_T] = \begin{pmatrix} h_a(y) \\ 0_{n-k} \end{pmatrix}. \tag{B.16b}$$

Thus, $[f_T, g_T] \in \Delta_T$ and hence Δ_T (and therefore, also Δ) is an involutive distribution.

II. Suppose Δ is an integrable distribution.

The proof is by induction in m, the dimension of the distribution.

If $m = 1$, then f is the only vector field and because of linear independence $f(x) \neq 0$, $\forall x$. Therefore, we can take the diffeomorphism T by the above lemma and based on it we can choose $\phi_i(x) = T_i(x)$, $i = 2, \ldots, n$. Then, in the transformed variables, $f_T(y) = (10 \ldots 0)^T$, $\forall y$, and $\phi_{iT}(y) = y_i$, $i = 2, \ldots, n$. Hence, the statement holds for $m = 1$.

Suppose by induction that the statement is true up to $(m - 1)$ vector fields. For simplicity, let again $x_0 = 0$. Let the m vector fields $f_1, \ldots, f_m$ satisfying the conditions and choose the diffeomorphism T so that f_{1T} satisfies $f_{1T}(y) = (10 \ldots 0)^T$, $\forall y$, where y is sufficiently small. Let us introduce the notation $g_i(y) := f_{iT}(y)$, $i = 1, \ldots, m$. Since the Lie bracket is invariant under coordinate transformations, hence $\{g_1, \ldots, g_m\}$ is also involutive. Thus, there exist smooth functions $\beta_{ijk} \in S(X)$ such that

$$[g_i, g_j] = \sum_{k=1}^{m} \beta_{ijk}(y)g_k(y), \quad \forall y. \tag{B.17}$$

Let us introduce the notation $h_i(y) := g_i(y) - g_{i1}(y)g_1(y)$, $i = 2, \ldots, m$ where $g_{i1}(y)$ is the first component of $g_i(y)$. Notice that $h_{i1}(y) = g_{i1}(y)(1 - g_{11}(y)) = 0$. Since the set $\{g_1, \ldots, g_m\}$ is linearly independent therefore $\{g_1, h_2, \ldots, h_m\}$ is also a linearly independent set by the construction. Denote $\bar{h}_i(y)$, $i = 2, \ldots, m$ the last $(n-1)$ components of $h_i(y)$. Then the set $\{\bar{h}_2, \ldots, \bar{h}_m\}$ is linearly independent in R^{n-1}. We have to show that $\{\bar{h}_2, \ldots, \bar{h}_m\}$ is involutive. Denote $\overline{X}$ the slice of X at $y_1 = 0$ then $\overline{X}$ is an open set in R^{n-1}. Because of linear independence, the form of $g_1(y)$ and $h_{i1}(y) = 0$ we know that $\exists \alpha_{ijk} \in S(X)$ such that

$$[h_i, h_j](y) = \alpha_{ij1}(y)g_1(y) + \sum_{k=2}^{m} \alpha_{ijk}(y)h_k(y), \quad \forall y$$

$$[h_i, h_j](y) = \sum_{k=2}^{m} \alpha_{ijk}(y)h_k(y), \quad \forall y$$

$$[\bar{h}_i, \bar{h}_j]\big((0\ \bar{y}^T)^T\big) = \sum_{k=2}^{m} \alpha_{ijk}\big((0\ \bar{y}^T)^T\big)\bar{h}_k\big((0\ \bar{y}^T)^T\big), \quad \forall \bar{y},$$

which shows that $\{\bar{h}_2, \ldots, \bar{h}_m\}$ is involutive in some neighborhood of $0_{R^{n-1}}$. Since the number of functions is only $m-1$ hence, by the inductive hypothesis, there are smooth functions $\bar{\phi}_{m+1}(\bar{y}), \ldots, \bar{\phi}_n(\bar{y}) \in S(\overline{X})$ and an open neighborhood $\overline{N}(0_{R^{n-1}}) \subset R^{n-1}$ so that $d\bar{\phi}_{m+1}(\bar{y}), \ldots, d\bar{\phi}_n(\bar{y})$ are linearly independent, $\forall \bar{y} \in \overline{N}$ and

$$\frac{\partial \bar{\phi}_i(\bar{y})}{\partial \bar{y}}\bar{h}_i\big((0\ \bar{y}^T)^T\big) = 0, \quad \forall \bar{y} \in \overline{N}, j = 2, \ldots, m; \ i = m+1, \ldots, n. \quad \text{(B.18)}$$

Let $\phi_i(y) := \bar{\phi}_i(\bar{y})$, $i = m+1, \ldots, n$, then $\phi_i(y)$ is really independent of y_1 and $\phi_{m+1}, \ldots, \phi_n \in S(X)$ thus we can choose $N = \{y \in X : \bar{y} \in \overline{N}\}$.

We have to show that

$$\langle d\phi_i, g_j\rangle(y) = 0, \quad \forall y \in N, \ j = 1, \ldots, m; \ i = m+1, \ldots, n. \quad \text{(B.19)}$$

Since $\{g_1(y), \ldots, g_m(y)\}$ and $\{g_1(y), h_2(y), \ldots, h_m(y)\}$ span the same subspace and only the first component of $g_1(y)$ differs from zero, furthermore $d\phi_i(y) = [0\ \partial\bar{\phi}_i/\partial\bar{y}]$, hence

$$\langle d\phi_i, g_1\rangle(y) = 0, \quad \forall y \in N, \ i = m+1, \ldots, n. \quad \text{(B.20)}$$

It follows

$$L_{g_1}\phi_i = \langle d\phi_i, g_1\rangle \equiv 0 \ \Rightarrow \ L_{[g_1, h_j]}\phi_i = L_{g_1}L_{h_j}\phi_i - L_{h_j}L_{g_1}\phi_i = L_{g_1}L_{h_j}\phi_i,$$

$$[g_i, h_j] = b_{j1}g_1 + \sum_{k=2}^{m} b_{jk}h_k, \quad b_{jk} \in S(X) \quad \text{(involutive property)},$$

$$L_{g_1} L_{h_j} \phi_i = L_{[g_1, h_j]} \phi_i = \sum_{k=2}^{m} b_{jk} L_{h_k} \phi_i,$$

$$\psi_{ij} := L_{h_j} \phi_i = \langle d\phi_i, h_j \rangle \in S(X),$$

$$L_{g_1} \psi_{ij} = \langle d\psi_{ij}, g_1 \rangle = \frac{\partial \psi_{ij}}{\partial y_1} = \sum_{k=2}^{m} b_{jk} \left((y_1 \, \bar{y}^T)^T \right) \psi_{ik} \left((y_1 \, \bar{y}^T)^T \right). \tag{B.21}$$

Now fix the index i and the vector $\bar{y} \in \overline{X}$. Then (B.21) is a linear vector differential equation in which the independent variable is y_1 and the unknown variable is the vector $(\psi_{i2}, \ldots, \psi_{im})^T$. By (B.18) and (B.21), the "initial condition" at $y_1 = 0$ is

$$\psi_{ij} \left((0 \, \bar{y}^T)^T \right) = \frac{\partial \bar{\phi}_i(\bar{y})}{\partial \bar{y}} \bar{h}_i \left((0 \, \bar{y}^T)^T \right) = 0. \tag{B.22}$$

Since the linear vector differential equation (B.21) is homogeneous and the initial condition is zero, it follows

$$\langle d\phi_i, h_j \rangle(y) = \psi_{ij} \left((y_1 \, \bar{y}^T)^T \right) = 0, \quad \forall (y_1 \, \bar{y}^T)^T \in N, \forall i, j, \tag{B.23}$$

which completes the proof.

$\square$

Theorem B.2 (Alternative Frobenius theorem) *Suppose* $f_1, \ldots, f_m \in V(X)$, $N(x_0) \subset X$ *is an open set and for each* $x \in N(x_0)$ *the set* $\{f_1(x), \ldots, f_m(x)\}$ *contains* k *linearly independent vectors. Then there exist smooth functions* $\phi_{k+1}, \ldots, \phi_n \in S(X)$ *such that*

(1) $\{d\phi_{k+1}(x_0), \ldots, d\phi_n(x_0)\}$ *is a linearly independent set.*
(2) $\exists V(x_0) \subset N(x_0)$ *open set such that*

$$\langle d\phi_i, f_j \rangle(x) = 0, \quad \forall x \in V(x_0), k+1 \le i \le n, 1 \le j \le m, \tag{B.24}$$

if and only if the distribution $\Delta(x)$ *spanned by* $f_1, \ldots, f_m$ *is involutive, that is, there exist smooth real functions* $\alpha_{ijl} \in S(X)$ *and an open set* $U(x_0) \subset N(x_0)$ *such that*

$$[f_i, f_j](x) = \sum_{l=1}^{m} \alpha_{ijl}(x) f_l(x), \quad \forall x \in U(x_0). \tag{B.25}$$

Remark The equations in (B.24) are partial differential equations, the solvability of which is based on the involutive property. The main problem is that for the applications in control engineering the partial differential equations have to be solved in symbolic form (i.e., formulas) which is seldom possible. However, the Frobenius theorem is the basis for many theoretical investigations.

B.3 Local Reachability and Observability

It is well known from the theory of continuous time linear time invariant (LTI) systems

$$\dot{x} = Ax + Bu, \quad x \in R^n, u \in R^m,$$
$$y = Cx, \quad y \in R^p \tag{B.26}$$

that such systems can be brought onto controllability staircase form. If $M \subset R^n$ is the controllability subspace then $A(M) \subset M$ and $B(R^m) \subset M$. In other formulation, the subspace M is A-invariant, if $A(M) \subset M$, and the system is (A, B)-invariant if simultaneously $A(M) \subset M$ and $B(R^m) \subset M$. If $M = R^n$ then the system is completely controllable. Notice that for continuous time LTI systems complete controllability (to the origin within finite time) and complete reachability (from the origin within finite time) are similar ideas. If the continuous time LTI system is completely controllable, then any state change can be performed within finite time.

Nonlinear input affine system class:
Let $X \subset R^n$, $f, g_1, \ldots, g_m \in V(X)$ and the nonlinear input affine system

$$\dot{x} = f(x) + \sum_{i=1}^{m} u_i g_i(x), \quad x \in R^n, u_i \in R^1, i = 1, \ldots, m,$$
$$y = h(x), \quad y \in R^p. \tag{B.27}$$

Locally reachable nonlinear system:
The nonlinear system is said to be locally reachable around a state $x_0 \in X$ if there exists a neighborhood $U(x_0)$ such that for every $x_f \in U(x_0)$ there exist a finite time $T_f > 0$ and a set of control inputs $\{u_i(t), t \in [0, T_f], 1 \le i \le m\}$ such that, if the system starts in the state x_0 at time 0, then it reaches the final state x_f at time T_f.

Invariant distribution of nonlinear system:
Let Δ a distribution on X and $f \in V(X)$. We say that Δ is invariant under f, or f-invariant, if $[f, h] \in \Delta, \forall h \in \Delta$. Briefly: $[f, \Delta] \subset \Delta$.

Remark This definition is an extension of the well known idea of invariant set for linear systems to nonlinear systems. Let namely $X = R^n$ and $M \subset R^n$ is a k-dimensional subspace which is A-invariant, i.e. $A(M) \subset M$, where for linear system yields $f(x) = Ax$. Let $\{v_1, \ldots, v_k\}$ be a basis for M and consider the distribution Δ generated by $\{v_1, \ldots, v_k\}$. Then

$$\Delta(x) := \text{Span}\{v_1, \ldots, v_k\} = M,$$

$$h \in \Delta \quad \Rightarrow \quad h(x) = \sum_{i=1}^{k} h_i(x)v_i, \quad h_i \in S(X).$$

I. First, assume that Δ is f-invariant.
The following properties are evident:

$$v_i \in \Delta,$$

$$[f, v_i] = \frac{\partial v_i}{\partial x} f - \frac{\partial f}{\partial x} v_i = -A v_i \in \Delta,$$

$$-A v_i = const, \quad \forall x \quad \Rightarrow \quad -A v_i \in \Delta(x) = M,$$

$$AM \subset M \Rightarrow M \quad \text{is } A\text{-invariant subspace.}$$

II. Now assume that the subspace M is A-invariant, that is, $A(M) \subset M$.
Since the Lie bracket is bilinear, it is enough to show for the case $h_i \in S(X)$, $h := h_i(x)v_i \in \Delta$ that $[Ax, h_i(x)v_i] \in \Delta$. But in this case

$$\left[Ax, h_i(x)v_i\right] = v_i \frac{\partial h_i(x)}{\partial x} Ax - A h_i(x)v_i = c_i(x)v_i - h_i(x) \sum_{j=1}^{k} \lambda_j v_j$$

$$= \sum_{j=1}^{k} \hat{h}_j(x)v_j \in \Delta, \quad c_i, \hat{h}_j \in S(X).$$

Lemma B.4 *Partition of function defining invariant distribution.*

Let $x_0 \in X$, $f \in V(X)$ and Δ a k-dimensional distribution on X. Assume $\exists U(x_0)$ open set such that the restriction of Δ on $U(x_0)$ is involutive and f-invariant. Then there exist a neighborhood $U_0(x_0)$ and a diffeomorphism T on $U_0(x_0)$ such that in the new coordinates $y = T(x)$

$$f_T(y) = \begin{pmatrix} f_a(y_a, y_b) \\ f_b(y_b) \end{pmatrix}, \quad \forall y = \begin{pmatrix} y_a \\ y_b \end{pmatrix} \in T(U_0) \tag{B.28}$$

where $f_a, y_a \in R^k$ and $f_b, y_b \in R^{n-k}$.
The proof uses the fact that Δ is involutive hence by the Frobenius theorem $\exists U_0, T$ such that in the new coordinates

$$\Delta_T = \{h \in V(X) : h_i(y) = 0, k+1 \leq i \leq n, \forall y \in T(U_0)\}$$

$$= \{h \in V(X) : h_b(y) = 0, \forall y \in T(U_0)\}.$$

Since Δ_T is also f_T-invariant, therefore the last $n - k$ elements of $[f_T, h_T]$ are zero. Using the partitions

$$f_T = \begin{pmatrix} f_a \\ f_b \end{pmatrix}, \quad h_T = \begin{pmatrix} h_a \\ 0 \end{pmatrix}, \quad y = \begin{pmatrix} y_a \\ y_b \end{pmatrix}$$

direct computation shows that the statement of the lemma is true.

Theorem B.3 *Reachability staircase form of nonlinear system.*

Let us consider the nonlinear system $\dot{x} = f(x) + \sum_{i=1}^{m} u_i g_i(x)$. Suppose there exist open set $U(x_0)$ and Δ k-dimensional distribution on $U(x_0)$ satisfying the following conditions:

(i) Δ *is involutive and x_0 is regular point of Δ.*
(ii) $g_i(x) \in \Delta, \forall x \in U(x_0), i = 1, \ldots, m$.
(iii) Δ *is f-invariant.*

Then there exist an open set $U_0(x_0)$ and a diffeomorphism T on $U_0(x_0)$ such that

$$f_T(y) = \begin{pmatrix} f_a(y_a, y_b) \\ f_b(y_b) \end{pmatrix}; \quad g_{iT}(y) = \begin{pmatrix} g_{ia}(y_a, y_b) \\ 0 \end{pmatrix}, \quad i = 1, \ldots, m, \quad \text{(B.29)}$$

where $y = \begin{pmatrix} y_a \\ y_b \end{pmatrix} = T(x)$ and $y_a, f_a, g_{ia} \in R^k, y_b, f_b \in R^{n-k}$.

In the proof the form of f_T follows from the previous lemma, and that of g_{iT} from the fact that $g_i \in \Delta \Leftrightarrow g_{iT} \in \Delta_T$.

Corollary y_b is not reachable because neither y_a nor u have influence on it. The goal is to find the smallest k and a distribution belonging to it satisfying the conditions of the theorem. For this purpose, iterative algorithms are available which finish at most n steps [152].

Now let us consider the observability problem. In case of linear systems, the observability condition was derived by using repeated differentiation of the output and the relations resulting from it.

Locally observable nonlinear system.
The nonlinear system is locally observable in the state x_0 if $\exists N(x_0)$ open set such that $\forall x \in N, x \neq x_0$ can be distinguished from x_0 based on input and output observations.

The nonlinear system is locally observable if it is locally observable $\forall x_0 \in X$.

Repeated differentiation of nonlinear system.

$$y_j = h_j(x),$$

$$\dot{y}_j = dh_j \dot{x} = dh_j f(x) + \sum_{i=1}^{m} u_i \, dh_j g_i(x) = (L_f h_j)(x) + \sum_{i=1}^{m} u_i (L_{g_i} h_j)(x),$$

$$\ddot{y}_j = d \underbrace{(dh_j f)\dot{x}}_{L_f h_j} + \sum_{i=1}^{m} \dot{u}_i \underbrace{dh_j g_i}_{L_{g_i} h_j} + \sum_{i=1}^{m} u_i \, d \underbrace{(dh_j g_i)}_{L_{g_i} h_j} \dot{x}$$

$$= d(L_f h_j)\left(f + \sum u_i g_i\right) + \sum \dot{u}_i (L_{g_i} h_j) + \sum u_i \, d(L_{g_i} h_j)\left(f + \sum u_s g_s\right)$$

$$= L_f^2 h_j + \sum_i u_i L_{g_i} L_f h_j + \sum_i \dot{u}_i L_{g_i} h_j + \sum_i u_i L_f L_{g_i} h_j$$

$$+ \sum_i \sum_s u_i u_s L_{g_s} L_{g_i} h_j,$$

from which it can already be seen that $y_j^{(k)}$ is the "linear combination" of terms like

$$(L_{z_s} L_{z_{s-1}} \ldots L_{z_1} h_j)(x), \quad 1 \le s \le k, \tag{B.30}$$

where $z_1, z_2, \ldots, z_s$ are vector fields from the set $\{f, g_1, \ldots, g_m\}$.

Observation space of nonlinear system.
The observation space $\mathcal{O}$ of the nonlinear system is the linear space of functions over the field R^1 of the form

$$L_{z_s} L_{z_{s-1}} \ldots L_{z_1} h_j, s \ge 0, z_1, \ldots, z_s \in \{f, g_1, \ldots, g_m\}, \quad 1 \le j \le p. \tag{B.31}$$

In the linear combinations of functions, the coefficients are real numbers not functions of x.

Lemma B.5 *Piecewise-constant approximation of the input signal.*
Let $u_1, \ldots, u_s \in R^m$ and consider the piecewise-constant input signal

$$u(t) := \begin{cases} u_1, & 0 \le t < t_1, \\ u_2, & t_1 \le t < t_1 + t_2, \\ \vdots & \vdots \\ u_s, & \sum_{k=1}^{s-1} t_k \le t \le \sum_{k=1}^{s} t_k. \end{cases} \tag{B.32}$$

Denote $y_j(x_0) = y_j(x_0, t_1, \ldots, t_s)$ the j-th component of the output signal of the system belonging to $u(\cdot)$ starting from the initial state x_0. Then

$$\left(\frac{\partial}{\partial t_1} \cdots \frac{\partial}{\partial t_s} y \right)_{t_k=0, \forall k} = (dL_{v_s} L_{v_{s-1}} \ldots L_{v_1} h_j)(x_0), \tag{B.33}$$

where v_k has the form

$$v_k = f + \sum_{i=1}^{m} u_{ki} g_i, \quad k = 1, \ldots, s. \tag{B.34}$$

By the lemma if the piecewise-constant input signal is applied and the duration of the "pulses" goes to zero, then $((\partial/\partial t_1, \ldots \partial/\partial t_s)(y))_{t_k=0, \forall k}$ is a special repeated Lie derivative. The proof can be performed by induction on k.

Theorem B.4 *Observability staircase form of nonlinear system.*

Denote $\mathcal{O}$ the observation space of the nonlinear system and for every $x \in X$ let $d\mathcal{O}(x) \subset (R^n)^*$ be the subspace which consists of the row vectors $d\alpha(x), \alpha \in \mathcal{O}$. Suppose there exists a neighborhood $N(x_0)$ such that $\dim d\mathcal{O}(x) = k < n, \forall x \in N$. Then there exists a diffeomorphism T on N such that using the coordinate transformation $z = T(x)$ and the notations $z = \binom{z_a}{z_b}, z_a \in R^k, z_b \in R^{n-k}$ the system can be decomposed in the form

$$f_T(z) = \begin{pmatrix} f_a(z_a) \\ f_b(z_a, z_b) \end{pmatrix}, \qquad g_{iT} = \begin{pmatrix} g_{ia}(z_a) \\ g_{ib}(z_a, z_b) \end{pmatrix},$$

$$h_T(z) = h(T^{-1}(z)) = h_T(z_a). \tag{B.35}$$

The proof is based on the fact that the observation space is closed under Lie differentiation.

Remark

(i) z_b is not observable since it does not influence z_a in any way nor is it reflected in the output signal.
(ii) A general decomposition similar to Kalman decomposition of linear systems is known also for nonlinear systems, see [106].

Brunovsky canonical form of linear system.
SISO linear controllable system can be brought onto "controller" form in which

$$\dot{x} = Ax + bu,$$

$$A = \begin{bmatrix} 0 & 1 & 0 & \cdots & 0 \\ 0 & 0 & 1 & \cdots & 0 \\ \vdots & \vdots & \vdots & \ddots & \vdots \\ 0 & 0 & 0 & \cdots & 1 \\ -a_0 & -a_1 & -a_2 & \cdots & -a_{n-1} \end{bmatrix}, \qquad b = \begin{pmatrix} 0 \\ 0 \\ \vdots \\ 0 \\ 1 \end{pmatrix}, \tag{B.36}$$

$$\det(sI - A) = s^n + a_{n-1}s + \cdots + a_1 s + a_0.$$

Applying the state feedback $u := v + (a_0 a_1 \cdots a_{n-1})x$ the state equation appears in SISO Brunovsky canonical form and the resulting system consists of n integrators in series if x_1 is considered as output:

$$\dot{x} = Ax + bu = v + (a_0 a_1 \cdots a_{n-1})x =: \hat{A}x + \hat{b}v,$$

$$\hat{A} := A + ba^T = \begin{bmatrix} 0 & 1 & 0 & \cdots & 0 \\ 0 & 0 & 1 & \cdots & 0 \\ \vdots & \vdots & \vdots & \ddots & \vdots \\ 0 & 0 & 0 & \cdots & 1 \\ 0 & 0 & 0 & \cdots & 0 \end{bmatrix}, \qquad \hat{b} = \begin{pmatrix} 0 \\ 0 \\ \vdots \\ 0 \\ 1 \end{pmatrix}, \tag{B.37}$$

$$\dot{x}_1 = x_2, \ldots, \dot{x}_{n-1} = x_n, \qquad \dot{x}_n = v.$$

Similarly, MIMO linear controllable system $\dot{x} = Ax + Bu$ can also be brought onto MIMO Brunovsky canonical form based on the following construction:

$$r_0 = \text{rank } B = m,$$

$$r_i = \text{rank}[BAB \ldots A^i B] - \text{rank}[BAB \ldots A^{i-1} B], \quad i \geq 1,$$

$$0 \leq r_i \leq m, \quad \forall i,$$

$$r_i \geq r_{i+1},$$

$$m = r_0 \geq r_1 \geq \cdots \geq r_{n-1}, \quad \sum_{i=0}^{n-1} r_i = n,$$

$(B.38)$

The Kronecker indices $\kappa_1, \ldots, \kappa_m$ are defined in the following way:

$$\kappa_i := \sum_{r_j \geq i} 1: \quad \text{the number of } r_j \text{ integers which are } \geq i,$$

$$\kappa_1 \geq \kappa_2 \geq \cdots \geq \kappa_m \geq 0 \quad \text{(after possible reordering)}, \quad \sum_{i=1}^{n} \kappa_i = n, \quad (B.39)$$

$$\sigma_i := \sum_{j=1}^{i} \kappa_j, \quad i = 1, \ldots, m.$$

If the MIMO linear system is controllable, then it can be brought on Brunovsky canonical form by using linear coordinate transformation and state feedback. The Brunovsky canonical form is defined as follows:

$$\dot{x} = \hat{A}x + \hat{B}u,$$

$$\hat{A} = \text{BlockDiag}\{\hat{A}_1, \ldots \hat{A}_m\},$$

$$\hat{A}_i = \begin{bmatrix} 0 & 1 & 0 & \cdots & 0 \\ 0 & 0 & 1 & \cdots & 0 \\ \vdots & \vdots & \vdots & \ddots & \vdots \\ 0 & 0 & 0 & \cdots & 1 \\ 0 & 0 & 0 & \cdots & 0 \end{bmatrix}_{\kappa_i \times \kappa_i}, \quad (B.40)$$

$$\hat{B} = [e_{\sigma_1} e_{\sigma_2} \cdots e_{\sigma_m}],$$

where e_{σ_i} is the σ_i-th standard unit vector.

For illustration of the above construction, consider the *linear system* having $A_{9 \times 9}$, $B_{9 \times 3}$ matrices in the state equation and apply a selection scheme to find the linearly independent columns of the controllability matrix:

$$
\begin{array}{ccc}
b_1 & b_2 & b_3 \\
Ab_1 & Ab_2 & Ab_3 \\
A^2b_1 & \circ & A^2b_3 \\
A^3b_1 & & \circ \\
\circ & &
\end{array}
$$

Then the above construction gives the following results:

$$r_0 = 3, \qquad r_1 = 3, \qquad r_2 = 2, \qquad r_3 = 1, \qquad r_4, \ldots, r_8 = 0,$$

$$\delta = 3,$$

$$\kappa_1 = 4, \qquad \kappa_2 = 3, \qquad \kappa_3 = 2,$$

$$\sigma_1 = 4, \qquad \sigma_2 = 7, \qquad \sigma_3 = 9,$$

$$m_3 = r_3 = 1,$$

$$m_2 = r_2 - r_3 = 2 - 1 = 1,$$

$$m_1 = r_1 - r_2 = 3 - 2 = 1,$$

$$m_0 = r_0 - r_1 = 3 - 3 = 0,$$

$$n = 1 \cdot m_0 + 2 \cdot m_1 + 3 \cdot m_2 + 4 \cdot m_3 = 9.$$

For comparison, consider the dimensions of the subspaces spanned by the corresponding distributions:

$$
\begin{array}{ll}
\dim \Delta_0 = 3 & \dim \Delta_3 - \dim \Delta_0 = 1 + 2 \cdot 1 + 3 \cdot 1 = 6 \\
\dim \Delta_1 = 6 & \dim \Delta_3 - \dim \Delta_1 = 1 + 2 \cdot 1 = 3 \\
\dim \Delta_2 = 8 & \dim \Delta_3 - \dim \Delta_2 = 1 \\
\dim \Delta_3 = 9 &
\end{array}
$$

Preparing block dimensions of Brunovsky form for nonlinear system.
If the nonlinear system can be brought to Brunovsky canonical form using nonlinear state feedback, then we expect to find it in the sequence $\{r_i\} \to \{\kappa_i\} \to$ Brunovsky form.
   Let the nonlinear system be

$$\dot{x} = f(x) + \sum_{i=1}^{m} u_i g_i(x), \tag{B.41}$$

$$X \subset R^n \text{ open}, \qquad 0 \in X, \ f, g_1, \ldots, g_m \in V(X).$$

Let us perform the following construction:

$$C_i := \left\{ ad_f^k g_j : 1 \le j \le m, 0 \le k \le i \right\}, \qquad 0 \le i \le n - 1,$$

$$\Delta_i := \operatorname{Span} C_i \quad \text{(generated distribution).} \tag{B.42}$$

Especially, $\Delta_0 = \mathrm{Span}\{g_1, \ldots, g_m\}$, $\Delta_1 = \mathrm{Span}\{g_1, \ldots, g_m, [f, g_1], \ldots, [f, g_m]\}$, etc. Now define the following integers:

$$r_0 := \dim \Delta_0 = m; \qquad r_i := \dim \Delta_i - \dim \Delta_{i-1}, \quad i \geq 1,$$

$$0 \leq r_{i+1} \leq r_i, \quad \forall i,$$

$$\delta := \max\{i : r_i \neq 0\},$$

$$r_\delta > 0, \quad \text{but } r_i = 0, \forall i > \delta,$$

$$m_\delta := r_\delta; \qquad m_i := r_i - r_{i+1} \geq 0, \quad i = 0, \ldots, \delta - 1,$$

$$r_i := \sum_{j=i}^{\delta} m_j,$$

$$\dim \Delta_i = \sum_{j=0}^{i} r_j,$$

(B.43)

$$\dim \Delta_\delta - \dim \Delta_i = \sum_{k=i+1}^{\delta} (k - i)m_k = m_{i+1} + 2m_{i+2} + \cdots + (\delta - i)m_\delta,$$

$$\kappa_i := \sum_{r_j \geq i} 1 : \text{the number of } r_j \text{ integers which are } \geq i,$$

$$\kappa_1 \geq \kappa_2 \geq \cdots \geq \kappa_m \geq 0 \quad \text{(after possible reordering)}, \qquad \sum_{i=1}^{n} \kappa_i = n,$$

$$\sigma_i := \sum_{j=1}^{i} \kappa_j, \quad i = 1, \ldots, m.$$

Formulation of the MIMO nonlinear state feedback problem:
Given:

$$\dot{x} = f(x) + \sum_{i=1}^{m} u_i g_i(x),$$

(B.44)

$$\kappa_1, \ldots, \kappa_m \text{ and } \kappa_1 \geq \kappa_2 \geq \cdots \geq \kappa_m.$$

Do there exist:

(i) $U(0) \subset R^n$ open set.
(ii) $q : U \to R^m$ smooth function.
(iii) $S : U \to R^{m \times m}$ smooth function, $\det S(0) \neq 0$.
(iv) locale diffeomorphism $T : U \to R^n$, $T(0) = 0$,

satisfying the conditions:

$$v = q(x) + S(x)u,$$
$$z = T(x), \qquad\qquad\qquad\text{(B.45)}$$
$$\dot{z} = Az + Bv,$$

A, B is of Brunovsky canonical form belonging to the indices $\kappa_1, \ldots, \kappa_m$.

Theorem B.5 *Solvability conditions for the nonlinear state feedback problem.*
Let the locally reachable nonlinear system be $\dot{x} = f(x) + \sum_{i=1}^{m} u_i g_i$. Suppose that

(a) $g_1, \ldots, g_m$ *vector fields are linearly independent at 0, $\Delta_0 = Span\{g_1, \ldots, g_m\}$ and dim $\Delta_0 = r_0 = m$.*
(b) 0 *is a regular point of the distribution Δ_i, $\forall i \geq 0$.*

Then the MIMO nonlinear state feedback problem has a solution if and only if the following two conditions are satisfied:

(i) dim $\Delta_\delta = n$.
(ii) Δ_{i-1} *is involutive if $m_i \neq 0$.*

Proof We consider only the construction (if) part.

(1) Since $\delta = \max\{i : r_i \neq 0\}$, dim $\Delta_\delta = n$ and dim $\Delta_{\delta-1} = \dim \Delta_\delta - r_\delta = n - m_\delta$ hence the codimension of $\Delta_{\delta-1}$ is m_δ and is also involutive, thus by the alternative Frobenius theorem $\exists U_0(0)$ open set and $\exists h_{\delta,i} \in S(X)$ smooth functions, $1 \leq i \leq m_\delta$, such that their differentials are linearly independent at $x = 0$ and

$$\langle dh_{\delta,i}, ad_f^l g_j \rangle(x) = 0, \quad \forall x \in U_0, 0 \leq l \leq \delta - 1, 1 \leq i \leq m_\delta, 1 \leq j \leq m. \tag{B.46}$$

These functions will be the parts of the coordinate transformation T. The matrix M_δ of type $m_\delta \times m$, defined by $(M_\delta)_{ij}(x) := \langle dh_{\delta,i}, ad_f^\delta g_j \rangle(x)$, has rank $M_\delta(x) = m_\delta$ (maximal), $\forall x \in U_0$.

(2) Next, consider the distribution $\Delta_{\delta-2}$: dim $\Delta_{\delta-2} = \dim \Delta_\delta - m_{\delta-1} - 2m_\delta = n - m_{\delta-1} - 2m_\delta$. Two cases are possible ($m_{\delta-1} = 0$ and $m_{\delta-1} \neq 0$).

If $m_{\delta-1} = 0$, then it can be shown that the differentials of the $2m_\delta$ functions $\{h_{\delta,i}, L_f h_{\delta,i} : 1 \leq i \leq m_\delta\}$ are linearly independent and annihilate the vector fields in $\Delta_{\delta-2}$.

If $m_{\delta-1} \neq 0$ then $\Delta_{\delta-2}$ is involutive and its codimension is $m_{\delta-1} + 2m_\delta$. By the alternative Frobenius theorem there exist an open set $U_1(0) \subset U_0(0)$ and $m_{\delta-1} + 2m_\delta$ exact differential forms which annihilate $\Delta_{\delta-2}$ at all $x \in U_1$. Now, from the set which annihilate $\Delta_{\delta-2}$ we already know $2m_\delta$ pieces, namely $\{dh_{\delta,i}, dL_f h_{\delta,i} : 1 \leq i \leq m_\delta\}$. Let us select further $m_{\delta-1}$ smooth functions $\{h_{\delta-1,i} : 1 \leq i \leq m_{\delta-1}\}$ such that the set

$$\{dh_{\delta,i}, dL_f h_{\delta,i} : 1 \leq i \leq m_\delta\} \cup \{h_{\delta-1,i} : 1 \leq i \leq m_{\delta-1}\}$$

is linearly independent and each vector in the set annihilates $\Delta_{\delta-2}(x)$, $\forall x \in U_1$. Define the $m_{\delta-1} \times m$ type matrix $M_{\delta-1}$ by

$$(M_{\delta-1})_{ij}(x) := \langle dh_{\delta-1,i}, ad_f^{\delta-1}g_j\rangle(x), \quad 1 \le i \le m_{\delta-1}, 1 \le j \le m,$$

and consider the $(m_\delta + m_{\delta-1}) \times m$ type $[M_\delta^T M_{\delta-1}^T]^T$ matrix. It can be shown that this matrix has maximal row-rank for each $x \in U_1$.

(3) The construction can similarly be continued, when it is finished, we will have the following functions:

$$\{h_{\delta,i}, L_f h_{\delta,i}, \ldots, L_f^{\delta-1}h_{\delta,i} : 1 \le i \le m_\delta\},$$

$$\vdots$$

$$\{h_{2,i}, L_f h_{2,i} : 1 \le i \le m_2\},$$

$$\{h_{1,i} : 1 \le i \le m_1\}.$$

The differentials of these functions are linearly independent in some open set $U(0)$ and annihilate all vector fields in Δ_0. The total number of functions is

$$\delta m_\delta + (\delta - 1)m_{\delta-1} + \cdots + 2m_2 + m_1 = \dim \Delta_\delta - \dim \Delta_0 = n - m = n - \dim \Delta_0.$$

(4) At the end, it is necessary to consider two cases ($m_0 = 0$ and $m_0 \ne 0$). In either case, we have $\sum_{i=0}^\delta m_i = r_0 = m$.

If $m_0 = 0$, then $\sum_{i=1}^\delta m_i = m$, $(\delta + 1)m_\delta + \delta m_{\delta-1} + \cdots + 3m_2 + 2m_1 = n$, thus the following set contains exactly n functions:

$$\{L_f^l h_{k,i} : 1 \le i \le m_k, 0 \le l \le k, 1 \le k \le \delta\}.$$

The differentials of these functions are linearly independent over some neighborhood $U(0)$.

If $m_0 \ne 0$, then there are just $n - m_0$ functions in the above set. We have to choose m_0 further functions (see the earlier construction): $\{h_{0,i} : 1 \le i \le m_0\}$ such that the differentials of the n functions in the set

$$\{L_f^l h_{k,i} : 1 \le i \le m_k, 0 \le l \le k, 0 \le k \le \delta\}$$

are linearly independent in some neighborhood of the origin. Form the following matrix:

$$M = [M_\delta^T M_{\delta-1}^T \cdots M_1^T M_0^T]^T,$$

$$(M_k)_{ij}(x) = \langle dh_{k,i}, ad_f^k g_j\rangle(x), \quad 1 \le i \le m_k, 1 \le j \le m.$$

Since the number of rows in this matrix is $\sum_{i=0}^\delta m_i = m$, hence M is square and, by the construction, nonsingular for all x in a neighborhood of 0.

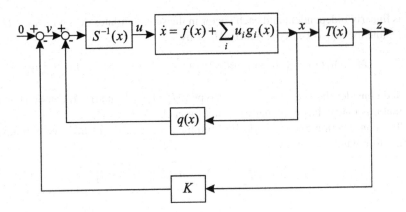

Fig. B.2 Compensation of nonlinear system with external and internal state feedback

Remark

(1) Amongst the integers m_i there are at most m strictly positive.
(2) Define $\kappa_i := i + 1$ for those values of i for which $m_i > 0$, then these are precisely the Kronecker indices. Therefore, $m_k > 0$ is valid only if $k = \kappa_i$ for some i so that the corresponding $h_{k,i}$ functions appear in the above list.

Define the following functions:

$$\phi_1 := h_{\delta,1}, \ldots, \phi_{m_\delta} := h_{\delta,m_\delta}.$$

If $m_{\delta-1} > 0$ then $\phi_{m_\delta+1} := h_{\delta-1,1}, \ldots, \phi_{r_{\delta-1}} := h_{\delta-1,m_{\delta-1}}$,

$$\vdots$$

If $m_0 > 0$ then $\phi_{r_1+1} := h_{0,1}, \ldots, \phi_{r_0} := h_{0,m_0}.$

Notice that we have exactly m of this functions ϕ_i. For every i the integer κ_i shows how many times the function ϕ_i was Lie-differentiated in the list. Define $z_i : U \to R^{\kappa_i}$ and the mapping $T : U \to R^n$ as follows:

$$z_i := \begin{pmatrix} \phi_i(x) \\ L_f \phi_i(x) \\ \vdots \\ L_f^{\kappa_i-1} \phi_i(x) \end{pmatrix}, \qquad z := \begin{pmatrix} z_1(x) \\ z_2(x) \\ \vdots \\ z_m(x) \end{pmatrix} = T(x).$$

Then T is a local diffeomorphism, $\phi_i(0) = 0$ (possibly after subtracting an appropriate constant), hence $T(0) = 0$. According to the construction, it yields

$$\frac{d}{dt} z_{i,1} = \langle d\phi_i(x), \dot{x} \rangle = \left\langle d\phi_i(x), \left[f(x) + \sum_{j=1}^m u_j g_j(x) \right] \right\rangle$$

$$= L_f \phi_i + \sum_{j=1}^{m} u_j L_{g_j} \phi_i = L_f \phi_i = z_{i,2},$$

$$\frac{d}{dt} z_{i,l} = L_f^i \phi_i = z_{i,l+1}, \quad 1 \le l \le \kappa_i - 1,$$

$$\frac{d}{dt} z_{i,\kappa_i} = L_f^{\kappa_i} \phi_i + \sum_{j=1}^{m} u_j L_{g_j} L_f^{\kappa_i-1} \phi_i.$$

Since,

$$L_{g_j} L_f^{\kappa_i-1} \phi_i = \langle d\phi_i, ad_f^{\kappa_i-1} g_j \rangle = (-1)^{\kappa_i-1} M_{ij}$$

hence

$$\frac{d}{dt} \begin{pmatrix} z_{1,\kappa_1} \\ \vdots \\ z_{m,\kappa_m} \end{pmatrix} = q(x) + S(x)u,$$

where

$$q_i(x) = L_f^{\kappa_i} \phi_i,$$

$$s_{ij}(x) = (-1)^{\kappa_i-1} M_{ij}(x),$$

$$v = q(x) + S(x)u,$$

$$\frac{d}{dt} z_{i,\kappa_i} = v_i.$$

Hence, using the nonlinear static feedback $u = S^{-1}(x)[-q(x) + v]$ the system is in Brunovsky canonical form. The characteristic equation of the resulting system is $s^{\kappa_i} = 0$, which can be stabilized by $v = -Kz$ outer linear feedback, see Fig. B.2.

B.4 Input/Output Linearization, Zero Dynamics

We have already seen that under some condition the *state equation* of the nonlinear system can be linearized using a relatively simple compensation. Now let us examine whether is it possible to linearize the full *input/output* relation. The problem will be investigated in a neighborhood $U(x_0)$ of the point x_0 where x_0 is typically an equilibrium point, that is, $f(x_0) = 0$. We consider here only the MIMO case and assume that the number of the inputs and the outputs are equal (but y may differ from the real system output). First, the vector relative degree has to be defined.

Formulation of the MIMO input/output linearization problem:
Given:

$$\dot{x} = f(x) + \sum_{i=1}^{m} u_i g_i(x),$$

$$y_i(x) = h_i(x), \quad i = 1, \ldots, m, \quad h = (h_1, \ldots, h_m)^T,$$

$$f, g_i \in V(X), \quad h_i \in S(X), \quad i = 1, \ldots, m.$$

Do there exist:

(i) $r = (r_1, \ldots, r_m)^T$ relative degree vector.
(ii) $U(x_0)$ open set.
(iii) $v = q(x) + S(x)u$.
(iv) $q : U \to R^m$ smooth function.
(v) $S : U \to R^{m \times m}$ smooth function, $\det S(x_0) \neq 0$,

such that

$$y_i^{(r_i)} = v_i, \quad i = 1, \ldots, m. \tag{B.47}$$

If the problem has a solution then, with internal control v, the system looks like m decoupled integrators plus an additional zero dynamics system. The integrators can be stabilized by outer linear feedbacks. The entire system is stable if the zero dynamics is also stable.

The basic approach to obtain a linear input/output relation is simply to repeatedly differentiate each output by the time. Differentiating y_i by the time yields

$$\dot{y}_i = \nabla h_i \dot{x} = \left\langle dh_i, f + \sum_{i=1}^{m} g_i u_i \right\rangle = L_f h_i + \sum_{j=1}^{m} u_j L_{g_j} h_i.$$

If each $L_{g_j} h_i$ is zero, then the differentiation is continued and we get

$$\ddot{y}_i = \nabla(L_f h_i) \dot{x} = \left\langle d(L_f h_i), f + \sum_{i=1}^{m} g_i u_i \right\rangle = L_f^2 h_i + \sum_{j=1}^{m} u_j L_{g_j}(L_f h_i).$$

Let r_i be the smallest integer for which

$$y_i^{(r_i)} = L_f^{r_i} h_i + \sum_{j=1}^{m} L_{g_j}\left(L_f^{r_i-1} h_i\right) u_j, \tag{B.48}$$

and for at least one j also yields $L_{g_j}(L_f^{r_i-1} h_i)(x_0) \neq 0$.

Relative degree vector:
The MIMO input affine nonlinear system is said to have the relative degree vector $r = (r_1, \ldots, r_m)^T$ at the point x_0 if $\exists U(x_0)$ open set such that:

(i) $L_{g_j} L_f^k h_i(x) \equiv 0, \forall x \in U, j = 1, \dots, m, k = 0, \dots, r_i - 2$.

(ii) The matrix $S(x) = [s_{ij}(x)]_{m \times m}$ is nonsingular at x_0 where $s_{ij} = L_{g_j} L_f^{r_i-1} h_i$, i.e.

$$S(x) = \begin{bmatrix} L_{g_1} L_f^{r_1-1} h_1 & \cdots & L_{g_m} L_f^{r_1-1} h_1 \\ \vdots & \ddots & \vdots \\ L_{g_1} L_f^{r_m-1} h_m & \cdots & L_{g_m} L_f^{r_m-1} h_m \end{bmatrix}. \tag{B.49}$$

Since $S(x_0)$ is nonsingular, hence $S(x)$ is also nonsingular in a sufficiently small neighborhood of x_0.

Lemma B.6 *Linear independence of the resulted differentials.*

If there exists the relative degree vector $r = (r_1, \dots, r_n)^T$, *then the set of row vectors* $\{dL_f^k h_i : 0 \le k \le r_i-1, 1 \le i \le m\}$ *is linearly independent in some open set* $U(x_0)$, *furthermore* $\sum_{i=1}^m r_i \le n$.

The proof is by induction. It can be shown step by step that in the linear combination $\sum_{i=1}^m \sum_{k=0}^{r_i-1} \alpha_{ik}(dL_f^k h_i)(x_0) = 0$ each coefficient is zero based on the fact that $S(x_0)$ has an inverse. As a consequence of linear independence, $\sum_{i=1}^m r_i \le n$ is evident.

Normal form of MIMO input affine nonlinear system.
If the relative degree vector $r = (r_1, \dots, r_n)^T$ exists, then the following coordinates can be chosen:

$$z_{11} = h_1(x), z_{12} = L_f h_1(x) \cdots z_{1,r_1} = L_f^{r_1-1} h_1(x),$$

$$z_{21} = h_2(x), z_{22} = L_f h_2(x) \cdots z_{2,r_2} = L_f^{r_2-1} h_2(x),$$

$$\vdots \tag{B.50}$$

$$z_{m1} = h_m(x), z_{m2} = L_f h_m(x) \cdots z_{m,r_m} = L_f^{r_m-1} h_m(x).$$

Let us introduce the notations $z_i = (z_{i1}, \dots, z_{i,r_i})^T \in R^{r_i}, i = 1, \dots, m$ and $z_o = (z_1^T, \dots, z_m^T)^T$. These coordinates may be parts of a coordinate transformation since the differentials $dL_f^k h_i(x), 0 \le k \le r_i - 1, 1 \le i \le m$ are linearly independent in some open set $U(x_0)$. Let $\rho = \sum_{i=1}^m r_i$ then $n - \rho$ additional functions $\eta_i(x) \in S(X)$ can be constructed such that the differentials $d\eta_1(x), \dots, d\eta_{n-\rho}(x)$ complete the above differentials to basis at the point x_0. Define $z_u := (\eta_1(x), \dots, \eta_{n-\rho}(x))^T$ then $(z_o^T, z_u^T)^T = T(x)$ is a coordinate transformation. Denote $T^{-1}(x)$ the inverse transformation and define the following internal feedback and system parts

$$q_i(x) := L_f^{r_i} h_i(x), \qquad q(x) := [q_i]_{m \times 1},$$

$$s_{ij}(x) := L_{g_j} L_f^{r_i - 1} h_i(x), \qquad S(x) := [s_{ij}(x)]_{m \times m},$$

$$v_i := q_i(x) + \sum_{j=1}^{i} s_{ij}(x) u_j, \qquad v := q(x) + S(x)u, \qquad \text{(B.51a)}$$

$$b_i(z_o, z_u) := q_i\left(T^{-1}(z_o, z_u)\right), \qquad b(z_o, z_u) := [b_i]_{m \times 1},$$

$$a_{ij}(z_o, z_u) = s_{ij}\left(T^{-1}(z_o, z_u)\right), \qquad A(z_o, z_u) := [a_{ij}]_{m \times m},$$

$$b_{u,k}(z_o, z_u) := L_f \eta_k\left(T^{-1}(z_o, z_u)\right), \qquad b_u := [b_{u,k}]_{(n-\rho) \times 1},$$

$$a_{u,kj}(z_o, z_u) := L_{g_j} \eta_k\left(T^{-1}(z_o, z_u)\right), \qquad A_u(z_o, z_u) := [a_{u,kj}]_{(n-\rho) \times m}. \qquad \text{(B.51b)}$$

Then the MIMO input affine nonlinear system appears in the following normal form:

$$\dot{z}_{i1} = z_{i2}, \qquad \dot{z}_{i2} = z_{i3}, \ldots, \dot{z}_{i,r_i-1} = z_{i,r_i},$$

$$\dot{z}_{i,r_i} = b_i(z_o, z_u) + \sum_{j=1}^{m} a_{ij}(z_o, z_u) u_j$$

$$= L_f^{r_i} h_i\left(T^{-1}(z_o, z_u)\right) + \sum_{j=1}^{m} L_{g_j} L_f^{r_i - 1} h_i\left(T^{-1}(z_o, z_u)\right) u_j$$

$$= v_i, \qquad i = 1, \ldots, m, \qquad \text{(B.52a)}$$

$$\dot{z}_o = b(z_o, z_u) + A(z_o, z_u)u = q(x) + S(x)u = v, \qquad \text{(B.52b)}$$

$$\dot{z}_u = b_u(z_o, z_u) + A_u(z_o, z_u)u.$$

Corollary

(i) *If the inner-loop nonlinear state feedback $u(x) = -S^{-1}(x)q(x) + S^{-1}(x)v$
is applied, then linear subsystems $\dot{z}_i = A_i z_i + b_i v_i$, $y_i = c_i z_i$ appear in
Brunovsky canonical form where $c_i = e_1^T \in (R^{r_i})^*$. The subsystems can be collected in MIMO Brunovsky canonical form belonging to the block dimensions
$r_1, \ldots, r_m$:*

$$\dot{z}_o = A_o z_o + B_o v, \qquad \text{(B.53)}$$

$$y = C_o z_o.$$

(ii) *Since z_u has no effect on z_o and y, hence z_u is unobservable. It is clear that z_o
is observable.*

(iii) *If x_0 is an equilibrium point, that is, $f(x_0) = 0$ and $h(x_0) = 0$, then, according
to the coordinate transformation, $z_o = 0$ belongs to x_0. On the other hand we*

can also choose $z_u = 0$ to x_0 and thus $(z_o, z_u) = (0,0)$ belongs to the equilibrium point x_0.

(iv) *The transfer function of the observable part belonging to the equilibrium point is $\text{diag}(1/s^{r_1}, \ldots, 1/s^{r_m})$ in accord to the Brunovsky canonical form that is a decoupled system in integrator sense and can be further compensated by outer-loop linear state feedback. However, the unobservable part can cause stability problem.*

(v) *Apply the inner-loop nonlinear state feedback*

$$u(x) = -S^{-1}\big(T^{-1}(z_o, z_u)\big)\big[-q\big(T^{-1}(z_o, z_u)\big) + v\big],$$

then the unobservable part can be written in the form

$$\dot{z}_u = b_u(z_o, z_u) - A_u(z_o, z_u)S^{-1}\big(T^{-1}(z_o, z_u)\big)q\big(T^{-1}(z_o, z_u)\big)$$
$$+ A_u(z_o, z_u)S^{-1}\big(T^{-1}(z_o, z_u)\big)v$$
$$=: f_u(z_o, z_u) + \sum_{i=1}^{m} g_{ui}(z_o, z_u)v_i. \tag{B.54}$$

(vi) *In case of outer-loop linear stabilizing feedback $v = -K_o z_o$, the observable part becomes $\dot{z}_o = (A_o - B_o K_o)z_o$ and the unobservable part is $\dot{z}_u = f_u(z_o, z_u) - G_u(z_o, z_u)K_o z_o$. The state matrix of the linearized system is*

$$\begin{bmatrix} A_o - B_o K_o & 0 \\ \star & \frac{\partial f_u}{\partial z_u}\big|_{(0,0)} \end{bmatrix}.$$

Using Lyapunov's indirect method, we obtain that the composite system is locally asymptotically stable if the eigenvalues of $\frac{\partial f_u}{\partial z_u}\big|_{(0,0)}$ have negative real parts.

Zero dynamics:
The system $\dot{z}_u = f_u(0, z_u)$ is called the zero dynamics of the input affine nonlinear system. The zero dynamics represents the dynamics of the unobservable part when the input is set equal to zero and the output is constrained to be identically zero. The zero dynamics is locally asymptotically stable if the eigenvalues of $\frac{\partial f_u}{\partial z_u}\big|_{(0,0)}$ have negative real parts.

Minimum phase nonlinear system:
The transportation zeros of continuous time LTI systems are strongly related to the behavior of the system if the output is identically zero. If the transportation zeros have negative real parts, then the LTI system is minimum phase, otherwise it is nonminimum phase. Since the zero dynamics is also defined for identically zero output, hence the input affine nonlinear system is called minimum phase if the zero dynamics is locally asymptotically stable.

Dynamic extension with integrators:
Assume the MIMO nonlinear system $\dot{x} = f(x) + G(x)u$, $y = h(x)$ does not have any well defined vector relative degree but the rank of $S(x)$ is constant $r < m$ if x is in the neighborhood of x_0. Then we can try to extend the system with integrators and amplifiers such that the extended system already has vector relative degree. The following algorithm formulates a procedure for it. The functions of the extended system will be denoted by "tilde".

Dynamic extension algorithm:
Step 1. Let $\tilde{n} := n$, $\tilde{x} := x$, $\tilde{x}_0 := x_0$, $\tilde{f} := f$, $\tilde{G} := G$, $\tilde{h} := h$, $\tilde{u} := u$, $\tilde{m} := m$.

Step 2. Computation of $\tilde{S}(\tilde{x})$ in the neighborhood $\tilde{U}$ of $\tilde{x}_0$.

  (i) If rank $\tilde{S}(x) = m$ on $\tilde{U}$ then stop, the system $(\tilde{f}, \tilde{G}, \tilde{h})$ has vector relative degree.
  (ii) If rank $\tilde{S}(x)$ is not constant over the neighborhood of $\tilde{x}_0$ then stop, the system cannot be extended such that the vector relative degree does exist.
  (iii) If rank $\tilde{S}(\tilde{x}) = r$, then let $\tilde{r} := r$ and continue with Step 3.

Step 3. Use elementary column operations to determine the matrix $\beta(\tilde{x})$ such that the last $(\tilde{m} - \tilde{r})$ columns of the matrix $\tilde{S}(\tilde{x})\beta(\tilde{x})$ become zero. Since the rank of $\tilde{S}(\tilde{x})$ is constant hence $\beta(\tilde{x})$ can be partitioned in the form $\beta(\tilde{x}) = [\beta_1(\tilde{x})\beta_2(\tilde{x})]$ so that $\beta_1(\tilde{x})$ consists of the first $\tilde{r}$ column.

Step 4. Let $\tilde{u} = \beta_1(\tilde{x})z_1 + \beta_2(\tilde{x})w_2$ and $\dot{z}_1 = w_1$ where $z_1, w_1 \in R^{\tilde{r}}$, $w_2 \in R^{\tilde{m}-\tilde{r}}$. This is equal to adding $\tilde{r}$ integrators and the amplifiers $\beta_1(\tilde{x})$, $\beta_2(\tilde{x})$. The new inputs will be w_1, w_2. The extended system is the following:

$$\begin{pmatrix} \dot{\tilde{x}} \\ \dot{z}_1 \end{pmatrix} = \begin{pmatrix} \tilde{f} + \tilde{G}\beta_1 z_1 \\ 0 \end{pmatrix} + \begin{bmatrix} 0 & \tilde{G}\beta_2 \\ I & 0 \end{bmatrix} \begin{pmatrix} w_1 \\ w_2 \end{pmatrix}, \qquad (B.55)$$

where $\bar{x} = (\tilde{x}^T, z_1^T)^T$ is the new state and $\bar{u} = (w_1^T, w_2^T)^T$ is the new input. The new $\bar{f}, \bar{G}$ can be read out from the above state equation. The output is not changed, that is, $\bar{h}(\bar{x}) = \tilde{h}(\tilde{x})$.

Step 5. Rename $\bar{x}, \bar{u}, \bar{f}, \bar{G}, \bar{h}$ to be the new $\tilde{x}, \tilde{u}, \tilde{f}, \tilde{G}, \tilde{h}$ and go back to Step 2.

The algorithm successfully finishes after finite many steps if in every steps of the algorithm the rank of $\tilde{S}(\tilde{x})$ remains constant in some neighborhood of $\tilde{x}_0$. This condition is called the *regularity condition*. If the algorithm successfully finishes, then the extended system has vector relative degree. The compensator is a dynamic system whose form is

$$\dot{z} = c(x, z) + d(x, z)w,$$
$$u = \alpha(x, z) + \beta(x, z)w. \qquad (B.56)$$

Especially if only one iteration step was necessary, then the extended system is

$$\dot{z}_1 = w_1,$$
$$u = \beta_1(x)z_1 + \beta_2(x)w_2. \tag{B.57}$$

The *resulting nonlinear system* is the original system precompensated with the dynamic compensator. Since the resulting system already has vector relative degree hence the above discussed MIMO nonlinear control methods can be applied for it.

References

1. Abraham, R., Mardsen, J.M., Ratiu, T.: Manifolds, Tensor Analysis, and Applications. Springer, Berlin (1988)
2. Ahmad, N.J., Ebraheem, H.K., Qasem, M.Q.: Global asymptotic stability for systems with friction and input backlash. In: Proceedings of the 13th Mediterranean Conference on Control and Automation, Limassol, Cyprus, pp. 212–218 (2005)
3. Al-Bender, F., Lampaert, V., Swevers, J.: The generalized Maxwell-slip model: a novel model for friction simulation and compensation. IEEE Trans. Automat. Control **50**(11), 1883–1887 (2005)
4. Allgower, F., Zheng, A. (eds.): Nonlinear Model Predictive Control. Birkhauser, Basel (2000)
5. Alonge, F., DIppolito, F., Raimondi, F.M.: Globally convergent adaptive and robust control of robotic manipulators for trajectory tracking. Control Eng. Pract. **12**, 1091–1100 (2004)
6. Andreev, F., Auckly, D., Gosavi, S., Kapitanski, L., Kelkar, A., White, W.: Matching, linear systems and the ball and beam. Automatica **38**(12), 2147–2152 (2002)
7. Armstrong-Hèlouvry, B.: Control of Machines with Friction. Kluwer Academic, Boston (1991)
8. Asada, H., Slotine, J.E.: Robot Analysis and Control. Wiley, New York (1986)
9. Asare, H., Wilson, D.: Design of computed torque model reference adaptive control for space-based robotic manipulators. In: Proceedings of the Symposium ASME Winter Annual Meeting WAM, Anaheim, USA, pp. 195–204 (1986)
10. Åstrom, K.J., Canudas-de-Wit, C.: Revisiting the LuGre friction model. IEEE Control Syst. Mag. **28**(6), 101–114 (2008)
11. Åström, K.J., Wittenmark, B.: Computer-controlled Systems. Prentice-Hall, New York (1997)
12. Awrejcewicz, J., Olejnik, P.: Analysis of dynamic systems with various friction laws. Appl. Mech. Rev. **58**, 389–411 (2005)
13. Barreiro, A., Banos, A.: Input-output stability of systems with backlash. Automatica **42**, 1017–1024 (2006)
14. Batista, A.A., Carlson, J.M.: Bifurcations from steady sliding to stick slip in boundary lubrication. Phys. Rev. E **57**(5), 4986–4996 (1998)
15. Beldiman, O., Bushnell, L.: Stability, linearization and control of switched systems. In: Proc. of the American Control Conference, San Diego, California, pp. 2950–2954 (1999)
16. Blue, P., Guvenc, L., Odenthal, D.: Large envelope flight control satisfying Hinf robustness and performance specifications. In: Proceedings of the American Control Conference, Arlington, USA, pp. 1351–1356 (2001)
17. Bona, B., Indri, M.: Friction compensation in robotics: an overview. In: Proceedings of the 44th IEEE Conference on Decision and Control, and the European Control Conference, Seville, Spain (2005)

B. Lantos, L. Márton, *Nonlinear Control of Vehicles and Robots*,
Advances in Industrial Control,
DOI 10.1007/978-1-84996-122-6, © Springer-Verlag London Limited 2011

18. Bonchis, A., Corke, P., Rye, D.: A pressure-based, velocity independent, friction model for asymmetric hydraulic cylinders. In: Proceedings. 1999 IEEE International Conference on Robotics and Automation, Detroit, Michigan, pp. 1746–1751 (1999)

19. Borner, M., Andreani, I., Albertos, P., Isermann, R.: Detection of lateral vehicle driving conditions based on the characteristic velocity. In: Proceedings of the IFAC World Congress, Conference CD, Barcelona, Spain, pp. 1–6 (2002)

20. Boubadallah, S., Noth, A., Siegwart, R.: PID vs LQ control techniques applied to an indoor micro quadrotor. In: Proceedings of the IEEE/RSJ International Conference on Intelligent Robots and Systems IROS, Sendai, Japan (2004)

21. Boubadallah, S., Siegwart, R.: Backstepping and sliding-mode techniques applied to an indoor micro quadrotor. In: Proceedings of the IEEE International Conference on Robotics and Automation ICRA, Barcelona, Spain, pp. 2247–2252 (2005)

22. Brandt, T., Sattel, T.: Path planning for automotive collision avoidance based on elastic bands. In: Proceedings of the 16th IFAC World Congress, Prague, Czech Republic, pp. 1–6, (2005). Paper TU–M16–TO/4 02,746.pdf

23. Breivik, M.: Nonlinear maneuvering control of underactuated ships. M.Sc. Thesis. Norwegian University of Science and Technology, Department of Engineering Cybernetics, Trondheim, Norway (2003)

24. Bretschneider, C.L.: Wave and wind loads. In: Handbook of Ocean and Underwater Engineering. McGraw-Hill, New York (1969)

25. Camacho, E.F., Bordons, C.: Model Predictive Control, 2nd edn. Springer, London (2004)

26. Caundas-de-Wit, C., Ollson, H., Åstrom, K.J., Lischinsky, P.: A new model for control of systems with friction. IEEE Trans. Automat. Control $40(3)$, 419–425 (1995)

27. Chen, L.-W., Papavassilopoulos, G.P.: Robust variable structure and switching-σ adaptive control of single-arm dynamics. IEEE Trans. Automat. Control $39(8)$, 1621–1626 (1994)

28. Chen, W.-H., Ballance, D.J., Gawthrop, P.J., OReilly, J.: A nonlinear disturbance observer for robotic manipulators. IEEE Trans. Ind. Electron. $47(4)$, 932–938 (2000)

29. Chen, Y.-Y., Huang, P.-Y., Yen, J.-Y.: Frequency-domain identification algorithms for servo systems with friction. IEEE Trans. Control Syst. Technol. $10(5)$, 654–665 (2002)

30. Chui, C.K., Chen, G.: Kalman Filtering. Springer, Berlin (1999)

31. Cook, M.V.: Flight Dynamics Principles, 2nd edn. Elsevier, London (2007)

32. Coza, C., Macnab, C.: A new robust adaptive-fuzzy control method applied to quadrotor helicopter stabilization. In: Proceedings of the Annual meeting of the North American Fuzzy Information Processing Society NAFIPS, Montreal, Canada, pp. 454–458 (2006)

33. Das, A., Subbarao, K., Lewis, F.: Dynamic inversion of quadrotor with zero-dynamics stabilization. In: Proceedings of the IEEE Conference on Control Applications CCA, San Antonio, USA, pp. 1189–1194 (2008)

34. Davrazos, G., Koussoulas, N.T.: A review of stability results for switched and hybrid systems. In: Proc. of 9th Mediterranean Conference on Control and Automation, Dubrovnik, Croatia (2011)

35. Dhamala, M., Lai, Y.-C., Kostelich, E.J.: Analyses of transient chaotic time series. Phys. Rev. E 64, 64–73 (2001)

36. Doupont, P.E.: Avoiding sick-slip through PD control. IEEE Trans. Automat. Control $39(5)$, 1094–1097 (1990)

37. Dupont, P., Armstrong, B., Altpeter, F.: Single state elastoplastic friction models. IEEE Trans. Automat. Control $47(5)$, 787–792 (2002)

38. Farrell, J.A., Barth, M.: The Global Positioning System & Inertial Navigation. McGraw-Hill, New York (1999)

39. Fliess, M., Levine, J., Martin, P., Rouchon, P.: Lie-Bäcklund approach to equivalence and flatness of nonlinear systems. IEEE Trans. Automat. Control $44(5)$, 922–937 (1999)

40. Fossen, T.I.: MATLAB Gnc Toolbox. Marine Cybernetics, Trondheim (2001)

41. Fossen, T.I.: Marine Control Systems. Guidance, Navigation, and Control of Ships, Rigs and Underwater Vehicles. Marine Cybernetics, Trondheim (2002)

42. Fossen, T.I., Sagatun, S.I., Sorensen, A.J.: Identification of dynamically positioned ships. Control Eng. Pract. $4(3)$, 369–376 (1996)

43. Freund, E., Mayr, R.: Nonlinear path control in automated vehicle guidance. IEEE Trans. Robot. Automat. **13**(1), 49–60 (1997)
44. Guo, Y., Hill, D.J., Jiang, Z.-P.: Global nonlinear control of the ball and beam system. In: Proceedings of 35th IEEE Conference on Decision and Control, vol. 3, Kobe, Japan, pp. 2818–2823 (1996)
45. Guo, Z., Huang, L.: Generalized Lyapunov method for discontinuous systems. Nonlinear Anal. **71**, 3083–3092 (2009)
46. Hartley, R.: An investigation of the essential matrix. Report GE-CRD, Schenectady, New York, USA (1993)
47. Hartley, R., Zisserman, A.: Multiple View Geometry in Computer Vision, 2nd edn. Cambridge University Press, Cambridge (2003)
48. Hauser, J., Sastry, S., Kokotovic, P.V.: Nonlinear control via approximate input-output linearization: the ball and beam example. IEEE Trans. Automat. Control **37**(3), 392–398 (1992)
49. Hensen, H.A., van de Molengraft, M.J.G., Steinbuch, M.: Friction induced hunting limit cycles: a comparison between the LuGre and switch friction model. Automatica **39**, 2131–2137 (2003)
50. Hensen, R.H.A., van de Molengraft, M.R.J.G.: Friction induced hunting limit cycles: an event mapping approach. In: Proceedings of the American Control Conference, Anchorage, AK, pp. 2267–2272 (2002)
51. Hirschorn, R.M.: Incremental sliding mode control of the ball and beam. IEEE Trans. Automat. Control **47**(10), 1696–1699 (2002)
52. Hung, N.V., Tuan, H.D., Narikiyo, T., Apkarian, P.: Adaptive control for nonlinearly parameterized uncertainties in robot manipulators. In: Proc. 41th Conference on Decision and Control, Las Vegas, USA (2002)
53. Ihle, I.A., Arcak, M., Fossen, T.I.: Passivity-based designs for synchronized path-following. Automatica **43**(9), 1508–1518 (2007)
54. Ioannou, P.A., Sun, J.: Robust Adaptive Control. Prentice Hall, Upper Slade River (1996)
55. Isherwood, R.M.: Wind resistant of merchant ships. RINA Trans. **115**, 327–338 (1972)
56. Isidori, A.: Nonlinear Control Systems I, 3rd edn. Springer, Berlin (1995)
57. Isidori, A.: Nonlinear Control Systems II. Springer, Berlin (1999)
58. Jafarov, E.M., Parlakcy, M.N.A., Istefanopulos, Y.: A new variable structure PID-controller design for robot manipulators. IEEE Trans. Control Syst. Technol. **13**(1), 122–130 (2005)
59. Jang, J.O., Son, M.K., Chung, H.T.: Friction and output backlash compensation of systems using neural network and fuzzy logic. In: Proceeding of the 2004 American Control Conference, Boston, Massachusetts, pp. 1758–1763 (2004)
60. Johansson, M.: Piecewise Linear Control Systems—a Computational Approach. Springer, Berlin (2003)
61. Keviczky, T., Balas, G.J.: Receding horizon control of an F-16 aircraft: a comparative study. Control Eng. Pract. **14**(9), 1023–1033 (2006)
62. Khalil, H.K.: Nonlinear Systems, 3rd edn. Prentice Hall, Upper Saddle River (2002)
63. Khatib, O.: A unified approach for motion and force control of manipulator robots: the operational space formulation. IEEE J. Robot. Automat. **3**, 43–53 (1987)
64. Kiencke, U., Nielsen, L.: Automotive Control Systems. For Engine, Drive Line, and Vehicle. Springer, Berlin (2000)
65. Kim, H.J., Shim, D.H.: A flight control system for aerial robots. Control Eng. Pract. **11**(12), 1389–1400 (2003)
66. Kis, L., Lantos, B.: Sensor-fusion and actuator system of a quadrotor helicopter. Period. Polytech. Electr. Eng. 1–12 (2010)
67. Kis, L., Prohaszka, Z., Regula, G.: Calibration and testing issues of the vision, inertial measurement and control system of an autonomous indoor quadrotor helicopter. In: Proceedings of the International Workshop on Robotics in Alpe-Adria-Danube Region RAAD, Conference CD, Ascona, Italy (2008)
68. Kis, L., Regula, G., Lantos, B.: Design and hardware-in-the-loop test of the embedded control system of an indoor quadrotor helicopter. In: Proceedings of the Workshop on Intelligent Solutions in Embedded Systems WISES, Regensburg, Germany, pp. 35–44 (2008)

69. Kiss, B.: Motion planning and control of a class of flat and liouvillian mechanical systems. Ph.D. Thesis, L'Ecole Nationale Superieure des Mines de Paris, Paris (2001)

70. Klein, V., Morelli, E.A.: Aircraft System Identification. Theory and Practice. American Institute of Aeronautics and Astronautics, Inc., Reston (2006)

71. Kogan, D., Murray, R.: Optimization-based navigation for DARPA Grand Challenge. In: IEEE Conference on Decision and Control, San Diego, USA, pp. 1–6 (2006)

72. Lampaert, V., Swevers, J., Al-Bender, F.: Modification of the Leuven integrated friction model structure. IEEE Trans. Automat. Control **47**(4), 683–687 (2002)

73. Lantos, B.: Theory and Design of Control Systems II. Akademiai Kiado, Budapest (2003) (In Hungarian)

74. Lantos, B.: Nonlinear model predictive control of robots, cranes and ground vehicles. In: Proceedings of the Workshop on System Identification and Control Systems, Budapest, Hungary, pp. 201–215 (2005)

75. Lantos, B.: Path design and receding horizon control for collision avoidance system of cars. WSEAS Trans. Syst. Control **1**(2), 105–112 (2006)

76. Lantos, B., Somlo, J., Cat, P.T.: Advanced Robot Control. Akadémiai Kiadó, Budapest (1997)

77. LaSalle, J., Lefschetz, S.: Stability by Lyapunov's Direct Method. Academic Pres, New York (1961)

78. Leine, R.I.: Bifurcations in discontinuous mechanical systems of Filippov-type. PhD thesis, Technische Universiteit Eindhoven (2000)

79. Leine, R.I., Nijmeijer, H.: Dynamics and Bifurcations of Non-smooth Mechanical Systems. Springer, Berlin (2006)

80. Levine, J.: Analysis and Control of Nonlinear Systems. Springer, Berlin (2009)

81. Lewis, F.L., Dawson, D.M., Abdallah, C.T.: Robot Manipulator Control. Dekker, New York (2004)

82. Liberzon, D.: Switched systems. In: Hristu-Varsakelis, D., Levine, W.S. (eds.) Handbook of Networked and Embedded Control Systems, pp. 559–574. Birkhauser, Boston (2005)

83. Liberzon, D., Morse, A.S.: Basic problems in stability and design of switched systems. IEEE Control Syst. Mag. **19**(5), 59–70 (1999)

84. Lindegaard, K.P., Fossen, T.I.: A model based wave filter for surface vessels using position, velocity and partial acceleration feedback. In: Proceedings of the IEEE Conference on Decision and Control CDC, Orlando, USA, pp. 946–951 (2001)

85. Loria, A., Panteley, E., Popovic, D., Teel, A.R.: A nested Matrosov theorem and persistency of excitation for uniform convergence in stable nonautonomous systems. IEEE Trans. Automat. Control **50**(2), 183–198 (2005)

86. Lyapunov, A.M.: Problem general de la stabilite de mouvement. Ann. Fac. Sci. Toulouse **9**(2), 203–474 (1907)

87. Maciejowski, J.M.: Predictive Control with Constraints. Prentice Hall, London (2002)

88. Madani, T., Benallegue, A.: Control of a quadrotor mini-helicopter via full state backstepping technique. In: Proceedings of the IEEE Conference on Decision and Control CDC, San Diego, USA, pp. 1515–1520 (2006)

89. Mahvash, M., Okamura, A.: Friction compensation for enhancing transparency of a teleoperator with compliant transmission. IEEE Trans. Robot. **23**(6), 1240–1246 (2007)

90. Martin, P., Rouchon, P.: Feedback linearization and driftless systems. Math. Control Signals Syst. **7**(3), 235–254 (1994)

91. Márton, L.: Robust-adaptive control of nonlinear singlevariable mechatronic systems and robots. PhD thesis, Budapest University of Technology and Economics (2006)

92. Márton, L.: Distributed controller architecture for advanced robot control. In: Proc. of IEEE International Symposium on Industrial Electronics, Cambridge, United Kingdom, pp. 1412–1417 (2008)

93. Márton, L.: On analysis of limit cycles in positioning systems near Striebeck velocities. Mechatronics **48**, 46–52 (2008)

94. Márton, L., Lantos, B.: Friction and backlash induced limit cycles in mechanical control systems. In: Proc. of European Control Conference, Budapest, Hungary, pp. 3875–3880 (2005)

95. Márton, L., Lantos, B.: Control of mechanical systems with Stribeck friction and backlash. Syst. Control Lett. **58**, 141–147 (2009)

96. Márton, L., Lantos, B.: Friction and backlash measurement and identification method for robotic arms. In: Proc. of 14th IEEE International Conference on Advanced Robotics, Munich, Germany (2009)

97. Márton, L., Lantos, B.: Control of robotic systems with unknown friction and payload. IEEE Trans. Control Syst. Technol. (2011). doi:10.1109/TCST.2010.2086458

98. Márton, L., Hodel, A.S., Lantos, B., Hung, J.Y.: Underactuated robot control: comparing LQR, subspace stabilization, and combined error metric approaches. IEEE Trans. Ind. Electron. **57**(10), 3724–3730 (2008)

99. Márton, L., Fodor, S., Sepehri, N.: A practical method for friction identifcation in hydraulic actuators. Mechatronics (2011). doi:10.1016/j.mechatronics.2010.08.010

100. Menon, K., Krishnamurthy, K.: Control of low velocity friction and gear backlash in a machine tool feed drive system. Mechatronics **9**, 33–52 (1999)

101. Merritt, H.E.: Hydraulic Control Systems. Wiley, New York (1967)

102. Merzouki, R., Davila, J., Cadiou, J., Fridman, L.: Backlash phenomenon observation and identification. In: Proceedings of the 2006 American Control Conference, Minneapolis, Minnesota, USA, pp. 3322–3327 (2006)

103. Murray, R.M., Li, Z., Sastry, S.: A Mathematical Introduction to Robotic Manipulation. CRC Press, Boca Raton (1994)

104. Naerum, E., Cornella, J., Elle, O.J.: Wavelet networks for estimation of coupled friction in robotic manipulators. In: Proceedings of the 2008 IEEE International Conference on Robotics and Automation, Pasadena, USA, pp. 862–867 (2008)

105. Sepehri, N., Sassani, F., Lawrence, P.D., Ghasempoor, A.: Simulation and experimental studies of gear backlash and stick-slip friction in hydraulic excavator swing motion. ASME J. Dyn. Syst. Meas. Control **118**, 463–467 (1996)

106. Nijmeier, H., van der Schaft, A.: Nonlinear Dynamical Control Systems. Springer, Berlin (1990)

107. Nordin, M., Gutman, P.-O.: Controlling mechanical systems with backlash—a survey. Automatica **38**, 1633–1649 (2002)

108. Olsson, H., Astrom, K.J.: Friction generated limit cycles. IEEE Trans. Control Syst. Technol. **9**(4), 629–636 (2001)

109. Ortega, R., Spong, M.W., Gômez-Estern, F., Blankenstein, G.: Stabilization of a class of underactuated mechanical systems via interconnection and damping assignment. IEEE Trans. Automat. Control **47**(8), 1218–1233 (2002)

110. Panteley, E., Ortega, R., Gafvert, M.: An adaptive friction compensator for global tracking in robotic manipulator. Syst. Control Lett. **33**, 307–313 (1998)

111. Parra-Vega, V., Arimoto, S., Liu, Y.-H., Hirzinger, G., Akella, P.: Dynamic sliding PID control for tracking of robot manipulators: theory and experiments. IEEE Trans. Robot. Automat. **19**(6), 967–976 (2003)

112. Peni, T.: Cooperative and constrained control. Ph.D. Thesis. Budapest University of Technology and Economics, Budapest (2009)

113. Peni, T., Bokor, J.: Formation stabilization of nonlinear vehicles based on dynamic inversion and passivity. In: Proceedings of the Workshop on System Identification and Control Systems, Budapest, Hungary, pp. 217–228 (2006)

114. Pettersen, K.Y., Gravdahl, J.T., Nijmeijer, H. (eds.): Group Coordination and Cooperative Control. Springer, Berlin (2006)

115. Pierson, W.J., Moskowitz, L.: A proposed spectral form of the for fully developed wind seas based on the similarity theory of S.A. Kitaigorodskii. J. Geophys. Res. **69**(24), 5181–5190 (1964)

116. Price, W.G., Bishop, R.E.D.: Probabilistic Theory of Ship Dynamics. Chapman & Hall, London (1974)

117. Putra, D., Nijmeijer, H., van de Wouw, N.: Analysis of undercompensation and overcompensation of friction in 1DOF mechanical systems. Automatica **43**, 1387–1394 (2007)

118. Quinlan, S., Khatib, O.: Elastic bands: Connecting path planning and control. In: Proceedings of the IEEE International Conference Robotics and Automation, vol. 2, Atlanta, USA, pp. 802–807 (1993)

119. Rantzer, A., Johansson, M.: Piecewise linear quadratic optimal control. IEEE Trans. Automat. Control 45(4), 629–637 (2000)

120. Rauw, M.: Fdc 1.4—A Simulink Toolbox for Flight Dynamics and Control Analysis. Draft Version 7. Fdc User Manual. Haarlem, Haarlem (2005)

121. Regula, G., Lantos, B.: Backstepping-based control design with state estimation and path tracking to indoor quadrotor helicopter. Period. Polytech. Electr. Eng. 1–10 (2010)

122. Ren, W., Beard, R.: Distributed Consensus in Multi-vehicle Cooperative Control. Springer, Berlin (2008)

123. Rodonyi, G.: Vehicle models for steering control. Technical Report No. SCL-4/2003. Computer and Automation Research Institute of Hungarian Academy of Sciences, System and Control Laboratory, Budapest, Hungary (2003)

124. Rostalski, P., Besselmann, T., Baric, M., van Belzen, F., Morari, M.: Hybrid approach to modeling, control and state estimation of mechanical systems with backlash. Int. J. Control 80(11), 1729–1740 (2007)

125. Ryu, J., Gerdes, J.G.: Integrating inertial sensors with global positioning system (GPS) for vehicle dynamic control. J. Dyn. Syst. Meas. Control 126(2), 242–254 (2004)

126. Sastry, S.: Nonlinear Systems. Analysis, Stability and Control. Springer, Berlin (1999)

127. van der Schaft, A.: L2-Gain and Passivity Techniques in Nonlinear Control. Springer, Berlin (2000)

128. van der Schaft, A., Maschke, B.M.: On the Hamiltonian formulation of nonholonomic mechanical systems. Rep. Math. Phys. 34(2), 225–230 (1994)

129. Shevitz, D., Paden, B.: Lyapunov stability theory of nonsmooth systems. IEEE Trans. Automat. Control 39(9), 1910–1914 (1994)

130. Skjetne, R., Fossen, T.I., Kokotovic, P.V.: Robust output maneuvering for a class of nonlinear systems. Automatica 40(3), 373–383 (2004)

131. Slotine, J.E., Li, W.: Applied Nonlinear Control. Prentice Hall, Englewood Cliffs (1991)

132. Slotine, J.-J., Li, W.: Adaptive manipulator control: a case study. IEEE Trans. Automat. Control 33(11), 995–1003 (1988)

133. Somlo, J., Lantos, B., Cat, P.T.: Advanced Robot Control. Akademiai Kiado, Budapest (1997)

134. Sooumelidis, A., Gaspar, P., Lantos, B., Prohaszka, Z.: Design of an embedded microcomputer based mini quadrotor UAV. In: Proceedings of the European Control Conference ECC, Kos, Greece (2007)

135. Sooumelidis, A., Gaspar, P., Lantos, B., Regula, G.: Control of an experimental mini quadrotor UAV. In: Proceedings of the Mediterranean Conference on Control and Automation MED, Ajaccio, France, pp. 1252–1257 (2008)

136. Spillman, M., Blue, P., Lee, L., Banda, S.: A robust gain scheduling example using linear parameter varying feedback. In: Proceedings of the IFAC World Congress, San Francisco, USA, pp. 221–226 (1996)

137. Spong, M.W.: The swing up control problem for the acrobot. IEEE Control Syst. Mag. 15(1), 49–55 (1995)

138. Stevens, B.L., Lewis, F.L.: Aircraft Control and Simulation. Wiley, New York (1992)

139. Suraneni, S., Kar, I.N., Murthy, O.V.R., Bhatt, R.K.P.: Adaptive stick-slip friction and backlash compensation using dynamic fuzzy logic system. Appl. Soft Comput. 6, 26–37 (2005)

140. Swevers, J., Al-Bender, F., Ganseman, C.G., Prajogo, T.: An integrated friction model structure with improved presliding behavior for accurate friction compensation. IEEE Trans. Automat. Control 45(4), 675–686 (2000)

141. Tafazoli, S., de Silva, C., Lawrence, P.: Friction estimation in a planar electrohydraulic manipulator. In: Proc. of the American Control Conference, Seattle, Washington, pp. 3294–3298 (1995)

142. Tao, G., Kokotovic, P.V.: Adaptive Control of Systems with Sensor and Actuator Nonlinearities. Wiley, New York (1996)

143. Tataryn, P.D., Sepehri, N., Strong, D.: Experimental comparison of some compensation techniques for the control of manipulators with stick-slip friction. Control Eng. Pract. **4**(9), 1209–1219 (1996)

144. Tien, L.L., Albu-Schaffer, A., Luca, A.D., Hirzinger, G.: Friction observer and compensation for control of robots with joint torque measurement. In: Proc. of the 2008 IEEE/RSJ International Conference on Intelligent Robots and Systems, Nice, France (2008)

145. Tjahjowidodo, T., Al-Bender, F., Brussel, H.V.: Quantifying chaotic responses of mechanical systems with backlash component. Mech. Syst. Signal Process. **21**, 973–993 (2007)

146. Tjee, R.T.H., Mulder, J.A.: Stability and control derivatives of the de Havilland DHC-2 beaver aircraft. Report LR-556, Delft University of Technology, Faculty of Aerospace Engineering, Delft, The Netherlands (1988)

147. Tomei, P.: Robust adaptive friction compensation for tracking control of robot manipulators. IEEE Trans. Automat. Control **45**(6), 2164–2169 (2000)

148. Tomlin, C.J., Sastry, S.: Switching through singularities. Syst. Control Lett. **35**, 145–154 (1998)

149. Torsethaugen, K.: Model of doubly peaked wave spectrum. Technical Report No. STF22 A96204, SINTEF Civ. and Env. Eng., Trondheim, Norway (1996)

150. Utkin, V.I.: Variable structure systems with sliding mode. IEEE Trans. Automat. Control **22**(2), 212–222 (1977)

151. Varga, R.: Gerschgorin and His Circles. Springer, Berlin (2004)

152. Vidyasagar, M.: Nonlinear System Analysis, 2nd edn. Prentice Hall, Englewood Cliffs (1993)

153. Vik, B., Fossen, T.: A nonlinear observer for GPS and INS integration. In: Proceedings of the IEEE Conference on Decision and Control CDC, Orlando, USA, pp. 2956–2961 (2001)

154. Villwock, S., Pacas, J.M.: Time-domain identification method for detecting mechanical backlash in electrical drives. IEEE Trans. Ind. Electron. **56**(2), 568–573 (2009)

155. Wang, D., Liu, Y., Meng, M.: A study of limit cycles and chaotic motions of a single-link robot manipulator during slow motion. In: Proceedings of the IEEE International Conference on Systems, Man, and Cybernetics, Tokyo, Japan, pp. 884–888 (1999)

156. de Wit, C.C., Siciliano, B., Bastin, G. (eds.): Theory of Robot Control. Springer, Berlin (1997)

157. van de Wouw, N., Nijmeijer, H., Mihajlovic, N.: Friction-induced limit cycling in flexible rotor systems: an experimental drill-string system. In: Proc. International Design Engineering Technical Conference, Long Beach, California, USA (2005)

158. Wu, Q., Sepehri, N.: On Lyapunov's stability analysis of non-smooth systems with applications to control engineering. Int. J. Non-Linear Mech. **36**, 1153–1161 (2001)

159. Yong Li, Z.C.F.: Bifurcation and chaos in friction-induced vibration. Commun. Nonlinear Sci. Numer. Simul. **9**, 633–647 (2004)

160. Yousefi, H., Handroos, H., Soleymani, A.: Application of differential evolution in system identification of a servo-hydraulic system with a flexible load. Mechatronics **18**, 513–528 (2008)

161. Yu, J., Jadbabaie, A., Primbs, J., Huang, Y.: Comparison of nonlinear control design techniques on a model of the Caltech ducted fan. Automatica **37**(12), 1971–1978 (2001)

162. Zweiri, Y.H., Seneviratne, L.D., Althoefer, K.: Identification methods for excavator arm parameters. In: Proc. of 43th SICE Annual Conference, Sapporo (2004)

163. Zyada, Z., Fukuda, T.: Identification and modeling of friction forces at a hydraulic parallel link manipulator joints. In: Proc. of 40th SICE Annual Conference, Nagoya (2001)

Index

2DOF robot, adaptive control, 156, 380
2DOF robot, backstepping control, 164
2DOF robot, dynamic model, 35, 163, 380
2DOF robot, inverse kinematics, 162

A

Acceleration calibration, 221
Active suspension, model, 33
Adaptive control, 152, 366, 375
Adaptive control, robot, adaptation law, 156
Adaptive control, robot, control law, 154
Adaptive control, ship, control law, 257, 258
Airplane, aerodynamic effects, 113
Airplane, angle of attack, 111
Airplane, dynamic model, 116
Airplane, gravitational effect, 112
Airplane, gyroscopic effect, 116
Airplane, kinematics, 108
Airplane, linearized dynamic model, 118
Airplane, longitudinal model, 200
Airplane, LPV model, 200
Airplane, sideslip angle, 111
Airplane, trust forces and moments, 119
Angular velocity calibration, 223
Angular velocity estimation, 227
Appell equation, 89, 404
Attitude estimation, 227, 252

B

Backlash, 314, 320, 377
Backlash identification, 346
Backlash simulation, 315
Backstepping control, algorithm, 61
Backstepping control, quadrotor helicopter, 230
Backstepping control, robot, 160
Backstepping control, ship, 259

Backstepping control, ship formation, 280
Ball and beam, 364
Ballast system, 126
Barbalat lemma, 48
Body frame, 108, 130
Brushless DC motor, dynamic model, 217

C

CAN bus, 213
Car, dynamic model, 101, 175, 270
Car, input affine model, 102, 175
Car, linearized model, 103
Car, side slip angle, 99
Car, state estimation, 189
Car, turning angle, 99
Cartan fields, 25
Cascade control, 139
Center of mass, 399
Chaotic behavior, 333
Characteristic equation, 103, 137, 145, 177, 180, 233, 331, 332, 337, 338
Collision avoidance system, 170
Completely integrable distribution, 423
Computed torque method, 144
Constrained motion, 15, 22
Constraint, command signal, 206
Constraint, command signal change, 206
Constraint, output signal, 207
Constraint force, 16, 17
Continuously differentiable function, 417
Control horizon, 72, 184, 204
Crane, cable driven, 57
Crane, flatness control, 57
Crane, receding horizon control, 74
Current control, 139
Curvature, 173
Curve, 418

B. Lantos, L. Márton, *Nonlinear Control of Vehicles and Robots*,
Advances in Industrial Control,
DOI 10.1007/978-1-84996-122-6, © Springer-Verlag London Limited 2011

Other titles published in this series (continued):

Soft Sensors for Monitoring and Control of Industrial Processes
Luigi Fortuna, Salvatore Graziani, Alessandro Rizzo and Maria G. Xibilia

Adaptive Voltage Control in Power Systems
Giuseppe Fusco and Mario Russo

Advanced Control of Industrial Processes
Piotr Tatjewski

Process Control Performance Assessment
Andrzej W. Ordys, Damien Uduehi and Michael A. Johnson (Eds.)

Modelling and Analysis of Hybrid Supervisory Systems
Emilia Villani, Paulo E. Miyagi and Robert Valette

Process Control
Jie Bao and Peter L. Lee

Distributed Embedded Control Systems
Matjaž Colnarič, Domen Verber and Wolfgang A. Halang

Precision Motion Control (2nd Ed.)
Tan Kok Kiong, Lee Tong Heng and Huang Sunan

Optimal Control of Wind Energy Systems
Julian Munteanu, Antoneta Iuliana Bratcu, Nicolaos-Antonio Cutululis and Emil Ceangă

Identification of Continuous-time Models from Sampled Data
Hugues Garnier and Liuping Wang (Eds.)

Model-based Process Supervision
Arun K. Samantaray and Belkacem Bouamama

Diagnosis of Process Nonlinearities and Valve Stiction
M.A.A. Shoukat Choudhury, Sirish L. Shah and Nina F. Thornhill

Magnetic Control of Tokamak Plasmas
Marco Ariola and Alfredo Pironti

Real-time Iterative Learning Control
Jian-Xin Xu, Sanjib K. Panda and Tong H. Lee

Deadlock Resolution in Automated Manufacturing Systems
ZhiWu Li and MengChu Zhou

Model Predictive Control Design and Implementation Using MATLAB®
Liuping Wang

Predictive Functional Control
Jacques Richalet and Donal O'Donovan

Fault-tolerant Flight Control and Guidance Systems
Guillaume Ducard

Fault-tolerant Control Systems
Hassan Noura, Didier Theilliol, Jean-Christophe Ponsart and Abbas Chamseddine

Detection and Diagnosis of Stiction in Control Loops
Mohieddine Jelali and Biao Huans (Eds.)

Stochastic Distribution Control System Design
Lei Guo and Hong Wang

Dry Clutch Control for Automotive Applications
Pietro J. Dolcini, Carlos Canudas-de-Wit and Hubert Béchart

Advanced Control and Supervision of Mineral Processing Plants
Daniel Sbárbaro and René del Villar (Eds.)

Active Braking Control Design for Road Vehicles
Sergio M. Savaresi and Mara Tanelli

Active Control of Flexible Structures
Alberto Cavallo, Giuseppe de Maria, Ciro Natale and Salvatore Pirozzi

Induction Motor Control Design
Riccardo Marino, Patrizio Tomei and Cristiano M. Verrelli